DIPLOMACY ON THE JORDAN: INTERNATIONAL CONFLICT AND NEGOTIATED RESOLUTION

NATURAL RESOURCE MANAGEMENT AND POLICY

Editors:

Ariel Dinar
Rural Development Department
The World Bank
1818 H Street, NW
Washington, DC 20433

David Zilberman
Dept. of Agricultural and
Resource Economics
Univ. of California, Berkeley
Berkeley, CA 94720

EDITORIAL STATEMENT

There is a growing awareness to the role that natural resources such as water, land, forests and environmental amenities play in our lives. There are many competing uses for natural resources, and society is challenged to manage them for improving social well being. Furthermore, there may be dire consequences to natural resources mismanagement. Renewable resources such as water, land and the environment are linked, and decisions made with regard to one may affect the others. Policy and management of natural resources now require interdisciplinary approach including natural and social sciences to correctly address our society preferences.

This series provides a collection of works containing most recent findings on economics, management and policy of renewable biological resources such as water, land, crop protection, sustainable agriculture, technology, and environmental health. It incorporates modern thinking and techniques of economics and management. Books in this series will incorporate knowledge and models of natural phenomena with economics and managerial decision frameworks to assess alternative options for managing natural resources and environment.

With water becoming a subject to dispute between riparian nations, we strive to better understand processes of negotiation over water agreements. Therefore, understanding the non-water and water background natures of the disputes, the structure of possible and actual negotiations, and patterns of cooperation among riparian nations is of great importance. This book analyzes the Jordanian-Israeli water affairs in the Jordan River Basin, using a wide and interdisciplinary perspective. It reveals many untold facts from the entire dispute and negotiation processes, starting from the 1950s and ending with the water accord in 1974.

The Series Editors

Recently Published Books in the Series

Easter, K. William, Rosegrant, Mark W., and Dinar, Ariel:
Markets for Water: Potential and Performance
Smale, Melinda:
Farmers, Gene Banks, and Crop Breeding: Economic Analyses of Diversity in Wheat, Maize, and Rice
Casey, Frank, Schmitz, Andrew, Swinton, Scott, and Zilberman, David:
Flexible Incentives for the Adoption of Environmental Technologies in Agriculture
Feitelson, Eran and Haddad, Marwan
Management of Shared Groundwater Resources: the Israeli-Palestinian Case with an International Perspective
Wolf, Steven and Zilberman, David
Knowledge Generation and Technical Change: Institutional Innovation in Agriculture
Moss, Charles B., Rausser, Gordon C., Schmitz, Andrew, Taylor, Timothy G., and Zilberman, David
Agricultural Globalization, Trade, and the Environment

DIPLOMACY ON THE JORDAN: INTERNATIONAL CONFLICT AND NEGOTIATED RESOLUTION

by

Munther J. Haddadin
Former Minister of Water and Irrigation
and
Jordan's Chief Water Negotiator

KLUWER ACADEMIC PUBLISHERS
Boston / Dordrecht / London

Distributors for North, Central and South America:
Kluwer Academic Publishers
101 Philip Drive
Assinippi Park
Norwell, Massachusetts 02061 USA
Telephone (781) 871-6600
Fax (781) 871-6528
E-Mail <kluwer@wkap.com>

Distributors for all other countries:
Kluwer Academic Publishers Group
Distribution Centre
Post Office Box 322
3300 AH Dordrecht, THE NETHERLANDS
Telephone 31 78 6392 392
Fax 31 78 6546 474
E-Mail <services@wkap.nl>

Electronic Services <http://www.wkap.nl>

Library of Congress Cataloging-in-Publication Data

Haddadin, Munther J.
Diplomacy on the Jordan : international conflict and negotiated resolution / by Munther J. Haddadin.
p. cm. - (Natural resource management and policy ; 21)
Includes bibliographical references and index.
ISBN 0-7923-7527-0 (alk. paper)
1. Jordan River-Water rights. 2. Water resources development—Law and legislation—Jordan River Valley. 3. Arab-Israeli conflict—1993- I. Title. II. Series.

KMC660.J67 H33 2001
333.91'6217'095694—dc21 **2001038671**

Printed on acid-free paper.

Printed in the United States of America

TABLE OF CONTENTS

LIST OF FIGURES

LIST OF TABLES

LIST OF APPENDICES

FOREWORD

The Jordan River has attracted the attention of people around the world. The natives call it "the Shari'a," or the Code [of God.] The religious importance of the river is known to mankind. Travelers and explorers sailed down its course from north to south. All have noted that its waters are turbulent, its path windy, and its meandering course steep.

Old civilizations flourished in the Jordan River basin. The territories surrounding it were known to have been ruled by the same political authority for centuries. Riparian States came to exist in its basin only after the First World War, when Britain and France shared the spoils of the eclipsed Ottoman Empire. Lebanon, Syria, Jordan and Palestine were created as separate political entities and became riparian parties on the Jordan. Palestine became the target of the Zionist Movement for the establishment of a national home for the Jews. They were supported by Britain, who issued the Balfour Declaration in 1917 pledging support for their objective. The Jewish State, Israel, was established in part of Palestine in 1948, and became another riparian party on the Jordan River.

The struggle between the Zionist Movement and the Palestinian Arabs over Palestine had water as a primary target since the beginning of the twentieth century. Palestine is arid in the south where, living space had to be prepared for the incoming Jewish immigrants. Water is needed to make the dry environment habitable. That struggle continued, and the Arab-Israeli conflict clouded the Middle East and, at times, the world for over half a century. Israeli-Arab wars broke out in 1948, 1956, 1967, 1973, 1982, and in each war Israeli forces advanced into neighboring Arab territories. In the last encounter of 1982, Israeli forces advanced and occupied part of Beirut, the Lebanese capital.

The United States was at the center of events, worried about Israel's security and peace in the region that supplies the world with much of its oil needs. Under her patronage, peace was worked out between Israel and Egypt in 1978 at Camp David. A more comprehensive Peace Process in the Middle East was kicked off in 1991 under the sponsorship of the United States and the Soviet Union. The process yielded a mutual recognition by and between the Palestinians and Israel in 1993 and a Peace Treaty between Jordan and Israel in 1994. The process is still slowly moving, but stagnant at time, to reach a final settlement between Israel and each of the Palestinians, Syria and Lebanon.

The author was born and raised in Jordan, in that environment of prevailing hostilities. His country suffered the consequences of confrontation. Jordan received the bulk of the Palestinian refugees who fled their homes during and after the war of 1948. She received more Palestinians who fled their towns and refugee camps in the West Bank as a result of the 1967 war. The territories of Jordan bordering Israel suffered more damages as a result of the military and paramilitary confrontation after that war. The Jordan Valley was all but destroyed, and its people evacuated it for safer refuge on the eastern plateau of Jordan.

The author became involved with the consequences of the confrontations in early 1973 after he was called back from the United States, where he was working in engineering research. He was given the post of "second in command" in a government agency set up for the rehabilitation and development of the Jordan Valley. He worked as Senior Vice President of the Jordan Valley Authority for nine years, and became its President and Chairman of the Board for five years after that. Throughout those fourteen years of senior service, he became aware of the intricacies of the water conflict with Israel, and developed to be the leading advocate of Jordan's water rights.

The author was called away from his private consulting engineering practice in 1991 to join the delegations of Jordan to the Middle East Peace Process. He became member of the delegation to the Bilateral Peace Negotiations (1991-1995), in which he led the team negotiating water, energy and the environment. He led the Jordan delegation to the Water Resources Working Group of the Multilateral Peace Talks (1992-1995), and led the Jordanian team to the Trilateral Economic Committee (Jordan, Israel and the United States). The author was also a member of the Jordanian delegation to the Steering Group of the Multilateral Peace Talks. He was later called upon, on 19 March 1997, to serve as Minister of Water and Irrigation until he resigned his post, on 10 August 1998.

From his various positions, the author was the Jordanian official in charge of water and water negotiations during a very sensitive period of the history of the Middle East. From his positions in the Jordan Valley Authority, he managed Jordan's international waters during eras of turbulence and hostilities. He later made the engagements with the Israelis in a prolonged peace process that yielded a peace treaty with a special annex assigned to water. As Minister, he managed the follow-up of the implementation of the treaty provisions.

Throughout his involvement, he was eager to write about it when the time allowed. This book has been the result of intensive work the author undertook after he resigned his post as Minister. The author's notes on various occasions provided some material for the book, especially the notes he kept of the peace negotiations. However, all of the official material is available at the Ministry of Foreign Affairs of the Hashemite Kingdom of Jordan. Access to it requires the approval of the competent authorities.

ACKNOWLEDGEMENT

The author likes to express the deepest gratitude to the memory of the late King Hussein of Jordan for selecting the author to be among the few Jordanians who were dispatched to the various forums of the Middle East Peace Process, and for selecting him to care for the development of the Jordan Valley, and represent Jordan in all matters pertaining to international waters. The author would like to acknowledge with gratitude and admiration the leadership role played by His Royal Highness Prince Al Hasan Ben Talal, in directing the Jordanian efforts for the rehabilitation and integrated development of the Jordan Valley; and in structuring and guiding the Jordanian delegations to the Bilateral and Multilateral negotiations of the Peace Process. The author benefited very much from the experience and wisdom of the visionary Hashemite Prince.

The author also would like to acknowledge with gratitude the role that his senior colleague and former superior, Mr. Omar Abdallah Dokhgan, the first President and Chairman of the Board of the Jordan valley Authority, played in introducing the author to the political and ideological sensitivities involved in Jordan's pursuit of the development of the Yarmouk River. His valuable guidance at the time, and his confidence and trust, were very influential in shaping the attitudes of the author toward international water issues that faced Jordan.

Thanks undoubtedly are due to the author's wife, Lexi Kludt Haddadin, who has suffered, with patience and rare courage, throughout most of her married life because of the involvement of her husband in the above critical and sensitive issues. Special thanks go to their offshoots: Sumaya, Yezan and Badia, for their sacrifice and courage, and for being as supportive of their father as they have been throughout. Similar thanks go to the author's immediate family: his brothers and sisters, who have been instrumental in enabling the author to face the multitude of diverse problems emanating from his involvement in the service of his country and the King.

Thanks are due to Raja' Kamel al Dawoodi, who served as executive secretary and office manager to the two presidents of the Jordan Valley Authority, and to the successive Ministers of Water and Irrigation, for her diligence and skills in filing and record keeping, and her assistance in filing and retrieving the author's personal records.

Thanks are due to the many friends in Jordan and the United States who have aided the author with their encouragement and follow-up on the progress of the author's work. Margaret Koivula of Spokane, Washington deserves recognition for her "hot pursuit" and her review of several chapters. Special thanks go to Dr. Ariel Dinar of the World Bank for his patience and diligence in reviewing and editing the manuscript, and for his key role in arranging for the publication of the book. Thanks are also due to the anonymous reviewers who supplied the author with crucial viewpoints that were duly incorporated in the manuscript.

Finally, the efforts of Ms. Shira Yoffe, Mr. Case Bowman and Mr. Greg Fisk, all of the Department of Geosciences at Oregon State University, in the final editing of the manuscript, deserve the author's thanks and recognition.

CHAPTER I

Plans for the Jordan River Basin

The Jordan River waters are essential for the development and well being of the people of its basin. Plans for the development of its water resources were considered since the beginning of the twentieth century. These water resources were at the center of plans drawn to populate Palestine with immigrants in addition to its indigenous population. The basin was thus a target for the World Zionist Organization in its quest for the creation of a state for the Jews in Palestine. The history of the development plans and the associated complications are reviewed in this chapter.

1. Geography

The Jordan River drains territories in Lebanon, Syria, Israel, Jordan and the West Bank. It discharges its waters into the Dead Sea whose shoreline is the lowest, dry contour on earth. Several tributaries discharge their waters into the River. Three tributaries constitute the sources of river flow in the north before it discharges into Lake Tiberias. The first is the Hasbani that rises in Lebanon and flows southward towards the Jordan Valley. About six kilometers of its course flows along the border between Lebanon and Syria before it enters Israel. The Hasbani is joined by a sub-tributary called Al Wazzani before it discharges into the River Jordan north of Lake Tiberias. The second is the Dan tributary that rises inside Israel and flows in a southwesterly direction to join the River. The third is the Banyas that rises in the Syrian Golan Heights, at the foothills of Mount Hermon, and flows in a westerly direction to join the Jordan River north of Lake Tiberias also. There are several minor side streams that discharge flood water into the River and others that feed the Lake directly (see Figure I.1).

South of the Lake, the River is joined by its major tributary, the Yarmouk River. This tributary runs close to the border between Jordan and Syria, and forms part of the border when the railroad track, running adjacent to the river, crosses it to the north. Actually, the border between the two countries in the Yarmouk gorge is demarcated by the Yarmouk or the railroad, whichever is closer to Jordan. Before the Yarmouk joins the Jordan River near Baqura, it forms the border between Jordan and Israel. As the Jordan River meanders southwards to the Dead Sea, it is joined by several side wadis from the west and the east, the largest of which is the Zarqa River that rises and flows totally inside Jordan.

Several cities and towns are located in the Jordan River Basin including Amman, the capital city of Jordan, Irbid, Zarqa and several other

Jordanian towns and cities. In it also lie several Lebanese, Syrian, Palestinian and Israeli towns and cities including population centers in the Jordan Valley.

The Jordan Valley is part of the Great Rift Valley that starts in southwestern Turkey and extends southwards into the Syrian Ghab plains, the Lebanese Baqaa' Valley, the Jordan Valley, the Dead Sea, Wadi Araba, and under the Red Sea waters to terminate in Kenya in East Africa. The Jordan Valley is rich with agricultural land and with the water resources of the Jordan River Basin. South of the Dead Sea, the water resources are not as abundant, and a few side wadis discharge base flows and flood flows that have been put to beneficial economic use. Wadi Araba consists of vast stretches of arid land, but contains very limited renewable groundwater resources, the quality of which gets to be more brackish as one moves towards the center, away from the alluvial fans at the foot of the eastern and western escarpments.

As such, the Jordan Rift Valley (JRV) straddles territories and territorial waters of several countries. These from north to south: Turkey, Syria, Lebanon, Israel, Jordan, Palestine, Egypt and Saudi Arabia. Of them Syria, Lebanon, Israel, Jordan and Palestine are riparian parties on the Jordan River Basin and its waters. Egypt and Saudi Arabia are riparian parties on the Gulf of Aqaba, in addition to Israel and Jordan. Figures I.1 and I.2 show the Jordan River Basin and the Jordan Rift Valley respectively.

The multiple national entities and states that share the Jordan River Basin make it an international watercourse. The riparian parties are Lebanon, Syria, Israel, Jordan and Palestine. Several plans and attempts have been made by the riparian parties to develop the water resources of the basin as shall be detailed later.

2. Historical Developments

In ancient times, the Jordan Rift Valley was an important populated region that enjoyed development and prosperity. Archaeological finds testify to early settlement of man in the north Jordan Valley. Discoveries in Pella and on the northern bank of the Yarmouk at Adasiyya show that man had settled in the Valley since the eighth millennium before the beginning of the Christian Era (C.E). Roman water systems are visible today on the eastern and western escarpments of the Valley. Two of the Decapolis, Pella in Jordan and Biesan in Israel, are located in the Jordan Valley. South of the Dead Sea and along its shores there is evidence of ancient settlements. There are ancient water systems credited by archaeologists to the Nabateans of Petra and built in the fourth century before the Christian Era.[1] On the mountain slope that

---

[1] The author is witness to these water systems at Wadi Tlah in northern Wadi Araba, and has supervised the uncovering of Nabatean water systems in Wadi Khneizeereh and Wadi Feifa south of the Dead Sea.

Figure (I.1)- Map of the Jordan River Basin

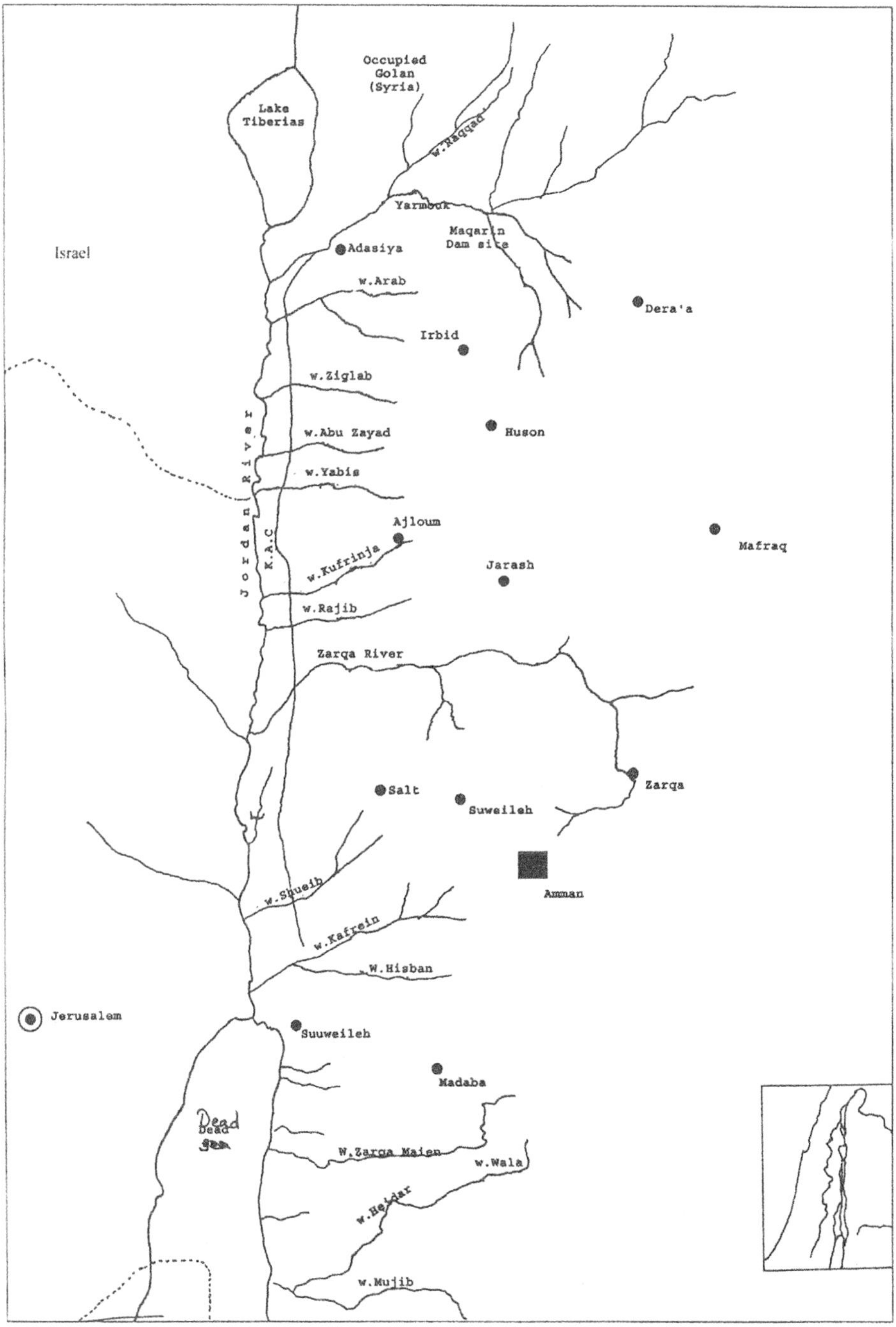

Figure (I.2)- Map of the Jordan Rift Valley

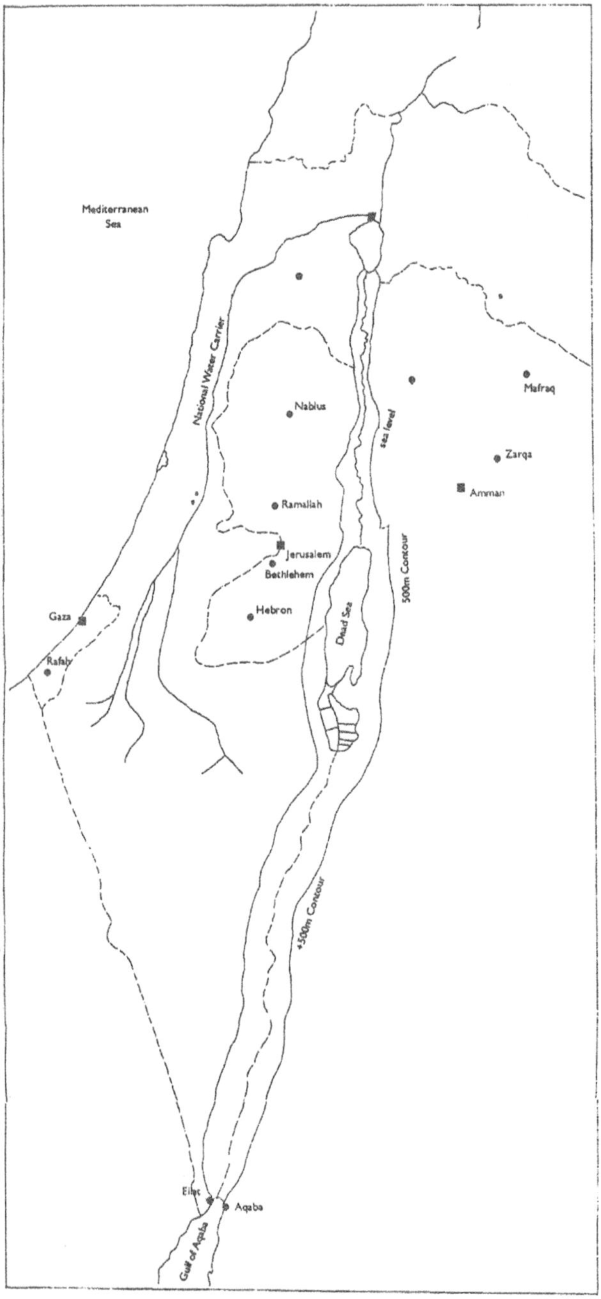

abuts today's town of Safi, south of the Dead Sea, Lot's cave has recently been discovered.

The Jordan Valley can be viewed from the eastern and western escarpments. John the Baptist baptized Jesus Christ in the Jordan River on its eastern bank. Bethany (Arabic Beit Anya) where John the Baptist baptized people "beyond the Jordan" has been uncovered recently at Wadi Safsaf (Kharrar) in Jordan in the vicinity of the baptism site. The Mount of Temptation, immediately west of Jericho, overlooks the Valley and so does Mount Nebo on the eastern escarpment. On the first, Jesus was tempted by Satan. The reward, if Jesus would kneel to him, was to be the kingdoms that Jesus could see from that mount. These were clearly in the Jordan Valley, which must have been well developed. On Mount Nebo, Moses stood and saw the "Promised Land", and it must have been well developed to be worthy of the divine promise. At the initial stages of Prophet Mohammad's preaching in the second decade of the seventh century C.E, his opponents pointed to what they thought was unrealistic exaggeration in the Prophet's preaching because, they said, he was promising the believers a paradise similar to the gardens of the Jordan Valley! In the Omayyad Era of the Arab Islamic rule, the Caliph, Hisham Ibn Abdul Malek, had a palace built for him in the Jordan Valley at Jericho, the remains of which still stand today. During the same era, a rebellion in the first decade of the ninth century, led by Abdul Rahman Ibn al Asha'ath against al Hajjaj, the grand governor of Iraq and the Orient, gathered most of its rebellious troops, said to have numbered 100,000, from the Jordan Valley, evidence of its heavy population at the time. During the Crusades (1096-1292), the Jordan Valley and the eastern escarpment formed a front line for decades. During the Ottoman rule (1516 -1916), the Jordan Valley was famed for producing sugar cane and sugar beets, and the remains of presses for producing sugar are visible there today.

3. Western Missions to Palestine

Palestine attracted the attention of Orientalists and exploratory Western trade missions since the early nineteenth century. Several missions were conducted by Orientalists from the West to explore the Near East and promote trade with it, and some were designed to better understand the Jordan River and its Valley. The region was then part of the Ottoman Empire with whom Europe had had frequent wars throughout the centuries, since the fall of Constantinople to the Ottoman Turks in 1453 C.E. Early European expeditions were led by Ireland's Christopher Costigan in 1835, and by Lieutenant Thomas Molyneux, a British Royal Navy officer, in 1847. In 1837, the *U.S.S. Constitution* landed American General Lewis Cass and several naval officers at Jaffa. They proceeded to the Dead Sea to survey it. Although

Cass and his party did not survey the Dead Sea, they did survive it. Several hours of floating in its waters produced skin irritation but no fatalities.[2]

A second American mission was led ten years later by Lieutenant William Francis Lynch, of the US Navy, who led 17 members in his company to explore the Dead Sea and the River Jordan in 1847. His aim was to "serve the cause of science" and was at "the service of American commerce with the region."[3] A description of the Jordan River and its valley as it looked at that time is revealing:

Lynch sailed down the course of the river Jordan from Lake Tiberias and had a land party parallel the boats. In the stretch between Lake Tiberias and the Dead Sea, the river Jordan contained no fewer than 27 rapids and was "more sinuous even than the Mississippi."[4] Nevertheless, the lower Valley, or Ghor, held other possibilities. His land party reported "an extensive plain, luxuriant in vegetation, and presenting ... a richness of alluvial soil, the produce of which, with proper agriculture, might nourish a vast population".[5]

As Lynch proceeded further along the Ghor, he noted a "most beautiful tract of alluvial soil located amid country "entirely destitute of cultivation."[6] To Lynch, this landscape could support small farmers and independent shopkeepers. In short, Lynch saw a fertile medium fit for what he called the duplication of American republicanism. Lynch's mission was subject to congressional criticism, and the unstable situation in America prior to the Civil War shelved his report and recommendations. It was concluded, however, that the Jordan River was not navigable, as it contained many rapids.

Many American writers visited the Holy Land in the nineteenth century. One was Herman Melville whose visit in 1856-57 inspired him to compose the epic poem *Clarel*. That poem records Melville's encounter with the desert which held for him the same fascination as the sea.[7] Ten years later, in 1867, another notable American author, Mark Twain, traveled to the Middle East, and wrote about the scenery in Palestine "...The noted Sea of

---

[2] W. Stanley Hoole, "A Visit to the Holy Land in 1837", *Journal of the American Oriental Society* 95 (October/December 1975).

[3] Robert E. Rook, "The 150th Anniversary of the United States Expedition to Explore the Dead Sea and the River Jordan", American Center of Oriental Research, Amman, Jordan, 1998.

[4] Lynch, *Narrative*, 265; cited in the above.

[5] Ibid.

[6] Ibid.

[7] See Daniel Hillel, "Rivers of Eden", Oxford University Press, 1994. Hillel quotes Melville's poem about Jerusalem and environs: "Here the city old/ Fast locked in torpor, fixed in blight/ Obscure, in monitory dream/ Oppressive, roofed with awful skies/ Whose stars like silver nail-heads gleam/ Which stud some lid over lifeless eyes" p.147.

Galilee was long ago deserted by the devotees of war and commerce, and its borders are a silent wilderness...Palestine is desolate and unlovely."[8]

4. Early Zionist Missions

In the late nineteenth century, Jewish activists gave momentum to a political movement, the Zionist movement, aiming at establishing a national state for the Jews to which they can all go and reside. Led by Theodore Herzl, a Russian Jew, the movement held the first Zionist Conference in the summer of 1897 in Basel, Switzerland, and Palestine was formally selected as the territory for that purpose. Encouraged by the Zionist movement, M. Abraham Bourcart, a German engineer, visited Palestine and stayed there for seven months to study its potential for development. In a handwritten letter[9] (Appendix 1) dated 18 February 1901 to Reverend John Wilkinson,[10] Bourcart sought the Reverend's council and help, to sell his ideas to the King of England. Bourcart had studied the Jordan River Basin and developed pioneering ideas that were later reiterated and detailed by other engineers. He proposed to divert the Litani River, a river flowing totally outside the Jordan River Basin through a 5 kilometer tunnel to the Hasbani and on to the Jordan River (Figure I.3). He concluded that power in the total of between 200,000 and 400,000 HP (Horse Power) could be generated from the combined flow of the Litani and the Jordan systems. The maps attached to Bourcart's letter indicated another tunnel to bring seawater from the Mediterranean at Haifa to the Jordan River at Biesan and on to the Dead Sea, an idea first advanced by Theodore Herzl, (Figure I.4). The idea was meant to generate power, and to raise the level of the Dead Sea by 20 to 50 meters. That depth of Mediterranean water in the Dead Sea would be fit for fisheries, boating and tourism (Figure I.4). Under the same project, the Jordan's freshwater would be diverted away from the river course as it exits Lake Tiberias to a parallel conduit running close to the River and west of it. The fresh water would be pumped from that conduit at three locations to reservoirs on the western escarpment, and water would be released from them to flow by gravity and irrigate the barren lands. The power generated by the project would be used to pump water and develop the lands of Palestine and transform barren lands to gardens. Bourcart suggested in his letter to Reverend Wilkinson that the King of England purchase the barren lands from the Ottoman Sultan, develop them, and then sell the land to the Jews to make money. Bourcart cited verses from the Old Testament (see Appendix 1) and requested that Jews be allowed to go

---

[8] Ibid, pp. 147.

[9] Copy of the letter (with sketches) is obtained from the Zionist Central Files.

[10] Founder of Mildmay Mission for the Jews in the late 1800's, and author of the book "Israel my Glory" published in 1889. He was close to the Monarch of Great Britain.

back to Palestine and live without fears. He summarized his conclusions as follows:

" So my conclusion would be:

1. Would not England have all interest to purchase the Holy Land out off the hands of the Sultan, cheap, barren as it is?
2. to let the Jews go back to Palestine, to protect them there and sell them land to higher but reasonable prices;
3. to recuperate thus of the money lost in the Transvaal war;[11]
4. to merit of God a new blessing;
5. to hasten the easing of the torments of the Jews, and the coming of our Lord."

Bourcart's plan for the development of the western Jordan Valley, and the rest of Palestine, although clouded with overestimated raw water flow data, was comprehensive, analytic and the first such plan on record to propose the diversion of the Litani River to the Jordan, and to connect the Dead Sea with the open sea at Haifa. His maps show that he also proposed the diversion of the Yarmouk River to Lake Tiberias (Appendix 1). The diversion would have two benefits: it would regulate the floods of the river, and would save it from being mixed with the seawater brought from the Mediterranean to the Jordan River to flow down to the Dead Sea. Several other plans followed that made similar proposals for diversion of the Litani, the Yarmouk, and for the connection of the Dead Sea with the Mediterranean.[12] Bourcart was influenced by verses in the Old Testament that made reference to the water in and around Jerusalem, and to the "healing" of the sea and the growing of fish (Appendix 1). His plans were the first, followed by several others laid down by the Zionist Movement, to develop the Lands of Palestine, increase its absorptive capacity and make room for Jews to immigrate and live there.

On the part of the indigenous population, the development of the Jordan Valley was a topic that attracted attention. In 1913, when Greater Syria (including Palestine) was still under the Ottoman Empire, George Franjieh, Director of Public Works in Palestine, proposed a scheme for the development of the Jordan River system for use in irrigation and power generation. The scheme envisaged the diversion of the Yarmouk tributary into Lake Tiberias, a canal of a 100 mcm annual flow capacity to irrigate the

---

[11]A war in which Great Britain fought against the Transvaal and Orange Free State, 1890-1902; also a war between Great Britain and the Transvaal, 1880-1881. The Transvaal is a province in the NE Republic of South Africa where Pretoria, the capital, is located. *Footnote inserted by the author.*

[12] The latest proposal for a Med - Dead Canal was advanced by Israeli parties in the course of the Middle East Peace Talks in August, 1994.

Jordan Valley, and two power plants to generate electricity.[13] The collapse of the Ottoman Empire at the end of the First World War terminated this scheme.

It can be seen that the efforts of the Zionist Conference to have plans prepared for the settlement of Jews in Palestine had seriously started before the indigenous population became aware of such plans. They were careful not to have those plans exposed to the public or to the Ottoman Sultan for fear of aborting them. The Ottomans were not sympathetic with Herzl's aspirations and did not support the establishment of a national home for the Jews in Palestine. The natives reacted to the Zionist plans much later by adopting competing plans prepared for the development of the Jordan Valley.

5. The British Mandate Period

The serious plans for Palestine came in the wake of the 1917 Balfour Declaration. A year earlier, Britain and France, the major allies of the First World War, agreed to divide the spoils of the Ottoman Empire between them, such that Britain would keep Palestine, Transjordan and Iraq under its mandate, and France would do the same to Syria and Lebanon. In the 1919 Paris Peace Conference, the World Zionist Organization tried hard to alter the northern boundary of Palestine and move it up into Lebanon to include the Litani River, and move the northeastern boundary into Mount Hermon and the Golan to include the sources of the Jordan River waters. The eastern boundary was to be moved from the Jordan River eastward to be close to the Hijaz Railroad track (Figure I.6). The World Zionist Organization submitted a statement to the Paris Peace Conference dated the third day of February 1919. Excerpts from that statement read under the title "Boundaries of Palestine" as follows:

"The boundaries of Palestine shall follow the general lines set out below:
Starting on the North at a point on the Mediterranean Sea in the vicinity South of Sidon and following the watersheds of the foothills of the Lebanon as far as JISR EL KARAON, thence to El BIRE, following the dividing line between the two basins of the WADI EL KORN and the WADI ET TEIM, thence in a southerly direction following the dividing line between the Eastern and Western slopes of the HERMON, to the vicinity West of BEIT JENN, thence Eastward following the northern watersheds of the NAHR MUGHANIYE close to and west of the Hejaz Railway.

In the East a line close to and west of the Hejaz Railway terminating in the Gulf of Akaba.

---

[13] "Water in the Middle East- Conflict or Co-operation", edited by Thomas Naff and Ruth C. Matson, Westview Press, London and Boulder, 1984, pp. 30.

Figure (I.3)- Bourcart's Proposal for Litani Diversion

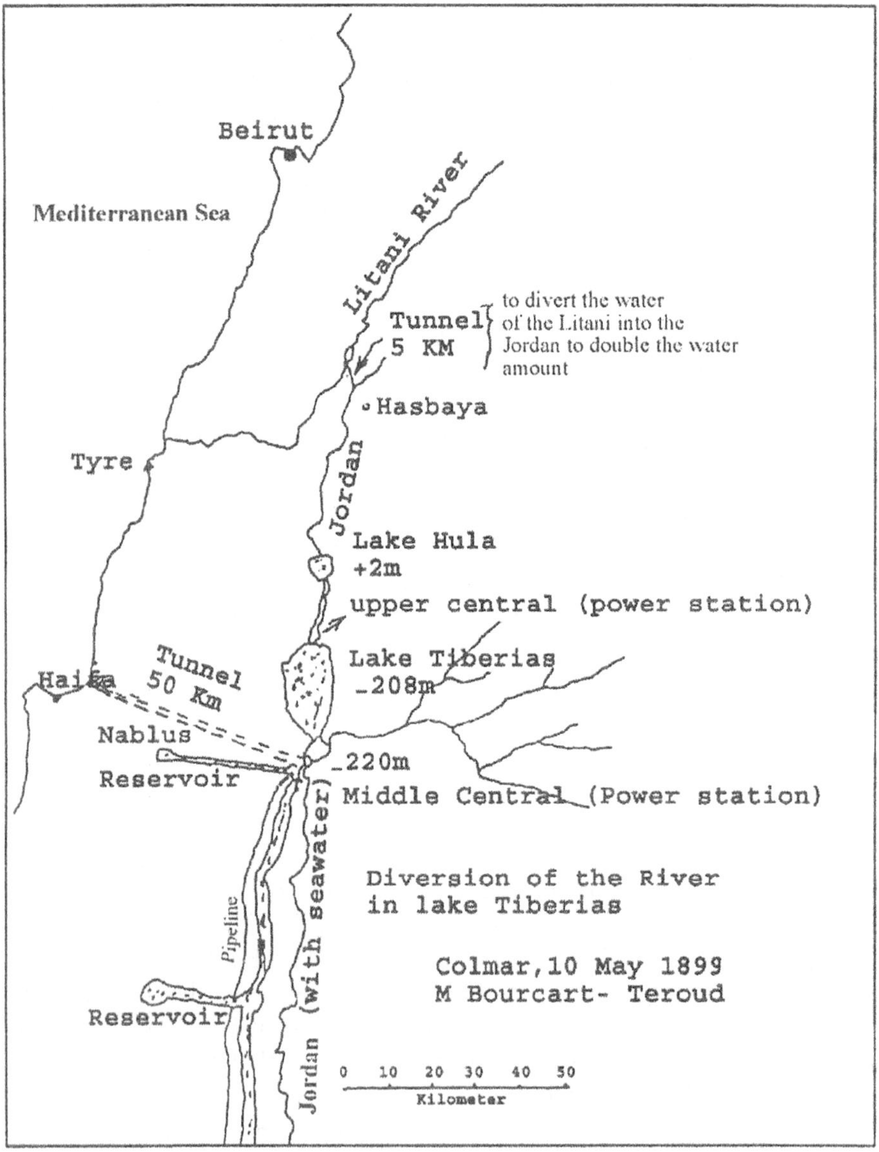

Figure (I.4)- Mediterranean- Dead Sea Connection (Bourcart)

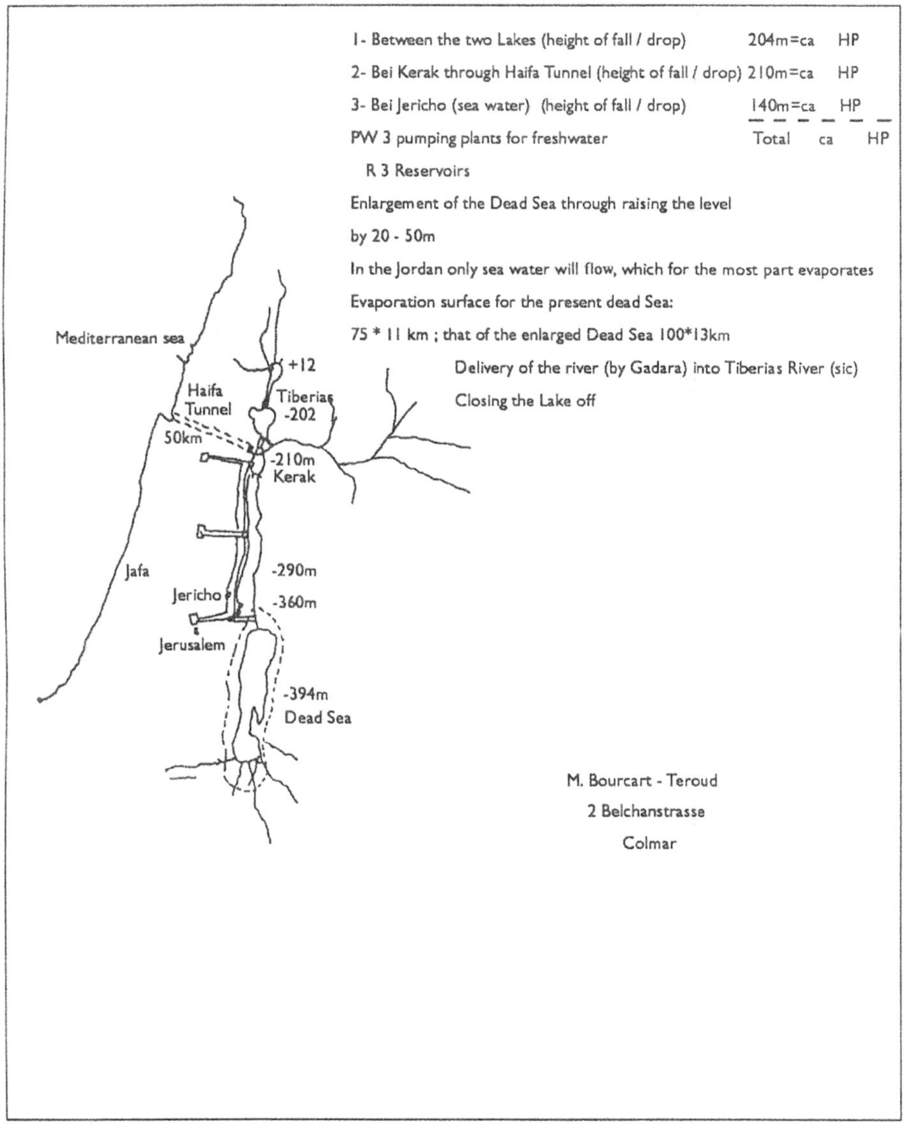

Figure (I.5)- Bourcart Plan Showing Biblical References

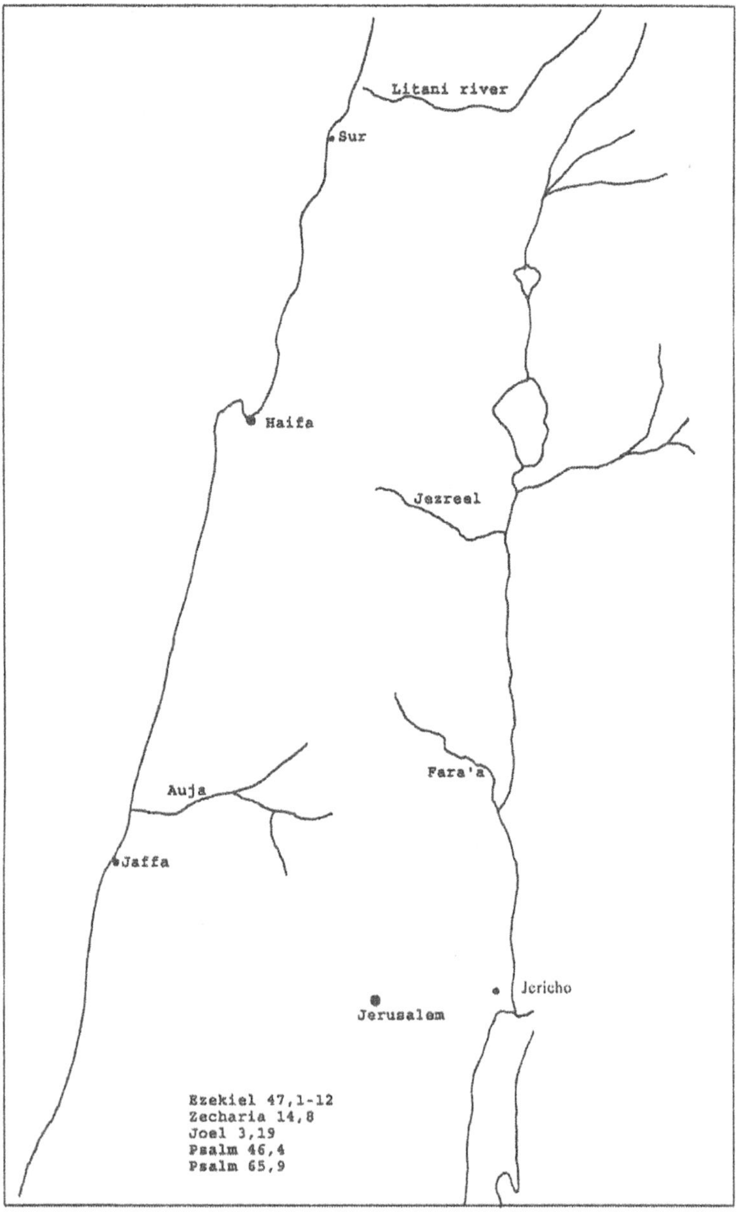

In the South a frontier to be agreed upon with the Egyptian Government.

In the West the Mediterranean Sea."[14]

The World Zionist Organization believed that its plans for settling Jews in Palestine would be hampered without an adequate and secure supply of water, thus the importance of the Litani and the upper tributaries of the Jordan River to their plans. The south of Palestine is a stretch of semi arid and arid territory in which agriculture would not be feasible without irrigation, and considerable amounts of water would be needed to develop irrigated agriculture there.

The Zionist attempts to alter the borders were frustrated by the reaction of France, from whose share of the territories a good part was proposed to be carved out and added to the British controlled territory. The borders of Palestine with Lebanon and Syria on the north and the northeast remained unchanged, and the arrival of Abdullah, the Hashemite prince, from Hijaz provided Transjordan with an advocate of her independence from the plans of Zionism, although Zionist attempts to settle Jews in Transjordan continued.

Palestine finally came under the mandate of Britain by decree of the League of Nations at the request of the Zionist Organization that included in its statement to the Paris Peace Conference a request that says:

"We ask that Great Britain shall act as Mandatory of the League of Nations for Palestine. The selection of Great Britain as Mandatory is urged on the ground that this is the wish of the Jews of the world and the League of Nations, in selecting a Mandatory, will follow as far as possible, the popular wish of the people concerned.[15]

The preference on the part of the Jews for a British Trusteeship is unquestionably the result of the peculiar relationship of England to the Jewish Palestinian problem."

This move would pave the way to honor the pledge made in the Balfour Declaration of 1917 to establish a national home for the Jews in Palestine, and was contrary to the earlier pledges made by Britain to the Sharif of Hijaz, Al Hussein Ben Ali, as he led the Arab Revolt against the Ottoman Turks in 1916 during the War. The Arabs became aware of the Declaration and of the 1916 secret Sykes-Pecot agreement after Russia broke away from the Allies in 1917, in the wake of the Bolshevick Revolution, when some leaks were authorized. The participation of the Sharif's son and envoy, Faisal, then proclaimed King of Syria, in the Paris Peace Conference

---

[14] Aaron T. Wolf, "Hydropolitics Along the Jordan River", the United Nations University Press, Tokyo, New York, Paris, 1995, Appendix II, pp. 224.

[15] Ref. 7 above, Appendix I, pp. 225.

did not help convince the victorious Allies to honor the pledges made to his father to recognize the Arab Kingdom in Greater Syria.

Figure (I.6)- Map of Zionist Proposed Borders for Palestine (1919)

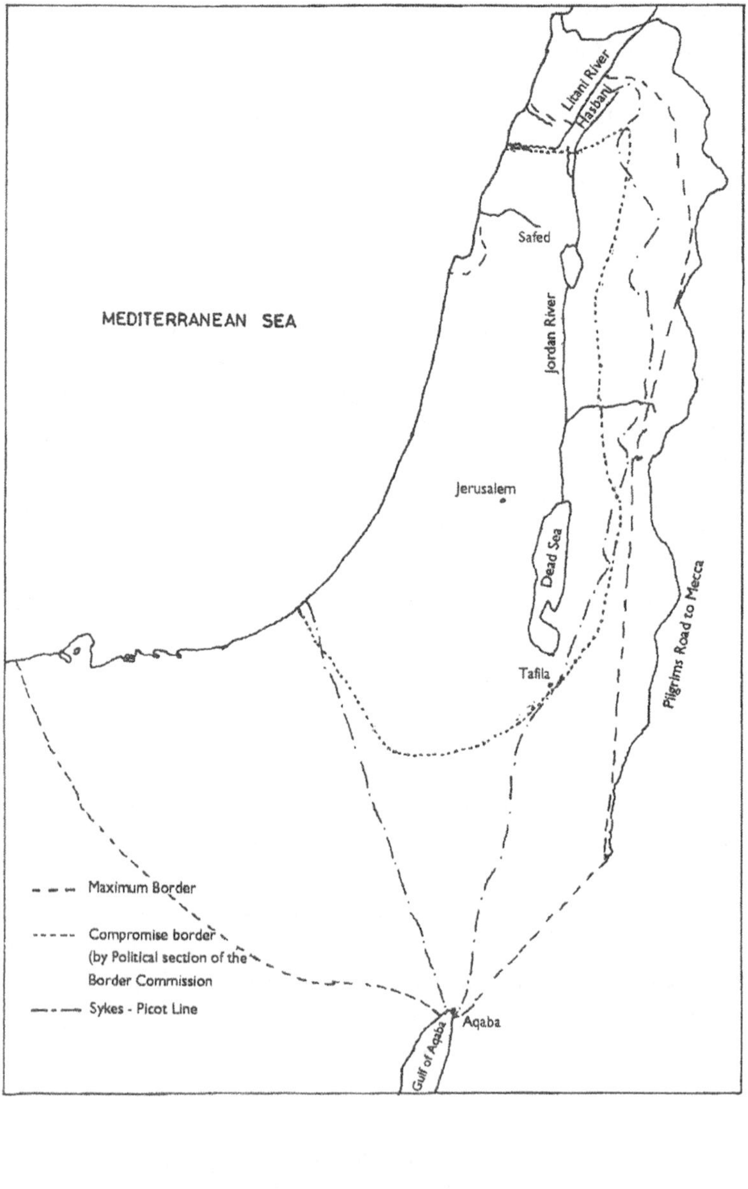

5.1 The Rutenburg Concession

The Jewish Agency, an arm of the World Zionist Organization, administered the plans for Jewish settlement in Palestine, and provided for their safety and wellbeing. In their efforts to evaluate the potential of Palestine for absorption, they realized the constraints that water resources might impose. They commissioned a Jewish civil engineer, Pinhas Rutenburg, in 1919, to carry out a study of the water resources of Palestine including those of the Jordan Valley. Rutenburg conducted a preliminary appraisal of the water resources of the Jordan River Basin for lack of measurements and data, and proposed an elaborate plan for the development of these resources including the Litani. He focused on the construction of dams on the tributaries and the installation of power plants, and on the development of irrigated agriculture. Rutenberg exaggerated the estimates of available flows in the basin and ended up with similar exaggeration in the areas that can be irrigated, and in the amount of electrical energy that can be generated. In the upper Jordan Valley he figured that an area of 22,000 hectars (ha) could be irrigated. In the Jordan Valley south of Lake Tiberias, he figured that 82,000 ha could be irrigated on the eastern slopes, and 48,000 ha on the western slopes. These estimates were over three times the area that was later found to be irrigable in the Jordan Valley. With the Litani included, Rutenberg figured that 40,000 ha could be irrigated in the plains of Tyre and Sidon in Lebanon, and 90,000 ha in the coastal plains of northern Palestine, over double the area that the Litani measured flow could irrigate. He had excess water to convey to areas south of the Dead Sea, through a canal running on the western shore, to leach salts from soils there and to irrigate 100,000 more hectars south of the Dead Sea or elsewhere. The amount of energy that could be generated in his calculations amounted to 3.168 billion HP.Hr annually. Exaggeration in water flows led him to exaggerate the amount of power that could be generated from these flows.

Rutenburg estimated that, away from the Jordan River Basin, the western slopes of the Palestinian mountains (sloping toward the Mediterranean) had enough water resources to develop and generate an equivalent amount of electrical energy and irrigate the whole coastal plain. That estimate was, again, an over-exaggeration. On the notion of the Jewish homeland, Rutenberg noted that: "Palestine will be Jewish, only if the entire work relative to the building up of Jewish life will be carried out by Jewish workers, however difficult or dangerous the work may be, and whatever sacrifices it may require. The re-building of Palestine by Arab labor would result in the creation of an Arab and not a Jewish Palestine, irrespective of the amount of Jewish capital drawn in."[16]

In his work, Rutenburg stated that "The fundamental principle underlying the plans of the electric part of the work has been that a relatively considerable quantity of energy should be supplied to centres chosen as points for intense immigration and consolidation, even before there is demand for such energy."[17] His ideas, and not figures, were to be reiterated by several subsequent studies on the subject.

Rutenburg, obviously acting on behalf of the Jewish Agency, was granted concessions by the Crown Agents through a Concession Agreement dated 21 September 1921 to develop hydropower in the Jordan Valley and on the Auja River (later renamed Yarqon in Hebrew) on the coastal plain of Palestine near Jaffa.[18] The concession was made transferable to a company that was yet to be established (The Palestine Electric Company was established later). "The Articles of Association and constitution of the said Company shall be subject to the approval of the High Commissioner for Palestine in agreement with the Jewish Agency referred to in the Mandate for Palestine."[19] Rutenberg was also given concession to drain the Huleh swamps north of Lake Tiberias to reduce evaporation losses, increase the flow velocity in the Jordan River for purposes of power generation down stream, and reclaim lands for cultivation. All components of the works were suggested in his above referenced study.

The Concession Agreement was ratified by the Foreign Minister of Britain and by the High Commissioner of Palestine in the same year (1921). The Prime Minister of Jordan, Hasan Khalid Abu al Huda signed an agreement with the Palestine Electric Company on 5 March 1927, and his government approved the concession on 8 January 1928.[20] [21]. Under that

---

[16] P. Rutenburg, "Water Resources of Palestine, I. Jordan Valley, Preliminary Project", a Confidential Report issued in Jerusalem, dated June 1920, bilingual (English and Hebrew), pp. 61.

[17] Ibid, pp. 47.

[18] The Crown Agents for the Colonies and Mr. Pinhas Rutenberg, "Agreement for the Granting of a Concession for the Utilization of the Waters of the Rivers Jordan and Yarmouk and Their Affluents for Generating and Supplying Electrical Energy", witnessed by Burchells Solicitors of Westminster, London; Herbert Oppenheimer Nathan and Vandyk, Solicitors, London; and Harry Sacher, Barrister at Law and Notary Public, Jerusalem, dated 21 September 1921.

[19] Ibid, pp. 2

[20] "Official Gazette No. 177", Princedom of Transjordan, 1928.

[21] The British Government asked Prince Abdallah Ibn al Hussein of Transjordan to accept the Concession and ratify the Agreement in 1924. The Prince declined on the grounds that such a concession is the concern of the people, and should be looked into by the Council of Deputies (Parliament). The British Resident Representative in

concession, a power house was built on the east bank of the Jordan at its confluence with the Yarmouk, known as the Rutenburg Project, and power lines transmitted electricity from that power house in Jordan to Palestine in the west. The project and the associated township were vandalized during the war of 1948, and it still stands today as a landmark in the northwestern corner of the Hashemite Kingdom.

5.2 More Water Plans for the Jordan

The water scarcity problem in Palestine was acknowledged by the mandatory powers in the early 1920's, most likely through Zionist lobbying. They wanted as much water as could possibly be made available to Palestine, to enhance its absorption potential. The issue of water was discussed in 1924 in a special section of the agreement between the French and the British regarding their mandates in the region. The agreement stated that "experts chosen by the governments of Syria and Palestine will work together to survey possibilities of exploitation of the waters of the upper Jordan River and Yarmouk River and their tributaries, for irrigation and power generation, to satisfy the needs of the Territories that are under the French mandates (in Syria). While this study is being done, the French government will give its representatives extremely liberal instructions regarding the use of the excess water for the benefit of Palestine."[22]

Jewish immigrants started to arrive in large numbers early in the British mandate, in implementation of the Balfour Declaration of 1917. The water availability for the indigenous population and the immigrants became an issue. To address that issue, the British Government commissioned a British engineer, Mavromatis, in 1922, to make an assessment of water resources and propose development schemes. He proposed an elaborate plan to irrigate the area around Lake Huleh and drain the swamps there, divert the Yarmouk into Lake Tiberias (earlier proposed by Bourcart and by Franjiyeh), construct two dams to generate electric power, and build irrigation canals down both banks of the Jordan River. Another engineer, Henriques, later commissioned by the British Government in 1928, proposed the irrigation of the Yarmouk Triangle, the area bound by the Jordan River to the west, the Yarmouk to the south, the Golan escarpment to the east and Lake Tiberias to the north.[23]

Amman of the High Commissioner in Jerusalem repeated the request to the Prime Minister on 15 March 1926.

[22] Nimrod, Y., "Angry Waters, The Controversy Over the Jordan River", Givat Haviva: Center for Arabic and Afro-Asian Studies.

[23] "Water in the Middle East - Conflict or Co-operation", edited by Thomas Naff and Ruth Matson, (above), pp. 30.

With the increase in Jewish immigration and settlement in the 1930's, the development of water resources to support immigrants became imperative. The Jewish Agency established a colonial company, Mekorot Water Company, for that purpose. During 1935 and 1936, Simcha Blass, a Jewish water engineer, suggested the first regional plan to supply water to western Galilee. It was the first plan to be implemented by the newly founded water company, Mekorot.

The Land of Israel Water Company, a private company established in 1932, was also active in water projects in the 1930's and 1940's. In 1945 the company presented its master plan, the Litani-Jordan-Negev Plan, that would have made possible the irrigation of 3.1 million dunums (a dunum is 1000 sq. meters) in Israel.[24]

5.3 The Parties Compete for Water

The door to Jewish immigration to Palestine was opened under the British mandate. The conditions in Palestine during that period reflected the concern of its population and their fear of Jewish immigration. Palestinian demonstrations were organized frequently, and violence erupted at times. The most famous was the strike of 1936 that lasted over six months, after which the British Government sent the Peel Commission, which appraised the situation and recommended the partition of Palestine between Palestinian Arabs and Jews. The Palestinians rejected the Commission's proposals and waged a new uprising in the summer of 1937. The British Government soon dispatched the Woodhead Commission in 1939, which appraised the practicality of the recommendations of the Peel Commission and introduced modifications to the partition plan. The Woodhead Report included a section on the limitations that scarce water resources placed on the possibility of population resettlement. In follow up of the Woodhead Report, His Majesty's Government convened a round-table conference to produce an agreed solution to Palestine's problems. Arab and Jewish delegates participated, but no agreement was reached. The British Government then issued the MacDonald White Paper in May 1939. The new directive restricted Jewish immigration to 75,000 during the following five years and made its further continuation dependent on Arab consent. It prohibited Jewish land purchases in predominantly Arab districts. Finally, it vaguely provided for a representative government in Palestine within ten years based on Arab demographic majority.[25] While the Arabs opposed partition and favored limitations or even termination of Jewish immigration, the Jewish Agency had opposed

---

[24]Elisha Kally with Gideon Fishelson, "Water and Peace", Praeger, Westport, Connecticut, London, pp. 6.

[25]Yoav Gelber, University of Haifa, "Jewish - Transjordanian Relations 1921 - 1948", Frank Cass Publishers, London, U.K and Portland, Oregon, USA, 1997, pp. 141.

limitations on Jewish immigration, and had reservations about partition. A 1945 *aide-memoire* on Palestine described Zionist reservations on partition: "With the sea in the West, the Jordan and Power and Potash concessions in the East, the chief water resources in the North, and the main land reserves in the South, any partition scheme seems bound to disrupt the country's economic frame, and wreck the chances of large scale development."[26]

The Zionist reservation was thus based on the geographic confines of the territories earmarked for the Jewish state. The natural resources of Palestine, the Zionists contended, did not support the idea of its partition. The Arabs had their own reservations about partition because it gave the Jews a lot more land than their numbers justified, and because the Palestinians did not accept to relinquish any part of their land to immigrant Jews.

Additionally, the Zionist Organization, acting through the Jewish Agency, made several intensive attempts to have Transjordan opened for Jewish immigration and settlement, but these attempts did not succeed despite the support that the Agency solicited from some tribal chiefs there.[27]

Soon after the Peel Commission report was published, the Government of Tranjordan initiated a study of Jordan Valley water resources. Mr. Michael G. Ionedis, a British engineer working for the Government of Transjordan and who had done work on the Tigris and Euphrates basins in Iraq, conducted the study. Ionedis published his findings in 1939. He estimated, for the first time, the available water resources and irrigable land in the Jordan Valley and concluded that the water resources were not adequate to support a Jewish state. His plan was based on water availability from the Jordan River and Lake Tiberias. According to his plan, both sides of the Jordan Valley would receive water through surface canals. Lake Tiberias would be used as the main reservoir for the collection of excess winter water including Yarmouk floods. Ionedis suggested conservation measures in the side wadis to improve existing irrigation schemes. He also proposed the diversion of the Yarmouk waters into a canal down the east side of the Jordan Valley to expand irrigation there, with the excess flood waters diverted for storage in Lake Tiberias. This idea was taken up seriously in the 1950's, as shall be seen later. It is interesting to note that diversion of Yarmouk floods to Lake Tiberias was recommended by Arab (Franjiyeh, Ionedis) and by Zionist (Bourcart, Rutenberg) plans.

About the same period of time, the United States Department of Agriculture appointed a land conservation specialist, Professor Walter C.

---

[26] 6 April 1945, cited in Weizmann Letters 1968, vol. XXII, pp. 299.

[27] Yoav Gelber, University of Haifa, "Jewish - Transjordanian Relations 1921 - 1948", Frank Cass Publishers, London, U.K and Portland, Oregon, USA.

Lowdermilk, to look into means of land conservation in the Near East.[28] In 1939, after a flight over the area, Lowdermilk conceived of the idea of the Jordan Valley Authority. His plan focused on the use of the Jordan River and the Auja (Yarqon) River, and the collection of winter floodwater. He proposed bringing seawater from the Mediterranean to the Dead Sea, an idea found in the analytic study of Bourcart in 1899 (see above). The transfer would compensate the Dead Sea for water diverted from the Jordan River for use outside the basin (irrigation of the Negev in southern Palestine). Similar ideas were concurrently advanced by Simcha Blass of Mekorot.

Lowdermilk's plan, published in 1944, reinforced the Jewish argument that proper water management would generate resources for 4 million Jewish immigrants in addition to 1.8 million Palestinians and Jews already living in Palestine. He included the Litani in his regional management scheme, and proposed the use of the Jordan and the Litani to irrigate the Negev, the diversion of the Yarmouk into Lake Tiberias, and gravity flow canals down the foot of the escarpments to irrigate the Jordan Valley, in addition to the Medditerranean-Dead Sea Canal. His recommendations were included in his book, "Palestine, Land of Promise," and were seriously taken up in the water planning of Israel in the early 1950's.

Concurrently, Simcha Blass prepared in 1944 what some viewed as the most comprehensive plan for solving the water shortage problem. An extension of his earlier work, the plan included the major water resources in the region: the Jordan, the Yarmouk, the Auja (Yarqon), water from springs, underground water, flood water, drainage water, and sewage water. Blass' plan also considered another problem that the diversion of the Yarmouk and the Jordan would cause, namely, the lowering of the level of the Dead Sea. To solve this problem, he recommended conveying water from the Mediterranean to the Dead Sea, which would also enable the generation of electricity - an idea Theodore Herzl had proposed eight decades earlier,[29] which was elaborated analytically by Bourcart in 1899 and reiterated by Lowdermilk in 1944.

The World Zionist Organization later commissioned two American engineers, Hays and Savage to prepare an all-inclusive study, "Water Resources in the Land of Israel", which would form the basis for the laying out of a national water network. The plan was published in 1948 as "TVA on the Jordan". They included Syria, Lebanon, and Transjordan and thereby the suggestion for a TVA duplication. If all the waters in the region, including the Yarmouk and the Litani were included, the Plan envisaged the expansion of

---

[28] Elisha Kally with Gideon Fishelson, "Water and Peace - Water Resources and the Arab - Israeli Peace Process", Praeger, Westport, Connecticut and London, 1993, pp. 6.

[29] Ibid pp. 6,7.

irrigated areas from 10,000 acres to 110,000 acres, and the supply of industrial and domestic water for about 4 million inhabitants in Palestine. The Plan mentioned the possibility of using a canal from the Mediterranean to the Dead Sea to maintain the Dead Sea level and to generate electrical power. In fact, the Hays - Savage effort served as a detailing of the Lowdermilk Plan.

In 1947, before the partition of Palestine was decreed by the United Nations, the special UN Commission on Palestine, UNSCOP, examined the Hays' Plan. The Commission received a British memorandum criticizing Hays for exaggerated estimates, and for not considering Jordanian rights nor the rights of the Palestine Electric Company (the Rutenberg concession). The British proposed a more limited alternative plan, which included the waters of the Litani, a multiyear water storage for irrigation and electricity generation in Lake Tiberias, and a reservoir for annual storage at Beit Natofa, a depression in northern Palestine, as suggested by earlier studies.

These have been the efforts exerted by the World Zionist Organization, by Jordan, by the U.S. Department of Agriculture, and by the Mandatory power, Great Britain. They all had the objective of utilization of the waters of the Jordan River Basin. The Zionist plans attempted to maximize the water resources available to Palestine by adding outside resources like the Litani. On their own, the Jordan Valley water resources were not sufficient to support a Jewish state alongside a Palestinian state. This was clearly demonstrated in the works of Ionedis on behalf of the Government of Transjordan, thereby the inclusion of outside resources by the Zionist plans. The ultimate beneficiaries of the various plans, however, were not the same, and confrontation was soon to follow, not over water, but over Palestine itself.

6. End of the Mandate- Creation of Israel

More Jews were brought into Palestine despite the provisions of the White Paper. Immigration intensified during the Second World War, as a result of the atrocities committed against the Jews in the European Continent. A three-way confrontation developed in Palestine: Jews against Palestinians, Jews against Government (British Mandate), and Government against Palestinians. Throughout the 1940's, Palestine witnessed more turbulence and violence than it ever had before. On water, especially that of the Jordan River Basin, Jews and Arabs had their own conflicting plans as advanced by their respective studies (see above).

As the turmoil continued, Britain took the matter to the newly established United Nations. The General Assembly voted in favor of the partition of Palestine between Jews and Palestinians (resolution 181 of November 29, 1947), similar to what the Peel Commission recommended in 1937. The Arabs rejected the partition resolution. Transjordan maintained army battalions of its own in Palestine by request of the Mandate government, as part of the Palestine Garrison. Violence re-erupted after the UN Resolution was passed, and Jewish engagements with the Arab Legion became frequent. Britain decided to end its Mandate on 15 May 1948.

As the evacuation date approached, the Arab countries consulted through the newly founded Arab League and agreed to have member states move troops into Palestine as soon as Britain departed, in order to prevent a Jewish take over and to protect the Palestinians against Jewish terrorism. The League also organized a force of volunteers under Fawzi Kawiqji. On May 15, the day Britain finished evacuation, the Jews announced the establishment of the Jewish State and called it Israel. A few hours later, President Truman of the United States, against the advice of his Secretary of State, extended recognition to the new state. The Soviet Union followed suit, and so did many other countries that were members of the United Nations.

The Arab states moved in military forces to protect Palestine and Palestinians against a Jewish takeover. Jordan sent in two more battalions in addition to the three battalions she had maintained in Palestine as part of the Palestine Garrison. Iraq sent troops through Jordan to Palestine, and so did Syria and Egypt. The Israeli forces, however, outnumbered the Arab forces combined, and were much better equipped and trained. The Arab Legion battled with Israeli forces in territories earmarked for the Palestinians in the UN Partition Resolution, and in Jerusalem. It was able to retain most of such territories. Iraqi troops helped retain the northern part of the West Bank, earmarked for Palestinians, in Arab hands. The Syrian forces occupied portions in northern Palestine earmarked for the Jewish state under the Resolution. Egypt, the largest and longest independent country, retained the Gaza strip. The 1948 military confrontations ended with a cease-fire, and Armistice Agreements were concluded in 1949 between Israel and the surrounding states: Egypt, Lebanon, Jordan, and Syria. The portions of Palestine held by the Syrian troops in the northeast were made the Demilitarized Zones under the Armistice Agreement between Syria and Israel. The Israeli forces occupied the rest of Palestine, more than was allocated to them under the Partition Resolution, and sent waves of Palestinians as refugees to the surrounding Arab territories. Many took refuge in the remaining part of Palestine (West Bank and the Gaza strip), and many others became refugees in Lebanon, Syria, Jordan and Egypt. Jordan hosted the highest percentage of Palestinian refugees. The creation of Israel at the expense of Palestinians, and the parallel creation of the problem of Palestinian refugees, marked the future of the Middle East and was the root cause of the domestic and regional turbulence that followed.

In 1950, the West Bank united with, and became an integral part of, the Hashemite Kingdom of Jordan (Figure I.7). None of the Arab states supported or recognized that unity, claiming that it would dissolve the

Palestinian entity. As a matter of fact, only Britain and Pakistan explicitly recognized the unity of the West Bank with the Hashemite Kingdom. However, when Jordan joined the United Nations in 1955, there was no reservation expressed by any member state as to the territories of the country that included the West Bank, which was tantamount to implied recognition. The effects of implied recognition are no different from those of express recognition. The international community at large dealt with Jordan thereafter in a way that cannot be construed except as recognizing Jordanian sovereignty over the West Bank. The multitude of agreements, protocols and conventions that Jordan entered into, without a single reservation by others as to their applicability to the West Bank, support international recognition.

7. Conflicting Water Plans

The early water planning displays the incompatibility of plans advanced by the Jewish Agency on the one hand, and the Government of Transjordan on the other. Competition over water became more serious after the creation of Israel. A state of war continued between her and the Arab states, and a sudden increase of the population in Jordan and the surrounding countries dictated urgent needs for accelerated economic development. Worst hit was Jordan, whose national economy was marginal. Jordan commissioned the British consultants, Sir Murdoch MacDonald and Partners, in 1949, to appraise potentials for the use of the waters of the Jordan River Basin. The consultants' work used a good part of the previous work of Ionedis (1939). Israel, on her part, published its seven years plan in 1953, expressing her intent to harness the waters of the Jordan River Basin. The Israeli efforts were based on previous work done by Rutenberg, Lowdermilk, Blass, Savage and Hays. There was limited additional field work.

7.1 The MacDonald Plan (Jordan):

The British consultants prepared their report by March 1951. MacDonald followed Ionedis' general scheme, projecting canals on both sides of the Jordan River, and allowed for the use of Lake Tiberias as a reservoir to store the excess flood waters of the Yarmouk and the Jordan rivers for use in the dry season, when the natural supplies are at their low ebb. A feeder canal from Lake Tiberias would convey water to the canal on the East Bank, which would also be fed by the Yarmouk River.

The basic principle of the MacDonald Plan has been adhered to by the Arab plans that followed. MacDonald asserted that "the general principle, which to our mind has an undoubted moral and natural basis, is that the waters in a catchment area should not be diverted outside this area unless the

Figure (I.7)- The Hashemite Kingdom of Jordan (1950-1988)

requirements of all those who use or genuinely intend to use the waters within the area have been satisfied."[30]

While the Arab states concerned approved of this principle, Israel totally opposed it because it would abort the Israeli plans, inherent in the works of Lowdermilk, Hays, and Blass, to transfer waters of the Jordan River outside its basin to the Negev. The priority of in-basin use constituted a basic difference between Arab and Israeli approaches to the utilization of the Jordan waters.

7.2 The Bunger Plan (Jordan)

The Jordan Government soon followed the MacDonald Plan with another in 1952. Mills E. Bunger, an American engineer working in Jordan for the Technical Cooperation Agency (TCA), the predecessor of the Agency for International Development (USAID), discovered a reservoir site on the Yarmouk River while he was on a flight from Amman to Beirut. A dam could be built at the mouth of that reservoir site where the river narrows and starts to acquire its name, the Yarmouk. Upstream of that dam location, several tributaries feed the river. Their confluence, Maqarin (the place where several conduits meet), forms the reservoir site. Each tributary has its own name (Shallala in Jordan, and Allan, Hreir, Dhahab and Zaydi, in Syria) and their joining (*qaran*) form the Yarmouk River. The part of the Hijaz Railway that connected Dara'a in Syria with Haifa in Palestine passes through the Yarmouk gorge, and has a station at Maqarin. From that railroad station on downstream, the Yarmouk or the adjacent railroad track, whichever is closest to Jordan, forms the boundary between the Hashemite Kingdom of Jordan and the Syrian Arab Republic. Some 25 kilometers downstream, the Syrian territories on the northern bank end, and the Palestinian territories, under British Mandate, start. The Armistice Agreement concluded between Syria and Israel in July of 1949 produced an "Armistice Line" separating the Israeli and Syrian forces. The areas lying between the Armistice Line and the international borders between Palestine and Syria were made demilitarized zones (Figure I.8). Israel had a frontage of about 7 kilometers on the Yarmouk, before the river joins the Jordan River at Baqura. The Yarmouk Triangle in Israel (Figure I.9), 2600 hectars in area, has frontage and water rights on the water bodies that adjoin it, namely, the Yarmouk, the Jordan and Lake Tiberias.

The United Nations Relief and Work Agency for Palestinian Refugees, UNRWA, established in 1949, moved in 1951 to aid the Jordanian Government in surveys for the "Bunger" dam on the Yarmouk at Maqarin.

---

[30] Sir Murdoch MacDonald and Parners, "Report on the Proposed Extension of Irrigation in the Jordan Valley", London, March, 1951.

Figure (I.8)- DMZ between Syria and Israel

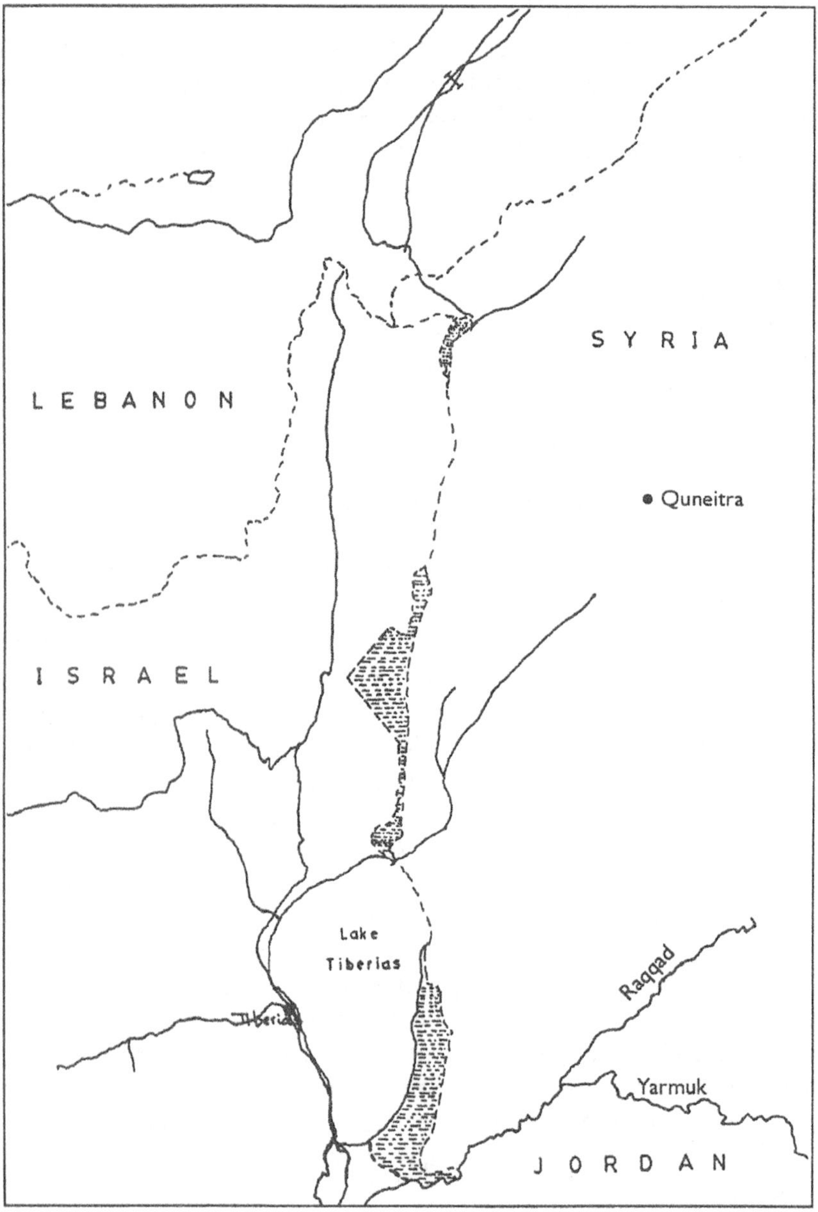

Figure (I.9)- The Yarmouk Triangle

The TCA (also known as Point 4) of the United States was attempting to shore up the economy of Jordan after the 1948 influx of the Palestinian refugees. The anxiety of TCA in this respect and the concern of UNRWA to make some of the refugees self-supporting promoted a pooling of efforts and resources directed at river development. Thus the TCA put up $929,000, UNRWA $856,000, and the Jordanian government some $200,000 (out of British loans) for the preliminary work on a dam at Maqarin, first nicknamed as the "Bunger Dam". It is to be noted that the Yarmouk flow at Maqarin at the time was about half of the entire flow of the River as measured at Adasiyya, shortly before it joins the Jordan River. The dam would thus regulate half the Yarmouk flow and the other half would flow downstream unregulated. The dam, therefore, would not deprive Israel of the waters needed to irrigate the Yarmouk Triangle. To the contrary, when the Plan was first conceived, it was stipulated that:

"Firstly, the proposal should aim at the maximum development of the Jordan Valley without involving international negotiations which might not be feasible at the present moment and, secondly, should comprise a scheme which could easily be fitted into any subsequent scheme derived from the use of the Lake Tiberias."[31]

The Bunger Plan was based on the utilization of the Yarmouk River waters for irrigation and the generation of electric power in such a way as not to impinge upon the rights of others or to hamper any other regional scheme that might be devised. In practical terms, the Plan would assure that enough Yarmouk water would still be available downstream of the dam for Israel's use. It also would avoid the need to get into Jordanian negotiations with Israel to obtain her acquiescence to implement the Plan. Such negotiations, even contacts, would not have been possible given the state of war between the two countries and the big issues at stake that involved the Palestinians and other Arab states. Harza Engineering Company of Chicago, Illinois, USA was commissioned to do the study and detail the Plan.

Subsequent to the Bunger Plan, Jordan and Syria concluded an agreement compatible with its provisions on June 4, 1953 for the utilization of the waters of the Yarmouk. Syria retained the rights to use the waters of the springs that emerge in her territories above elevation 250 meters above sea level, would benefit from 75% of the power generated at the dam, and would pay 5% of the cost of the dam construction. The Bunger Plan called for:

1. The building of a dam, 160 meters high on the Yarmouk River at Maqarin with a storage capacity of about 500 mcm for use in irrigation in the Jordan Valley in Jordan.

---

[31] T. R. Welling, Director TCA/Jordan, *Statement on the Yarmouk-Jordan Project,* presented before the Jordan Development Board, July 1, 1952.

2. A canal to lead from the dam on the south bank of the Yarmouk to Adasiyya, where a diversion dam would divert water from the Yarmouk River and the Maqarin canal southward into an East Ghor Canal almost to the Dead Sea.
3. A small dam across the Jordan River shortly below the Jordan - Israel cease-fire line to facilitate pumping of the water from the Jordan River to a West Ghor canal leading to Jericho.
4. A power plant at the dam site at Maqarin and another at Adasiyya to utilize the drop from the Maqarin canal to the Yarmouk at that location. The power output was estimated at 281 million kilowatts.hr annually.

The Bunger plan was the first to propose a site for storage of the Yarmouk floods other than Lake Tiberias, which was considered by the previous plans as the natural reservoir. The site was convenient to Jordan both technically and politically. It would enable the generation of power, and would maximize water storage on Arab lands rather than storing Yarmouk waters in Lake Tiberias, and the Arabs thought it unwise to store their water in a reservoir under the control of their enemy.

After the studies of the Maqarin/Bunger dam project were done, UNRWA, in agreement with the Jordan Government on March 30, 1953 and December 1953, allocated $40 million from its rehabilitation funds for the dam's construction. The total cost was estimated at $70 million. In July 1953, UNRWA, TCA and the Jordan Government appropriated funds for the initial expenditures, and the recruitment of workers for the scheme was begun.[32] It was at this point that the Plan was suddenly dropped and the United States Government, which had given her official endorsement, proposed another plan instead. It turned out that Israel had protested to the State Department of the United States and to the United Nations, against the Plan in the joint effort. Israel claimed that her rights were jeopardized. The implementation of the Bunger Plan, along whose lines Jordan and Syria drew their joint cooperation to develop the Yarmouk, was terminated. Bunger's continued promotion of it irritated American officials, as it caused difficulties to subsequent mediation efforts by the United States (see next chapter). Mills Bunger apparently fell in disfavor and was transferred to Brazil. He preferred to resign than to accept a transfer made against his will.[33]

---

[32] Georgiana Stevens, "The Jordan River Valley", International Conciliation, No. 506, New York, January 1956.
[33] American Friends of the Middle East, Inc., "The Jordan Water Problem: an Analysis and Summary of Available Documents", pp. 43.

7.3 The All Israel Plan (Israel)

Concurrent with the efforts that Jordan was undertaking to develop the Yarmouk, Israel was also crystallizing plans for the development of water resources, some of which were the subject of Jordan's plans for development. Since her statehood, Israel became active in preparing water plans to cope with the high demands of incoming European immigrants, and to confirm the commitment to agriculture inherent in the ideology of Zionism.

By 1950 the government of Israel had appointed a commission to plan the National Water Carrier (NWC), envisaged in the Blass and Lowdermilk plans and in the subsequent detailing by Hays and Savage. J. S. Cotton, an American engineer working for the government of Israel, was given the task of converting the Hays-Savage plan into an All Israel Plan that could be carried out between 1951 and 1955. The All Israel Plan included the draining of the Huleh marshes, an idea first advanced by Baron Rothchild in 1910 and included in the Rutenberg Concession, and the Jordan-Negev master plan that would divert the Jordan River water to the Negev (Lowdermilk- Blass ideas). The Plan included the NWC and branches for distribution, and regional water works. The NWC was to start with an intake on the Jordan River at a point south of and close to Jisr Benat Ya'qoub, cross the western escarpment, and deliver the water to a reservoir in Beit Natofa Valley. From there, the NWC would proceed through the coastal plain to the Negev (Fig I.10). The implementation of the NWC started in 1951 with work on the Yarqon (Auja) - Negev section of the Carrier. When Israel attempted to start work on the intake structure on the Jordan River at Jisr Benat Ya'qoub, it was met with Syrian opposition expressed in complaints to the United Nations Truce Supervision Organization (UNTSO), and to the Security Council, and, occasionally, in shelling from the overlooking Syrian Golan Heights. The intake point for the diversion was located in a demilitarized zone between Syria and Israel whose final status was not decided. Clashes of this type took place as Israel attempted, earlier that year, to drain part of the Huleh marshes located in yet another demilitarized zone there. Israel's work was terminated by a decision of the U.N. Security Council and by U.S. pressure voiced by the Secretary of State, John Foster Dulles, who threatened to cut aid to Israel if she did not abide by the Security Council Resolution (see below).

An Israeli technical report produced in 1951 indicated that the Beit Natofa Valley was not suited for a reservoir of the type the NWC required, and recommended that Lake Tiberias be used for the purpose instead. With the above military complications arising at Jisr Benat Ya'qoub and the geological conditions at Beit Natofa, Israel moved the intake structure down to a point on the northwest shore of the Lake at Tabgha.

When Jordan announced in 1951 her intentions to divert Yarmouk water to irrigate the Jordan Valley under the UNRWA-supported Bunger Plan, Israel announced she would close the gates of an existing dam built on the Jordan River shortly after it leaves Lake Tiberias. The dam was built under the 1921 Rutenburg Concession (see above) to draw water from the Jordan River as it exits the Lake to the power house at Baqura by means of a canal. The closure of the gates transforms Lake Tiberias into a huge reservoir and prevents water from flowing past the Lake into the Lower Jordan. The moves made by the two adversaries, Jordan and Israel, were in effect a mutual tit for tat. Israel's dependence on the Yarmouk paralleled Jordan's dependence on the Jordan. Action by Jordan on the Yarmouk would prompt Israeli action on the Jordan. As it stood at the time, Israel drew water from the Yarmouk to irrigate the Yarmouk Triangle, and was counting on the Yarmouk flow to be made available to help irrigate her West Jordan Valley (Beit Shean). Jordan, on the other hand, depended on the flow of the Jordan River to irrigate some 6800 hectars on the East and West banks of the Jordan River. Should these unilateral moves be implemented, Jordan and Israel would need replacement water to sustain the agriculture in their respective territories that depended on Jordan River and Yarmouk waters.

Financing of Israel's water plans had to rely on outside assistance, as was the case with the Jordanian plans, and the United States was the most likely donor that the two parties would approach. This fact enabled the United States to see the conflicting plans, and made her withdraw support both the Bunger Plan and the Israeli plans, since they were not coordinated and one could be harmful to the other.

7.4 The Syrian - Lebanese Plan

In 1951, and again in 1959, Syria and Lebanon entertained ideas aiming at diverting the flow of the Hasbani and Banyas tributaries to the Litani in the Beqa'a Valley of Lebanon. The scheme was an out-of-basin transfer that would have deprived other downstream riparian parties (Israel and Jordan) of parts of their rightful water shares. It would have contradicted the principle advocated in the MacDonald Report and endorsed by the Arabs.

The Plan should be looked upon in the context of the conflict between the Arab states and Israel, and the non-recognition by the Arabs of Israel's claims to land or water.

8. Unilateral Implementation of the Plans

Jordan and Israel were quick to have their plans prepared for the development of water resources of the Basin. Jordan's efforts to implement the Bunger (Maqarin) Plan were aborted by withdrawal of financial support soon after their take-off because of Israeli announcements to close off the dam on the Jordan River, and because of her protests to the United States and the United Nations. The facts on the ground and the state of war prevailing between the two countries did not permit the preparation of integrated or joint plans, nor even initiate any level of coordination. No water-sharing agreements were considered either.

The support by UNRWA and the United States was necessary to have the Jordanian Bunger Plan implemented. The support was justified by the need to buttress the economy of Jordan and promote her development efforts on the one side, and, on the other, to assist in alleviating the plights of the Palestinian refugees that poured into Jordan in the wake of the 1948 Israeli war with the Arabs. Actually, Israel's protests had to do with ignoring her as a rightful riparian on the Yarmouk, when, in that capacity, she had to be included in the development effort. Jordan could not afford politically to start any talks with Israel or strike any deal with her at a time when a state of war between Israel and the Arab states (including Jordan) prevailed. The withdrawal of support by the United States and UNRWA to the Plan was a blow to the Jordanian development efforts that had just started to take-off, and triggered worries in the Jordanian leadership as to how far Jordan could go in the development of Yarmouk resources without having to succumb to Israeli demands.

Israel started work on her All Israel Plan that included the drainage of the Huleh marshes in 1951. Part of the work had to be done inside a demilitarized zone (DMZ), delimited by the Syrian-Israeli Armistice Agreement of July 20,1949 (Figure I.8). Syria considered Israel's work an infringement on the DMZ and Israel's action provoked a number of military clashes. Syrian forces overlooked the Huleh valley from the hilltops of the Golan to the east, and targets in the valley were within the range of Syrian artillery. Israel, on the other hand, had Syrian citizens in the DMZ and close-by areas within the range of their riffles. In the spring of 1951, there was shooting in the DMZ, with Israel pushing out Syrian citizens from their villages of Bekara, Naymeh and Mazra'at el Khouri, which triggered Syrians to shell the Israeli equipment. The Israeli Air Force bombed al Himmeh, a Palestinian town on the northern bank of the Yarmouk that came under Syrian administration after the 1948 war. The United Nations intervened and the Security Council ordered a return of Syria citizens to their homes in the DMZ, but their villages and homes had been razed. During those and the following years, some incidents were designed to protect property, others were designed by Syria to interfere with the implementation of the water projects, and other incidents were triggered by Israel to remove unwanted Syrian population. However, these incidents kept the United Nations Truce Supervision Organization (UNTSO), observers busy reporting to headquarters, and the Security Council looked into several complaints by Syria and counter-complaints by Israel. Israel succeeded in draining and reclaiming some 6,000 hectars of the Huleh marshes and saved about 60 mcm of water that were heretofore being lost to evaporation, using the water to irrigate the reclaimed land.

Figure (I.10)- National Water Carrier of Israel

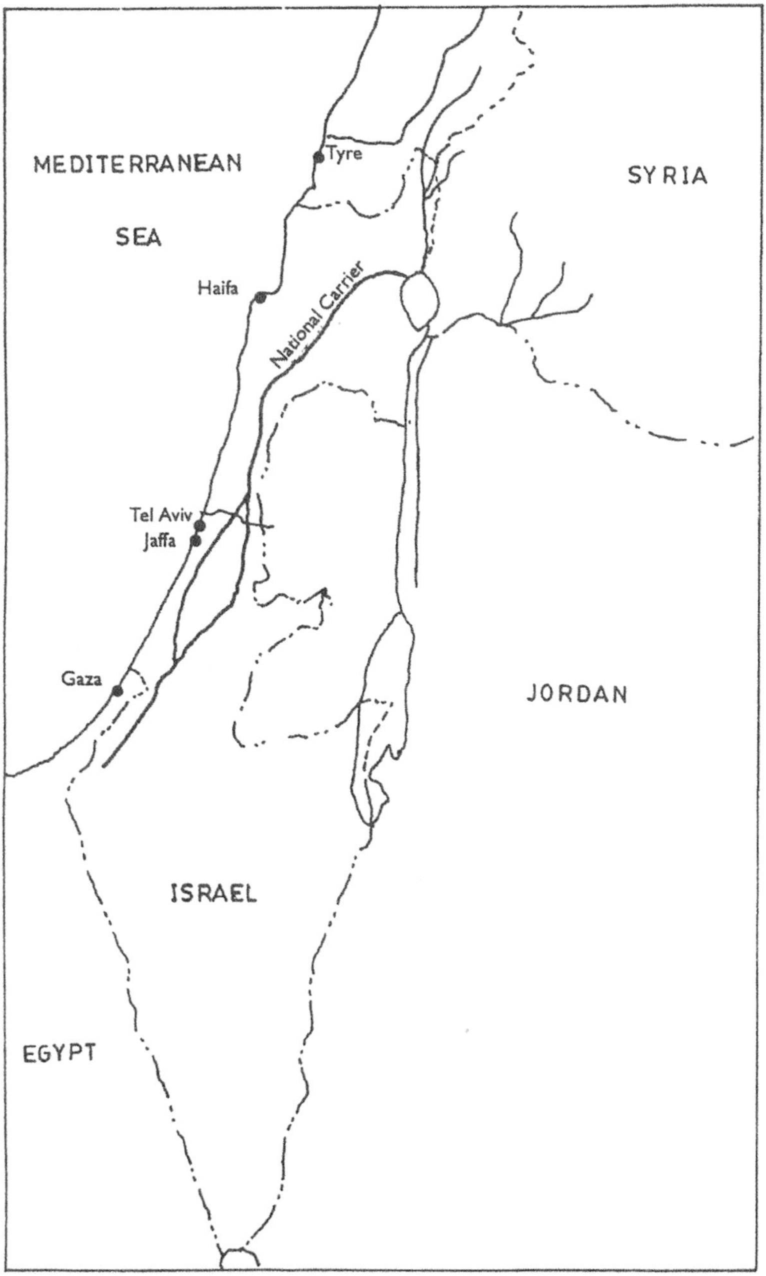

Figure (I.11)- Southern and Northern DMZ Between Syria and Israel

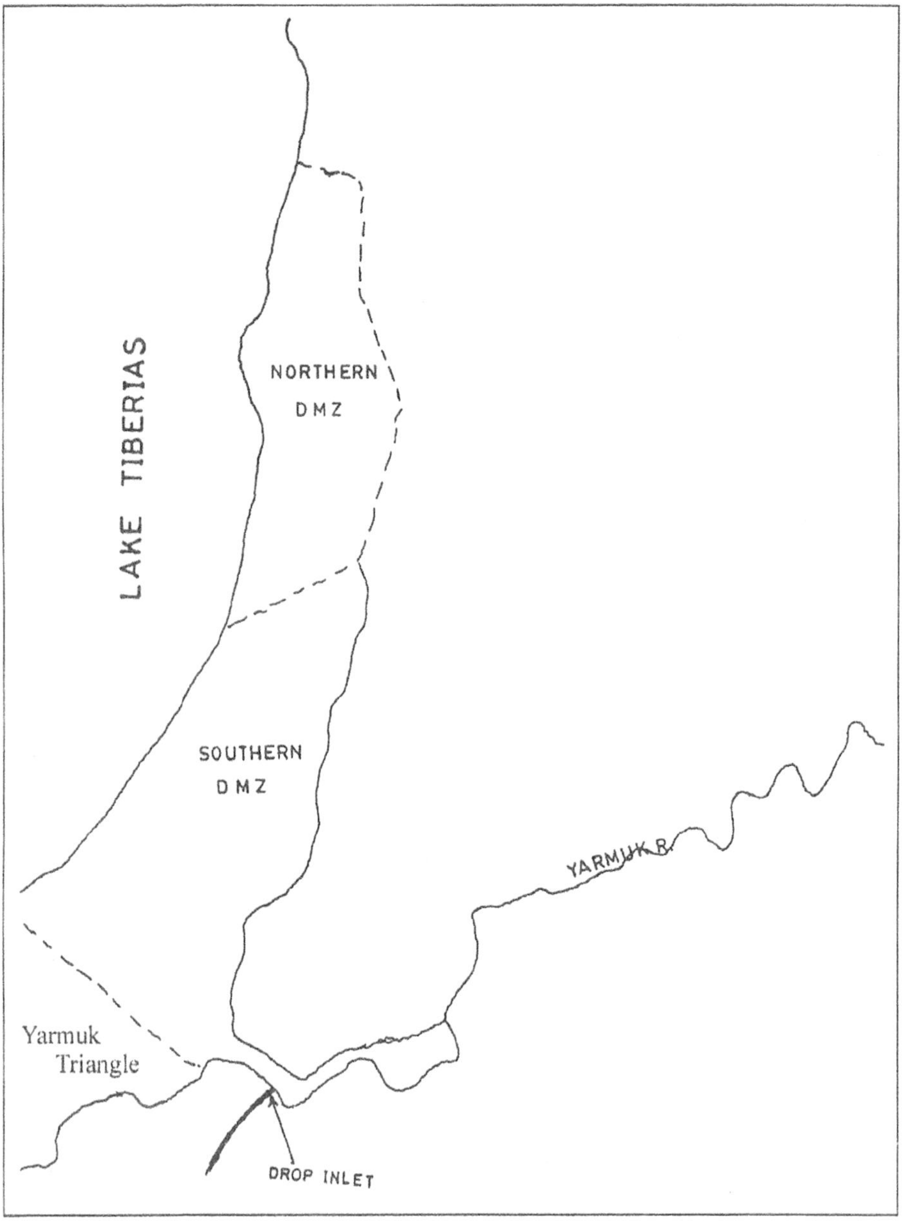

In 1953, Israel decided to initiate work on the diversion project and moved construction equipment to the diversion site in the DMZ immediately south of Jisr Benat Ya'qoub. The same pattern of clashes was repeated. In March 1953, incidents and shooting in the DMZ became serious and frequent. Throughout the period, these incidents threatened to shatter the Syrian - Israeli Armistice Agreement. In a letter dated October 16, 1953,[34] the Permanent Representative of Syria at the U.N. complained to the Security Council that Israel had started to divert the Jordan River into a new channel in the demilitarized zone with a view to making it flow through her own territory, in violation of Syrian-Israeli Armistice Agreement. Major General Bennike, Chief of Staff of UNTSO, made a decision that the Israeli diversion, which began on September 2, 1953, was contrary to the provisions of the General Armistice in relation to the demilitarized zone,[35] and requested the Israeli Government to cease working on the west bank of the River Jordan in the DMZ. In the Security Council, an extensive debate ensued. When the question was raised as to whether Israel was bound by the 1922, 1923 and 1925 agreements on the Jordan waters,[36] the Israeli Representative stated that Israel refused to consider herself bound by the obligations assumed by the Government of Palestine concerning the Jordan River because Israel did not inherit the international treaties signed by the United Kingdom as Mandatory power.[37] On October 27, 1953,[38] the Security Council unanimously adopted the following resolution:

"The Security Council

Having taken note of the request of the Chief of Staff of the Truce Supervision Organization dated 23 October 1953 (S/3122)

Desirous of facilitating the consideration of the question, without however prejudicing the rights, claims or position of the parties concerned,

---

[34] U.N. Document S/3108/Rev. 1.

[35] UN Document S/3122.

[36] League of Nations Treaty Series, Vol. XXII, p.366. The exchange of notes between France and Britain stipulated that any existing rights over the use of the waters of the Jordan by the inhabitants of Syria should be maintained unimpaired.

[37] Israel changed her arguments on this concept with time. In the bilateral peace talks with Jordan in August 1994, Israel cited the Rule of Succession of States as the basis for upholding border changes between Jordan and Palestine in 1946.

[38] Five days earlier, Ambassador Eric Johnston had started his trip to the Middle East to mediate an agreement over the Jordan River basin (see below).

Deems it desirable to that end that work started in the demilitarized zone on 2 September 1953 should be suspended during the urgent examination of the question by the Security Council

Notes with satisfaction the statement made by the Israel representative at the 631st Meeting regarding the undertaking given by his Government to suspend the works in question during the examination,

Requests the Chief of Staff of the Truce Supervision Organization to inform it regarding the fulfillment of that undertaking."[39]

It is to be noted that in this regard Israel did not abide by the Security Council resolution until the United States Government made serious statements to suspend economic aid to Israel if she did not abide.[40]

To avoid these complications, and as a result of technical findings pertaining to the geological conditions of the reservoir in Batofa Valley and its unfitness for storage of the NWC water, Israel moved the intake for the diversion to the northwest corner of Lake Tiberias at Tabgha. The Lake replaced the Batofa as a storage reservoir for the NWC, and the new diversion site was a fortified structure built inside the rocky mountain, immune to damage by Syrian shelling or bombardment.

9. Involvement of the United States

The United States was not without historical interest in the region. As narrated above, United States' missions came to the Jordan River basin and the Dead Sea as far back as 1837. The United States, at the time of President Woodrow Wilson's administration, sent the King-Crane Commission in 1918 to get acquainted with with the region and its problems. The United States was informed of the intentions of Britain and France to have the League of Nations issue a decree asking each to exercise, on behalf of the League, mandate powers over their respective territories as agreed between their foreign ministers in 1916. It was also appraised of the Balfour Declaration and the intentions of the British Government to have a national home for the Jews established in Palestine, for which the United States granted support. The King-Crane Commission reported the Zionist argument for a Jewish National Home in Palestine:

"...There is no need of displacing the present population, for with afforestation, modern methods of agriculture, utilization of water power,

---

[39] U.N. Document S/3128.

[40] The U.S. eagerness to cool the situation was more likely meant to improve the cold atmosphere with which Ambassador Eric Johnston, the personal envoy of President Eisenhower, was received in the Middle East on his first trip there.

reclamation of waste lands, scientific irrigation, and the like, the land can contain several times its present number of inhabitants".

The United States introduced herself directly to the Palestinian issue in 1946 when a joint United States-United Kingdom mission came to Palestine to appraise the situation before Britain relinquished the Mandate over Palestine to the United Nations. Finally, the United States, during the Truman Administration, was the first country to recognize the State of Israel upon its proclamation on 15 May 1948.[41] It soon became the source of major support for Israel, and was hopeful to maneuver to have Israel accepted in the region, resolve the conflict and to make peace between Israel and the Arab countries.

The fallout that resulted from the failure of the Arab states to protect Palestine and the Palestinians against Zionist designs affected all Arab states. The domestic front in the recently independent Syria after the 1948 war was not very stable. A military coup, the first in a series of such coups, led by Colonel Husni Zaim, and covertly sponsored by the United States,[42] seized power in late March 1949. Zaim was reported to have sent a secret message to David Ben Gurion, the first Prime Minister of Israel, in April 1949, offering to sign a separate peace treaty with Israel, settle some 300,000 Palestinian refugees in Syria and establish a joint militia in return for some border modifications along the cease fire line, and half of Lake Tiberias.[43] Ben Gurion declined the offer and proceeded to have an Armistice Agreement signed that year. In less than six months, Zaim was overthrown and killed by another military coup led by his subordinates, Sami Hinnawai, Fawzi Silu and Adeeb Shishakly. A third coup was mounted by Shishakly and ousted Hinnawi. A fourth was mounted against Shishakly in 1954 by two majors, Mustafa Hamdoun and Abdul Ghani Qannout that prompted Shishakly to resign and flee the country.

In Egypt, a military coup ousted King Farouk in July 1952, and replaced him with his young son, Fuad. A year later, the Monarchy was abolished and Egypt was proclaimed a republic with a president, General Mohammad Najib, heading a Revolutionary Command Council. The Egyptian

---

[41] A review of the U.S. position and of the respective officials was broadcast by CNN in an interview with the late Clark Clifford, former Secretary of Defense and advisor to President Truman. The interview was rebroadcast on the day Mr. Clifford passed away in October 1998. He said in the interview that the White House, against the advice of the Secretary of State, prepared the statement of recognition before they knew what the name of the Jewish state would be.

[42] Miles Copeland, "The Game of Nations", Weidenfeld and Nicolson, London W1,1969; pp. 42.

[43] Aaron Wolf, "Hydropolitics Along the Jordan River", United Nations University Press, 1995.

confrontation with Britain started in order that British troops vacate the Suez Canal Zone and because of conflict over the Sudan. The United States became aware of the Egyptian coup shortly before it was mounted, and maintained contacts with it through a delegate of the "Free Officers". It was not surprising that some tacit support was given to the coup after it was mounted, especially in the face of British plans to abort it using the British forces stationed in the Canal Zone.

In Lebanon, the strong political party of the Syrian nationalists, led by Antoine Sa'adeh, was routed and their leader executed in July 1949. The Party retaliated two years later by assassinating the Lebanese Prime Minister, Riyadh al Sulh while on an official visit to Jordan. President Bishara al Khouri resigned his post and was succeeded by Mr. Camile Chamoun.

In Jordan, the founder of the Kingdom, King Abdullah Ibn al Hussein, was assassinated in the Aqsa mosque in Jerusalem on 20 July 1951. His son, Talal, ascended the throne only to abdicate less than a year later, and his son, al Hussein became king at the age of 17.

Elsewhere in the world, the post war era carried with it more challenges to the United States' leadership. Europe started to recover from the six years of extreme turbulence that the War brought. The North Atlantic Treaty Organization, NATO, was established in 1949, and West Germany was allowed democratic elections and the formation of the Federal Republic. It successfully started economic recovery under the Marshall Plan. In the Asian subcontinent, two independent countries, India and Pakistan, emerged in 1947 after gaining independence from Britain. In China, the nationalists, led by Chiang Kai-shek were forced to flee to Taiwan and a communist takeover established the Peoples Republic of China in 1949; Japan entered an era of reconstruction and rehabilitation after she surrendered to the commander of the American seventh fleet in 1945; Korea plunged into a civil war and the United Nations interfered under American leadership to prevent a communist takeover, and the country was eventually divided along the 42nd parallel into a communist North Korea and a western oriented South Korea. In Southeast Asia, France was on a retreat, giving way to the communists in Vietnam. In Iran, the leftist supported government of General Musaddaq nationalized the oil industry but was soon ousted and the Shah returned from exile. These world developments posed huge burdens on the United States, the leader of the Western World.

The above events were taking place in the Middle East while the Cold War was escalating and the Korean War was on. The British and French influence in the region was gradually fading due to the economic post-war conditions of these two powers and the new power structure in the world. To the Western World, an important strategic region like the Middle East cannot be left in a vacuum for Soviet adversaries to fill, nor could it be left for the peaceful spread of communism that, the West thought, flourished on the want and frustration of people. The United States had to act to protect Western interests in the region, which concerned basically oil, security of Israel, check of the spread of communism and denial to the Soviets of a foot-hold on the coasts of the warm waters of the Indian Ocean, the Gulf, the Arabian Sea, the Red Sea, or the Mediterranean Sea.

The United States decided to wage a two-pronged offensive against the potential spread of Soviet influence in the region. It made plans for military alliances that would confront the Soviets and solicited support for them from Middle Eastern states. An alliance in the Middle East similar to the North Atlantic Treaty Organization would serve that objective. The other component of United States' action was aimed at improving the standard of living of people in the region through social and economic development. The underdeveloped Arab societies, with their increasingly nationalistic trends, and the disenchanted Palestinians, were considered prime targets for communism. The hostility between Israel and the Arabs was the main obstacle to waging an all-out war on want, and mounting a collective effort for accelerated economic and social development. Overcoming that obstacle was viewed necessary to achieve the objective of such development and thus abate the potential threat of the expansion of communism. Water, it appeared, would serve such an objective and more. On the one hand, it could stop the violent clashes that were taking place over it, and, on the other, it could be used as a vehicle for some sort of cooperation between the Arab States and Israel, and an all out offensive against want and poverty could thus be waged. These objectives motivated the United States to get involved.

However, the fundamentals underlying the Arab-Israeli conflict were not confronted in the approach the United States followed. The Arabs viewed Israel as an invading society that had set up a state in Arab lands at the expense of Palestinians. They could not bring themselves to terms with accepting that invasion or any of its results. On top of this concern was the case of refugees and their rights to repatriation and compensation. Any notion of cooperation with the "Zionist entity" was not thinkable, nor was it acceptable before resolving the basic issues of conflict. Postponing that difficult task, the United States had to move with extra care to promote a Unified Plan for the development of the Jordan Valley and the use of the Jordan River waters.

CHAPTER II

Resorting to Diplomacy

The conflicting plans advanced separately by Israel and the Arab countries for the development and use of the Jordan River waters quickly became the source of added tension in the region. The United States had to move to deeper involvement, especially after that it had disassociated itself in 1953 from the Bunger plan that Jordan was hoping to implement with support from the TCA of the United States and UNRWA. The U.S.. had to ease the tensions, and to come up with a workable plan for the development of the basin to which all parties could subscribe. By then, the Marshall Plan in Germany had gained credibility, and defeated Germany appeared on her way towards a peaceful recovery and progress. Plans to promote peace and cooperation in the region would be welcome and supported by the American people. The incentives for a deeper U.S. involvement were: a) the attempt to forge the integration of Israel in the Middle East; b) the assurance of Israel's continued existence and security; c) the resettlement of Palestinian refugees; d) the closure of the Middle East to communism and the influence of the Soviet Union, and, e) the protection of Western influence and interests in the extensive oil reserves of the region. With regard to the refugees, there was mounting congressional pressure in 1953 to resolve the Palestinian refugee problem,[44] and the United States was prepared to contribute towards the financial cost of their resettlement in the host countries and to urge the United Nations and the Western countries to do the same.

To the Arab peoples, none of the above incentives were attractive. They were actually disincentives. Israel, viewed as the result of flagrant occupation of Arab lands at the expense of the Palestinian Arabs, was a body that was foreign to the region, and her acceptance in the region was beyond the convincing ability of any leader. Any attempt to consider the acceptance of Israel was reason to charge its advocate with treason. The intention of the United States to assure her continued existence and security ran against the aspirations of the Arab peoples. The resettlement of the Palestinian refugees in the host countries was unthinkable. Their cause was championed by political leaders of government and opposition alike, as it gave credibility to those champions. Many Arabs favored approaches to the Soviet Union, albeit not to communism, as an alternative to the West, which had been influential in the loss of Palestine to Zionism. Finally, the desire of the United to protect

---

[44] Georgiana Stevens, "Jordan River Partition", Hoover Institution Studies, 1965, pp. 17.

the oil reserves caused Arab concern over the loss of natural resources to monopoly by the West. Again, many Arabs viewed the Soviet Union as a potential ally that could help protect Arab oil interests.

On the other hand, there were Arab leaders who viewed alliance with the West as the strategic choice. The Western societies, unlike the communist Soviets, are believers in the same God worshiped by Arab Moslems and Christians; the West owned modern technology and wealth that the Arab peoples needed to advance and improve their living standards; and, the Arabs had the experience of historical contact with the West in war and in peace, and therefore a taste of what it was like to deal with western societies. Finally, the West came out victorious from two World Wars, and was fighting a major war in the Far East (Korea) against the atheist communists to prevent their takeover of East Asia. In that respect, the West was viewed as a protector of the monotheistic religions against communist take-over.

Motivated by the military clashes between Syria and Israel over the water works and by strategic incentives, President Dwight David Eisenhower appointed Mr. Eric Johnston, on 7 October 1953, with an Ambassadorial rank as his personal envoy to the Middle East. Johnston's mission was to try and work out a unified plan for the development of the Jordan Valley to which Arabs and Israelis could subscribe. He was confronted with a huge task in which he was supposed to bring enemies who were in a state of war to adopt a cooperative plan. The challenge was immediate, because Syria and Israel had just clashed militarily over Israel's work in the DMZ to drain the Huleh marshes, and Israel had defied the decision of the Chief of Staff of the Truce Supervision Organization of 24 September 1953 by refusing to stop work in the DMZ. The issue was taken up by the Security Council, which supported the decision of the Chief of Staff on 27 October (see above). It was at this point that the U.S. Secretary of State, John Foster Dulles, threatened to cut off United States aid to Israel if Israel did not promptly comply with the Security Council resolution. Israel did so the following day, on October 28, 1953. This move on the part of the U.S. helped reduce Arab apprehension to the mission that Ambassador Johnston had started in the Middle East a week earlier.

1. The Chas T Main/ TVA Plan

Prior to Johnston's appointment, and in parallel to the U.S. slowdown of support to the Bunger Plan, UNRWA, upon recommendation of the British Foreign office and with United States approval and tacit participation,[45] moved to propose a plan for the sharing and utilization of the waters of the Jordan River Basin. The objective was that projects resulting therefrom would

---

[45] Georgiana Stevens reported in 1965 that the TVA sponsored study was under UNRWA and U.S. auspices.

assist in alleviating the difficulties facing the Palestinian refugees, especially in Jordan, by creating a viable means of living in the Jordan Valley that could facilitate their resettlement. Additionally, a shared plan would assure coordination, if not cooperation, among the riparian parties in the basin and exclude the likelihood of one riparian inflicting damages on another, willfully or accidentally, by implementing water projects using the same water that a co-riparian was already using. The underlying purpose of UNRWA was spelled out in a letter addressed to the Tennessee Valley Authority (TVA), a government agency of the United States, excerpts of which are quoted below. The TVA, on its part, contacted the American engineering consulting firm, Chas T. Main, for that purpose. A desk study, without field investigations, was conducted utilizing the previous plans for the basin, and Chas T. Main came up with what was called the Unified Development Plan of the Jordan Valley, or Main/TVA Plan. Over two years this Plan was subjected to amendments during the diplomatic work of Ambassador Eric Johnston, until it took its final form in what came to be known as the Unified Plan, or, at times, the Johnston Plan.

In her letter to Gordon Clapp, Chairman of the Board of TVA, the Acting Director of UNRWA, Leslie J. Carven, wrote in November 1952:
" Towards the end of the last year it was already clear that the United Nations Relief and Works Agency for the Palestinian Refugees in the Near East would shortly be entering into a Program Agreement with the Government of the Hashemite Kingdom of Jordan to reserve a very large sum of money for a scheme to use the water of the Jordan and Yarmouk Rivers for irrigating the Jordan Valley and for establishing the refugees there.

Although the main outlines of the scheme had begun to take shape, it was thought proper (in view of the importance of the funds to be committed) to examine all previous plans - of which there are many- for utilizing the waters of the Jordan River and its tributaries, and to determine the extent to which the projects which the Agency might be called upon to finance were really economical and would not be rendered nugatory by other projects undertaken by other interests in the same watershed".

UNRWA's goal was clear. It was to irrigate the Jordan Valley and to establish the Palestinian refugees there, an objective shared by the United States and resented by the Arabs as noted above. In its transmittal letter submitting its report to UNRWA, the TVA mentioned that it had the study conducted with the consent of the State Department. Both UNRWA and the United States had earlier joined in supporting Jordan's Bunger Plan aiming at partially regulating the Yarmouk River at Maqarin, and at irrigating the Jordan Valley. UNRWA wanted to be sure that the project it was supporting would be sustainable and not be rendered nugatory by some future projects in the basin that could be implemented by Israel. Actually, this concern of UNRWA stemmed from the announcement by Israel that she would close off the Jordan flow out of Tiberias if the Jordanian project denied Israel Yarmouk water she had been using (see Chapter I above). In a separate assignment, UNRWA had contracted the services of a legal expert, Mr. Sefit, to conduct a study on the legal aspects of the development of the Jordan Valley.

The essence of the Chas T. Main report is expressed in the letter forwarded to UNRWA's Acting Director. The following excerpt sums it up: "As a problem of engineering, the most economic and the quickest way to get the most use from the waters of the Jordan River System requires better organization of the headwaters on the Hasbani and in the Huleh area to serve the lands by gravity flow within that part of Jordan watershed, and use Lake Tiberias as a storage reservoir for the flood flows of the Jordan and the Yarmouk Rivers. From Lake Tiberias these waters would be made available by gravity flow to irrigate lands on the east and west sides of the Jordan Valley to the south.

Use of the natural reservoir afforded by Lake Tiberias takes advantage of an asset already at hand. There is no known alternative site, at any cost, for a reservoir that would effectively regulate and store the flood flows of the Jordan and its main tributary, the Yarmouk.... Thus the report describes the elements of an efficient arrangement of water supply within the watershed of the Jordan River System. It does not consider political factors or attempt to set this system into the national boundaries now prevailing."[46]

The Main/TVA Plan was submitted to UNRWA and to the Government of the United States on 31August 1953, and, a month after that, to the Arab states concerned. The major features of the Plan were not new but were suggested in earlier plans (Bourcart, 1901; Mavromatis, 1922; Ionedis, 1939; Bunger, 1952). Those features included small dams on the Hasbani, Dan, and Banias, and the channeling of their waters in open canals to irrigate areas around Huleh and in lower Galilee. The Plan called for draining of the Huleh marshes and preparing them for agriculture, the construction of a medium-size power dam at Maqarin on the Yarmouk (58 meters high raisable to 95 meters), a diversion dam at Adasiyya, additional storage in Lake Tiberias, and gravity flow canals down both sides of the Jordan River. A few power plants were also included to generate an average total of 210 million kwh annually (Figure II.1). It excluded the Litani River because it is not part of the Jordan watershed. The Plan further described only in-basin use of the Jordan River, although it alluded to the out-of-basin potential uses by stating that "It is recognized that each of these countries may have different ideas about the specific areas within their boundaries to which these waters might be directed."[47]

---

[46] Miriam Lowi, "Water and Power," Cambridge University Press, London, 1993.
[47] Main, Chas T. Inc. "The Unified Development of the Water Resources of the Jordan Valley Region", Knoxville, Tennessee Valley Authority, 1953.

The Main/TVA Plan was the kind of homework needed before the United States could start her active involvement in attempting to resolve the conflict over Jordan Valley water resources, and to achieve what she could of her strategic objectives listed above. The views of the competing Arabs and Israelis of the Plan would be revealed to the Presidential envoy after he started his diplomatic mission. The envoy whom President Eisenhower appointed for the task, Ambassador Eric Johnston,[48] was the Chairman of the International Advisory Board of the Technical Cooperation Agency (TCA). He had made his fortune through trade in electrical appliances, a business he started in Spokane, Washington. He also served three terms as President of the U.S. Chamber of Commerce, was Chairman of the Association of Movie Industries and was active in promoting the sale of American films abroad. Upon his appointment, Mr. Johnston was equipped with the Main/TVA Plan on his first visit to the Middle East in October of 1953. Some had this to say about the Main Plan and Johnston:

"The plan, designed to tempt the Arabs into at least limited cooperation with the Israelis, was a third-rate idea with at least a second-rate chance of success because it had a first-rate negotiator, Eric Johnston, to advocate it. Its only disadvantage was that it made sense."[49]

The Plan gave annual allocations to Israel (394 mcm) to irrigate 416,000 dunums, Jordan (774 mcm) to irrigate 490,000 dunums, and Syria (45 mcm) to irrigate 30,000 dunums, and did not allocate any quota for Lebanon. The framework of Johnston's mission included several important points,[50] communicated to him before departure:
(1) The distribution of water should conform as closely as possible to the recommendations of the TVA (Main) report. As a minimum, Jordan should receive a substantially greater volume than it would from unilateral development of the Yarmouk alone, and Israel should renounce all rights to divert for irrigation more than a specified volume of water from the Jordan River and tributaries; (2) Armistice line adjustments should be effected so that physical control over Lake Tiberias, the outlet from the Lake, and the diversion canal from the Yarmouk would not be exercised exclusively by Israel; (3) demilitarized zones in the Jordan Valley should be eliminated; and, (4) the development of the Litani River must not be considered in the present

[48] Born in Washington D.C. in 1895, raised in Spokane, WA; attended University of Washington Law School in Seattle, WA, but had his education interrupted by the First World War when he was drafted and served with the Marines in China. Served under President Truman, Eisenhower and Kennedy. Died at George Washington Hospital in Washington D.C., in August 1963.
[49] Miles Copeland, "The Game of Nations", 1970.
[50] Miriam R. Lowi, "Water and Power," Cambridge University Press, 1993, p.86.

context.[51] Ambassador Johnston was accorded full latitude to advocate variations in the TVA/ Main Plan within the above terms of reference, and was instructed to use future levels of military and economic aid, plus UNRWA contributions, as bargaining chips with the riparian states.

Johnston's mission addressed several political objectives that promoted American interests, but were not attractive to the Arab side. On Jordanian-Israeli matters, the mission would advocate higher allocations to Jordan primarily because of the requirements to resettle the Palestinian refugees there, and to give incentives to Jordan to subscribe to a cooperative plan with Israel rather than follow the "unilateral" plan for the development of the Yarmouk in cooperation with Syria. A cooperative plan with Israel should have in it more benefits to Jordan than the unilateral development. Additionally, Jordan's economy needed to be boosted to alleviate the pressures brought about by the influx of refugees.

On the Syrian-Israeli front, the mission was to attempt to move bilateral issues further. Control of Lake Tiberias was to be reconsidered to avoid Israeli monopoly, and thus make the Lake an acceptable reservoir for the Basin's water. That meant the mission should attempt to effect adjustments in the Armistice lines to give Syria a part of Lake Tiberias, and to eliminate the demilitarized zones.

Finally, Ambassador Johnston was given the latitude to use American aid to achieve his goals. That would impact the attitudes of the riparian parties, especially Jordan and Israel, whose economies were being shaped and could use all the assistance they could get.

2. The Johnston Shuttle Diplomacy

Long before Secretary of State Henry Kissinger, in the wake of the October 1973 war in the Middle East, embarked on what came to be known as "shuttle diplomacy", Ambassador Eric Johnston had initiated that kind of diplomacy. He exercised it between 1953 and 1955 in four rounds of shuttling between the Arab states and Israel.

Mr. Johnston's first trip to the Middle East began in Beirut, Lebanon on 22 October 1953 and lasted until 3 November. It covered the four riparian states (Lebanon, Syria, Jordan, Israel) and Egypt. Copies of the Main/TVA Plan had been forwarded to the parties through diplomatic channels before Johnston started his trip.

His mission was met with a cool reception in the area. It came at a time when feelings ran high on both Arab and Israeli sides over Israel's plans to drain the Huleh marshes and to divert the Jordan River, and just after a

---

[51] Secretary of State J. F. Dulles to Eric Johnston, Foreign Relations of the United States, 1952-1954, vol. 9, 13 October 1953, no. 686, pp. 1348-52. Quoted by Lowi.

serious Israeli raid into the West Bank town of Qibyah, of the Hashemite Kingdom of Jordan. Fifty Jordanian villagers were killed in the raid that Israel mounted in retaliation against infiltrators across the Armistice Line. Ambassador Johnston, in his first approach, was faced with the task of gaining a hearing among Arab officials. The Arabs, especially Jordanians, were apprehensive and wondered why the United States was favoring the Main/ TVA Plan when it had withdrawn its support from the Bunger Plan. Three elements characterized the Arab attitude. First, they viewed Johnston's mission as yet another manifestation of American pro-Israel policy. Johnston himself was, among other posts he occupied, a vice chairman of the American Christian Palestine Committee, a pro-Zionist philanthropic organization. Second, they viewed America's intention not as a humanitarian gesture to improve socio-economic conditions throughout the Jordan River basin, but rather as an attempt to promote Arab cooperation with Israel and seduce Arabs into recognizing the Jewish state. Of primary concern to the Arabs was their commitment not to recognize the State of Israel nor to ever co-operate with her at any time before the causes of conflict were adequately resolved, not the least of which was the issue of Palestinian refugees and their right to return to their homes. They feared that the Plan would aim to resettle those refugees permanently in the host countries without regard to their right of return as decreed by the United Nations General Assembly in resolution 194 of 1948.

It took Johnston the utmost persuasion, in the name of President Eisenhower, to secure a hearing in Beirut, Amman, and Damascus. Only upon being assured that they need not contemplate any direct dealings with Israel over water were these Arab governments persuaded to listen to the Johnston proposals. However, Ambassador Johnston gained official hearing in Egypt. He met with General Mohammad Najib, Chairman of the Revolutionary Command Council of Egypt, and with Colonel Jamal Abdul Nasser, member of the same Council. The reaction of Egypt was cordial and a committee was formed to study the Johnston proposals (the Chas T. Main/ TVA Plan).

In addition to the allocation of water resources to the riparian parties, the implementation of the Plan would hinge on using Lake Tiberias as a storage reservoir from which water would be drawn to irrigate lands in Jordan and in Israel. The Plan ignored political boundaries and proposed the construction of two canals running from the Lake down both banks of the Jordan River. One would irrigate the East Jordan Valley under Jordan's jurisdiction, and the other would irrigate the West Jordan Valley whose northern part is within Israel, and the southern part is part of the West Bank, then an integral part of the Hashemite Kingdom. Thus the two countries would rely on a common reservoir and a common canal, a matter that would induce cooperation between them. Syria, on the other hand, would draw water from the Yarmouk and the upper Jordan before those water bodies discharged into the common reservoir, Lake Tiberias. Her withdrawal would affect water availability to both Israel and Jordan downstream. Lebanon was not allocated water, but Lebanese withdrawal from the Hasbani or the Wazzani would likewise affect water availability downstream. Thus the water sharing formula and adherence to it were basic in assuring the sustainability of downstream projects in Jordan and Israel, and that required a form of commitments and supervision to assure adherence to that formula. In the Plan, hydropower generation on the Hasbani was proposed. Finally, the water quality was not a central issue, although it was emphasized later in the mission. The lower Jordan River, south of Lake Tiberias, would be transformed into a drain, because the freshwater resources feeding it would be consumed before they discharged into it. The lands in Jordan and Israel that were irrigated from it would shift their sources to the two main canals running down from Lake Tiberias. No consideration was made of the religious heritage the River has to Christianity as a site where John the Baptist baptized Jesus at "Bethany beyond the Jordan".

From the beginning of Johnston's talks with Arab governments, it became clear that any water scheme involving Arab acquiescence would have to be carried forward under neutral or United Nations authority. Any other course implying cooperation with Israel or breaking the Arab boycott of that country was totally unacceptable. With such an Arab stand and atmosphere in their ranks, Mr. Johnston did not have an easy task at hand. He was able to open discussions on the only possible basis, namely, as a neutral moderator, and to induce all concerned to study the Plan he carried, and to have them make comments and counterproposals of their own.

2.1 The Reaction in Jordan

The reaction in Jordan was the most indicative of the Arab response to Johnston's mission. His reception in Jordan was very conservative. A private meeting with the Prime Minister, Dr. Fawzi al Mulki, was arranged at his house. The reaction to Johnston's visit can be detected in the contents of a letter forwarded by the Foreign Minister, Mr. Hussein Fakhry al Khalidi, to the Prime Minister. The Foreign Minister referred to the Plan that was carried by Mr. Johnston, a copy of which had earlier been forwarded to the Ministry by the American Embassy in Amman. He also referred to a report submitted by Mr. Sefit, the expert whose services were contracted by UNRWA to study and report on "The Legal Aspects of the Project for the Development of the Jordan Valley." The Foreign Minister made reference to a letter that Johnston carried, dated 14 October 1993 addressed to the Prime Minister, a copy of which had been forwarded to the Ministry before Johnston's arrival. An English re-translation of President Eisenhower's letter from Arabic reads:

" I am sending Mr. Eric Johnston who is the Chairman of the International Consultative Council as my personal representative with the rank of

Ambassador to discuss with you issues of interest to some countries of the Near East. I believe that the situation in the Near East has reached a stage that prompts us to focus our efforts to create situations to proceed forward and implement projects to harness the resources of this region. When I appointed Mr. Johnston for this important mission, I assured him of my full dependence on him and have authorized him to convey to you my personal views. In light of the many benefits that I believe will accrue from our joint attempts to find satisfactory solutions to the standing problems in this region I am confident that you will contribute in every way possible towards the success of the mission of Mr. Johnston.

Signed (President Eisenhower)"

The Foreign Minister wrote in November 1953 to the Prime Minister a detailed memorandum in which he appeared to distance himself from Johnston's mission, and expressed criticism of UNRWA for its role in the Jordan Valley Plan that Johnston carried with him. Because of its significance and importance, excerpts are translated and quoted below. The views and sentiments expressed in the letter reflected not only those of the Foreign Minister, but also of the Arab parties in general. Mr. al Khalidi wrote:[52]

"Mr. Johnston arrived in Jordan after he visited Lebanon where he met with Prime Minister Abdallah el Yafi. Upon his arrival here he met with Your Excellency a private meeting, and then met with Your Excellency again in my presence and the presence of the Minister of Education, Mr. Ahmad Toukan, and Mr. Lynch, the American Charge de Affairs. I may summarize what was discussed in the meeting that lasted for about two hours after exchange of views and participation from all".

Putting such developments on official record had not been customary in Jordanian tradition of inter-cabinet transactions. The responsibility of government action, by Constitution, is collective and is not assumed by any single minister or prime minister. The fact that al Khalidi opted to put this on record is evidence of his disagreement with the objectives of the Johnston mission. The memorandum displays Jordan's concern with the attitudes of other Arab countries, and her attempt to act in conformity with her Arab sister states.

The Arabs viewed not only the Plan as evidence of America's siding with Israel, but also the mediator, Eric Johnston, was suspected of harboring Zioinist connections and sympathies. In his summary, the Foreign Minister mentioned that Johnston "confessed that he heads the Movie Association of America, and that Jews were the majority of members in that Association. But he strongly denied that he harbored any inclinations or sympathy towards the Jewish cause or the Zionist movement, and insisted that he is an American in

[52] Memorandum to the Prime Minister from the Foreign Minister, Records of the Prime Ministry, Amman, Jordan (in Arabic), translation by the author.

the first place and has the American interests as his priority.... He also denied that he had any connection or acquaintance with Mr. Lowdermilk, the author of the project in 1937 to develop the Jordan basin, and who resides at this time in Israel and works as a consultant to the Israeli Government in the field of water and irrigation".

The memorandum went on to read:

"11. I find myself obliged to put on record some of what happened to Mr. Johnston in the rest of the Arab countries. His Excellency Abdallah el Yafi, the Prime Minister of Lebanon has informed us, as Your Excellency may recall, that he had met with Mr. Johnston in Beirut before he arrived in Jordan and had discussed with him his mission. Mr. el Yafi informed Mr. Johnston, in no ambiguous terms, that Lebanon and the Arab countries utterly refuse the discussion of any joint project with Israel to develop water resources, or to enter in direct or indirect negotiations with her about this subject. As to the Iraqi Government, it has refused by words of her Prime Minister, Mr. Fadil al Jamali, to discuss this subject with Mr. Johnston, or even allow him to enter Iraq because of the evidence that the Iraqi Government has evidence proving that this emissary was sympathetic with the Zionist cause in the past, and has strong connections and contacts with its leaders. The Iraqi Prime Minister held on to this opinion until the end of the meetings of the Political Committee [of the Arab League] in Amman.

12. You recall, Excellency, that Prime Minister al Jamali has raised this subject in the meetings of the Political Committee in Amman and insisted on his opinion despite the fact that the other representatives of the Arab countries did not quite agree with his theory, especially that Mr. Johnston had already visited Lebanon, and was about to enter Jordan carrying a message from President Eisenhower to Your Excellency. Your view was that the requirements of diplomatic courtesy dictate that you receive him even on private basis, especially that he carries a message to you from the President of the United States. And you know, Excellency, that Mr. Johnston has hopped over to Israel and met the officials there, and that he has visited Egypt and met with His Excellency President Mohammad Najib. He then visited Syria and met with His Excellency President Shishakli and important members of his government; and visited Lebanon again and met with President Chamoun. Additionally, it has been rumored that the Government of Iraq will lift the order preventing his entry to Iraq, and that he will meet the ruling officials of this brotherly country.

13. It is beneficial to mention at this stage of my report what the Political Committee of the Arab League decided about the subject in its last meeting in Amman that convened on 23/10/1953. After the Political Committee discussed and listened to the various opinions about the subject, it issued its decision number 7 stating:

"The Political Committee recommends that the three sisterly Arab States that have common interests in the waters of the Jordan basin [sources and tributaries] utterly refuse consideration of any joint project to utilize the waters of this river with the enemy Israel. This decision should be made crystal clear to the western countries involved in the matter".

14. It was clear after the Political Committee issued its decision that the Jordan Government should refuse to discuss this matter with Mr. Johnston in an official manner, but as I mentioned earlier, Mr. Johnston had already arrived in Amman, and you found, Excellency, that the duties of courtesy require that he be accorded a private meeting. This meeting was made with Your Excellency in the presence of the aforementioned, and it had its private character. It does not, in my opinion, contradict with the above decision of the Political Committee mentioned in the previous paragraph.

15. The decisions taken by the Political Committee were presented to the Council of Ministers of Jordan who approved them all as Your Excellency recalls. The Council authorized me to undertake correspondence with the concerned parties to implement the recommendations mentioned above. I did that at the time and, in implementation of the recommendations and the decision of the Jordanian Council of Ministers, informed all the concerned Arab States, the other countries involved in the matter, and the General Secretariat of the Arab League either directly, or through memorandums I sent to our embassies and consulates abroad, and the embassies of the western countries involved in the matter."[53]

It is interesting to note how the Jordanian Foreign Minister was distancing himself away from the decision to meet with Ambassador Johnston. He went on to express in that memorandum his views after he had read the Main/TVA report. He protested UNRWA's decision to look for another unified project, thereby ignoring the Yarmouk (Bunger) project and the investments made in it by the UNRWA, Point 4 (the U.S. TCA), and the British loan. UNRWA also ignored the agreement between the Jordanian and Syrian governments of June 1953 concerning the joint development of the Yarmouk waters. The Minister protested a claim by UNRWA that the new undertaking was approved by the UNRWA's Consultative Council at a time when the representatives on the Council of Syria, Egypt and Jordan were never consulted, and that the matter was kept secret among only the four foreign

---

[53] Note how the Foreign Minister is tactfully reminding the Prime Minister that his meeting with Johnston was not in line with the decision of the Council of Ministers that endorsed the decision of the Political Committee of the Arab League taken a few days earlier. It is clear that the Foreign Minister was not comfortable by the actions of the Prime Minister, and reiterates that those actions had a private character.

Figure (II.1)- Map of Main/TVA Plan

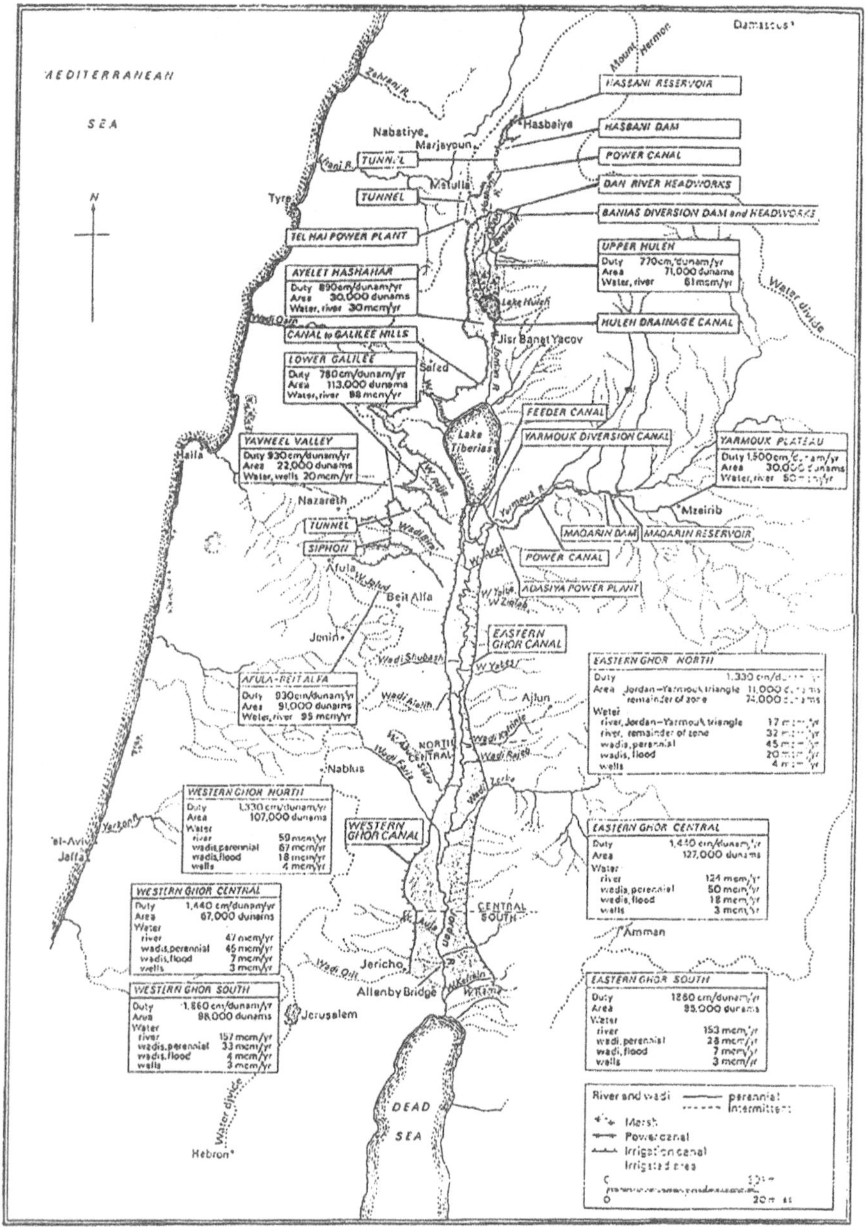

Council members, who were the representatives of the United States, Britain, France and Turkey. The same secrecy applied to the other study conducted by the legal expert, Mr. Sefit, who submitted his report on "The Legal Aspects of the Project to Develop the Jordan Valley". The Foreign Minister of Jordan added several other points in his review of the TVA and the Sefit reports, as follows:

"20. The Director of UNRWA, when he sent his request [to the TVA] undoubtedly at the instigation of high political references,[54] was completely aware of his duties and authorities, and of the purpose for which UNRWA was established. The mandate and operations of UNRWA are confined to two parts: the first is to provide for the livelihood of Arab refugees in the Arab countries, and the second is to create jobs and projects for the settlement of those refugees in order that they become self supportive whereby their names can be deleted from the lists of UNRWA dependents. After UNRWA in 1951 surrendered its care for twenty thousand of those Arab refugees who remained in Israel, it had nothing to do with Israel as it left the destiny of those miserable refugees to the mercy of the Jewish authorities in Israel.

21. It is therefore strange that UNRWA would pay such a visible attention to the resources of the entire Jordan basin, and to instruct the TVA to ignore and forget the current political boundaries. It appears as though UNRWA was created and founded to attend to the new Jewish refugees who pour into the occupied part of Palestine so that they, too, would benefit from the unified irrigation projects proposed by UNRWA, and [to serve the interests of] the other authorities that used it to consider the matter from all its aspects for the benefit of the population in the basin regardless of the scope of its mandate. The basic aim of the preparation of this project, and of forcing it on the Arab countries at these critical times, is so clear that it does not need any proof.

22. I now proceed to the second introduction written by Mr. Gordon Clapp, Chairman of the TVA, and he was the chairman of the committee named after him that was sent by the Reconciliation Committee two or so years ago, to study the optimum temporary solutions to create jobs for Arab refugees and their settlement. His committee submitted the known report that recommended the terracing for soil conservation, and the projects of afforestation, and the utilization of some limited water resources in the Arab host countries.

23. The date of this report [TVA] is 31 August 1953, and is submitted to the Director General of UNRWA. The introduction mentions that the TVA is a government organization, and it has done this study upon the request of

[54] This is an obvious reference to the role of the U.S. State Department in the formulation of the Plan.

the United States Department of State and after its approval. It is clear that the study has been done with the knowledge of the U.S. State Department, and may be by its order. That is why the UNRWA intended to have the political boundaries ignored, and to overlook their immense importance for the negotiations and agreements that have to be made among the parties concerned before the utilization of such a river basin."

The memorandum of the Foreign Minister of Jordan continues to present a critical review of the Main/TVA Plan (Unified Plan), and compares the benefits that Israel would get with those expected for the Arab riparian parties. The impact of the project on the level of the Dead Sea was also mentioned, and the negative effect this would have on any future Jordanian projects to extract Dead Sea salts was outlined. The memorandum raised the transport of Mediterranean Sea water to the Dead Sea to stabilize its level. The impact of raising the level of Lake Tiberias was highlighted, in that the raising would submerge important Christian holy places around the Lake. The Foreign Minister reiterated his astonishment at the involvement of UNRWA in the Unified Plan, as it fell outside UNRWA's geographic jurisdiction. He linked the Plan with the settlement of new Jewish immigrants in Israel, a task outside UNRWA's mandate.

The Foreign Minister proceeded to review the report of the legal expert, Mr. Sefit. In his memorandum, the Foreign Minister made clear that the UNRWA had started thinking about the Unified Project in the middle of 1952, and that it contracted the services of the TVA on 5/11/1952 at a sum of U.S. $ 55,000. Despite those moves by UNRWA, it entered into a second agreement with the Jordanian Government on 30/3/1953 for the Yarmouk Project, knowing that it was negotiating another separate project with a different authority (TVA). He noted that Mr. Sefit had recommended the implementation of small, labor-intensive projects that would not raise complicated water sharing issues like the Unified Plan does.

"You see, Excellency," wrote the Foreign Minister, "that the expert was very frank in the introduction to his report that he submitted to UNRWA. It was supposed to take up the recommendations of its expert, and not to adopt a Unified Project that was forced on it by the Government of the United States of America, a project that has nothing to do with UNRWA, and is not consistent with its mandate".

"...this dual policy (of UNRWA) reveals a skew behavior not becoming of an agency or commission representing the United Nations and its stature and authority. This policy shows clearly that the Unified Project is not meant to benefit Jordan or the Arab refugees, but is intended to pressure and influence the Arab countries to share economic benefits with Israel, and eventually urge these states to negotiate peace with the Jews."

Mr. al Khalidi continued to appraise the legal findings of Mr. Sefit:

"The author of the [Sefit] report points out in its conclusion to the decision taken by the Jordanian Council of Ministers on 31/8/1953 that was duly ratified by the Regents Commission on 15/9/1953 to cancel the Concession granted to Rutenburg. The expert notes that the consequences of such a cancellation are limited to a fair compensation to the owner of the concession. The expert does not discuss the legality of the unilateral cancellation decreed by the Council of Ministers, which indicates that the cancellation was proper and is the right of the Jordan Government to make it. Otherwise, the expert would have discussed it and shown his arguments about it".

The reference to the Rutenburg Concession was important because the proposed Unified Plan would allow the diversion of the Yarmouk waters to its three different riparian parties (Syria, Jordan and Israel) at locations before the Yarmouk gets to the Rutenberg power house. In the case of the Jordan River, its waters were proposed for use before it exited from Lake Tiberias, and none was allocated for the Rutenberg Project. This proposal meant the permanent denial to the Rutenberg project (which stopped operations operational in the wake of the 1948 war) of Yarmouk and Jordan waters needed for its operation. The cancellation of the Concession by the Council of Ministers of Jordan was thus not incompatible with the thinking behind the Main/TVA Plan. Both would disable the powerhouse of the already stalled Rutenburg Project.

2.2 Reaction in Israel

Mr. Johnson proceeded to Israel from Jordan. In talking with the Israelis, he had to try to persuade them to harness their own extensive water plans to some internationally approved formula for sharing of the Jordan waters, in lieu of unilateral diversion of the Jordan River. This linkage was essential to prevent violence over the Jordan. The Israelis had their objections to the Main/TVA Plan on the grounds that it excluded the Litani, and that it favored the in-basin use of the waters, which would deny them the implementation of their out-of-basin transfer plan to irrigate the Negev with Jordan River water. The political reaction of Israel was first expressed in a "Background Note" issued by the Israeli Embassy in Washington D.C. in October 1953, which stated that while Israel stood ready to negotiate an interstate agreement on water, "it cannot be expected to freeze all its development in the hope that the other states may some day be prepared to hold such discussions".

In a broadcast on November 30, 1953, the Israeli Foreign Minister, Mr. Moshe Sharett (later Prime Minister), who headed the Israeli negotiating team, announced the position of the Government of Israel:

"As to the question of regional water planning, we were the first who declared our willingness to sit together with our three neighbors who may have joint water interest with us... for the purpose of discussing a regional arrangement based on a just distribution of water resources. As long as no arrangement of this sort exists because the neighbor states refuse to meet with Israel, we consider ourselves free to use the water of the rivers which flow in our country as our property. We are convinced that the undisturbed continuation of Israeli development work constitutes perhaps the most effective method to insure regional cooperation. In the same manner, however, as Israel does not give up its freedom of policy and independent opinion in exchange for American aid… it cannot be attracted by immediate foreign aid to such an extent that it would give up the prospects of its future development and consolidation."[55]

The public announcement clearly alluded to the incentives that Johnston carried with him, which were in the form of American aid. It stressed Israel's intention not to let such aid sway her policies, nor to lure her into accepting projects that could jeopardize her future prospects for development. However, within Israel, Sharett appeared to favor serious consideration of the Johnston proposals on political grounds. Two possible advantages were: (1) improved relations with Arab neighbors that might result from joint use of Jordan waters; and, (2) the international financing of water development Mr. Johnston was able to offer.

On his part, Mr. Johnston maintained a low profile. In an interview with Columbia Broadcasting Service (CBS), On 1 December 1953, Mr. Johnston described the objective of his trip to the Near East by saying: "I did not go to the Near East carrying a specific plan. I had in my little bag a proposal. This proposal urges attention that the idea aims at coordinating the development of the Jordan basin. I did not expect a [yes or no] answer to these proposals; to the contrary, I feel that an answer given without a comprehensive study will not be mature."[56]

3.The Arab Counterproposals- The Arab Plan (1954)

Mr. Johnston's first success was to stimulate the formulation of counterproposals to the Main/TVA Plan. In the case of the Arab states, it was more difficult to promote such reaction. The Arab states had multiple apprehensions caused by a spectrum of reasons. These reasons ranged from their non-recognition of Israel on the grounds of the plight her establishment inflicted on the Palestinians and their basic rights, to the recent withdrawal by the United States and UNRWA of support to the Bunger Plan, with many other reasons in between. The November memorandum of the Jordanian Foreign Minister displayed the reservations in Jordan, which were also shared by the other Arab countries. However, Johnston proceeded to Egypt from

---

[55] Georgiana Stevens, above, pp. 20.

[56] Publication of the U.S. State Department, 28 December 1953, pp. 892.

Israel and was given a helping hand by the new Egyptian revolutionary government. The Egyptian officials were anxious to assert their role as leaders of the Arab countries, and also were eager to show friendliness to the United States at the start of their assumption of power and proclamation of the Republic that summer (1953). Johnston met with President Mohammad Najib and with Jamal Abdul Nasser, and the Egyptian leadership expressed its preparedness to have the proposals studied and formed a committee, on December 3, 1953, to study the utilization of the waters of the Jordan and its tributaries and to make a careful review of the Main/TVA Plan. The committee was expanded six weeks later, however, to include other Arab parties rather than confine it to Egypt.

On January 12, 1954, the Arab League Political Committee, in its meeting in Cairo, and upon the recommendation of Egypt, decided to establish an Arab Technical Committee to continue the study of the utilization of the waters of the Jordan and come up with an Arab counter-proposal to the Main/TVA Plan. The Arab Technical Committee was composed of the members of the above Egyptian Committee and representatives of the Syrian and Lebanese Governments. It is interesting to note that no representative of the Jordanian Government, the primary Arab beneficiary from any plan to develop the river basin, was appointed on that Technical Committee. This could be related to the resentment the members of the Arab League harbored towards the unification of the West Bank with the Hashemite Kingdom in 1950. All Arab League members opposed the move, and Jordan, not yet a member of the United Nations, had to fight her battle inside the Arab League alone. Other reasons could be related to the warm ties Jordan maintained with the West, or to the diminished political weight Jordan had after the sudden changes in her leadership. Jordan's founder, King Abdallah, was assassinated on 20 July 1951, and his successor, King Talal abdicated less than a year after he ascended to the throne. The young prince, al Hussein, was proclaimed King before he became of age. Later, however, in April 1954, two Jordanian members were added to the Arab Technical Committee.

The Committee held meetings in Cairo from January 14 to January 16, 1954, and laid down specifications for an Arab Project to counter the Main/TVA Plan. It made the following guidelines in the form of recommendations to the Council of the Arab League[57]

1. The Arab Project for the development of the resources of the Jordan River should preserve the irrigation of cultivated areas in the sources and basin of the Jordan River and its tributaries, and these areas should benefit fully from the utilization of electric powers generated from these waters.

---

[57] Omar Z. Ghobashy, "The Development of the Jordan River", Information paper number 18, Arab Information Center, New York, November 1961, pp. 16.

2. The Arab project should safeguard the full utilization of the Jordan River waters for the benefit of the Arabs.

The Committee determined the areas which could be irrigated and the quantity of water necessary for their irrigation from the river and its tributaries in Jordan, Syria and Lebanon. It also estimated the amount of electric power that could be generated from waterfalls. Moreover, the Committee evaluated the area under Israeli control that was dependent on Jordan water resources and the quantity of water necessary for the sustainability of irrigation of these areas, with no provisions for expansion.

On January 19, 1954, by resolution number 600, the Arab League Council adopted the above recommendations and requested the Technical Committee to pursue its technical work. The Arab Technical Committee had another meeting, this time in Amman, Jordan, on February 13, 1954, followed by a visit to the River basin and its tributaries in Jordan, Syria and Lebanon, examining all local projects and collecting data. The Committee was joined by the American engineers who were responsible for the American proposals, including the renowned Dr. Wayne D. Criddle, and by members of UNRWA. The Jordanians were not happy, because they were not represented on the Committee and their views were neither heard nor heeded. They made their unhappiness known to the Technical-Political Committee in its meeting the following month, when representatives of Jordan were named on the Technical Committee.[58] The Arab League Council adopted resolution no. 756, dated April 5, 1954, approving the recommendation of the Political Committee to form an engineering-military committee with representatives from Jordan, Syria, Lebanon and Egypt to study the project from all its aspects. This was the first opportunity for Jordan to contribute to the Arab effort in that field.

The Arab Plan as proposed by the Arab Technical Committee included:[59]

1. Utilization of the Yarmouk water for the purpose of irrigation and of hydroelectric power.

---

58 Handwritten notes of the minutes of meetings on 19/4/1954, in which the Jordanian representatives (Hamad el Farhan and Izzeddin Yunis) raised the haste in which the Arab Plan was put together. They also wondered why the meeting was held after, and not before, the Technical Committee had formulated the Arab Plan, had it translated into English, and forwarded to the U.S. Embassy in Cairo so that Johnston would be invited to discuss it the following month. The meeting was chaired by Mahmoud Riyadh and attended by Ahmad Farag, both of Egypt, and by Kazim Kharraz and Ibrahim Istuwani from Syria, and Ibrahim Abdul Aal from Lebanon. In that meeting, Lebanon and Jordan opposed the idea of storage in Lake Tiberias.

59 "The Arab Project for the Development of Water Resources in the Jordan River Basin and its Tributaries", Arab League, Cairo, Egypt, March, 1954.

2. Utilization of waters of the Jordan River and its tributaries north of Lake Tiberias for the purpose of irrigation and production of hydroelectric power.
3. Utilization of waters of the Jordan River and its tributaries south of Lake Tiberias.
4. Utilization of waters of the plains, and of wells.

The Plan was completed in March 1954 and revised a few months later for submission to the Arab League. In its final form it provided for the following measures (Figure II.2):

3.1 Utilization of the Yarmouk River

The basic features of this part of the Arab Plan, from which Jordan was the primary beneficiary, were as follows:

a. The construction of a storage dam on the River, either at Maqarin[60] or at Wadi Khalid, and a hydroelectric power plant at the same point.
b. The construction of a canal from the above dam to Adasiyya, where a hydroelectric plant would be built.
c. A dam for diversion and regulation to be built at Adasiyya in order to divert the waters into the East Ghor Canal for irrigation purposes in the Jordan Valley and a small amount (60 mcm) into Lake Tiberias for storage.[61]
d. An East Ghor Canal to be built along the east side of the Jordan from Adasiyya southward almost to the Dead Sea. This canal would be fed by the diversion at Adasiyya and a feeder canal from Lake Tiberias.[37] Shortly below where the Armistice Line between Jordan and Israel meets with the Jordan River, a siphon would convey water from the East Ghor Canal to the West Ghor for irrigation of that part of the Jordan Valley lying on the West Bank.
e. The two power plants would be used for the mutual benefit of Syria and Jordan.
f. Of the Yarmouk flow of 450 mcm, 330 mcm was allocated to Jordan and 90 mcm to Syria. (The rest of the flow would go to Israel (17 mcm) and to spills (13 mcm)).

It is worth noting that the Arab Plan had more stress on hydropower generation than the TVA/Main Plan did, and opted to store Yarmouk water on the river course with partial diversion of its floods to Lake Tiberias. These

---

[60] A dam at that location would regulate half of the annual flow of the river as measured at Adasiyya. At Wadi Khalid, the flow is increased over that at Maqarin, because of the contribution of Wadi Raqqad, a tributary draining eastern Golan in Syria.

[61] Omar Z. Ghobashy, ref. 57 above, pp.17.

[37] Ibid.

were the flood flows that would not be captured by a dam either at Maqarin or at Wadi Khalid. In this respect, the TVA/Main Plan also relied on Lake Tiberias as a common reservoir in which the Yarmouk excess waters would be stored. The Arab Plan proposed the conveyance of water to the West Bank through a siphon from the East Ghor canal which, in turn, would be fed from the Yarmouk and from Lake Tiberias. This arrangement removed West Bank dependency on an irrigation canal running down from Lake Tiberias through Israeli controlled territories as TVA/Main Plan proposed. Thus a facility common between Jordan and Israel was eliminated, and the importance of the other common facility (Lake Tiberias) was minimized.

3.2 Utilization of the Jordan River Waters North of Lake Tiberias

The Arab Plan called for the utilization of the Jordan River north of Lake Tiberias for the benefit of all riparian parties including Israel. Its primary features were:

a. Construction of a dam on the Hasbani tributary about 12 km upstream from its confluence with the Jordan River.
b. Construction of a carrier canal from the Hasbani dam to irrigate Lebanese lands in the Hasbani Valley, using 35 mcm of water annually, and to generate hydroelectric power from the fall of water at the end of this canal back into the Hasbani at el-Ghajar village.
c. Construction of two canals from the Banyas tributary for the purpose of irrigating Syrian lands on both banks of the tributary. These canals would use 20 mcm of water annually.
d. Construction of a canal from the Banyas tributary in Syria, crossing the Jordan River and proceeding southward to below Lake Huleh, feeding branch canals irrigating Israeli lands. This canal would use some 66 mcm of water to irrigate 78 thousand dunums of Israeli land in upper Huleh. Israel would also receive 30 mcm directly from the River to irrigate 30 thousand dunums in Hashahar north of the Lake, and 22 mcm (of which 3 mcm are from springs and 19 mcm from wells) to irrigate 22 thousand dunums to the northwest of Lake Tiberias. The total allocation for Israel would be 118 mcm, of which 96 mcm were from the River and the Banyas tributary, 3 mcm from springs and 19 mcm from groundwater.
e. The utilization of 22 mcm of Jordan River water to irrigate the Syrian district of Boteiha, northeast of Lake Tiberias. Added to the 20 mcm of allocation to Syria from Banyas, the total allocation to Syria was 42 mcm.

The Arab Plan, in this section, recognized the rights of existing irrigated areas in Israel to be sustained, allocated water to those lands and named the sources. No allocation, however, was made for further Israeli uses. The Arab allocation to Israel was far below the level that Israel wanted, but was, nonetheless, a starter in the sense that the Arabs recognized water rights of lands under Israeli jurisdiction.

3.3 Utilization of the Jordan River Waters South of Lake Tiberias

The remainder of the flow in the Jordan River would be discharged into Lake Tiberias for storage and for use in irrigating lands south of the Lake for the benefit of Jordan and Israel. The plan provided for the following utilization of water:

a. A feeder canal between Lake Tiberias and the East Ghor Canal in Jordan would provide Jordan with a total of 428 mcm, of which 368 would be drawn from the Jordan River and 60 mcm from the Yarmouk floods stored in Lake Tiberias as mentioned above. Jordan would add to these quantities a flow of 270 mcm from the Yarmouk base flow and Maqarin storage, 213 mcm from springs and side wadis, and 64 mcm from wells (groundwater). Jordan's total share would thus be 911 mcm from the River and its tributaries and 64 mcm from groundwater, totaling 975 mcm. This total would irrigate an estimated area of 490 thousand dunums, of which 296 thousand and 194 thousand are located in the East Bank and the West Bank respectively. These lands would respectively require 574 mcm and 401 mcm of water, of which groundwater would contribute 36 mcm and 28 mcm respectively.

b. A canal off-taking at the southwest tip of Lake Tiberias to Biesan, almost to the Armistice Line between Jordan (West Bank) and Israel to provide Israel with 41 mcm of Jordan River water to which a quantity of 85 mcm would be added from springs and side wadis. In addition, a total of 45 mcm was reserved for irrigation of the Yarmouk Triangle south of Lake Tiberias in Israel. Israel's allocation to irrigate areas south of Lake Tiberias added up to 171 mcm annually. Added to 118 mcm allocated to Israel north of Lake Tiberias, the total quantity that Israel would thus receive was 289 mcm per year.

c. The total usable resources of the basin thus add up to 1,431 mcm, made up of 1,047 mcm of River water, and 384 mcm of springs, wells and side wadis.

The Arab Plan accounted for the irrigation of lands inside the River basin, including existing irrigated areas inside Israel. It is interesting to note that the Arab Plan allocated to the Yarmouk Triangle of Israel water from Lake Tiberias. The total allocation for the Triangle was 45 mcm of which 17 mcm would be from the Yarmouk and the balance (28 mcm) from Jordan River water withdrawn from the Lake. The Plan opposed the transfer of water outside the River basin. Its water allocations are summarized in Table 2.1 below.

The Arab Plan accounted for a total of 1,348 mcm of surface water (including springs), and 83 mcm of groundwater extracted from wells. It is to be noted that the Main/TVA Plan accounted for 1,213 mcm of surface water only.

In summarizing its views, the Arab Technical Committee, before preparing its counterproposal, pointed out that the Main Plan ignored political boundaries and allocated too small an amount of water to the Arabs, considering the location of the major sources of the Jordan River in Arab lands. It did not benefit Lebanon despite the flow of the Hasbani in Lebanese territory. Similarly, Syrian rights to the maximum use of the waters in her territory were not fully recognized. The Committee observed, moreover, that the Johnston Proposals would be of immediate benefit to Israel, and of delayed benefit to Jordan.

On the proposal for storage of Yarmouk waters in Lake Tiberias, the Committee observed that this would make Jordan completely dependent upon Israel, and also found technical difficulties with such storage. In the first place, the degree of salinity in Lake Tiberias exceeded that of the Yarmouk River, and since the Ghor (Jordan Valley) area in Jordan suffered from soil salinity, the use of water with a high degree of salinity would damage Jordanian agriculture. Secondly, it pointed out that there would be a higher rate of evaporation in Lake Tiberias, which is broad and shallow, than in a reservoir at Maqarin, which would be narrow and deep. The influence of the Bunger Plan was visible in the Arab counterproposal. Johnston blamed that on its author, Mr. Bunger, requested Bunger's transfer from the TCA office in Amman.

The Arab Plan addressed the important issues encountered in the utilization of the water resources of the Basin. These were: 1) allocation; 2) water quality; and, 3) storage. The Arab counter proposal increased the river water allocation to the Arab parties. Jordan's share, for example, was increased from 774 mcm as proposed by the Main/TVA Plan to 911 mcm. The plan to maximize storage of Yarmouk waters in the river gorge at Maqarin or Wadi Khalid, away from Israeli control in Lake Tiberias, which was proposed by the Main/TVA Plan. Portions of the Yarmouk floods that could not be so regulated would be diverted to Lake Tiberias or any other economic location. The water quality issue was linked to the Yarmouk storage in Lake Tiberias. Because the salinity of the Lake water was higher than the salinity of Yarmouk water at the time, the increase in Yarmouk water salinity by storage in the Lake would exacerbate the adverse agricultural environment in the Jordan Valley, a result of existing soil salinity and increased water salinity.

Figure (II.2)- The Arab Plan (1954)

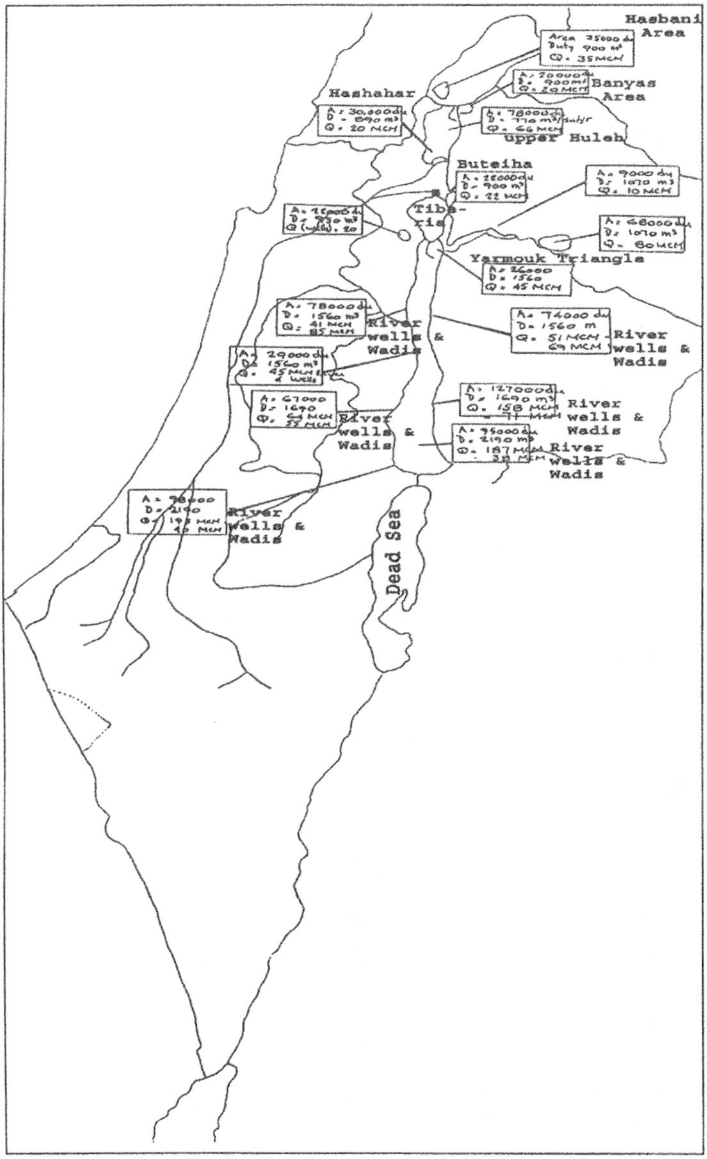

Table 2.1: Details of Water Allocation in the Arab Counterproposal (Arab Plan, 1954)

Riparian	Site of Irrigation	Area '000 dunum	Water Mil M3	From River	From Side Wadis	From Ground
Lebanon	Hasbani	35	35	35	-	-
Total Lebanon		(35)	(35)	(35)	-	-
Syria	Banyas	20	20	20	-	-
	Boteiha	22	22	22	-	-
	Upper Yarmouk	68	80	80	-	-
	Lower Yarmouk	9	10	10	-	-
Total Syria		(119)	(132)	(132)	-	-
Israel	Upper Huleh	78	66	66	-	-
	Hashahar	30	30	30	-	-
	Bavneil	22	22	-	3	19
Sub Total		130	118	96	3	19
	Yarmouk Triangle	26	45	45	-	-
	Biesan Ghor	78	126	41	85	-
Sub Total		104	171	86	85	-
Total Israel		(234)	(289)	(182)	(88)	(19)
Jordan	N.E Ghor	74	120	51		
	M.E Ghor	127	229	158		
	S.E Ghor	95	225	187		
Sub Total	East Bank	296	574	396	142	36
	N.W Ghor	29	49	45		
	M.W Ghor	67	119	64		
	S.W Ghor	98	233	193		
Sub Total	West Bank	194	401	302	71	28
Total Jordan		(490)	(975)	(698)	(213)	(64)
Grand Total	All Riparians	878	1431	1047	301	83

4. The Israeli Counterproposals - The Cotton Plan (1954)

Israel did not lack ready plans to present Johnston with a counterproposal. She had already published her Seven Years Plan for the development of water resources. However, in her response to the Main/ TVA Plan, Israel dug into her Zionist "water heritage", rich with plans discussed in the previous chapter. The American engineer, John. S. Cotton, who worked as a consultant to the Israeli Government, reviewed those plans and came up with an "upper bound" version based heavily on the Lowdermilk- Blass- Hays and Savage plans. This version was known as the Cotton Plan.

The Cotton Plan had a shortcoming that the Arab Plan did not. Contrary to the in-basin principle on which the Arab counterproposal was based, the Cotton Plan included out-of-basin water resources, and out-of-basin lands for irrigation from basin waters. It called for inclusion of the waters of the Litani and a diversion of the Jordan through a canal at Jisr Benat Ya'qoub (on which work was suspended by the Security Council resolution of 27 October 1953), to a reservoir in the Batofa Valley near Nazareth, from which the National Water Carrier would off-take to convey water through pipes and tunnels to the Negev. Parts of the Carrier were under construction between Auja (Yarqon) River near Jafa/Tel Aviv and the Negev. Under the Cotton Plan, the salinity of water in Lake Tiberias would be reduced by the introduction of fresh water from the Litani River. The water needs of the Jordan Valley in the Hashemite Kingdom of Jordan were to be met by drawing Yarmouk water through a canal system under Jordan's control.

The Cotton Plan would, in effect, draw on the headwaters of the Jordan system, that is, the Hasbani, the Dan, and the Banyas for fresh water, and would reach north outside the Jordan basin to the Litani in Lebanon for further sources of fresh water. In Israeli eyes, this ambitious plan was a natural sequel to the plans of the early Zionists, who had foreseen that the viability of the Jewish state in Palestine would depend on her ability to take water resources from surrounding northern territories southwards toward the dry regions of Palestine.

Cotton's estimates of the water resources available for use, including the Litani, amounted to 2345.7 mcm. The allocation to different riparian countries was 450.5 mcm to Lebanon (from the Litani), 30 mcm for Syria, 575 mcm for Jordan, and 1290 mcm for Israel. Besides the quantitative objection Israel had to the Main/ TVA Plan, she objected to the involvement of the United Nations in the operation of joint water resources. It is worth noting that, despite the headache that the American engineer, John S. Cotton, caused to Johnston, he was not transferred out of Israel as Bunger had been transferred out of Jordan.

Table 2.2 compares the allocation to different riparian parties under the Main Plan and under each of the counterproposals.

Table 2.2: Annual Allocations Under the Different Plans (mcm)

	Lebanon	Syria	Jordan	Israel	Total
Main/TVA Plan	Nil	45	774	394	1,213
Arab Plan[62]	35	132	975	289	1,431
Cotton Plan[63]	450.5	30	575	1,290	2,345.7

The differences among Arabs and Israelis in attitudes, objectives, allocated quantities, water storage and quality issues, supervision of implementation, and means of water conveyance were significant, as evidenced in the preceeding text, and the reconciliation of these differences was quite a challenge to the diplomatic skills of Ambassador Johnston, and a technical challenge to his engineering escorts. He obviously was determined, however, to proceed with his mission, and aimed at success by providing a solution to the resettlement of the refugees, at least in Jordan.

---

[62] Includes 64 mcm of groundwater in Jordan and 19 mcm of groundwater in Israel.
[63] Includes all of the Litani River water.

CHAPTER III

Attempts at Reconciliation - Johnston's Shuttle Diplomacy

Ambassador Johnston received the counterproposals from both the Arab and Israeli sides through diplomatic channels, and they were worlds apart. The difficult task of reconciliation would soon commence, and it was done through more contacts, visits and more intensive diplomacy. The Arabs would not meet the Israelis face to face, because none of the Arab states recognized Israel and a state of war still prevailed since the day of Israel's creation. A host of issues clouded the atmosphere, of which, the cause of the Palestinian refugees and their right of return ranked high. The United Nations General Assembly passed resolution 194 on 11 December 1948 asserting in paragraph 11 the right of the Palestinian refugees to repatriation or compensation. The political environment in the Middle East and the fast development of events had important impacts on Johnston's mission. A brief general overview of the situation in the region is presented below, along with a rundown on the topics of negotiations. This overview is not meant to be an academic, historical review of that era, but is meant to refer to the political events and issues that affected the mood of the peoples and governments of the region. Those, in turn, impacted the talks that Johnston conducted. It should be noted that the diplomatic skills for which Mr. Johnston was famed made a real difference. Interesting also was the moderating stand of Egypt, which helped materially in achieving substantial progress. After Mr. Johnston received the responses of the parties to the Main/TVA Plan, he set out for his second round of negotiations in the region in June 1954.

1. Negotiating with the Arab Side- Second Round

1.1 Public Mood

On the eve of Johnston's second round in June 1954, tensions between Israel and the Arab states were running high. Throughout that month, there had been numerous shooting incidents across the Israel-Jordan Armistice lines, culminating at the end of the month in the outbreak of fighting in Jerusalem. There were incidents on the Armistice lines with Syria as well, and Egypt and Israel quarreled over freedom of navigation in the Suez Canal and the Straits of Tiran. Moreover, Colonel Adeeb Shishakli of Syria, and General Mohammad Najib of Egypt had recently been ousted from power. "An atmosphere of internal dissention prevailed in all the countries of the region."[64]

Moreover, the general political atmosphere was not favorable to open alliance with the West. Britain and France were viewed by the Arabs as the primary parties responsible for the Arabs' misfortunes since the departure of the Ottoman Turks from Arab lands in 1917. The immediate recognition of Israel by the United States' and the Soviet Union in 1948 stunned the Arab public. The United States efforts in establishing UNRWA and in financing it on the one hand, and her entry in the region to help its economic and social development on the other, helped form a favorable public image of the United States. The communist party, organized in several Arab countries including the riparian parties on the Jordan and in Egypt, was spreading negative propaganda against imperialism and colonialism and was portraying the United States as the leader of the camp of imperialism. However, with Arabs being believers in God, the atheist communist party was fighting an uphill battle to gain grounds in their midst. The popular mood did not feel comfortable with atheist doctrines and favored countries whose peoples believed in God. The unqualified support that the United States gave to Israel was not made public, but the donations and assistance that she provided to the Palestinian refugees and to the Arab countries were duly publicized.

By the time of Johnston's second visit, the talks about a regional military alliance in the Middle East had intensified, accompanied by similar talk about the "vacuum" in the Middle East. Public sentiments in the Middle East were opposed to the western idea of forging a regional military alliance in the form of a "Middle East Defense Organization, MEDO" designed after the North Atlantic Treaty Organization (NATO). The relentless efforts of the nationalistic regimes in Syria and Egypt publicly stood firmly against that proposed alliance. Public sentiment was not ready to entertain any alliance with the West, primarily because of the West's role in creating and supporting Israel. Public awareness of the details of the Cold War was not very deep, and the activities of the leftist political parties, overt or underground, enhanced the political momentum of anti-western sentiments.

1.2 Political Environment

The political environment in the region did not display stability. Egypt, run by a military junta since the revolution in 1952, started to witness domestic confrontations. Premier Jamal Abdul Nasser, the real leader of the revolution, had ousted the President of the new republic, General Mohammad Najib. He displayed cautious friendliness to the United States, as he conducted a campaign to have Britain evacuate the Suez Canal zone. Nasser

---

[64] Yoram Nimrod, "The Jordan Angry Waters," New Outlook vol. 8, no. 5 (July-August,1965), p. 24.

was also hoping for economic aid and for military assistance from the United States. The Americans, on their part, supported a peaceful revolution that could rid Egypt of its ills and start a march toward democracy. They wanted the emergence of a leader who would be so popular and strong as to take unpopular decisions and survive to implement them. They also wanted to promote the formation of MEDO, concerning which Nasser had his own reservations. In a meeting with American officials, Nasser wondered about the enemy whom the new alliance would target. On the one hand, the enemy of the United States in the area was communism and the Soviet Union, and, on the other, the enemy of the Arabs was clearly Israel.

"The Arabs will say you are trying to get them to unite to fight *your* enemy," Nasser said to the American officials who approached him about the matter, "while they know that if they show any intention of fighting *their* enemy you would quickly stop all aid. Any regional military agreement which did not take this attitude into account would be a fraud."[65]

Nasser had his own plans for leadership, and realized that a friendly United States on his side would be a lot more convenient than having her on the side of Britain and France against him.[66] Israel, on the other hand, developed worries about that intimate relationship between Nasser and the United States and opted to try and foil it. In 1954, an Israeli covert operation was waged in Cairo under which the American Embassy came under an attack that Israel hoped would be blamed on the Egyptians, thus damaging the friendly Egypt-United States relations. The attempt failed and the fiasco was blamed on, and named after, the Israeli Defense Minister, Yitzhak Lavon. He resigned from the Israeli Cabinet of Moshe Sharett and was replaced by another hawkish politician, David Ben Gurion.

The United States' courting of Nasser prompted him to make every effort to facilitate the mission of Ambassador Johnston. Nasser, having met with Eric Johnston on his first trip, was behind the formation of the Arab Technical Committee of the Arab League, under Egyptian chairmanship, to study the Johnston's proposals, and Egypt was helpful in moderating the more extreme Arab positions.

Jordan's King Hussein had assumed his constitutional powers on 2 May 1953, and was eager to initiate and accelerate economic and social

---

[65] Conversation between Nasser and Amer, his Chief of Staff of the Egyptian Army, with Colonels Albert Gerhard and Wilbur ("Bill") Eveland of the Pentagon, at the suburban home of Hassan Touhami, Nasser's senior aide, in October 1954. See Mile Copeland's "The Game of Nations," pp. 123-124.

[66] The U.S. CIA made contacts with the "Free Officers" group that Nasser headed, before they mounted their revolution in 1952. The U.S. was careful to keep good relations with the real man behind the movement of the Free Officers. See, "The game of Nations," Ibid.

development of the country. He was dismayed at the writing off of the Maqarin Dam project and was willing to look into the alternative, the Unified Development of the Jordan Valley that Ambassador Johnston was promoting. The Palestinian problem and the case of the Palestinian refugees haunted him as it did all the politicians in his country. Jordan had hosted the highest percentage of Palestinian refugees in 1948, and had united in 1950 with the West Bank. A substantial percentage of the Kingdom's population had thus become of Palestinian origin. However, that unification was not formally recognized by any of the Arab states, and was explicitly recognized only by the United Kingdom and Pakistan. The Kingdom had proclaimed its first constitution under the second King, Talal Ben Abdullah, and the constitution allowed the formation of political parties, guaranteed the rights of citizens, and assured a democratic process. Jordan was not yet admitted to the United Nations, and was seriously considering joining the planned military alliance, MEDO. When contacts to form that regional defense organization did not materialize into a consensus among the concerned states, partial regional arrangements were seriously considered. Contacts were started to form the Baghdad Pact with membership of Turkey, Iraq, Iran and Pakistan. Jordan was a strong candidate to join that Pact, and many Jordanian politicians favored the idea, to protect the country against any Israel military threat, especially on its West Bank. The recent incidents across the Armistice lines there were evidence of such dangers and were a continuous reminder of Israeli designs to take over the West Bank. However, the campaign waged by Egypt and Syria against the Baghdad Pact made it difficult for Jordan to join, and so it did not.

Syria, independent since 1946, had been going through a series of military coups. The first was instigated and orchestrated by the U.S. Central Intelligence Agency and was staged by Husni Za'im on 30 March 1949. He lasted in power for 136 days, before a group of his military subordinates, fronted by Sami Hinnawi, mounted another coup on 14 August 1949, killed Za'im and buried him in the French cemetery. Exactly four months later, colonel Adib Shishakli mounted a coup, jailed Hinnawi and ran the country himself through a succession of civilian front men until 21 November 1951, when he came into the open as Syria's strong man. Shishakli fled the country in February 1954, in the face of yet another coup led by colonel Mustafa Hamdoun and colonel Abdul Ghani Qannout, two military officers enlisted by the Arab Socialist Ba'th Party. They managed to place authority in the hands of civilians who had been displaced by Shishakli, and elections were made for a parliament.

Johnston's first visit started in October 1953 while Shishakli was in power in Syria, but the regime changed by the time Johnston made his second visit in June 1954. Syria was opposed to MEDO and very much suspected the motives behind its formation. Syria's relations with Turkey have never been very friendly, since the time of the Arab Revolt of 1916 in which Syrians rallied around al Hussein, the Sharif of Mecca, and helped drive out the Ottoman Turks. The Syrian's immediately proclaimed Faysal Ben al Hussein as King of Syria in the wake of that revolt. Syria's relations with Turkey became particularly clouded in 1939, when France, the mandate power, carved out Alexandretta, the northwestern Syrian province, and gave it to Turkey to keep her from allying with Germany in the Second World War. Joining a pact like MEDO with Turkey was unthinkable to the minds of Syrian politicians. On the domestic front, Syria was not in a very stable condition by virtue of the successive military coups since independence. It had hosted a sizeable number of Palestinian refugees in 1948, and was vocal in nationalistic and pan Arab slogans.

Lebanon, independent since 1943, had a new President, Mr. Camile Chamoun, who took office in 1952 upon the resignation of his predecessor, Bisharah Al Khouri. Her Prime Minister, Riad al Sulh, had been assassinated in Jordan the year before, in retaliation for the execution of Antione Sa'adeh, the chief of the Syrian Nationalist Socialist Party in 1949. The country was managing a delicate ethnic balance in accordance with a National Charter agreed upon shortly before independence. Lebanon had a sizeable community of Palestinian refugees who had fled the violence in Palestine in 1948 and took refuge in Lebanon where they were, as were their peers in Syria, without citizenship, and were treated as resident aliens.

Iraq, proclaimed independent since 1933, had a more stable environment. It had a Hashemite monarch, King Faysal the Second, and a "democratic" system of government including an elected Parliament. It had participated in the war with Israel in 1948, and her troops were credited with holding onto the northern part of the West Bank. The above five countries, mostly newly independent from Britain and France, plus Saudi Arabia and Yemen founded the Arab League in 1945.

This, in brief, was the political environment in the concerned Arab countries at the time of Johnston's shuttle diplomacy. On the level of the people, there were a few political parties that were allowed to be active, and a few others operated underground. The parties that had Arab extensions were primarily the Moslem Brotherhood, the Communist Party, the Ba'th Party, the Arab Nationalists Party, and the Syrian Nationalist Socialist Party. These parties were licensed in some countries and were banned in others. The banned parties had sympathizers and were more influential on the streets and in the ranks of the Armed Forces than the licensed ones. All parties in the Arab countries resented the establishment of Israel, and their ideologies, with the exception of the communists, excluded the acceptance of the state of Israel in the Middle East.

Of primary concern to the Arabs was the implicit, or explicit, potential of cooperating with Israel before an acceptable solution to the

Palestine problem was reached. Another concern was the likelihood of diffusing the political issue of Palestinian refugees through the Unified Plan, and having them settle permanently in the host countries. Simultaneously, there was Arab concern for the well being of the refugees and an intention to improve their living conditions. These issues, along with other important technical issues would surface in Johnston's negotiations with the Arabs.

Israel, on the other hand, was busy building the infrastructure of a modern state. She had come out victorious from her war with the Arabs and had signed Armistice agreements with them. The new state was opened to immigration of Jews and many poured into it from Europe and from Arab countries. The challenge she faced was immense. She had to buttress her defense capabilities to assure her security against renewal of hostilities with the Arabs, and to pursue accelerated economic and social development to cater for the immigrants' needs. Water was needed for expansion in land reclamation, especially in the dry south, to make room for more population. The State was established on the principles of western democracy (not applicable to her Arab population), and was run by a democratically elected government dominated by the founding figures led by David Ben Gurion, the most prominent of the founding fathers at the time.

1.3 Talks with the Technical Committee

Despite the adverse internal conditions in the visited countries, Johnston was anxious to proceed and obtain agreement, at least in principle, to the Unified Plan. The strategy he adopted was to be neutral, to appear sympathetic with the party the Mission members were talking to, and to emphasize the benefits his proposals would bring to them. Referring to the incompatibility of Syrian and Jordanian objectives from the uses of the Yarmouk, a member of the Johnston's mission, Mr. Ludlow, wrote on August 5, 1955 to his colleague, Oliver Troxel, what was revealing of Johnston's negotiations strategy:

"...Given the above I don't think that we want to put ourselves in the position of either paying Syria's price or fighting Jordan's battle with Syria for her. I don't think we want to take eithers side (except perhaps with that side we're talking to at the time!). I think tactically we will want to propose plans which will appear to steer a fair middle course between the interests of the two governments and then be prepared to agree to what both sides may agree upon."

The Arab Technical Committee prepared the "Arab Plan" and had it transmitted to Johnston through diplomatic channels (the U.S. Embassy in Cairo). Ambassador Johnston appreciated the fact that the Plan acknowledged the right of existing irrigated lands in Israel to continue to receive water from the river system and allocated water to them (289 mcm/year). This principle was a plus to begin with. However, the Arabs expressed objection to Lake

Tiberias as a main storage reservoir for Yarmouk floods, and wanted maximum storage on the river course through a dam like Bunger had proposed in 1951. They based their objection, as mentioned above, on political and technical grounds. On the one hand, they did not feel comfortable with storing in Israeli territory the water on which Jordan would be dependent, and, on the other, such waters would have an increased salinity after they were mixed with Lake Tiberias waters. Coupled with the salinity of soils in the Jordan Valley, increased water salinity would be constraining on Jordanian agriculture. Another technical reason was the high evaporation loss from the surface of the Lake (165 sq. km in surface area).[67] If the surface area of the Lake is increased with Arab water storage, there would be a corresponding increase in the evaporation quantity. Evaporation, at any rate, increases the salinity level of the mix because it takes away pure water.

There were other objections by the Arab Technical Committee to Johnston's proposed Unified Plan. It noticed that the Plan did not allocate any water share to Lebanon, and that it allocated only a small percentage to Syria (45 mcm/yr). It considered that the Plan allocated too high a percentage of water (394 mcm/yr) to Israel, a share that comprised 33% of the total resources when the Israeli territories contributed only 23% of the flow. The Committee felt that the flows of the Hasbani and Banyas should be used in power generation and irrigation upstream in Arab lands, instead of leaving them to flow for use in similar purposes downstream in Israeli territories. Another basic objection expressed by the Committee was the implied use of the waters of the basin outside its territories, a notion that served the purposes of Israel who wanted to divert Jordan River water to irrigate the Negev. Finally, there was disagreement over the total water resources of the basin. The Unified Plan estimated the total river flow at 1,213 mcm per year, and the Arab Plan estimated the resources at 1,431 mcm that included 83 mcm of groundwater per year (See Chapter II).

The first breakthrough that Johnston achieved, with the help of Egypt, was the change in the Arab mood that once ran high toward rejecting Johnston's mission and proposals. After the first meeting of the Technical Committee of the Arab League in mid-January 1954, Mahmoud Riyadh, of Egypt's Ministry of War, reported that the Arab states did not want to reject a regional scheme outright. Nonetheless, they could not accept it as it stood, since "Israel would have complete control of the waters on which Jordan

---

[67] The evaporation factor was actually immaterial in its effect on the quantity of Arab water stored, because the evaporation would take place with or without diversion of Yarmouk flood for storage, and it was taken into account in the resource accounting and allocation.

would be dependent..."[68] The above reservations expressed by the Technical Committee were also cited.

Johnston's meetings with the Arab Committee began in Cairo on 13 June and lasted until 17 June 1954.[69] The issues to be discussed with the Arabs pertained to a) storage of Yarmouk floods as it impacted strategic factors and also water quality. The Arabs were not comfortable with storage of these floods in Lake Tiberias, as the Americans favored; b) water allocation to the parties that was based on the areas in the basin within the territories of each riparian party that could be irrigated, and not on areas outside the basin; c) water duty, which is amount of water needed to irrigate a unit area of arable lands, and which impacts allocation; and, d) supervision of implementation of the Unified Plan.

The Arab Plan was discussed, and agreement was reached on four crucial points.[70] (1) Storage: in principle, storage of the Yarmouk floods would be made in Arab lands for purposes of irrigation and power generation. (2) Allocation of water: it was agreed that quantities allocated to Syria and Lebanon would be the same as indicated in the Arab Plan without affecting the share allocated to Jordan as stipulated in the Unified Plan. In effect, the additional shares the Plan allocated to Syria and Lebanon would have to come out of the Israeli allocation under that Plan. (3) After agreement with the American side on the general framework of the Project, the concerned Arab States would take the necessary measures at the United Nations to form the International Committee stipulated in the Unified Plan for supervision of the Plan and to define its scope of work. (4) The American side suggested postponing a decision on the power plant on the Hasbani at el-Ghajar until a detailed study of that proposal was completed.

Points of disagreement remained as follows. (1) Water duty: the American side wanted to have the figures of the Unified Plan adopted, while the figures of the Arab Plan were about 15% higher. (2) Israel's share: the American side insisted that Israel be allocated approximately the overall share as stipulated in the Unified Plan. (3) In-basin use: the American side insisted that the use of water by Israel not be limited while the Arab side insisted that Israel use the water inside the basin. (4) Yarmouk storage: the American side

---

[68] Report on Conversation with Mahmoud Riyadh, 15 January 1954, United States National Archives 683.84a322/1-1554, as reported by M. Lowi. See her "Water and Power," Cambridge University Press, 1993.

[69] Attendants were Hamad Al Farhan and Izzeddin Yunis (Jordan), Dr. Ibrahim Al Ustuwani, Dr. Subhi Mazlum and Tawfiq Shatella (Syria), Fuad Ammon, Ibrahim Abdel Aal and Nadim Dimashquiyyeh (Lebanon), and General Mahmoud Riyadh, Dr. Mohammad Ahmad Salim and Ahmad Ali Farag (Egypt).

[70] "The Johnston Negotiations- Minutes of Meetings,"Records of the Prime Ministry, Amman, Jordan.

insisted that the additional cost associated with the Yarmouk storage must not exceed $10 million over what was estimated in the Unified Plan, which specified 95 meters for the height of the Yarmouk dam. (5) Adasiyya diversion: the American side insisted hay the Adasiyya diversion be built promptly, while the Arab side insisted that no implementation should be made in the Yarmouk basin until the all-basin project is decided. The Adasiyya diversion was in the form of a weir built across the Yarmouk River near Adasiyya to divert Yarmouk water to Jordan and to Lake Tiberias in Israel.

The Arab Technical Committee recommended, at the end of this negotiation round, that the negotiations be continued to try and formulate a project that would bring the maximum benefit to the Arab countries. Failure to continue the negotiations, they added, would give Israel the chance to implement her own project, one which would divert all the waters of the Jordan river from a point north of Lake Tiberias outside the basin to the coastal plain and the Negev.

In essence, the points that needed to be ironed out with the Arab side were not easy. Maximizing the Yarmouk storage on the river course would minimize Yarmouk flows that could be stored in Lake Tiberias. To discourage that, Johnston set a ceiling on what could be spent with American assistance on Yarmouk storage over and above what the Unified Plan estimated. That was going a few steps toward the Arab objective because this ceiling obviously meant that the Arabs could build a higher dam provided that the additional funds would not be provided by the U.S. The water duty issue defined the total amount of water that the arable lands needed for irrigation, and thereby determined how much water each riparian party should be entitled to. The Arab side which had a greater amount of arable land within the basin used higher water duty to maximize their entitlement to basin waters, and continued to insist that the use of the waters be made only within the basin. That condition limited Israel's entitlements. They made the construction of the Adasiyya weir contingent upon reaching an agreement on the Plan as a whole. In that respect, the Arab side missed an opportunity to have the weir built. They could have postponed the construction of the diversion facilities from the weir Tiberias, thus avoiding the use of Lake Tiberias, but allowing water diversion to Jordan. The weir proved to be crucial to Jordan as shall be revealed later.

Despite the differences, Johnston was optimistic. In a telegram to the Department of State from Cairo, on the eve of his departure for Israel, Eric Johnston concluded by saying: "I leave Cairo much encouraged at the prospect of obtaining Arab cooperation in integrated Valley development". A significant achievement here had been the relaxation of Arab opposition to diversion of Yarmouk floods for storage in Lake Tiberias, an undertaking, Johnston rightfully thought, would be conducive to cooperation among adversaries.

2. Negotiating with the Israelis- Second Round

The Government of Israel had been studying the Main /TVA Plan that was transmitted to them shortly before Johnston's first visit in October 1953, and devising a counterproposal as well. Government appointed experts expressed their disagreement with the Plan's principal premise that excluded the Litani River, and with the conclusion on water allocation to the different riparian parties. Contrary to what the Arabs contended, the Israelis maintained that Israel's share was too little, and that without the inclusion of the Litani none of the riparian states would receive her water needs. Actually, the comment of Moshe Sharett, Israel's Foreign Minister, to Eric Johnston on his first visit in the fall of 1953 was to that point. He stated that Israel's attitude towards the Unified Plan would be determined in large part by the distribution of the Jordan's waters and by the possibility for the inclusion of the Litani waters in the scheme. In contrast with what the Arabs insisted on, the Israelis criticized the notion of exclusive, in-basin use of basin waters, because it would mean the waste of water that could be put to beneficial use outside the basin, and because these notions neglected the development needs of such areas as the Negev.

The Israeli Government then commissioned the American engineer, John S. Cotton, former advisor on the planning of Israel's National Water Carrier system, to prepare a "truly regional scheme," incorporating the Israeli criticisms and suggestions outlined above. The Cotton Plan, approved by the Israeli Cabinet in early June, had as its guiding principles the aim "to provide for the full irrigation of all irrigable lands in the Kingdom of Jordan, and all irrigable areas in Southern Lebanon as well as Syrian lands in the Upper Jordan basin, leaving for Israel all the surplus water from the Jordan-Yarmouk system and the Litani flow."[71] The Plan included the diversion of the Litani into the Jordan system, an idea first advanced by Bourcart in 1901. It envisaged separating Jordan's irrigation network from Israel's by constructing dams in Jordan to impound the major part of the Yarmouk waters for Jordan's use. It advocated the resumption of the Jordan River diversion scheme at Jisr Banat Yacoub, and the construction of two carriers in Israel, one to the northern Negev, and another from Lake Tiberias to Biesan in the West Jordan Valley. The Litani would provide fresh water to Lake Tiberias, to compensate for the increased salinity caused by the diversion of the Jordan River upstream. The total flow of the water resources in the Cotton Plan was twice that in the Unified Plan because of the inclusion of the Litani, and the

---

[71] Brecher, "Israel's Foreign Policy," p. 197.

allocation to Israel was increased threefold; Jordan's allocation was cut by 25 percent and Syria's by 30 percent. The Cotton Plan became the Israeli counter-proposal to the Unified Plan, as the Arab Plan was the counter-proposal of the Arabs.

Negotiations with Israel began on 20 June 1954. The situation Johnston faced in Israel was no less complicated than the one he confronted with the Arabs. Perhaps some excerpts of the minutes of his meetings with the Israeli side, headed by Prime Minister Moshe Sharett,[72] better describe the situation. On his second visit, the minutes, as written by the American side,[73] begin with Johnston's opening remarks:

"States [Johnston] he brings best wishes of the President of the United States, that we are proud of the part we have played in development and growth of Israel. Believes future development depends on partnership between the two countries. Stresses that neither we nor the Government of Israel can be "islands" and cites the hope expressed to Henry Byroade by Ben Gurion that U.S. relations with the Arabs improve because that would benefit all of us in the area."

Johnston went on to say, "Economic development is essential in the region as a whole, not in one area alone. We think the political climate of the Near East cannot be changed without changing the economic climate. One of the factors in such an attempt is the utilization of the waters in the area. Even though our discussion will probably confine itself to the Jordan River, I want to assure you that we in the United States are interested in other programs for economic development of the area, and we want you, as people of courage and vision who have carved out this nation, to join us in carrying out the economic development of this area. And if you achieve fuller wants, further irrigation is necessary. Men like you must take the leadership in these efforts. It requires statesmanship of the highest order; it requires sometimes denying yourselves the things to which you think you are entitled, sometimes doing the things that may not seem expedient. These are always the responsibility of leadership... We trust that you will approach this whole program from that viewpoint."

"In discussing river development, we have before us three plans: 1) the Unified Plan is not <u>the</u> plan, but <u>a</u> plan, subject to modification as engineering data accumulate;[74] 2) so called Cotton Plan, based mainly, I

---

[72] "U.S. - Israeli Discussions-- First Meeting--Jerusalem, June 20, 1954, at 4 p.m.," Minutes of meeting, State Department Records, from the National Archives of the United States, Washington D.C.

[73] State Department Records, from the National Archives of the United States, "US-Israeli Discussions--First Meeting--Jerusalem, June 20,1954 - 4 pm".

[74] Ambassador Johnston was referring to the data then being gathered and accumulated by the Joint Venture of Michael Baker, Inc. and Harza Engineering Co.

understand, on the data of another engineer, Mr. Hays; 3) the Arab Plan, discussed in detail in Cairo."

Ambassador Johnston then had the following argument to present to the Israelis about the Litani, the Lebanese river that Mr. Cotton included in his plan:

"... Now the difficulty in considering the Cotton Plan, so far as we are concerned, is that it takes into consideration a source of water not in the watershed and utilizes that resource. The Unified Plan would have been a different plan had it taken into account the Litani River. But we have consistently said, and we still maintain, that the Litani is a stream wholly within Lebanon, a national asset of Lebanon. We can no more require that it be given up for the benefit of the region than we can require that you give up the Yarmouk-Jordan Triangle for the benefit of the region. We have steadily refused to do that, despite great pressures. Therefore we cannot consider the inclusion of the Litani River in the development of the Jordan Valley. The only way that river can be attained for use and benefit of the region is by persuading Lebanon that it should be included within the area. It seems to us that that matter rests almost entirely within your hands and depends on the action and attitudes that you take in the matter."

The exclusion of the Litani at this early stage of Johnston's mission was a positive starter, because the Arab side would never agree to its inclusion. In the same meeting, Ambassador Johnston had the following to say to the Israelis in defense of the Unified Plan. Johnston's presentation to the Israelis is self explanatory, and his words describe in detail the virtues of the Unified Plan to Israel:

"Now in considering the three plans --and we must exclude the Litani-- we come back to the program which we presented to you last Fall as perhaps being the only one that does embody the wants and desires-- taking account of the political difficulties-- of the four states included in the Jordan watershed.

There are five features in this plan and its implications: (1) Division of the waters. We think that the division as outlined in the Unified Plan is probably a fair one and we would urge you to adopt approximately the amount of water indicated in that plan- or approximately 400 mcm's annually. (2) Utilization. Now, we have consistently stated, both here and publicly in

---

The Joint venture was employed by the Hashemite Kingdom of Jordan under finance from the United States Technical Co-operation Agency (TCA). Their work was authorized on June 24, 1953. The Joint Venture submitted an appraisal report of existing data in January 1954. Interim findings were submitted in a report in January 1955. Following conferences on the river development in the Spring of 1955, the Joint Venture proceeded to prepare the Master Plan, and submitted it on July 15, 1955.

the United States, that we thought that when that water was once allocated to a nation, that nation could use the water as it saw fit. And we have consistently stated that if a certain amount of water were allocated to Israel, she could take that water to the coastal plain. That is a very serious stumbling block to the Arab States. They have consistently stated that the waters must be used in the watershed. We still adhere to the principle I have outlined to you. (3) Storage. Again we feel that the Unified program is perhaps the best as regards provision for storage. This also involves some type of international control. (4), and we feel that once a division has been made and a determination made of what shall be done in the area, then there should be some type of international authority over allocation (not the use) of the waters. (5) Surveys. We think there must be access to the Yarmouk-Jordan Triangle for surveys. We have not yet been permitted into that area with engineers. We would like to start this work as soon as possible so that the Jordan- Yarmouk waters will not continue to run wasted into the Dead Sea."

"Thus, we would like to start dirt flying in this area as soon as possible, which we feel can be very soon if we can reach a general agreement here. And therefore we would like to consider what suggestions you have for modification of the Unified Plan."

In conclusion, Johnston cited new scientific inventions, e.g., atomic power, that may revolutionize economic development in the future and bring tremendous changes in the next 50 years; that these tools were within Israel's grasp; that he believed they could aid in preserving and enlarging freedoms through improving the economic situation. Johnston hoped that "new horizons of tomorrow" would figure in those discussions, and believed the world was watching.

The Israeli side responded with equally diplomatic language. The minutes of the same meeting of June 20, 1954 continue:

"Sharett: Thanks Ambassador Johnston for 'most interesting, stimulating and inspiring statement'. Feels his allusion to a partnership between the Governments of Israel and the United States is a 'very great compliment,' although there is such disparity between the size, wealth and power of the two countries. Wishes to emphasize that the Government of Israel cannot feel so detached, however".

"The problem we are discussing is the very essence of our future life. We are not rejecting the idea of leadership in the area by any means, but long before we can aspire to that exalted position, we have to assure something that is much nearer and more tangible and compelling to us -- our survival. We have to ensure this against very heavy economic burdens, against pressure of the countries by which we are immediately surrounded, by which, in fact, we are harmed; a pressure that is to this day inexorably hostile."

"...Now we have that little River Jordan.... Every drop of it is precious; every drop of it is part of our future. We realize that geography has ordained that this river cut through another territory, creating certain rights. On the other hand, a main tributary, the Yarmouk, forms our boundary for a certain distance. This gives us certain rights. We cannot be blamed for trying to make the most of those rights and for making the most of our geographic position which has placed us in control of a part of that stream. We need it in order to make our own life possible. Our population even today is very far from being self-supporting, physically or financially. This population will grow; and it will have to grow without bursting its boundaries-- that is our determination, all talk to the contrary not withstanding."

Sharett went on to say that Israel "may have to be a haven for others in the future; e.g., from North Africa, where Jews are living on the brink of a volcano. The haven of the United States for Jews is a source of 'inestimable moral and material strength to us' but that is not the fate of many other Jewish communities."

He proceeded to defend the inclusion of the Litani, against the convictions of the American delegation and, of course, against the will of Lebanon and the Arab League Committee. He viewed the Litani as basic to regional development. In a way, he considered the Litani inclusion in the Plan as a Lebanese contribution to Jordan's water needs:

"We do not feel we can be called upon to limit ourselves to any single river or system of rivers. If planning is regional, <u>all the land and all the waters</u> of the region must be encompassed. We quite realize you cannot force anyone-- Lebanon or us. We can only help in producing a promising plan. But if you contemplate regional planning, it is unthinkable to us that the Litani should be excluded: you cripple the chances of development in this area by taking it out."

"Thus of all the plans that have so far been drawn up, the only one deserving the title of 'regional plan' is the Cotton Plan. We don't say it is ideal. But I say that, if it is not a matter of practical politics that Lebanon should present Israel with "X" mcm's of water a year, and if Israel is called on to give up "X" mcm's for Jordan's benefit-- water she badly needs and which she could demonstrate would be of such advantage to herself-- why should not Lebanon be invited to supply that volume of water for Jordan's benefit as a matter of inter-state assistance within the Arab world?"

Sharett added that he did not think that "Israel can be blamed for resisting giving up waters for Jordan's benefit any more than Lebanon can be blamed for resisting giving up the luxury of letting the Litani run into the sea."

Sharett's response was not quite in line with the geography of the basin. Jordan has a right to the Jordan River's water, just as much as Israel has. Both are riparian parties along with Syria and Lebanon. Neither Jordan nor any other party can claim rights on the Litani, an exclusively Lebanese river. If Lebanon gives up any part of the Litani, she would be doing so out of her own will because she is under no obligation to do so. In the case of Israel and the Jordan River, there is obligation on Israel to let the Jordan River flow downstream to get to Jordan so that she can access her rights on the river. It is also to be noted that Lebanon needs the waters of the Litani to develop the southern part of Lebanon, where poverty had taken root. It is very crucial to improve the standard of living of people there in order to serve the cause of stability and peace in that area.

Ambassador Johnston's response to the Prime Minister in that meeting seemed to indicate a retreat from the position he had expressed in his own presentation a short while before. He said, "I would like to say that the Cotton Plan envisages that use of the Litani waters would be years and years away. This is playing on your side."

Sharett continued to advocate the Cotton Plan and stated that the Unified Plan was not regional and that it had some organic defects, and insisted on keeping the inclusion of the Litani alive.

Ambassador Johnston resorted to his skills in diplomacy. He complimented Sharett on his excellent exposition.

"We are not so very far apart. At no time did we discuss allocation of the Litani. 'yes' or 'no'. We simply said we were discussing the Jordan Valley. The Litani is something in which you have time on your side; if you can integrate your country into the area, this will take care of itself. In the immediate future, all of the water that you can possibly use is available to you from the Jordan River. And it is this that we want to discuss with you, because it is an international stream, arising and flowing through four countries. Anyone doing construction within it without the riparian rights having been determined will inevitably bring about more bitterness and conflict. Well-qualified engineers will have to go into this. But we want to get started on this river with a greater degree of unanimity for the benefit of Israel."

"As regards the relation of the Arab League to our recent Cairo meeting. So far as we know this was not an Arab League meeting. The attitude that the Arab States took through the leadership of Egypt has been extremely salutary; without that leadership, we could never have made the progress we made. You will recall that last fall two of the Arab States told me unequivocally they would have nothing to do with any program giving Israel one drop of these waters. Rather than being an Arab League meeting, with all the connotations this suggests, this meeting under the Egyptians' leadership resulted in real progress."

Sharett: "There is one figure in the Main Plan that leaps to the eye: that the Plan gives 774 mcm's to the Kingdom of Jordan, the Cotton Plan gives 575 mcm's. Now, that is not the only difference. But supposing that the Kingdom of Jordan really needs 774 mcm's-- it then falls short of its requirements by 200 mcm's. Why should it claim that water from Israel when it can get it from Lebanon? Lebanon is not using that water. Israel undeniably can use that volume of water with great advantage to itself and the human race..."

Ambassador Johnston, "I appreciate this. Lebanon, also, has ideas of her own regarding use of the Litani waters. I think, Mr. Prime Minister, that in all fairness, until Lebanon knows what she can do with her waters, she will not give them up -- any more than you would give up the Negev on the grounds that you have insufficient population to fill it."

Ambassador Johnston spent three days negotiating with the Israeli team led by the Prime Minister, Mr. Moshe Sharett, and including Levi Eshkol, Aharon Wiener, Simch Blass, and others. Johnston's company included Jones, Colburn, Gardiner and Barnes. The points discussed in this second round were the Litani issue, the water division, and the use of Lake Tiberias for storage of the Yarmouk floods. Ideas of stage implementation of the development scheme were discussed. Ambassador Johnston used the preparedness of the United States Government to finance the Plan as leverage to lure the Israeli side into making progress. He also used his potential success with the Arabs as another incentive for Israel to be positive and forthcoming. An important chip in Johnston's hand was the fact that the development of the Jordan Valley entailed the settlement of Palestinian refugees, something that Israel, the United States and the 'whole world' wanted to see progress in. In short, Johnston enumerated the pluses of the scheme for Israel:

"(1) You have a measure of Arab cooperation immediately, if we can arrive at a formula that is acceptable to both sides.

(2) In the immediate construction in the Valley, you help to solve the refugee problem, one in which you are as much interested as we are.

(3) If we can agree on a framework, a water program construction can start immediately.

(4) Israel under this program gets all the water she can use now and leaves to a later date and future consideration the exact allocation of the water.

(5) Israel gets the continued interest of the United States in her further development".

On the minuses, Johnston enumerated to the Israeli side two points:

"(1) That Israel gives up the immediate prospect of the Litani-- which, incidentally, you can not use for 15 years anyhow, so you are not really giving up much, and

(2) Israel gives up what is a relatively minor quantity of water that she cannot use in any event."

Johnston tried to convince the Israelis that a water-sharing agreement with the Arabs would clear the political air, and, for that reason alone, it was important to make concessions. He urged the Israeli team to accept three principles: (1) a prior claim to water lay with the Kingdom of Jordan for the resettlement of refugees; (2) Lake Tiberias would be used as a common storage reservoir; and, (3) a neutral body would supervise the distribution of water. Israel had a negative response to the first two principles, and a highly qualified positive one to the third.

Israel rejected the idea of using Lake Tiberias as a regional reservoir, something that the Arab side was also uncomfortable with. They also did not accept the division of water stipulated in the Unified Plan and neither did the Arabs. The motives of each party to reject each of the two ideas were different. Nonetheless, progress was made during the second round that left Johnston encouraged.

3. American Evaluation of the Second Round

In a memorandum on the status of Jordan Valley negotiations after Johnston's departure from the region, the situation was described as hopeful. Although no agreement had been reached, both sides had indicated acceptance of the principle of unified development. Johnston was optimistic that the differences could be reconciled. The most difficult issue outstanding was the division of the waters. The other issues were not quite easy either. His task at this point was to formulate a new proposal on allocations that would be acceptable to both sides.[75] This task obviously demanded more diplomatic efforts, and the continuation of shuttle diplomacy.

The division of water depended much upon the land areas to be irrigated and on the water duty, which in turn relates to evapo-transpiration of the crops, the method of irrigation, and conveyance efficiencies. These factors are normally defined by the use of survey data and scientific formulae. The Mission was still handicapped by the lack of real data. In Jordan, the Government had employed in June 1953 the Joint Venture of Michael Baker Inc. and Harza Engineering Company, under finance from TCA of the United States, to prepare a Master Plan for the development of the Jordan Valley in the Kingdom of Jordan. It was collecting valuable data that would definitely be needed before a final proposal could be made for the unified development of the Jordan Valley. Data was also lacking on the Hasbani in Lebanon, and Johnston suggested that the United States finance a study to have data collected on that sub-basin. In the mean time, he had to accept whatever data given to him by the Lebanese on the Hasbani, and the Syrians on the Banyas. Dr. Wayne Criddle, of Johnston's mission, made the field trips and conducted talks with the technical people in the countries of the riparian parties.[76] He was assisted by local government technicians and technicians from the United

---

[75] Memorandum prepared in the Department of State, "Current Status of Jordan Valley Negotiations," Foreign Relations of the United States, 20 December 1954, no. 938, pp. 1727-30.

[76] Meeting on 18 June 1984 between the author and Mr. Wayne Criddle at Harza Engineering Company, Chicago, Illinois.

States (Baker-Harza). He spent the period from 5 to 12 October 1954 in Lebanon, the 12th to 16th in Syria, the 16th to 27th in Jordan; and from 27 October to 30 November in Israel, and helped prepare data needed to update the Unified Plan.

4. Johnston's Third Round

4.1 New Data Availability

The third round of shuttle diplomacy began on 27 January 1955. Three issues remained to be resolved. The first was water allocation to the various riparian states; the second was the use of Lake Tiberias for storage of the Yarmouk floods in addition to regulation of the Jordan River flow, and the issue of international supervision. Each issue was basic by itself, and posed a challenge to the abilities of the American diplomat.

Interim findings of the American consulting engineers, Baker and Harza, were transmitted in a report to their employer, the Government of Jordan, on 14 January 1955. The draft of that report was completed sometime during the Fall of 1954, and was made available to the Johnston Mission at that time. The above mentioned field visits of Dr. Wayne Criddle in October - November 1954 reinforced the findings of the study. The final report containing the Master Plan for the development of the Jordan Valley in the Hashemite Kingdom was submitted in ten volumes on 15 July 1955, in time for Johnston's use in his fourth and final round. The Interim Report of January 1955 had results of soil classification and a survey of water resources. It indicated that there was more arable land in the Jordanian portion of the Jordan Valley than had been assumed by any of the development plans at hand, which meant greater water needs. It also indicated that the water duty (irrigation water needed to irrigate a unit of land area) was less than what previous estimates had indicated, which meant less total water needs. These findings induced a net reduction of 69 mcm in water needs to irrigate Jordanian lands, and thus enabled Ambassador Johnston to have the figures revised and to offer Israel more water than he had done in the second round. This off was made at the expense of the Jordanian share that was indicated in the Unified Plan as prepared by Chas T. Main.

4.2 Political Developments

The political developments in the region posed a threat to the success of the Johnston Mission during its third round. The Baghdad Pact was very much on the agenda of the countries of the region, splitting the Arab Middle Eastern states into two opposite camps, one that supported the Pact (Iraq, Lebanon and Jordan), and another that opposed it (Egypt and Syria). The mood on the streets of the pro-Pact states was not in line with official positions. Demonstrations against the Baghdad Pact took to the streets of Amman, Baghdad and Beirut. The government of Faris Al Khoury of Syria, a pro-Iraq politician, was brought down and replaced by a government headed by the pro-Egypt politician, Sabri Al Asali. Both the pro-Pact camp and the anti-Pact camp were opposed to any arrangements for cooperation with Israel. The political mood was reflected in Johnston's reception in each of the capitals that he visited, and prospects of success appeared to be slim.

Israel, on the other hand, was also unhelpful helpful either. Mr. Teddy Kollek, Director General of the Prime Minister's Office and member of the Israeli team negotiating with Johnston, gave a lecture in Tiberias, Israel on 5 January 1955, just before the third round started. He was quoted as saying:

"A solution in principle on the distribution of Jordan and Yarmouk waters between Israel and Jordan will be obtained, and as a result thereof it is possible that refugees will be settled on the eastern Jordan bank in accordance with the Johnston's Program."

"Israel will agree to Mr. Johnston's proposal on condition that she will be given the possibility of transferring Jordan water from the drainage area to other places, i.e., the Negev. She would, however, under no circumstances favor international control of the water sources." Mr Kollek added that Israel would in any case object strongly to control of Lake Tiberias as proposed by Syria and Jordan and to the use of such waters for irrigation of the Arab areas."[77]

Announcements of that type helped to generate in Arab ranks resentment of Johnston's Mission, especially when it was inferred that Johnston's efforts aimed at the final resettlement of Palestinian refugees in their host countries. The announcement coincided with the hot political environment in Arab streets, already hostile to the West and the Baghdad Pact. Moreover, the Israeli official position changed course and could place obstacles in Johnston's way.

4.3 Coordination with Britain and UNRWA

Before Mr. Johnston embarked on his third visit, he met with officials from the British Foreign Office on 21 January 1955. He briefed the British on what he hoped to achieve in his forthcoming round of negotiations. His points included the hope to reach agreement on:

(1) Shares of water in light of the Interim report of Baker-Harza that indicated Jordan's need was 760 mcm as opposed to 829[78] mcm in the Main/TVA Plan

---

[77] Foreign Service Despatch from the U.S. Embassy, Tel Aviv, to the Department of State, Washington D.C., dated 7 January 1955, quoting Haboker newspaper report of 6 January.

[78] Includes 55 mcm of groundwater and 774 mcm of surface water.

(or the 975[79] mcm estimated by the Arab Plan). This new information along with the hydrological survey explained in the report would leave 454 mcm for Israel of which some 150 mcm could be diverted out of the basin; (2) storage of water on which Johnston indicated that the Interim Report rejected Maqarin as a storage site and preferred a downstream site at the Wadi Khalid confluence with the Yarmouk. Two dam heights were possible. A 40 meters high dam to store 47 mcm and cost $11.6 million, and a 60 meters dam to store 118 mcm and cost $17.5 million. The rest of the flood flows would be stored in Lake Tiberias; and, (3) there would be some kind of international control of the division of water. "The United States," Johnston told his British audience, "would have the position that once the allocations to the riparian states had been agreed upon, any of the sovereign states concerned could use the allocated water anywhere it desired in its territories". That assertion helped support the Israeli plans for using Jordan River water outside its basin within the confines of her own share.

On the issue of benefit to the refugees from the scheme, Johnston pointed out that the issue was outside his own frame of reference and that the plan was for UNRWA to finance the works in Jordan under arrangements whereby significant benefits to refugees would emerge. He was to meet with Mr. Labouisse, UNRWA's Director, in Athens to discuss this particular aspect of the matter.[80]

Johnston shifted to the more reasonable of positions regarding sovereignty over water resources. Israel, as expressed by Sharett, talked about her freedom to do whatever she pleased with the waters that flow in her territories. This stance was in line with the provisions of the Harmon doctrine, under which a state is entitled to exercise absolute sovereignty over the waters flowing in her territories. The Arab side, on the other extreme, advocated the principle of "Absolute Integrity" of the river basin, under which waters of a basin can be used only inside that basin. In between came the approach adopted by Ambassador Johnston. In many ways, his approach carried with it certain aspects of both principles. Absolute sovereignty would be exercised by a state on her own share of water resources. The shares of the various riparian parties would be defined according to the irrigation needs of the arable lands in the basin that fall within the territories of the concerned riparian. This accounted for the need to satisfy the needs of the basin first, a principle enshrined in the doctrine of absolute integrity of the river basin.

---

[79] Includes 64 mcm of groundwater and 911 mcm of surface water.

[80] Memorandum of Conversation, Department of State, Washington D.C., January 21, 1955. Attendants from the British Foreign Office were Mr. Evelyn Shuckburg, Undersecretary for Middle East Affairs, Mr. G. G. Arthur, and Mr Ronald Bailey, First Secretary, British Embassy, Washington D.C.

The briefing to the British Foreign Office was indicative of the role of Britain in the whole process. It was by recommendation of Britain that the UNRWA initiated the Jordan River study in the first place, and her role since the Balfour Declaration had been central. Caring for the Palestinian refugees and their resettlement, and for the security of Israel, remained on the British agenda.

4.4 Talks with the Israelis

The Israeli delegation to the talks of the third round was headed by Mr. Moshe Sharett, Prime Minister.[81] As a result of the data update of the Interim Baker-Harza report, Johnston made the Israelis an offer at the beginning of his meetings with them, in the morning of 27 January. Johnston said, "It was now possible to suggest 448 mcm as Israel's share of the river water. This amount adds 90 percent of the difference between the original proposal (346 mcm) and the amount Israel had suggested (462 mcm)". In the afternoon meeting, Prime Minister Sharett was candid and informed Johnston that the Israeli Government experts were not satisfied with the figure of 462 mcm which Mr. Johnston had indicated was his understanding of Israel's requirement, as stated in the second round on June 28, 1954. Other items must be added to this total and, in any case, the Israeli experts were not speaking for the Israeli Government when they advanced their suggestions. He pointed out that Israel could not agree to squandering precious water to appease the Arabs and that Israel would use all the water she possibly could obtain and more. This change in Israeli attitude was the first hardship the Israelis threw in Johnston's way in this round. The second concerned Yarmouk storage in Lake Tiberias.

"As far as Lake Tiberias storage was concerned," Sharett said, "my Government understood the technical reasoning behind its inclusion in the plan, but Tiberias must be considered in the light of present international circumstances. The Government of Israel opposed the proposal to use Tiberias for storage, certainly in the first instance. The most the Israelis could do will be to keep the question open."

"It should be appreciated that Israel is young, small and hemmed in by enemies. The nation would be jealous in safeguarding its territory and its

---

[81] The Israeli delegation was comprised of: Mr. Levy Eshkol, Minister of Finance, Mr. Teddy Kolek, Director General, Prime Minister's Office, Mr. Jacob Herzog, Chief, American Section, Foreign Ministry, Mr. M. Blass, Director, Water Planning for Israel, Mr. Pinhas Sapir, Director General, Finance Ministry, and Mr. Aharon Wiener, Assistant Director, Water Planning for Israel.
The American delegation with Johnston was comprised of Mr. Arthur Z. Gardiner, Mr. George Barnes, Mr. Wayne Criddle, and Mr. Oliver L. Troxel. Mr. Ivan B. White, Counselor at the US Embassy in Tel Aviv, attended the meetings.

sovereign rights. The use of Tiberias as an international reservoir would lay the basis for territorial claims. The full exercise of sovereign rights is essential to Israel's interest."

It became obvious that the two sides, Arabs and Israelis, were in agreement in this respect, but each for their own reasons. The Arabs did not want to use Tiberias as a storage reservoir because of water quality and evaporation issues and because they did not want any of their waters to be controlled by their enemy; and the Israelis did not want the Arabs to use the Lake for fear of their laying sovereignty claims over it. Additionally, Israel did not want to release any Tiberias water to any Arab riparian because she wanted all the waters of the Lake. Johnston, on the other hand, favored the use of the Lake for common storage, because it would trigger and promote co-operation between Arabs and Israelis, at least in water matters.

Mr. Johnston, in his response to the Prime Minister, paid respects to Mr. Sharett's candor but thought he was misinformed on certain vital points. In the first place, Johnston wished to make it clear that "allocations which are proposed for the Arabs were not made for theatrical effects but would be used to provide a living for the Arab refugees in whom all concerned have a deep interest." Mr. Barnes of the Johnston party explained that "the figures we now presented happened to approximate the June figures but this was a coincidence as they were based on the requirements for Jordan as determined by the professional team which had just completed its work there, and that, following satisfaction of Jordan's needs, the residue of the water, after Syria and Lebanon's claims had been met, was allocated to Israel."

Mr. Aharon Wiener of the Israeli team indicated that Israel's needs amounted to 568 mcm, up 23% from their requested amount of 462MCM in the second round, and up 27% from the updated offer Johnston was making in that session. Wiener added that Jordan's needs should be calculated on the basis of deducting 20% of the area of arable land, and on the basis of availability of groundwater on the Jordanian side of the Jordan Valley, which he estimated in the order of 200 mcm per year based on experience on the Israeli side of the Jordan Valley. Johnston and his team defended the calculation of Jordan's requirements on the grounds that Wiener's estimate of groundwater availability on the Jordanian side was over any reasonable estimate, and that their estimates of availability would not exceed 50 mcm that were, in any case, needed to irrigate lands that the canals would not reach. Mr. Criddle pointed out that Jordan's needs were based on an irrigation efficiency of 65% as compared to only 60% used in Israel. The irrigation efficiency, along with the water duty, the area of arable lands that could be irrigated in each season, and the cropping pattern, all determine the quantity of water needed to satisfy the requirements of the riparian party.

It is to be noted here that Jordan's water needs were squeezed from several directions. The Baker-Harza estimate of those needs (760 mcm) effected a net reduction of 69 mcm or 8% of the amount allocated to her by the Main/TVA Plan (829 mcm), by simultaneously reducing the water duty and increasing the irrigation efficiency. Israel, the advanced country in comparison with Jordan, was assigned lower irrigation efficiency than less-advanced Jordan (60% compared to 65%). Additionally, Wiener tried to cut Jordanian arable land by 20%, to effect another drastic reduction in Jordan's water needs. As will be mentioned later, the fourth round of Johnston's Mission imposed further cuts on Jordan's share, following the Israeli argument as advanced by Wiener.

Ambassador Johnston stressed the interests of the Arab refugees and that, if the theory of allocating on the basis of economic benefit to Jordan was accepted, there was no other solution. The Israelis appeared to have accepted this theory. The storage in Lake Tiberias, although entertained by previous plans, was inconvenient for Israel on technical grounds alone, because other sites in Israel (the Batoufa depression), once thought fit for storage, were found not to be so. Israel needed for her own purposes the storage capacity of the Lake. Mr. Blass of the Israeli team pointed out the minimum of minimums that could be accepted by Israel as a solution to the problem. "First", he said, "Israel must have all the Yarmouk water which it now uses for existing settlements [Mr. Blass indicated on another occasion that this quantity was 45 mcm]. Secondly, no Israeli government could take the responsibility of permitting a drop of water from Tiberias to go to another nation. Israel requires the entire flow of the upper Jordan for her own purposes."

In the meeting of 31 January, the Israeli delegation presented Johnston with a written counterproposal containing essentially the above Israeli position on water sharing, storage in Lake Tiberias, international supervision, and salinity in Lake Tiberias. They contended that Jordan's share should be based on the irrigation of 400,000 dunums in lieu of the 513,000 indicated in the Baker-Harza report, and on the use of groundwater there. Johnston was outraged at the Israeli position. Regarding the 400,000 dunum area in Jordan, Johnston protested, "How can we know? How can we know? Have you been on the other side? All of your figures are based on 400,000, and we simply cannot sell anything of the kind. We simply couldn't sell this to the United States, or to the Arabs."

4.5 Meetings with the Arab Committee

Preliminary meetings were held at the Egyptian Ministry of Foreign Affairs in Cairo for members of the Arab Technical Committee on 1 and 2 February 1955 in preparation for the meetings with Eric Johnston.[82] The

---

[82] Participants were General Mahmoud Riyadh, Saleh Abdul Rahman Mahmoud, Dr. Mohammad Ahmad Salim and Ahmad Ali Farag (Egypt); Mr Khalousi El Kheiry,

Committee reviewed the cropping pattern in Jordan and decided to amend the one presented in the Interim Baker-Harza report. Mr. Khalousi el Kheiry, Minister of Economy of Jordan, reported that the consultants never discussed the cropping pattern with their employer, the Jordan Government.[83] The Committee reviewed the water requirements and decided to have them amended for Jordan in light of the amendment of the cropping pattern. They affirmed the need to have the Yarmouk floods stored by a high dam at Maqarin with 400 mcm storage capacity of which 100 mcm would be for dead storage and for power generation. The Committee also decided to discuss with Wayne Criddle, the engineering arm of Ambassador Johnston, his report on the utilization of the water resources of the Jordan River basin. They thought his water duty was very low.[84]

Meetings with the Johnston Mission started at 4:30 p.m. on 2 February 1955.[85] Johnston's effort focused on establishing the Interim Baker-Harza report of 14 January as a basis for proceeding with the Jordan Valley development. The Arab side had apprehensions about the report findings and about the report submitted by Dr. Criddle. 1) They wondered why Criddle did not give his own estimate of irrigable lands in Israel, when he did that for the Arab countries. They viewed this ommission as a sign of bias toward Israel. 2) They expressed that they were 100% against the small storage dams suggested in the Interim report by Baker-Harza, and wondered why the consultants ignored the idea of Arab engineers to have a high dam built at Maqarin, as conveyed to Baker-Harza by an Arab instruction on 30 September 1954. 3) They disagreed with the cropping pattern adopted by the consultants and thought that it was copied from the cropping pattern suggested by UNRWA, in the summer of 1954, which emphasized irrigation as opposed to power. The cropping pattern was not accepted by the Arab side, which considered it inadequate for the economy of Jordan. 4) A point of heated discussion was the irrigation efficiency, thought by the Arabs to be on the high side as adopted by the Interim Report. 5) Another point of

---

Hamad El Farhan, and Izzeddin Yunis (Jordan); Ibrahim Al Ostuwani, Dr. Subhi Mazlum, and Lt. Col. Bakri Kuttrash (Syria) and Ibrahim Abdel Aal (Lebanon).

[83] Note that the draft Interim Report was made available by the American Consultants to the Johnston Mission in the fall, but was never discussed with their employer who received the report in January.

[84] Dr. Wayne Criddle had developed with his colleague the famous Blane-Criddle formula that depicts crop water requirements.

[85] On the Arab side, the attendants were those indicated in the previous footnote plus Lt. Col Salah Gohar, and Mr. Mohammad Taher Shaah (Egypt), Kazim Jazzar (Syria), Nadim Demashkia and Fuad Ammoun (Lebanon). On the Johnston Mission the attendants were Amb. Johnston, Arthur Z. Gardiner, George Barnes, Oliver Troxel, and Wayne Criddle, with Howard Elting from the U.S. Embassy in Cairo.

disagreement was the water duties. 6) A last point was the amount of water allocated for Syrian use from the Yarmouk.

Five of the above points raised by the Arabs impacted the water shares (points 1,3,4,5,6), and point 2 dealt with storage. The Arab side put more weight on the allocation of water than they did on storage. They feared that more water allocated to Israel meant a higher capacity to absorb Jewish immigrants and thereby a stronger enemy than the one they had at hand. Naturally, they also wanted more water for their own benefit.

Johnston complained that he was accused in Israel of appeasing and siding with the Arabs on the one hand, while on the other the Arabs suspected that he was concealing sympathy for Israel. He expressed willingness to agree with the Interim Report findings, and noted that the net land area that could be cultivated in any season, as indicated in the report, was a higher percentage than what would normally be acceptable. Baker-Harza accounted for only 3% of the gross area to be uncultivated in any one season, leaving some 97% of the gross arable land irrigated. He finally succeeded in obtaining an agreement from the Arab side to the gross land area that could be irrigated in the Jordan Valley, and the debate on net irrigated area continued. The new allocations to Syria (52 mcm) would be corrected to be 80 mcm, and the storage issue would be discussed with two engineers from Harza who were called to Cairo to defend their choice of alternative dams and sites and the report at large. There were intense discussions regarding the cropping pattern. The Undersecretary of Agriculture in Jordan was called to the negotiations with the Harza people.[86] Comparisons were made of the average water duty adopted by Baker-Harza (1,480 m^3/dunum) and the average water duties in the Ghouta of Damascus (1,700 m^3/dunum), in the Ghab of Syria (1,200 m^3/dunum), and in the Imperial Valley of California (1,350 m^3/dunum). The Arab side contested the criterion of land classification in Israel, which was developed by Lowdermilk in 1947, then adviser to the Israeli Government. Consequently, the arable land areas in Israel were contested. The Arab side preferred to have the water division by percentages of the total water available, not by absolute quantities, and qualified their proposal by the fact that the percentage division is clearer and would be equitable in all conditions of water availability. General Riyadh noted that the acceptance by the Arabs, expressed the summer before, of an international supervision of water distribution was linked to a distribution by percentage. Mr. Johnston explicitly stated that the maximum amount of water he could get for Jordan would be the amount mentioned in the Baker-Harza Interim Report-- 760 mcm per year. He stressed that he could not sell this figure to Israel, and

[86] Mr. Nasouh Al Taher, Undersecretary of Agriculture, joined the meetings of the main committee, and Messrs. Davis and Johns of Harza Engineering Co. joined the discussions of the subcommittee.

reiterated that an agreement to a unified development plan would be for the benefit of the Arabs because failure to come to an agreement would give Israel the excuse to implement her diversion scheme unilaterally. In the same meeting, there was debate over the water that would be reclaimed from draining the Huleh marshes. The American side stressed their understanding that whoever invested to drain the marshes [i.e., Israel] was entitled to the reclaimed water.

On the issue of storage, the debate followed a similar pattern to that in the second round. The Arab side wanted maximum storage on the Yarmouk, with only leftover floods being diverted to Tiberias, and the American side insisted on the economics of the undertakings. Perhaps the remarks in the meeting with the Arab Committee, made by Mr. Gardiner of Johnston's company on this matter are revealing:[87]

"Plans provide for more storage of Yarmouk water in Tiberias than on the Yarmouk. Reasons have been reached on grounds of technical and economic feasibility. Would like to inject an element of humor into the situation. We went to Israel and found resistance to storage in Lake Tiberias, the reason being that they realized the Yarmouk is still largely Arab and would like to avoid any Arab claim to use of Tiberias and avoid the corollary necessity of a control authority that would deal with Tiberias. The Arabs would like as high a dam as possible. We are in the position of having to justify our action to the American people, and we have some problems which you may not understand."

In this respect, earlier reports of Syrian desires to have a presence in the Lake reinforced Israeli apprehensions. Colonel Husni Zaim of Syria expressed his willingness to make unilateral peace with Israel in exchange for half of Lake Tiberias.[88] The Armistice Line drawn between Syria and Israel lay only ten meters away from the northeast shoreline of the Lake, but Syrian soldiers often bathed in the water. The Israelis also discovered that they wanted all the storage capacity of the Lake for their own purposes after geological investigations revealed that the Batoufa reservoir was unfit for the storage they were hoping to establish there. However, Jordan would be unable to irrigate the lands indicated in the Baker-Harza Interim Report without a draft from Lake Tiberias. It was thought at the time that no matter how high

---

[87] Quote from the minutes of meeting on the morning of February 6, 1955 at the Foreign Ministry in Cairo, pp. 12.

[88] Shalev, Ayre, "Cooperation in the Shadow of Conflict," 1989 (in Hebrew), quoted from Aaron T. Wolf, "Hydropolitics Along the Jordan River," United Nations University Press, 1995, p.44.

the dam on the Yarmouk was at Maqarin, it could not impound all the Yarmouk floods.[89]

The negotiations culminated in a memorandum given to the Arab side by Ambassador Johnston for forwarding to their respective governments. In that memorandum, dated 8 February 1955, Ambassador Johnston mentioned two alternatives to water sharing: the first, that he preferred, defined the final shares of each riparian party; the second, proposed by the Arab side, stated that a partial allocation would be made first, followed by another allocation for Jordan and Israel three years later.[90] The memorandum contained an arrangement for neutral supervision by which the United States Government, or the Secretary General of the United Nations, would propose a panel of 20 or 30 distinguished and qualified persons, none to be a national of the Arab States or Israel. From the panel, a board of three experts would be appointed. The Arab riparian parties would chose one, Israel would chose another, and the two persons thus selected would choose a third to serve as Chairman of the Board.

4.6 Lobbying for Support in Jordan

The Johnston Mission then proceeded to visit Jordan. They arrived on Thursday, February 10th, and used Friday the 11th, which is the Moslem weekend to visit the Yarmouk and the proposed dam sites. Discussions during the course of the field trip with Jordan's Minister of Economy, Khalousi el Kheiry, and Minister of Finance, Anstas Hanania, were particularly profitable. Their interest and understanding were increased to the point where they agreed to give a certain amount of support [in the Cabinet] to the project.[91] Ambassador Johnston called on the King, the Prime Minister, and lesser ranking officials and emphasized to each the benefits to Jordan from the Unified Plan as follows:

" 1. The proposed plan will meet all of Jordan's water needs according to

---

[89] The Mukheiba site, discovered in 1962, was not yet under consideration. A dam built at Mukheiba would impound the remaining floods with no surplus to be stored in Lake Tiberias

[90] Partial allocation idea was introduced by Dr. Subhi Mazloum of Syria. Under it, Syria and Lebanon would get 132 mcm and 35 mcm respectively, and Jordan and Israel would get 375 mcm and 350 mcm respectively. This allocation would leave about 200 mcm of river water to be divided between Jordan and Israel after certain unresolved questions are clarified.

[91] In an interview on 19 April 1999 with Tawfig Marar, former President of the Jordan Engineers Association, Marar told the author that Khalousi Kheiry, in the Arab Engineers meeting near Jericho in the Spring of 1955, responded to a comment from an Arab engineer criticizing the storage of Yarmouk floods in Tiberias. "Why do you envy us over that storage?" Kheiry was convinced it was to Jordan's advantage, and so does the author.

the best engineering advice. The very basis of the proposal is that priority will be given to Jordan's needs.

2. The project will greatly increase Jordan's wealth and degree of economic self-sufficiency. Experts estimate that it will increase ten-fold the value of agricultural production in the Jordan Valley. More immediately, construction work will result in large-scale employment, valuable local contracts, and heavy purchases from local suppliers.
3. The plan, if accepted by Israel, will establish a limitation on Israel's withdrawals from the river system. Because of Israel's geographic advantage, unless such an agreement is made, Israel cannot be prevented in the long run from taking as much Jordan water as she pleases.
4. In addition to direct agricultural benefits, the development of the Jordan Valley would lay a sound basis for industrial development. Industrial crops can be grown which would be processed in Jordan, and industries and services would spring up offering employment to many thousands beyond those actually engaged in agriculture.
5. Refugees would become useful and productive members of Jordan's economy through the agricultural and industrial opportunities created."[92]

On Water allocation, Johnston briefed the Jordanian monarch and officials that, "at present, Israel is using over 60%[93] of the water being used within the basin, and she is rapidly continuing to develop an even larger share of the total resources. Under the proposal for dividing the total water supply, Israel would receive only about 36%, with Jordan receiving 52% and Lebanon and Syria the balance. In the absence of an agreement, Jordan would be unable to irrigate more than about two-thirds of her irrigable Valley lands. Water from Tiberias is absolutely essential; the Yarmouk flow alone is not enough to meet Jordan's needs. Jordan's requirements were carefully computed with long term development in mind, allowing for maximum crop flexibility to meet future needs."[94]

Ambassador Johnston advocated the construction of a 40 meters high dam at Wadi Khalid. "Only 92 mcm would be diverted on average into Lake

---

[92] "Discussions at Amman, February 10-13, 1955," State Department Records, U.S. National Archives, Washington D.C.

[93] Estimated at 170 mcm per year. See alternative B of Johnston's proposal to the Arab Technical Committee on 8 February 1955. The alternative was based on an idea advanced by Dr. Subhi Mazloum, Director of Irrigation in Syria.

[94] "Discussions at Amman, February 10-13, 1955," State Department Records, U.S. National Archives, Washington D.C.

Tiberias, while Jordan would draw far more from the Lake."[95] He refuted the "ill founded" concept of "security storage" advocated by the Arab negotiators. "The highest dam possible to construct on the Yarmouk could not conserve sufficient river water to irrigate all lands in Jordan for as long as one agricultural year if Tiberias flows were cut off. To irrigate all lands in the Valley, Jordan must rely on Tiberias storage; no possible dam on the river can change this physical fact of life."[96]

Ambassador Johnston was clearly soliciting support for his proposals. He used the dual argument of flagging the disadvantages of an agreement to Israel and emphasizing the benefits of such an agreement to Jordan. An agreement would limit Israel's usage and bring numerous benefits to Jordan and, in the absence of such an agreement, Israel would pump from the river as much water as she pleased.

However, Ambassador Johnston's proposal did not obstruct the construction of a high dam. If Jordan felt she should go for a high dam, for whatever reason, the United States would not object. The American financing for a high dam would not be more than $10 million over the cost of the 40 meters dam at Wadi Khalid, making American financial support of the dam limited to $21.6 million as a ceiling. The King appreciated Johnston's efforts and advice and invited him and several members of the Cabinet to lunch. In an effort to crystallize Jordanian thinking and provide a written basis on which negotiations with the other parties might best proceed, Ambassador Johnston gave to Mr. el Kheiry, at lunch on 13 February, a draft proposal for a communication that the Jordan Government might give to the Johnston Mission. In essence, the communication would inform Ambassador Johnston of the approval of his proposal by the Jordanian Government.

During that same day, the Lebanese Government sent a telegram to the Jordanian Government (and to the Syrian and Egyptian Governments as well) requesting that no final decision be taken without further consultation among the Arab states concerned, and inviting them to Beirut for consultation. The Jordanian Government agreed to the Lebanese request, explaining to Johnston that they would otherwise be in an untenable position. By that time, harsh criticism was levied against Iraq for taking unilateral action in the Turko-Iraqi-Iranian pact (the Baghdad Pact). A unilateral action on Johnston's proposals by Jordan, a pro-Pact country, would only make Jordan a target of similar criticism and would promote linkage between accepting Johnston's proposals and a desire by Jordan to join the Baghdad Pact. Jordan then, in fact, missed a valuable opportunity to provide Johnston with a solid base on which he could stand in his negotiations with Israel, and deliver on his promise that Jordan's water needs had priority in allocation. As his mission

---

[95] Ibid.
[96] Ibid.

developed, Jordan's water share became less, and as time progressed, it was even further diminished.

It was therefore the political pressure that stopped Jordan from giving Johnston the memorandum of approval that he sought. The heat that Iraq came under over the Baghdad Pact intimidated the Jordanians. The unilateral action taken by Iraq's Prime Minister, Nuri al Sa'id, to join the new West-sponsored defense alliance with Turkey, Iran, and Pakistan brought wide criticism of nationalist Arabs and radicals. Jordan preferred to stay within the Arab ranks and responded positively to the Lebanese request for more consultations before a deal was struck with Ambassador Johnston.

4.7 Follow Up Discussions In Syria

After Amman, Johnston traveled to Damascus. Due to the pending change of Cabinet and the absence of instructions from a responsible Government, the Syrian Committee (Jazzar, Mazloum and Khatib) presented technical views only. They demanded a share for the Boteiha farm of 22 mcm from the Jordan River and the provision of enough flow to have 50 Kilowatts of power generated.[97] The Syrians also reiterated their claims to 20 mcm from the Banyas Springs to enable lands in the area to be served by gravity. On the issue of storage, the Syrians favored more storage on the Yarmouk. They presented that a 300 mcm dam could be built at Wadi Khalid at a cost of only $6.0 million more than the cost estimated for the storage of a 200 mcm dam at the same site. They, however, preferred the Maqarin site because of the added power that could be generated there. They estimated a 55,000 kilowatt installation with 200 million kilowatt hours annually at Maqarin, as compared to 40,000 kilowatt installation with 140 million kilowatt hours annually at Wadi Khalid, for the same storage capacity of 300 mcm. They estimated a doubling in the electricity demand in Damascus over the following three years. The American side pointed out that the Jordanian need for irrigation releases from the dam would not coincide with the Syrian needs for power releases. The Jordanian and Syrian interests in releases from the dam, the Americans contended, would be opposite. They also contested the cost of the higher dam and estimated it would cost about $30 million.

The meeting then turned to the status of the Johnston negotiations. The subject involved four topics: 1) the question of division of water for irrigation purposes. As Syria and Lebanon had already accepted certain quantities, the only issue remaining on the Arab side appeared to be the allocation to the Hashemite Kingdom of Jordan, which would be a decision

---

[97] The American side (Gardiner, Troxel, Criddle and Locling from the US Embassy in Damascus) wrote in their Memorandum of Conversation: "It appeared that the Syrians are reconciled to the Israeli diversion at Jisr Banat Yaqub".

not involving Syria and Lebanon, but Jordan only;[98] 2) the question of storage of floods in Tiberias and the need to draw water from Tiberias. On this issue there was no disagreement on the Arab side, if the problem is stated in general terms; 3) the issue of control authority where substantial agreement in principle had been reached by all sides; and, 4) the question of the size and location of the reservoir on the Yarmouk, and the question of a power installation. These issues were impossible to resolve before Johnston's return to the United States and before the establishment of a governmental policy by Syria. In view of this fact, the Mission suggested that the Arabs agree on the first three points of Mr. Johnston's proposal, made in Cairo, and postpone decision on the fourth until a later date. The Syrian experts appeared to agree that this approach might prove constructive and undertook to advise their ministers accordingly.

4.8 Meeting with the Arab Ministers in Lebanon

Ambassador Johnston and his company proceeded to Lebanon for a landmark meeting with the Arab side, whose concerned ministers held a high-level meeting there by invitation from Lebanon. Johnston's first meeting was held in Beirut on February 19, 1955 at the office of the Prime Minister. The Arab representation was high powered and, because of the importance of the meeting and the attendants, the participants are listed below:

Lebanon, the host country
H.E. Sami Bey El Solh, Prime Minister.
H.E. Alfred Naccashe, Foreign Minister.
Fuad Bey Ammoun, Director General of the Ministry of Foreign Affairs.
Mr. Ibrahim Abdel Aal, Director of the Ministry of Public Works, Lebanon's representative on the Arab Technical Committee.

Egypt
Brig. General Mahmoud Riad, Charge d' Affairs, Egyptian Embassy, Damascus, Head of the Arab Technical-Political Committee to negotiate with Johnston.
Dr. Mohammad Ahmad Salim, Secretary General, National Production Board, Head of the Arab Technical Committee.
Mr. Ahmad Farag, Technical Adviser, National Production Board, a representative of Egypt on the Arab technical Committee.

Jordan
H.E. Walid Salah, Minister of Foreign Affairs.
H.E. Khalousi el Kheiry, Minister of National Economy.

---

[98] "Memorandum of Conversation, Damascus, Syria - February 17, 1955," State Department records, U.S. National Archives, Washington D.C.

Hamad el Farhan, Under Secretary, Ministry of National Economy, Jordan's representative on the Arab Technical- Political Committee.
Mr. Izzedine Mufti, Under Secretary, Ministry of Finance.
Mr. Izzedine Yunis, Engineer, Department of Irrigation, Jordan's representative on the Arab Technical Committee.

Syria
H.E. Khalid Ali el Azm, Foreign Minister.
Mr. Kazim Jazzar, Director General, Ministry of Public Works, Syria's representative on the Arab technical Committee.
Dr. Subhi Mazloum, Ministry of Public Works, Syria's representative on the Arab Technical Committee.

Government of the United States
Ambassador Eric Johnston.
Mr. Arthur Z. Gardiner.
Mr. George Barnes.
Mr. Oliver Troxel.
Mr. Wayne Criddle.

American Embassy, Beirut
Mr. Arman Meyer, Charge d'Affaires.
Mr. John D. Temlinson.

In this first meeting, the Minister of Foreign Affairs of Jordan asked if the Yarmouk diversion to Tiberias in the amount of 90 mcm would help control the salinity if Israel diverted the Jordan River at Jisr Banat Yacub; and the Syrian Foreign Minister asked if the increase in Tiberias salinity arose from causes resulting after the year 1948, when Israel was established. Mr. Criddle answered both and said that the Israelis cannot possibly divert the entire flow of the Jordan north of Lake Tiberias in view of the irregular flow; and that the supplies from the Yarmouk were not necessary to keep the salinity in the Lake at tolerable levels. He added that there had been no material rise in the saline content of the Lake's water since 1948. Mr. Gardiner continued by pointing out that "it was not a condition to store Yarmouk water in Tiberias. If it were not thus stored the waters would be lost, and the water would be lost to Jordan rather than any other party."[99] Discussions covered the issue of storage in Tiberias, its water salinity, the possibility of Israel cutting off supplies to Jordan from Tiberias whereby, the Jordanian Foreign Minister said, Jordan would have to compensate her

[99] Ibid.

farmers. He asked if Jordan would be compensated in return. Ambassador Johnston alluded to the American financing of a dam on the Yarmouk for security storage in an amount equaling $10 million over the cost of an economic dam. The meeting covered the other issues of water allocation and supervision.

The meeting turned to a discussion of a draft memorandum, which had been handed to the representatives of the Arab countries during the course of the afternoon. Various amendments were considered, and as a result of the discussions agreement was reached on the following text of a confidential memorandum and on a text of a non-committal communiqué to the press.

MEMORANDUM

Beirut, Lebanon
February 19, 1955

"As a result of discussions between representatives of the Governments of Jordan, Lebanon, Syria and Egypt, and of the Government of the United States, relative to the development of the Jordan River Valley, preliminary understanding has been achieved with respect to major elements of the proposed plan.

The purpose of this preliminary memorandum of understanding is to set forth the points on which substantive agreement has been reached. It is understood that this formation is not conclusive and that certain questions remain to be discussed and resolved at a later date.

Principles and elements of the proposed program upon which tentative agreement now exists are the following:

I. Storage

Water of the Jordan and Yarmouk Rivers will be stored and regulated (a) primarily through reservoir facilities to be constructed on the Yarmouk River and (b) through the operation of Lake Tiberias as a reservoir.

The importance to Syria and Jordan of the full utilization of the Yarmouk through the construction of a storage dam on the Yarmouk creating a reservoir with a capacity of up to 300 mcm was recognized. In view of the relationship between storage and security for Arab crops and lands, Ambassador Johnston will recommend that the United States contribute $21,800,000 toward the cost of this dam. Flood waters exceeding the storage capacity of this reservoir and irrigation needs will be spilled into Lake Tiberias for release to the Hashemite Kingdom of Jordan.

Arrangements concerning the control of the waters shall in no way alter existing territorial rights and claims.

II. Supervision

A neutral body acceptable to all parties will be established to oversee withdrawals and releases of water. Details will be the subject of further discussion.

III. Division of the Waters

(a) Of the requirements of the Hashemite Kingdom of Jordan, 537 mcm of water will be drawn from the Jordan and Yarmouk Rivers. This is in addition to internal resources of wells, springs and wadis within the Kingdom of Jordan.
(b) The requirements of Syria totaling 132 mcm will be withdrawn from the rivers up to the following amounts- 20 mcm from the Banyas, 22 mcm from the Jordan and 90 mcm from the Yarmouk. Provisions will be made for the protection of established irrigation and power interests at Boteiha.
(c) The requirements of Lebanon will be met through withdrawals of 35 mcm annually from the Hasbani River.
It is understood that all quantities expressed above are based upon average annual supply as indicated by existing records of the discharges of the Yarmouk and Jordan Rivers, and may be subject to adjustment depending upon hydrologic conditions of these two rivers."

That was the confidential memorandum of understanding. Jordan's share was indicated to be 537 mcm, excluding water from wells, springs and wadis estimated at 240 mcm annually. However, an annual evaporation quantity of 17 mcm was deducted from the total so that the annual share of the Hashemite Kingdom was 760 mcm. Such explanation was made in a separate joint communique for the press.

In his cable to the Secretary of State after the agreement, Ambassador Johnston emphasized its "tentative" nature. "In brief summary," Johnston wrote, "Arabs have on this basis accepted division of waters proposed by me, with Jordan taking deduction of 16 MCMs from rivers based on present estimated supply from wells. ... Mission leaves for Jerusalem today, for talks there and in Tel Aviv in which I understand Ben Gurion will participate." By that time, Ben Gurion had joined the Israeli Cabinet as Defense Minister, replacing Lavon who became unpopular after the Lavon Affair in Cairo (see above). Lavon was opposed to Johnston efforts aimed at sharing the waters with the Arabs.

On 14 February, the Director General of UNRWA who was visiting East Jerusalem, Jordan, made a press conference in which he was quoted on 15 February by the Jordanian paper, Filistine, as saying that "UNRWA was not concerned with the politics and that should Mr. Johnston fail in his Mission, the agency would, providing it proved feasible, participate in the execution of the Yarmouk project to the extent to which it would benefit the refugees." UNRWA's staement undoubtedly reinforced Arab fears of the real intents behind the Unified Plan and its connection with the final resettlement of Palestinian refugees.

4.9 Discussions In Israel- American Pressure

The Johnston Mission then proceeded to Israel. The Technical group started their meetings on 20 February 1955 at 4:00 p.m. in the first floor committee room of the Prime Minister's office in West Jerusalem. Dr. Criddle went out of his way to have the flow figures used by the Israelis and by the American delegation match, but the Israelis[100] had their reservations about the American figures, which were taken from the Baker-Harza Interim Report. The Israelis had their figures from the Cotton Plan and from Baker-Harza.[101] Mr. Kollek and Mr. Herzog had sent the American delegation in Cairo a cable dated February 2, through the facilities of the US Embassies in Tel Aviv and Cairo, stating that Israel's depletion allocation[102] would be: 62 mcm for use in the Huleh basin, 300 mcm diversion at Jisr Banat Yacub, and 130 mcm for use around and below Tiberias. The total Israeli depletion amounted to 492 mcm. In the same cable, they quoted a message from Prime Minister Sharett "No proposal subjecting to slightest limitation our freedom [to] use water allocated to us wherever and in whatever quantities we see fit can even be discussed." Johnston reported this tough Israeli position to Secretary of State, John Foster Dulles, from Cairo, to which Secretary Dulles replied by a cable dated 7 February with copies to United States Embassies in Amman, Tel Aviv, and Cairo, and to the United States Consulate in Jerusalem.[103] The Secretary agreed with Johnston that the crux of his negotiations was the Banat Yaqub diversion problem. He advised him to have Sharett be informed

[100] Kollek, Sapir, Herzog, Blass, Wiener and Farski.

[101] This meant that Israel was given a copy of the Interim Report by some party despite the fact that, as the American delegation always asserted, the report belonged to Jordan. Also, the Americans obtained its draft in the Fall of 1954, before Jordan laid eyes on it in January 1955.

[102] Depletion is the net quantity of water that the stream loses after use of its water. It is equal to the quantity diverted for use less the quantity that returns to the stream after that use (return flow). The out-of-basin use does not contribute any return flow to the stream. Further, the flow of the Jordan below Lake Tiberias would not be useable because the diversion of all the average fresh flow of the river and tributaries was proposed to be made before that flow reached the Lower Jordan.

[103] The U.S. never closed its Consulate in Jerusalem, and it was operating independently from the U.S. Embassy in Amman, a symptom of non-recognition by the U.S. of the status of Jerusalem as it existed at the time. This fact could be a manifestation of the attitude of the U.S. towards the unity of the West Bank with the Hashemite Kingdom.

through the United States Embassy in Tel Aviv that the Banat Yaqub diversion can proceed if the Government of Israel (GOI) accepted Johnston's plan. If Israel proceeded with the diversion unilaterally, the United States would endorse a Security Council resolution that would order Israel to stop work. The United States would do that regardless of whether or not General Burns (of the UN Truce Supervision Organization, UNTSO) would issue an order for Israel to stop, and would over-ride any decision by General Burn that might give a green light for the diversion works to be resumed. “If Security Council debate started,” wrote Dulles, “it certain that Burn's last report to SC [Security Council] on status DZ [Demilitarized Zone] would be considered fully, particularly matter of Israel policing in DZ. Such SC action we wish avoid but are confident it would be forthcoming if GOI acted on project in absence agreement with you. If on the other hand GOI accepts your plan we could support Gen. Burns if and when he indicated project could proceed whether or not Syria objected. Thus acceptance your proposal would make re-commencement Banat Yaqub matter for settlement between GOI and Burns with U.S. sympathetic support.” The cable further suggested to Johnston to assure GOI that the storage in Lake Tiberias “will prejudice neither GOI's claims (rpt [repeat] claims) to sovereignty over DZ nor GOI's right to negotiate territorial settlement with Syria.”

The American reaction to Israel's demands was that of the stick and the carrot, in the hope that it would convince Israel to agree to Johnston's proposals. It is clear from the cable of the Secretary of State that Israel had been claiming sovereignty over the DMZ between her and Syria.[104] Israel may have based her claims on the fact that the DMZ areas were part of Palestine and were allocated to the Jewish State by the UN General Assembly Partition Resolution 181 of 29 November 1947.

In the meeting of the technical group in Israel on February 20, the spills that can be diverted from the Yarmouk to Tiberias were found to be 99 mcm (compared to the 90 mcm figure used by the Arabs). Finally, it was demonstrated that the net difference between the two sets of figures (Israeli and American) for the Jordan-Yarmouk system as a whole was only 21 mcm.

The Israelis stated that they could make an off-the-river diversion at Banat Yaqub of about 300 mcm and Mr. Criddle made the point that, although theoretically possible, this diversion was not feasible on any reasonable economic grounds. The Israelis also said that they required virtually all of the storage space available in Lake Tiberias for Israel's use, and that the limitation on Lake Tiberias usefulness for storage was the danger of increased saline spring inflow, if the Lake was drawn down more than a few meters. They continued to be convinced that the diversion of 30 mcm of

---

[104] These claims were reiterated by General Ariel Sharon, Minister of National Infrastructure, to the author, in their meeting of 18 February 1998.

saline springs away from the Lake was necessary, especially in light of the Banat Yaqub diversion.

The technical working group then moved at 8:00 p.m. to the Cabinet room for discussion by full delegations. Mr. Johnston chaired the American side, and Mr. Eshkol chaired the Israeli side.

In that meeting, Johnston informed the Israelis that he had reduced the figure he proposed in January for Jordan's withdrawal from the two rivers to 520 mcm,[105] down from the January figure of 540 mcm. He thus was able to increase Israel's depletion from 375 mcm, proposed in January to 400 mcm, excluding the savings from the Huleh drainage. Following an exchange of questions and answers concerning figures used, Mr. Eshkol replied by rejecting, in effect, the principle of allocating any Jordan River water to Jordan, re-asserting that "Israel's grave doubts concerning the use of Lake Tiberias as a reservoir." Mr. Eshkol's rejection of allocating any Jordan River water to the Hashemite Kingdom entailed a tacit political message, in addition to its technical significance. It was known that the allocation from the Jordan River to Jordan was destined to irrigate the West Bank Ghors. Eshkol's rejection could be construed as a rejection of Jordan's sovereignty over the West Bank, exercised since their unity was accomplished in 1950.[106]

The Israeli delegation stated again the need to divert 30 mcm of saline spring flow away from Lake Tiberias. Ambassador Johnston said that he was willing to consider a step that would go far toward meeting the Israeli position, although he was by no means certain that it could be made acceptable to the Arabs. This step was to include the 30 mcm of saline springs, if diverted away from Tiberias, in the Jordan River water that would be delivered to Jordan at a certain point, JX.[107] He indicated that this delivery might be possible because the saline spring water, when mixed in the canal with the unusually pure water of the Yarmouk, would result in water well within the salinity tolerance for irrigation. Mr. Criddle pointed out that there was serious doubt that, in any event, the diversion of the saline springs would be required, since he questioned whether Israel could divert so much of upper

---

[105] In the Beirut Memorandum, Jordan's share was set at 537 mcm, but that included 17 mcm that would be lost to evaporation. The 520 mcm Johnston referred to must have been the "net" allocation to Jordan from the Yarmouk and the Jordan rivers.

[106] Later, in the peace negotiations in January 1992, the Israeli head of delegation asserted in the negotiation room that Judea and Samaria (as he called the West Bank) had been "occupied" by Jordan before Israel liberated them in 1967.

[107] JX was a point on the alignment of the Jordan irrigation main canal as designed by Baker-Harza (now the King Abdallah Canal) near the town of Adasiyya. The canal level drops at that point to make it fit to receive water from Lake Tiberias by gravity.

Jordan water directly from the stream at Banat Yaqub, rather than mixing it with Lake Tiberias and taking it out by pumps.[108]

One notices here how Ambassador Johnston ignored Eshkol's rejection of water allocation to Jordan from the Jordan River. The Ambassador threw a proposal that carried double benefit to his Israeli negotiators by: (a) suggesting that their request of saline spring diversion, an idea he had not raised with the Arabs, was being entertained; and, (b) further increasing their allocation at the expense of the fresh water portion he earlier had proposed as Jordan's share. In this regard, the jeopardy to Jordan was more than doubled. He reduced her share by 20 mcm and replaced another 30 mcm of Jordan's fresh water proposed share with saline water. This shift compromised Johnston's thesis that Jordan's water allocation had priority, and as such, Jordan should get all the water she needed.

Mr. Eshkol reiterated forcefully the earlier Israeli position that Israel should take all the Jordan, except for the agreed upstream diversions (Lebanon's 35 mcm and Syria's 42 mcm), and Jordan and Syria should take all the Yarmouk, implying that Israel might be prepared to modify her claims to Yarmouk River water intended for the Yarmouk-Jordan Triangle. Again, this separation of allocation between the riparian parties (Israel and the Arabs) would not have been conducive to co-operation, an objective that the U.S. Government targeted. The Arab riparian parties on the Yarmouk would have gladly accepted Eshkol's proposal that enabled them to keep the entire river, but would have insisted getting a rightful share of the Jordan to satisfy Jordan's needs. It was apparent that neither the Israelis, nor the Arabs, were after joint co-operation, as evidenced by the Eshkol proposal and by the Arab and Israeli negative attitudes towards storage in Lake Tiberias.

Mr. Troxel argued with the Israelis that, stripped of complications, Johnston's proposal called for the delivery to the Hashemite Kingdom of only 106 mcm of water from the Jordan on a net basis, i.e., in addition to the amount of Yarmouk water merely stored in Tiberias, and making no allowance for Yarmouk water for Israel. Should Israel require Yarmouk water, such a share would have to be added to the above net flow to the Hashemite Kingdom. If 25 mcm was to be allocated to Israel from the Yarmouk, then the amount to be transferred to the Hashemite Kingdom from the Jordan River (Tiberias) would be 131 mcm. He emphasized that since the Israeli technicians themselves claimed that they could not use the 30 mcm of saline spring water, Israel was giving up only 76 mcm[109] net of sweet water in consideration for the settlement of the problem and to assist in getting started on the Palestine refugee problem. Mr. Eshkol replied that instead of getting

---

[108] This was how the Israeli diversion was eventually implemented in 1964.

[109] The net transfer from the Jordan River to the Hashemite Kingdom then becomes 106 minus 30, or 76 mcm.

Jordan water, the Arabs should *pay* water for the privilege of storing their flood flows in Israel's territory, and he again reiterated Israel's objection to the use of Lake Tiberias. Ambassador Johnston pressed the Israelis "to make a concrete proposal for Jordan water in sufficiently realistic terms to permit him to go back to the Arabs with some hope of negotiating successfully."

On the next day, February 21, the two sides resumed their meetings. The Israeli delegation emphasized the point that Israel was using 39 mcm of water in the Jordan-Yarmouk triangle (a figure of 45 mcm was cited by them on a previous occasion). Mr. Gardiner forcefully made the point that, by Mr. Blass' own statement, much of that water was being wasted. Mr. Criddle demonstrated that that usage indicated such a high water requirement per unit of area that it would make the United States' position untenable to argue with the Arabs - who had been asked to accept a smaller water requirement just across the river - that the Israeli claim was justified. In that meeting, Mr. Eshkol and Mr. Kollek made a proposal, the "Eshkol Plan", based on two assumptions: (1) that Israel's rights to 40 mcm in the Jordan-Yarmouk triangle would be protected; and, (2) that Israel would be permitted unrestricted use of the Jordan River, less diversion allowances of 35 mcm in Lebanon on the Hasbani, 20 mcm in Syria on the Banyas and 22 mcm in Syria in the Boteiha area. Under these conditions, Israel would consider: (1) making Lake Tiberias available for the storage of some 300 mcm of water from the Yarmouk; (2) making arrangements under which Israel would guarantee unconditionally the delivery of 50 mcm (40 of which would have a salinity no higher than the general average in Lake Tiberias) to Jordan at point JX, in addition to the average of any diversions from the Yarmouk into the Lake; and, (3) guaranteeing the delivery of additional water up to a maximum of 50 mcm at point JX, with the understanding that that share would be decreased by 1 mcm for every 2 mcm of new useable well water (i.e., in excess of the 16 mcm then estimated to be available) discovered in Jordan during an adequate well drilling program conducted over a period of, say, 5 years' time.

The Eshkol Plan moved closer to Johnston's proposals. It accepted Johnston's figures of allocations to Syria and Lebanon and allocated a guaranteed quantity of 50 mcm of Jordan River water to Jordan, and another ceiling of 50 mcm, adjustable downwards as more groundwater was found to be available for use in Jordan. It accepted providing a storage facility for the Yarmouk floods in Lake Tiberias, and specified Israel's share of the Yarmouk at 40 mcm per year. Compared to Johnston's proposal before the injection of the saline springs allocation, Eshkol's plan allocated 46 mcm less water to Jordan. Mr. Gardiner advanced another proposal that became known during the negotiations as the "Gardiner Plan." It would formalize Israeli control over the Jordan, setting forth Israel's obligation to deliver specific flat amounts to Syria at Boteiha and Jordan at point JX, and would formalize Jordanian-Syrian control over the Yarmouk. Ambassador Johnston stated that if Israel would agree, under this formalization, to deliver 75 mcm of Jordan water "net" to Jordan (compared to the 106 mcm "net" requirement), he would try to secure Arab acceptance. The Israeli delegation, however, expressed doubt whether they could accept such a proposal. The two delegations discussed the partial allocation alternative, but did not come to any conclusions. The delegations spoke of increasing the storage in Lake Tiberias from the estimated 700 mcm to about 1,000 mcm, making the additional storage available for Arab use provided the Lake would be confirmed as an Israeli lake. Mr. Blass privately indicated to Mr. Gardiner that the Israelis were bound to follow United States' wishes. The division of water, in Blass' opinion, would require the Arabs to yield 40 mcm from the Yarmouk to Israel for the Jordan-Yarmouk triangle and the Israelis to provide 70 mcm to Jordan. In some of the meetings of the technical group, representatives of Harza Engineering Company and Michael Baker Inc. participated.

It is ironic that the Jordanians were advocating the memorandum reached in Beirut on 19 February, while Ambassador Johnston, only a day later, was diminishing Jordan's water shares in his talks with the Israelis. In Jordan, Minister of Economy, Khalousi el Kheiry told American Embassy officials on February 23rd that he would go to Damascus around March 7th "to defend the Johnston Plan on political grounds; the technical arguments having already been accepted." The Minister of Defense and Education, Anwar Nuseibeh, told the U.S. Army Attache that "Jordan would accept Ambassador Johnston's proposals because they are in this Kingdom's interest," but that he doubted that Israel would accept them. The Minister of Agriculture had let Embassy officers know that, in the important Cabinet meeting of February 12th, he was a staunch advocate of the Johnston's proposals."

After the end of the third round, with the above achievements and away from the field, Eshkol, Kollek and Sapir of the Israeli Government called on Johnston in his office in New York on 14 March. Discussions resulted in Eshkol's advancing a proposal to increase releases from the Lake to Jordan from 50 mcm to 65.5 mcm and then to 75 mcm. Johnston's reaction was that the Arabs could not possibly be brought to agree to less than 100 mcm, and Eshkol said he would present that to his Government. As an incentive, Eshkol asked Johnston to secure United States' assistance in developing a project to desalinate Mediterranean Sea water, an idea Johnston referred to at the beginning of the second round of talks with the Israelis.

4.10 American Assessment of the Third Round

The assessment of the State Department of the results of the third round were very positive on the Arab side, and negative on the Israeli side. The appraisal noted that "the tentative agreement with the Arabs would enable Israel to proceed unhampered with her diversion proposal at Banat Yaqub and to use water diverted at this point outside the Jordan Valley in accordance with Israel's wishes, thus removing a formidable issue from the docket of the Security Council." The appraisal went on to say, "In general, it is a fair conclusion that the forces in Israeli politics that look to unilateral development of the Jordan River, for Israel's benefit alone, are in command of the situation at the present time, and those recommending a conciliatory gesture to the Arabs on the waters are in a minority. Important opinion in Israel favors a course that will result in Israel's proceeding in her own way at Banat Yaqub, probably before Israeli elections to be held in July 1955, with the expectation that delays and confusion on the Arab side, and in the West, will work in Israel's favor."[110] The appraisal recommended that the United States use available leverages on Israel to have her conform to the Johnston proposals. These leverages were: (1) the supply of agricultural products under Public Law 480; (2) limiting the aid for fiscal 1955 to the amount allotted without further increases; and, (3) the size of aid to Israel in fiscal 1956. Ambassador Johnston, appreciative of the solid support he received from the State Department, was against the use of economic pressure on Israel, for purposes related to the success of his mission.

Secretary Dulles was apprised of the progress made in a meeting attended by the Secretary, Governor Stassen, Ambassador Eric Johnston, Assistant Secretary Allen, and Mr. Arthur Gardiner, Director of the Near East-Asia Bureau at the State Department. Johnston praised the cooperation extended to him by King Hussein of Jordan and said that "it was probably his influence which had led the Jordanians to take a reasonable and constructive view in later negotiations which culminated in a tentative agreement with the Arab States in Beirut." Secretary Dulles asked about the cost of the Jordan Valley Project, and Johnston replied that the estimates were in the order of $90 million inside Jordan alone, $45 million in Israel, and that it was intended to have the project implemented in Jordan through UNRWA. He indicated hope for contributions from other countries to diminish the U.S. burden. The additional $60 million which Ambassador Johnston was authorized to discuss, intended for Lebanon, Syria and Jordan, had been used by Ambassador Johnston only in part. Mr. Allen pointed out that the best argument for the plan lay in refugee resettlement. "This plan is the best one we have to help solve the refugee problem, and it would go far to break the log jam posed by the attitude of many of the refugees." Governor Stassen observed, "if some constructive action is not taken, there might well be a war for the Jordan waters."

---

[110] "Policy Toward Israel in Relation to Jordan River Development," a confidential memorandum to the Secretary of State dated 14 March 1955 from Assistant Secretary, Mr. Allen.

Secretary Dulles determined that "we should allocate the remaining 1955 aid funds of approximately $10 Million to bring Israel's total for the year up to $40 Million, that we should move slowly on the surplus food program as there appeared to be no urgency in this case, and that as far as aid for 1956 was concerned Israel's allotment should be secret and buried in a regional total."[111]

4.11 Briefing the United Nations Secretary General

On March 21st, 1955, Ambassador Johnston briefed Mr. Dag Hammarskjold, the Secretary General of the United Nations. In response to a question as to the effect of the general political climate on his negotiations, Mr. Johnston said that he was hesitant to make an evaluation that could be better made by others. He recalled a conversation with president Chamoun of Lebanon, in which Mr. Chamoun had frankly anticipated a Syrian-Iraqi union and had indicated that Lebanon's main problem would then be one of maintaining her independence. He spoke briefly of King Hussein of Jordan, stating that he found him to be courageous and intelligent, and gave him a great deal of the credit for the success of the negotiations through his positive stand, when many of his advisors were vacillating. In response to another question concerning Mr. Fadhil Jamali, the Prime Minister of Iraq, and his influence on the negotiations, Mr. Johnston replied that he had lunched with him while in the area. Mr. Jamali had apologized for his statement at the time of the first round of negotiations, in which he had said that Mr. Johnston was *persona non grata* in Iraq, explaining that he had been incorrectly informed as to Mr. Johnston's proposals.

Mr. Johnston said that perhaps the most significant aspect of his trip was the constructive attitude displayed by top ranking Arab leaders throughout the area. He contrasted this attitude with what he had experienced on his first trip to the Near East in October 1953. He emphasized that on several occasions he had told Arab leaders that the Jordan Valley negotiations, although extremely important in and of themselves, were really the first step toward the development of peace and stability in the area. He said he had received no adverse reaction to these statements, and that genuine beginnings could be found to the acceptance of Israel's existence. Mr. Hammarskjold commented that the Arabs often get to a certain point and then balk. Mr. Johnston agreed. The group then discussed the likelihood of Israel beginning her diversion works at Banat Yaqub, and concluded that pressure could be brought to bear on the Israelis not to proceed. The United States would join in supporting General Burns in his position against the diversion.

---

[111] "Memorandum of Conversation," Department of State, March 16, 1955.

4.12 Briefing the British Government

The British Government was kept informed of the developments in the negotiations. Mr. Anthony Eden, the Foreign Secretary, had sent messages to the Jordanians and the Israelis urging their cooperation with Mr. Johnston. Those messages, by Mr. Gardiner's account, had been helpful. A meeting was held at the Foreign Office in London on May 18th, 1955 in which Mr. Arthur Gardiner, Director of the Near East-Asia Bureau at the U.S. State Department and a member of Johnston's Mission, briefed the British officials. He mentioned two interesting points: (1) that the Israelis preferred that the Arabs build a high dam on the Yarmouk to fully regulate its flow and stay away from storing any Yarmouk waters in Lake Tiberias. The American delegation rejected that proposal on the grounds that a high dam would cost around $60 million; and, (2) if storage in the Lake was a must, they wanted it deferred until the year 1960. The reason for this was the failure of the Israelis to find adequate alternate reservoir sites at Sahl al Batouf and elsewhere in Israel. It is interesting to note how alike the views of the Israelis and the Arabs were, both shying away from mutual cooperation as implied in the storage in Lake Tiberias and deferring it as long as possible. The Americans, on the other hand, preferred storage in Tiberias and were willing to finance the raising of its shores at an estimated cost of $5 million. Storage in Lake Tiberias would induce cooperation, something that would help open the door for formal acceptance by the Arabs of the existence of Israel.

4.13 Public Reactions in Jordan

The public mood in Jordan displayed suspicion, mostly against Johnston's proposals. The Jordanian negotiator, Mr. Hamad al Farhan, Undersecretary of the Minstry of Economy, delivered a lecture on the Yarmouk-Jordan development on April 9th, 1955, before some 200 persons in al Hussein College, the Government boys' high school in Amman.[112] The setting was not the most propitious, since al Hussein College was believed to have supplied some of the rioters for the October 16th, 1954 disturbances in Amman.[113] Though the Bunger project was described by al Farhan as the "most preferred", the lecturer ended his description of the Johnston project, as amended by the Arab Technical Committee, as the best of all existing projects because it maximized the water available for Jordan, and promised to increase the national income by 40 percent and to give employment to 120,000 workers. The public mood could be detected from the questions asked after the lecturer delivered his speech:

---

[112] The author was in ninth grade at that highschool and attended the lecture.

[113] The author participated in those riots against the Baghdad Pact, the Johnston Plan, and the Eisenhower Doctrine.

Question: Why does the Hashemite Kingdom engage in talks with Johnston? It involves negotiating with the Jews.

Answer: If Jordan had the money with which to finance its projects, we could have told Johnston: "Thank you; we will finance our projects and determine what we will do". But we do not have the money to finance the projects. When we asked the Arab countries to provide the money, some of them apologized and the rest did not reply.

Question: Does not Ambassador Johnston represent the viewpoint of the Jews?

Answer: Johnston conveys the views of the State Department in Washington. In the beginning he represented the views of the Jews but later he approved many of the Arab views. He is a clever man who made his own fame and fortune.

Question: What do you think of the Baker-Harza Company?

Answer: It is a well-known, international company, but it is composed of men who are not infallible.

Question: Do people of Jordan support the Government in favoring the Johnston plan?

Answer: The people have not been consulted, but it is taken for granted that the Government represents the people and defends their rights.

Other questions went along similar lines. There was sympathy for the Bunger Plan and its originator, and lots of suspicion regarding the Johnston plan.

4.14 Memorandum with the Israelis

On the Israeli front, contacts between Johnston and the Israelis continued in Washington and New York. On 5 July 1955, Johnston arrived at a Memorandum of Understanding with representatives of the Government of Israel. It was parallel to the one he concluded with the Arabs in Beirut, with some differences. The major difference was in the share allotted to Israel from the Yarmouk, which was set at 40 mcm per year, and the share of the Hashemite Kingdom from Lake Tiberias, which was set at 100 mcm, including 30 mcm of water from saline springs. The Syrian and Lebanese shares remained the same, and Israel would use the remainder of the Jordan River flow. The Israelis agreed to provide storage facilities for Yarmouk floods in Lake Tiberias with certain conditions, and agreed to the formation of an engineering board for supervision.[114] This Memorandum would form the reference point for subsequent Israeli positions, even after the Johnston

[114] State Department Records from the National Archives of the United States, Washington D.C., declassified on 10/8/1994.

Mission was concluded. The text of the Memorandum is contained in Appendix 2.

4.15 The Palestinian View of the Johnston Mission

On another, yet unaccounted for front, the Arab Higher Committee for Palestine, stationed in Beirut, Lebanon, and headed by the veteran Palestinian, Haj Mohammad Amin al Husseini, submitted a memorandum dated August 18th, 1955 to the Prime Minister of Lebanon expressing opposition to the attempts by Eric Johnston to have the Unified Plan accepted. "The above makes it clear that the Johnston Plan was set up to serve the Israeli interest above all other considerations. It constitutes a wide effort for the liquidation of the Palestine question and establishment of cooperation between the Arabs and the Jews which will lead to peace, the recognition of the Jewish state and all that it brought about of aggression, violation of the integrity of the Arab nation...". Haj Amin al Husseini concluded his memorandum by stating his request: "... the Arab Higher Committee appeals to the Arab States concerned in this imperialistic-Zionist plan to reject it outright."

5. Johnston's Fourth Round

5.1 Talks in Jordan

Ambassador Johnston made his fourth visit to the region on 24 August 1955. He started with Jordan, whose Cabinet was changed on 30 May 1955. The new Prime Minister, Mr. Sa'id al Mufti, formed a committee to study the Unified Development of the Yarmouk-Jordan water system. The consultants, Baker and Harza, had submitted their report on the Master Plan for the Jordan Valley Development on 15 July. The Committee made a field trip to the proposed dam sites on the Yarmouk.[115] The trip had considerable educational value for the Committee. It revealed inclination on their part for unilateral as opposed to unified development. Upon arrival, Ambassador Johnston met with the Jordanian Ministerial Committee[116] whose members

---

[115] The party consisted of the Prime Minister, Sai'd Al Mufti, Minister of Economy, Sai'd Ala' Eddine, Minister of Public Works, Nai'm Abdul Hadi, Minister of Reconstruction and Development, Sama'an Daoud, British Ambassador, C. B. Duke, and officers of Point IV in Jordan.

[116] The Committee was chaired by Mr. Said Ala' Eddin, Minister of Economy, and members were Mr. Hazza' al Majali, Minister of the Interior, Mr. Naim Abdul Hadi, Minister of Public Works, Mr. Ali Hindawi, Minister of Agriculture, and Mr. Samaan Daoud, Minister of Reconstruction and Development. The technicians were the same as the previous rounds: Messrs. Hamad Al Farhan, Nasouh al Taher, and Ezziddine Yunis.

were different from the ministers with whom Johnston had negotiated on the previous rounds. The first meeting was held on Thursday 25 August at the Ministry of Economy, and Johnston briefed the new ministers on the status of his negotiations, referring primarily to the Beirut Memorandum of 19 February 1955. Johnston pointed out that Jordan had become the central party to make or break his proposals. He had been informed by the Lebanese president, Mr. Camille Chamoun, when he met him in February, that Lebanon did not object to his project if Jordan accepted it. President Chamoun repeated the same opinion to the U.S. Ambassador the week before. Johnston said that Prime Minister Nasser of Egypt had expressed the same opinion when he met with him.

The second meeting was on the following working day, Saturday, 27 August. On Friday, Secretary John Foster Dulles announced in a speech before the Council on Foreign Relations that in order to secure a lasting and stable peace in the Middle East, the United States would: (1) pay adequate compensation to the Arab refugees; (2) underwrite some of the expenses for a regional water development project; and, (3) guarantee new and permanent political boundaries, which would replace the old Armistice lines.[117] In effect, he was pushing for a comprehensive Arab-Israeli political settlement, with compensation paid to the Palestinian refugees, and made a link between that and the Jordan Valley development with which Mr. Johnston had been busy. Johnston asked the Jordanians, before the meeting was open on Saturday, if they were aware of Dulles' announcement that was broadcast the night before, and asked for their impressions. When the Jordanians wondered why such an announcement was made in such delicate times, Mr. Johnston answered that he "had asked the Secretary to make that announcement at the beginning of his talks in Jordan in order that there would be no doubts, after he left Jordan, as to the good intentions of the United States toward the problem."[118] The statement by the Secretary renewed Arab fears that had prevailed at the outset of Johnston's mission in 1953, and helped generate opposition to his goals.

Ambassador Johnston presented the Jordanians with his latest proposals. The consultants, Baker and Harza, had shown in their Master Plan Report of 15 July 1955, after consultations with the Jordanians, that a high dam could be built at Maqarin with a capacity of 460 mcm (height approximately 147 meters). Johnston expressed agreement to have the United States finance the cost of a 300 mcm capacity dam, and that Jordan would be free to raise the dam in cooperation with Syria at no further expense to the United States. The Jordanians wanted to retain the option of not storing water in Tiberias and Johnston agreed, provided that the alternative would be

---

[117] Miriam Lowi, "Water and Power," Cambridge University Press, 1993, p.100.

[118] Minutes of Meetings, Johnston's fourth round, Prime Ministry, Amman, Jordan (in Arabic).

another economic storage site, or an economic use, so that the un-stored water would not be wasted. He argued that should any of the Yarmouk water be allowed to go to the Dead Sea by Jordan, Israel would build a case to deny Jordan any share of the Jordan River. In effect, Johnston moved closer to the Arab position with regard to storage.

Johnston's surprise came in that meeting of Saturday, 27 August, when his proposal for Jordan's share from the Jordan River was reduced to 100 mcm. His justification for that was the updated figures of water resources that proved to be less by 67 million cubic meters than the previous estimates, and by the fact that the area in the Jordan Valley that would not be irrigated should be more than that estimated by Baker-Harza. Obviously, Johnston was trying to build into his proposal the opinion expressed by Aharon Wiener of the Israeli team, during the third, round when Wiener proposed a 20 percent ratio of land to be normally left fallow in any given season. The Jordanians reacted with protest to that reduction, and contested the rationale behind the diminishing of Jordan's share. Ambassador Johnston pointed to the improvements in his proposals regarding supervision and storage, and noted to the Jordanians that, "life is the choice of the lesser evil."

Dulles' announcement triggered more suspicions in Arab ranks about the Johnston Mission than what had been there already. Syria denounced the proposals in Dulles' announcement as an attempt to legitimize a situation the Arabs never accepted. The announcement made Johnston's task doubly difficult, as it threw more political shadows on his mission than did the economic and technical issues combined. After the announcement, Johnston's preaching of the economic benefits and his insistence that the political status quo would be unaffected by his plan fell on deaf ears.

At an informal meeting on 30 August 1955 with the Prime Minister of Jordan, the latter told Johnston that his country would be prepared to support the proposals on economic grounds only, provided that: (1) the dam on the Yarmouk River at the Maqarin site would be built to allow for a future raising; (2) a greater amount of water from the Upper Jordan River would be released to Jordan from Lake Tiberias; and, (3) more specific guarantees would be given as to the actions to be taken in the event of violations. In a cable sent to Secretary Dulles, Johnston wrote, "In my judgement, the Hashemite Kingdom of Jordan seriously wants the project but it must surmount several difficulties. On the one hand, these involve the substance of the proposals and internal and public relations problems, while, on the other, they relate to Jordan's ability to enlist the support of the other Arab countries... The Jordanian Cabinet appears willing to try to isolate the project from the question of general political accommodation in the area and secure the support of Lebanon, Syria and Egypt." Johnston left with the Jordan Government a memorandum on 1 September containing his updated proposals, as discussed with the Ministerial Committee on his visit. The

Memorandum approved a dam on the Yarmouk with 300 mcm storage, with diversion of the excess flow to Tiberias to be decided in five years, a share for Jordan from Tiberias of only 100 mcm that could include up to 30 mcm of salty spring water, and that Jordan would have the entire flow of the Yarmouk, less up to 90 mcm to Syria and a delivery of 25 mcm to Israel.

5.2 Decision of the Arab Technical Committee

From Amman, Johnston proceeded to Cairo, Damascus and Beirut to make every effort to have his Plan endorsed. He was met with stern suspicion. At a meeting with the Arab Technical Committee on 8 September 1955, in Beirut, Johnston presented them with his latest: the Revised Unified Plan that he had forwarded to the Government of Jordan on 1 September 1955. It differed from the Beirut Memorandum on three issues: (1) Johnston accepted a higher dam with 300 mcm capacity (126 meters high) at Maqarin and the United States would finance up to $32,650,000 instead of the previous $21,600,000. Johnston also agreed to have the dam raised at the expense of the Arab side; (2) the storage in Tiberias can be deferred for five years; and, (3) the Jordan share from the Jordan River and the Yarmouk River was decreased from 537 mcm in the Beirut Memorandum to 477 mcm, i.e, by 60 mcm. Thus the share to be drawn from Tiberias would fall from 160 mcm as specified by the recent Baker-Harza study (or 140 mcm in the 1953 TVA/Main plan) to 100 mcm. The American side suggested that it would become necessary, to avoid loss of water, to compensate Jordan an amount of 30 mcm of saline water for the decrease in its share. This amount was later reduced to 15 mcm and was to be counted not as compensation to Jordan, but as part of her share of 100 mcm from Lake Tiberias. On the financing aspects, the American side offered to contribute $250,000 towards a study of the Hasbani River and the possibility of irrigation in Lebanon benefiting from its waters to the extent of 35 mcm per year. The American side pointed out that the works required for electric power plants on the Yarmouk would not be included in the American financing.

The Arab Technical Committee, in its 9 September report, recommended that, to the best extent possible, Jordan's share from Lake Tiberias at 160 mcm be maintained without reduction, and that no salty water be given to Jordan as part of her share, in whatever proportion. It left it for the Lebanese Government to decide on the American contribution toward study of the Hasbani, in light of Resolution no. 5440 of the Lebanese Chamber of Deputies that had rejected the Johnston Plan.

Despite profound reservations, the Arab Technical Committee[119] decided to approve the Revised Unified Plan. In late September, the Committee recommended to the Arab League to accept the Revised Plan.

Another Arab Committee met in Damascus on 3-4 October 1955, to look into the possibility of utilizing the Yarmouk River for the benefit of Jordan and Syria outside the Unified Plan, because of the political obstacles facing that Plan. The Foreign Ministers of Syria, Jordan and Lebanon[120] attended the first meeting. The ministers instructed the technical experts to study the possibility of utilizing the Arab rivers, Yarmouk, Hasbani and Banyas, in an independent manner for the benefit of the Arab countries. The technical experts[121] met that day and the following, and came up with a proposal as follows:

"The Yarmouk, the joint Syrio-Jordanian Agreement of June 1953 stipulated the construction of a dam at Maqarin and a diversion dam in the lower Yarmouk. But since the opposite bank of the Yarmouk at Adasiyya comprised a DMZ the access to which entailed political hardships, the experts suggested the construction of a diversion dam at Mukheiba, where the other bank was Syrian territory, and construction of a conveyance canal in Jordanian territory from that dam to the beginning of the irrigation canal in the Jordan Valley."

However, in addition to the higher cost involved in the above proposal, a substantial quantity of the Yarmouk base flow from the Himmeh and Mukheiba springs, estimated at 80 mcm per year, would remain downstream of the diversion dam, and would have to be pumped or diverted to the said canal for an additional cost. Otherwise, that quantity would flow past Adasiyya toward Israel and the Jordan River.

The Hasbani and Banyas, the experts proposed the construction of a dam on the Hasbani, and the diversion of the excess water from the Hasbani to the Litani inside Lebanon. Surplus water of the Banyas, after irrigation of Syrian lands in its gorge, would be diverted to Lebanon.

---

[119] Attendants were (1) From Egypt: Dr. Mohammad Ahmad Salim, (2) From Syria: Mr. Kazem al Jazzar and Dr. Subhi Mazloum, (3) From Jordan: Mr Izzedine al Mufti, deputized by the Finance Ministry, and Mr. Hamad al Farhan, deputized by the Ministry of National Economy, (4) From Lebanon: Mr. Ibrahim Abdul Aal.

[120] Mr. Said al Ghazzi was Syria's Foreign Minister and Prime Minister; Mr. Said al Mufti was Jordan's Foreign Minister and Prime Minister; and Mr. Salim Lahoud was Lebanon's Foreign Minister. Mr. Shawkat Shuquair, Syria's Army Chief of the General Staff, attended the meeting.

[121] Participants were Ibrahim Abdul A'al, representative of Lebanon; Kazim al Jazzar and Subhi Mazlum, representing Syria, and Hamad al Farhan, Izzed Din Mufti and Sweilem Haddad, representing Jordan.

The position taken by this group of technical experts was not the position taken by the Arab Technical Committee a few weeks back. The latter accepted Johnston's proposals with recommendations to increase Jordan's share, but the former was recommending another scheme that was along the lines of the Syrian-Lebanese Plan (Chapter I above).

5.3 Johnston with Premier Jamal Abdul Nasser

Johnston then proceeded to Cairo for a special appeal to Prime Minister Jamal Abdul Nasser. At the invitation of the Prime Minister, Johnston dined on the evening of 7 October 1955, at Nasser's home in Manshiyyat el Bakri on the way to Heliopolis.[122] A condensed report of the conversation at dinner read:[123]

"During the day (October 7, 1955), Nasser had met the incoming Arab Foreign Ministers at the airport. He felt that the Arab states were not in a mood to accept the Jordan Valley proposals. He said the Lebanese had brought a new Arab plan which would divert the Hasbani and Banyas Rivers by tunnel to the Lebanese coast for irrigation and power, and dump the balance into the Mediterranean. This plan would include a dam at Maqarin with a diversion dam above Adasiyya and a tunnel through the hill to the East Ghor[124].... Nasser stated that he had called Dr. Salim [head of the Arab Technical Committee] to his residence yesterday afternoon and spent an hour with him going over the Lebanese proposal. Dr. Salim had advised him that the proposal was completely uneconomic and infeasible. Nasser said he would therefore oppose it.

"Nasser felt the best way of handling the Jordan Valley project was to divide it into two parts- economic and political. He would try to do this in the Foreign Ministers meeting. He was sure they would approve the plan from the technical viewpoint, since the Arab Technical Committee was in agreement that it was the most practical and sensible plan. Once this was done, the Foreign Ministers would consider the political phases of the project, weighing its advantages and disadvantages. Nasser said he would have to handle this his own way. To urge the Arab states to accept the plan politically would

---

[122] Present were, in addition to Nasser and Johnston, Minister of Interior Zakaria Mohyiddine, Minister of War and Commander of the Armed Forces Abdul Hakim Amer, Ambassador Hussein (Egypt's Ambassador to the U.S.), Major Hasan Touhami, Nasser's close advisor, Kermit Roosevelt and Miles Copeland, both of the Central Intelligence Agency (CIA) of the United States.

[123] "Meeting Between Ambassador Johnston and Prime Minister Nasser, Cairo, October 8, 1955," a memorandum of conversation kept by the American side; State Department Records, National Archives of the U.S., Washington D.C.

[124] Nasser was obviously referring to the Plan proposed by the Lebanese, Syrian and Jordanian experts as a result of their October 3-4 meetings in Damascus.

probably drive them to oppose it. He therefore planned to allow them to argue it among themselves. They would then come to him privately and individually to find out his position, and he would urge them to accept the plan. He said this would be difficult but he felt sure it could be accomplished. He further stated that Lebanon was the leader of the opposition and that this was the reason he had sent for Prime Minister [Rashid] Karami to come to Cairo. Syria was neutral. Jordan wanted the project but like all of these governments, the Jordanian Government was very weak and fearful of being over-turned.

"Ambassador Johnston asked the Prime Minister [Nasser] if they wouldn't be over-turned just as easily by the opposition who could point out that under the new Arab Plan they were willing to lose nearly 200 mcm, 100 mcm from the Jordan (Jordan's share under Johnston's plan), and 80 mcm from the Yarmouk which they would be unable to capture. He said that this was true in the long run but that the Arab Governments were looking to the short run. Nasser brought out a large map with detailed engineering data in Arabic. The plan was called the Arab Plan. He and Ambassador Johnston spent some time discussing its engineering features. Nasser said it was the "Johnston Plan" with an Arab label and indicated that it would be the plan the Technical Committee would approve.

"After dinner General [Abdul Hakim] Amer in conversation with [Miles] Copeland asked for a copy of the Jordan Plan stating that he must study it "because we are going to see that it is adopted."

"Coming home, Major Touhami remarked that Nasser made very few promises, but he had promised Johnston that his proposal would be adopted. Touhami said it would be very difficult but that Nasser would do it. He pointed out that in spite of his many duties, the Prime Minister had devoted most of his day to the Jordan project, including giving a dinner for Ambassador Johnston when many Arab dignitaries were in town.

The Arab Foreign Ministers were to meet at 6 o'clock this afternoon. The Technical Committee met and presumably approved Johnston plan without modification. Nasser appeared genuinely anxious to have water plan accepted now and willing to make every effort to secure agreement of other Arab states. He seemed friendly and cooperative".

This record shows that the Arab Plan, in the judgment of Dr. Mohammad Ahmad Salim, the Chairman of the Arab Technical Committee, as quoted by Nasser, was not at all feasible and that it would make the Arabs lose about 200 mcm compared to the Arab allocations under the Johnston plan. Above all, it shows that Nasser, the most publicly respected Arab leader, and the emerging champion of Arab nationalism, was actually favoring the Johnston Plan, and undertook to have it accepted by the Arabs.

5.4 Johnston with the Concerned Foreign Ministers

Johnston was given the chance to advocate his plan with the Arab Foreign Ministers of the concerned states (Lebanon, Syria, Jordan and Egypt) in their meeting of 8 October 1955 in Cairo. He said in his presentation:[125]

"I should like only to say that this Plan represents a combination of the best thinking of Arab and American engineers.

"It is largely based on data presented in a 10-volume report covering two years of scientific study by private engineering companies employed by the Government of Jordan.

"It incorporates most of the ideas contained in the "Arab Plan" developed by the Arab Technical Committee.

"It embodies major modifications to meet the Arab point of view.

"It represents, in our opinion, as good a plan, as far a plan, and as workable a plan as it is possible for anyone to devise."

Mr. Johnston added:

"In my recent discussions with you, the point has been made that the plan has two aspects- one technical, the other political.

"I have been given to understand that the plan is now technically satisfactory to the Arab states.

"I have also been told that the principal problem facing the Arab states in deciding whether to accept or reject it is a political problem.

"From the very beginning, I have made it clear that I am aware of this political problem. It is for this very reason that I have proposed a plan which has no political implications of any kind.

"It does not call for relations between the Arab states and Israel.
It involves no greater recognition of Israel than has already been given by the Arab states in many other ways.

"It does not in any way impair or prejudice the rights of the refugees to a final settlement of their claim against Israel.

"It does not in any way affect the sovereignty of any of the Arab countries.

"It offers merely a modus vivendi-- a means of moving ahead at once with sorely needed advances in the social and economic field which will mean homes and jobs for many thousands of the Arab people.

"Its acceptance, in short, will have no effect whatever on the ultimate solution of the political problems between the Arab countries and Israel.

"In fact, the United States would not be prepared to support a plan which was based upon political rather than economic considerations.

"We have maintained from the beginning that we were willing only to support an economic undertaking entirely outside the context of political issues in the area.

---

125 "Notes from E.J [Eric Johnston]." dated October 8, 1955, State Department records, United States National Archives, declassified 10/8/1994.

"We are not willing now to consider any amendment of this plan or any other plan which has its basis in political rather than economic considerations.

"I feel, and the United States Government feels, that after two years of discussion, we have reached the time for decision.

"I think you should know that I shall be obliged to report to the President on the results of my conversations with you here in Cairo.

"I must now tell him whether the Arab states accept or reject the offer he authorized me to make two years ago.

"It is certainly not our intention to press you to accept this offer. If you do not want to participate in such a plan, we do not urge you to do so. We ask only that you understand the Plan itself and recognize the values which can be gained- or lost- by your decision.

"This offer, made with the best of intentions and in all good faith, reflects the sincere desire of the President to contribute to the progress and prosperity of the Arab nations.

"It represents a major step in the implementation of his stated policy of friendship and cooperation with the Arab peoples. We hope that this policy will bear fruit in terms of economic growth and social progress in the Arab states."

Ambassador Johnston then turned to the Israeli factor in the Plan:

"I have heard it said that the Plan will benefit Israel to the detriment of the Arab states.

"On the face of it, this is not true.

"Consider these facts:

1. Acceptance of the Plan will immediately place a ceiling on the amount of water Israel may take from the River Jordan. Except by force, I know of no other way to impose a limitation on Israel.
2. The Plan will assure the Arab states and particularly Jordan of the right to store Yarmouk floods in Lake Tiberias. We know - as your Arab engineers know- that without this guaranteed accommodation in Tiberias, a large volume of water indispensable to Jordan's development will be lost.
3. Israel's water allowance will be far less than she could otherwise take because of her geographical position on the River.
4. The fact is that the benefits of this plan are overwhelmingly on the Arab side.

Finally, let me emphasize the essential character of this Plan."

Ambassador Johnston then made a presentation of the features of the Plan before he ended his advocacy of it. He was eager to gain the approval of the Arab states so that work on its implementation could start. It did not sound like he would be having difficulty in getting the Israelis to approve it.

5.5 Decision of the Political Committee of the Arab League

The representatives[126] of the states concerned debated the issue. It is worth reviewing what went on in the meeting of the representatives of the four concerned countries. After the Johnston briefing, Dr. Salim made a presentation of the "Arab Plan" and referred to the field trip that was made in February 1954. The Jordanians had some corrections to make in the figures presented by Salim. They felt that the best solution was the Unified Plan.[127] But Syria did not agree.

"We would not accept a Unified Project," said the Prime Minister of Syria, Mr. Said al Ghazzi, "the parliamentary debate in Syria showed us that not any Syrian government can accept a Unified Project. The Parliament of Lebanon decided a similar decision. We further think that in Jordan there are nationalistic men who reject the Unified Project. We therefore hope that the technicians would continue studying the independent Arab project and submit its details and cost in full."[128] This intervention by the Prime Minister of Syria, and the lack of arguments by other delegates to the contrary, proved to be the stumbling block that stopped the Arab League from accepting the recommendation of the Arab Technical Committee to accept the Unified Plan. Any argument to the otherwise would stain its advocate with the intention to co-operate with the enemy. Because of that, it was hard to face the argument presented by Syria.

The Foreign Minister of Lebanon said that his delegation was committed to the decision of the Lebanese Parliament, which rejected the Plan, but that his delegation came with open hearts and open minds, and that they would take back with them whatever decision the ministers made and try to convince the Parliament.[129]

The meeting then issued the following decision:

"Representatives of the Arab States concerned, namely Jordan, Syria, Lebanon and Egypt, have studied the Arab Plan for the Jordan Valley and its tributaries, developed by the Arab Technical Committee, and have come to the conclusion that, in spite of the efforts exerted, certain important points still need further consideration.

---

[126] Attendants were: (1) From Jordan: Said al Mufti, Prime Minister, Said Alaa Eddin, Minister of Economy, Izzeddin Mufti, and Hamad al Farhan, (2) From Syria: Said al Ghazzi, Prime Minister, Ostowani, Shawkat Shuqeir, Chief of Staff, al Jazzar, Mazloum, (3) From Lebanon: Salim Lahoud, Foreign Minister, Fuad Ammon, Taqi Eddin, and Abdel Aal, (4) From Egypt: Khayrat, Deputy Foreign Minister, Mahmoud Riyadh, Mohammad Ahmad salim, and Ahmad Farag, (5) From the Arab League: Hassounah, Secretary General, and Abdul Muniem.

[127] Handwritten notes taken by one of the Jordanian delegates attending the meeting; found in the Jordanian file on the Johnston negotiations, Prime Ministry, Amman, Jordan.

[128] Ibid, sheet 3.

[129] Ibid, sheet 4.

"It was, therefore, decided that the experts be asked to pursue the mission with which they have been entrusted until an agreement safeguarding Arab interests is reached."

The above decision was then forwarded to the Political Committee of the Arab League, composed of the Foreign Ministers of the eight member countries.[130] Their resolution was communicated to Ambassador Johnston by the Secretary General of the League. The letter read:

"Cairo, the 15th October 1955

Dear Mr. Ambassador,

I have the honour to enclose herewith the text of the resolution adopted on October 11, 1955 by the Political Committee of the League of Arab States, approving the decision taken by the four States concerned with the Plan for the development of the Jordan Valley and its tributaries.

The Political Committee has asked me to express to Your Excellency its greatest appreciation of the efforts exerted in this respect, both by yourself and by your assistants as well as by the Arab experts.

The Political Committee would like, furthermore, to express its hope that the studies would lead to a satisfactory result.

I avail myself of this opportunity to renew to Your Excellency the expression of my highest consideration.

Secretary General
[Signed]
Abdul Khaliq Hassouneh"

5.6 Reflections on the Arab Reaction

The diplomatic skills of Ambassador Johnston were demonstrated beyond any shadow of a doubt. His defense of the Revised Unified Plan in the meeting of the Arab Ministers on 8 October was a good example of an excellent diplomat at work. He insisted that the Plan would not prejudice the right of the Palestinian refugees to a final settlement with Israel. That was true, but true also was that the attraction in the Plan to Israel, as repeatedly stated by Johnston himself to them, was the resettlement of Arab refugees in the Jordan Valley. To be able to "float" in-between and have the audience, Arabs and Israelis, believe that the Plan served their respective contradictory interests was in itself a supreme innovation by a skilled diplomat. Ambassador Johnston asked the Arab ministers to recognize the values that could be gained or lost by their decision, and that nothing but force would stop Israel from diverting a lot more water from the Jordan River than the

[130] The members were the then independent Arab states: Lebanon, Syria, Jordan, Iraq, Egypt, Saudi Arabia, Yemen, and the Kingdom of Libya.

Plan would authorize to them. This point in itself, was a reminder to the Arabs that the Plan would secure them their water rights peacefully. He represented that the Plan was a manifestation of the desire of the President of the United States to contribute to the well being of the Arab states, and that he would not urge them to accept the Plan if they did not want to. His emphasis on the economic and resource benefits of the Plan was a good way to allay Arab fears.

The meeting showed that Syria, and not Lebanon as Nasser told Johnston, was actually spearheading the opposition to the Unified Plan. Interestingly, the Foreign Minister of Egypt did not attend that important meeting, and sent his deputy instead. He might have been preoccupied with other obligations, but the meeting provided the opportunity for Egypt to influence the Arab representatives and could have, at the least, averted the decision that was taken and, subsequently, endorsed by the Political Committee. That decision, however, was not a rejection of the Johnston Plan. One notes that the ministers of the four Arab states referred to the "Arab Plan" for the development of the Jordan Valley, and not to the Unified Plan or the Johnston Plan. That was meant to emphasize the work of the Arab Technical Committee and its recommendation to have the Plan accepted. The conclusion reached by the end of Johnston's fourth round was an acceptance by the Arab Technical Committee of the Revised Unified Plan, as finally presented by Johnston. There was no decision by any competent body of the Arab League, or the concerned Arab governments, that rejected the Unified Plan.

The negative Syrian attitude cannot be disconnected from the political context. There was tension in the area because of the progress made towards establishment of the Baghdad pact. The radical Arabs viewed the Pact as a creature in an American dress, wearing British shoes and a Turkish hat! Syria had had four military coups in five years, and her political leanings were toed towards either alliance with Egypt or alliance with Iraq. Since Iraq had gone into an alliance with Turkey, with whom Syria had several bones to pick, including the issue of Alexandretta, Syria did not feel comfortable. Syria and Egypt, who publicly contested the Baghdad Pact and Western influence in the Middle East, could not come out supporting the American engineered and sponsored Unified Plan, which was viewed with suspicion by the people in the region. Perhaps a cable sent to the Secretary of State, Mr. John Foster Dulles, by the United States Ambassador in Baghdad, best describes the political atmosphere that prompted the decision of the Political Committee:

"Control: 7743
Rec'd: October 15, 1955, 11:53 a.m.

From: Baghdad

To: Secretary
No: 356, October 15, 2 p.m.
SENT DEPARTMENT 356, REPEATED INFORMATION AMMAN 45, BEIRUT 73, CAIRO 82, DAMASCUS 67, JIDDA 39, TRIPOLI 31

According to Foreign Office Undersecretary Yusuf Gailani report by Foreign Minister [Burhan Eddine] Bash A'yan from Cairo states Arab League Foreign Ministers reached "nearly complete agreement" on acceptability Johnston plan. Representatives said to have agreed plan is equitable conforms with past Arab League resolutions and does not (repeat not) involve recognition Israel. In view of adverse Arab public opinion, however, conference decided only way out of "dilemma" was to announce postponement decision pending further study. Report states that Chairman [of the] meeting Egyptian Vice Foreign Minister conveyed conference's agreement postpone decision to Ambassador Johnston who said to have been very disappointed. Johnston's reaction prompted conference chairman explain circumstances of postponement and express hope American offer would remain open. Ambassador Johnston reported to have expressed understanding and assurances.

When queried re above hopeful development against background earlier discouraging signs, Gailani says Saudis were principal objectors who later became more amenable after conversations with Nasser following representations to latter by Bash A'yan.

GALLMAN"

It appears as though other Arab representatives were refusing the Revised Plan. For Saudi Arabia to be principal objector must have been an overbid on the Syrian position. Iraq was a moderate monarch and tried to pacify possible Egyptian opposition! But the Saudis conformed after talking to Nasser. It was important not to appear conforming to the wishes of the West at a time when tidal waves, led by Nasser, were, on the surface, rising against the West. It is also interesting to note that the chairman of the meeting of the Arab Political Committee, who deferred the decision on the Plan, had asked Ambassador Johnston to keep the American offer open.

Johnston left Cairo for Israel, where he and his delegation met with the Israelis before he headed for the USA via Rome, on the morning of October 14, 1955, ending his fourth and last visit to the region.

5.7 Israel's Position on the Johnston Plan

The position of Israel on the Plan is described in a cable to the Secretary of State from Tel Aviv:

"CONFIDENTIAL

Control: 7963
Rec'd: October 15, 1955, 8:28 a.m.

Form: Tel Aviv
To: Secretary of State
No: 356, October 14, 7 p.m.

PRIORITY

SENT DEPARTMENT 356, PRIORITY ROME 24, REPEATED INFORMATION LONDON 117, PARIS 75, CAIRO 142, DAMASCUS 64, AMMAN 43, BEIRUT 10
ROME FOR AMBASSADOR JOHNSTON
Ambassador Johnston's talks with Israel Prime Minister and Finance Minister and their staff went very well. On technical side the two ministers indicated, although reluctantly and after some discussion, that the compromise proposal of dividing the salt water 50-50 between the Arab States and Israel would be acceptable to them although, in view of the uncertain status project on the Arab side, they do not (repeat not) plan formally to submit it to new Cabinet at this time. It thus appears that informally both Israel and the Arab States have reached a common basis on a water settlement formula and that the sole but formidable remaining obstacle is to obtain Arab concurrence at the political level.

In the earlier discussion yesterday, both Sharett and Eshkol took position that, having reached agreement with Ambassador Johnston, and two years having elapsed since negotiations commenced, Israel should regard itself as free to proceed with its work at Bnat Yaacov. After Johnston had discussion political implications of this and the assurances which had been given by Nasser and other political leaders on Arab side that Arab states could be brought within two or three months to point of agreeing to plan, the Israel Ministers finally agreed to defer resumption of work for temporary period. Finance Minister Eshkol talked in terms of similar two or three months period, but in subsequent conversation with Sharett, Johnston was assured the work at Bnat Yaacov would not (repeat not) be resumed until March 1 and that if an agreement with Arabs was a real prospect at that time, resumption of work might be deferred for short additional period. Both Sharett and Eshkol expressed skepticism regarding sincerity of Arab desires for water settlement and emphasized the firm intention of GOI to proceed with Israel's water program should Arabs fail to give political concurrence to technical understanding within period indicated. Eshkol expressed hope that USG would give moral and material assistance to Israel in proceeding with its unilateral water development should Arabs fail to accept the plan within time limitation. Johnston replied he would submit Finance Minister's request to Secretary and other US Government officials upon his return to Washington. Following telegram transmits joint communique of 2 delegations which was issued last night upon conclusion conversations. Johnston left Israel by plane this morning en route to New York and Washington.

LAWSON"

Thus the Israelis accepted to share the 30 mcm of saline water equally with the Arabs, as well as the figures for water division. The share to Jordan from Lake Tiberias would thus be 100 mcm including 15 mcm of saline water. Israel's share from the Yarmouk would be 25 mcm, as stipulated by the Johnston's proposal, but this figure was contested by the Israelis later on and they insisted that their share was 40 mcm as stipulated in their Memorandum of Understanding of 5 July 1955.

The Israeli acquiescence to the Plan was clear. It would be to their advantage either way, but more so if the Arabs did not approve the Plan. They would then be free to resume their work on the diversion project that had been suspended since October 1953 at the beginning of Johnston's shuttle. The reading of the Israelis of the likely Arab reaction to the Plan proved to be correct. However, the Israelis would not be as blamed for acting unilaterally should the Arabs refuse the Unified Plan as they would be in the absence of such a plan. They even expressed the hope that the United States would give moral and material support to their unilateral diversion undertakings.

With this Israeli position, and the presentation to the Arabs by Ambassador Johnston, it would be hardly a surprise to the Arabs if Israel went ahead with her unilateral diversion of the Jordan River waters. The Israeli intention and the Arab awareness of it is important to point out, at this time, in order that the notion of "Israeli stealing of Arab waters," a phrase frequently used by the Arab public and even intelligentsia, could be better understood in the proper context.

The detailing of the Jordan Valley Plan and its justification and advocacy as advanced by the American side[131] is presented in Appendix 3. The text was most likely so documented by the American side in order to provide answers to any questions regarding the efforts of the Johnston's Mission.

Appendix 4 presents a detailing by the author of the water allocations to Jordan as they changed over the course of Johnston's negotiations, and a division between the allocations made to the East Bank and the West Bank.

---

[131] Text given to the author by Dr. Wayne Criddle, and also a declassified document of the State Department Records, National Archives of the United States, Washington D.C.

The final requirement for irrigation of the East Bank Ghors was settled finally at 479 mcm, of which 175 mcm would come from the East side wadis, 8 mcm from groundwater wells, and the remainder, 296 mcm, from the Yarmouk River. The West Bank share was settled at 241 mcm, of which 8 would come from wells, 52 from side wadis, 81 from the Yarmouk and 100 mcm from the Jordan drawn from Lake Tiberias. It is to be noted that the Mission inflated the contribution of the Yarmouk was inflated by assuming higher return flows from Syrian uses, and that this inflation served to reduce the contribution to the Hashemite Kingdom (its West Bank) from the Jordan River. The combined allocation from international rivers (the Yarmouk and the Jordan) for use in the West Bank was settled finally at 181 mcm.

CHAPTER IV

Preventive Diplomacy

1. The Arm's Race

The developments in the Middle East accelerated after the end of Johnston's fourth round. Jordan had been approached to join the Baghdad Pact, which appeared to the Arab public as a creature with an American dress, wearing British shoes and a Turkish hat. In Jordan, a new Cabinet was formed, headed by Hazza' al Majali, on 15 December 1955. Majali was a known proponent of the Pact. Riots broke out in the streets of Amman and General Templer, of Britain, who was on mission to Jordan to pave the way for her admission to the Pact, left without success. Under popular pressure, the new Cabinet resigned only five days after it was formed. Another Cabinet was formed on 21 December, to last until 7 January 1956. A third cabinet was formed and lasted about four months before a fourth Cabinet took over for one more month. Finally, a transition Cabinet was formed, on 1 July 1956, and conducted general elections. A parliamentarian government was formed on 29 October 1956.

The situation in Iraq was not much better; rioters took to the streets of Baghdad about the same time as they did in Amman. Anti riot police opened fire and killed a few rioters. Syria and Egypt were campaigning against the Baghdad Pact, and public sentiments in the region were going their way. The Pact, signed between Turkey and Iraq in February 1955, and by Britain three months later, was described by some Western, Middle East specialists as a "traumatic experience for Syria," but that was an understatement.[132] They noted that "the Baghdad Pact so shook the Arab world that, for a while, it seemed beyond our competence, for all the economic aid at our disposal, to save any sort of position for the West in the Middle East."[133]

Denied the provision of military assistance from the United States, Nasser of Egypt turned to the Soviets and worked out a deal with them to have arms supplied to Egypt from Czechoslovakia.[134] The United States was aware of the Soviet-Egyptian deal. In fact it was reported[135] that two American officials[136] helped draft the speech that Nasser delivered at the Airforce Academy in late September 1955, in which he made a public announcement about the Czech arms deal. It boosted his popularity in Egypt

---

[132] Miles Copeland, "The Game of Nations," Weidenfeld and Nicolson, London, 1969, p.176.
[133] Ibid.
[134] Czechoslovakia also happened to be a major supplier of arms to Israel!
[135] Ibid, pp. 133-141.
[136] The U.S. officials were Kirmit Roosevelt and Miles Copeland of the CIA, Ibid, pp. 133-134.

and the Arab world beyond limits. The anti-Western sentiments intensified, especially as the border incidents on the cease-fire line between Egypt and Israel became more frequent. Thus, the public mood in October of 1955, when Ambassador Johnston was advocating the Jordan Valley Plan to the concerned Arab Foreign Ministers, was not conducive to entering into a unified project with Israel, nor to promoting Western programs or plans. Nasser asked for some time to have the Arab states approve the Unified Plan.[137] The Israelis, on their part, watched the situation very carefully and informed Johnston that they would wait for an adoption of the Unified Plan until the end of February 1956, after which date they would be free to resume their work on the diversion at Jisr Banat Yaqub.

To keep record of his Mission and agreements with the parties, Mr. Johnston suggested that the results of his deliberations and negotiations with the concerned parties be documented. This request was made through an "unofficial" memorandum that was distributed to the concerned parties by Oliver Troxel, a member of the Johnston Mission and an official of the U.S. State Department, on 31 January 1956. The memorandum contained the text of the Unified Plan, as finally agreed to between Johnston and the Arab Technical Committee on the one side, and by the Israeli team on the other.[138] The memorandum documented the agreements that were reached through Johnston's shuttle diplomacy, which started in October 1953 and ended in October of 1955.

In Israel, David Ben Gurion replaced Moshe Sharett as Prime Minister in early 1956. This change in Israel was indicative of the rapid deterioration of the Arab-Israeli situation. Ben Gurion represented the "activists" in the military and foreign affairs, while Sharett represented the "pacifist" line that favored a non-aggressive, conciliatory attitude which could lead to peace. Ben Gurion, however, believed that Israel should always retaliate with force against Arab acts of hostility or breaches of the Armistice. The developments in the implementation of the water plans were not encouraging either. The Syrian Foreign Minister received a report from the Syrian Embassy in Washington D.C. informing him that the Israeli Embassy in D.C. had told the State Department that Israel was determined to resume work for diversion of the Jordan River waters by the end of February, 1956. The Syrian Prime Minister called in the American Ambassador in Damascus and handed him an *aide-memoire* stating the Syrian position concerning the matter. It remained as was expressed in her complaint to the Security Council

---

[137] "Meeting Between Ambassador Johnston and Prime Minister Nasser, Cairo, October 8, 1955," a memorandum of conversation kept by the American side; State Department Records, National Archives of the U.S., Washington D.C.; see footnote 51 in the previous chapter.

[138] See Text of the Jordan Valley Plan in Appendix 3.

on 16 October 1953, at the onset of the first Johnston visit to the region. In essence, Syria's position considered that the diversion of the Jordan would remove a natural barrier, which would lead to Israeli military gain in violation of paragraph 1, Article 2 of the General Armistice Agreement concluded on 20 July 1949; that the diversion constituted an exercise of sovereignty on the DMZ, which conflicted with the characteristic of the area; and that it would deprive owners of Arab land irrigated by the River of a natural and exercised right. In addition, the last paragraph of General Bennike's decision, of 23 September 1953, asked Israel to instruct the authority that began to work in the DMZ on 2 September 1953 to stop work in the area. Subsequently, on 27 October 1953, the Security Council issued a unanimous resolution supporting Bennike's decision and providing that work in the DMZ should stop until issuance of a final decision by the Security Council on the basis of the Syrian complaint. No change of the status of that matter occurred and the Israelis were supposed to abide by the decision of the Security Council so long as it was effective, and it still was.

Despite Syrian fears to the contrary the first of March 1956 passed without the Israelis resuming work on the diversion scheme in the DMZ. The "advice" rendered to Israel by the U.S. to that effect was very helpful. In fact, the United States, after the end of the Johnston Mission, was planning another initiative to be launched by the U.N. Secretary General, Mr. Dag Hammarskjold. On another development, Jordan's King Hussein dealt a blow to British influence in his country when, on March 1, 1956, he ordered the dismissal of the Army Chief of General Staff, General John Baggot Glubb, and his deportation along with other leading British officers who had command posts in the Army and Air Force. That move gained the king immense popularity, both domestically and in Arab countries.

The region, however, had embarked on an arms race. Nasser needed to arm his forces to justify his revolt against the monarchy regime that, among other drawbacks, had supplied the Egyptian Army with deficient ammunition and weapons in the 1948 war. His deal with the Soviets through the Czechs triggered a fury in Israel and a campaign to obtain arms from the West. Prime Minister Ben Gurion had made a statement to the effect that the chances of war coming by mid-summer (1956) depended upon sales of arms to Israel. France harbored hostile attitudes toward Nasser, especially after the start of the Algerian armed revolt against France on November 1,1954. Nasser provided material and logistical support to the Algerian Liberation Front, and provided refuge for their fugitives and revolutionaries. It was not surprising to see France side with Israel against Nasser. However, France did not want to be the first and only Western country to engage in the arms race in the Middle East. In a meeting between Mr. Abba Eban, Israel's Ambassador to Washington, with the Assistant Secretary of State for Near Eastern Affairs, Mr. George Allen, on 23 April 1956, Eban told Allen that France had told the

Israelis on April 20, 1956 that they had earmarked twelve Mystere IV fighter planes for delivery to Israel, but that the United States would have to deliver something significant before the French would ship them. Eban recalled that the Secretary of State had spoken to him on this matter on 30 March, and that almost a month had passed and no arms were forthcoming. The meeting discussed other matters of concern to Israel, such as the close relationship the United States had with the Baghdad Pact and the Jordan River diversion scheme Israel intended to implement.

The United States did not want to appear to be siding with one party against the other, especially that she was attempting to forge cooperation in the region through the implementation of the Unified Plan, and in view of her planning to have another Mission launched by the United Nations for that matter. Mr. Allen explained to Eban that there were factors to be considered by the United States with regard to the supply of arms to Israel. Among such factors was the forth coming U.S.-Saudi Arabian negotiations on the Dhahran Airfield in the eastern province of Saudi Arabia, home to all the oilfields in that country. A sale of U.S. fighter aircraft to Israel during this time could prejudice these negotiations. On the Baghdad Pact, and in response to Israeli concerns over close U.S. relations with the Pact, Mr. Allen assured Mr. Eban that the Baghdad Pact helped, rather than threatened, Israel. On the Israeli diversion scheme, Mr. Eban hinted that ruling circles in Israel were willing to discuss an arrangement whereby, in return for a loan, for Israel's water development and an assurance of eventual United States' support for Israel's diversion of Jordan water at Jisr Banat Ya'qub, the Israelis would forgo resumption of work in the demilitarized zone (DMZ) for a certain period. Mr. Eban stressed the urgency of the matter and the hope that discussions could begin promptly. Israel, through Eban's diplomatic skills, made the linkages between armaments, security and the water issues in the Jordan basin, in a maneuver to gain support from the United States on some, if not all, fronts.

2. The Lahoud Plan for the Jordan Basin

Under those volatile circumstances and fragile political environment, the Foreign Minister of Lebanon, Mr. Salim Lahoud, made an overture to the U.S. Ambassador to Lebanon, Mr. Heath, on 28 April 1956. Lahoud expressed desire of the Arab states, especially Lebanon, to avert war with Israel, but that cease fire (on the Syrian - Israeli lines) and diversion of Jordan River waters were tied together. The Minister asked to seek Israel's commitment not to divert Jordan waters until agreement for their use could be reached. Such use would not involve, at first stage, a formal accord between Israel and Arab states, but would mean that Arab states would commence work on their side and in the demilitarized zone, if necessary, and Israel would independently undertake work on their side and in the demilitarized zone.[139] The Minister had earlier asked the Secretary General of the U.N., during his recent visit to Lebanon and the region, to press Israel for such a commitment.

In his dialogue with the American side[140], Minister Lahoud recognized the validity of the argument that a general Israeli commitment not to divert Jordan River waters would give the Arabs an indefinite veto. Lahoud indicated that he would be able to propose a specific period for arriving at an accommodation, if Lebanon were the only interested party. Lahoud recognized the inseparability of the large-scale construction envisaged under the Jordan Valley/Johnston Plan and an understanding regarding Lake Tiberias storage. He believed, however, that modest assistance for some limited water operations could be undertaken, leaving major undertakings that require accord between the Arabs and Israel for later. Examples of limited operations would be a diversion dam on the Yarmouk at Adasiyya, partial development of the East Ghor in Jordan, expansion of Boteiha farm in Syria, field studies of the Hasbani in Lebanon, etc. In view of Israeli readiness to start digging for their diversion scheme, Lahoud hoped Arabs might begin their projects slightly ahead of the Israelis. He asked that Ambassador Johnston and his experts come up with a plan reflecting these ideas, which he might present to his fellow Arabs.

The Johnston team responded and formulated a plan. It contained the following justification and proposals:

"1. The Arab states are capable of planning and working out their own destiny, and they are determined to guide their future with their own hands. This applies to decisions respecting the development of their water resources, as well as to the other decisions facing the Arab world.

2. As a first stage in the Arab development of the Jordan River basin, a diversion dam should be constructed at Adasiyya; a canal should be built down the East Ghor in Jordan; Lebanon and Syria should speed up the exploitation of the Hasbani and Banyas Rivers before they flow into Israel.

3. The next stage would be the construction of a high dam on the Yarmouk at Maqarin. This dam would store the waters of the Yarmouk on Arab soil, and would provide power for industrial use in Syria and Jordan. When this dam has been built, it will be possible to complete the canal network in the Ghor and irrigate the Jordan Valley clear to the Dead Sea. Arab rights to store

---

[139] Cable no. 1359 dated April 29, 1956, 1 p.m. from the U.S. Embassy in Beirut to Secretary of State.

[140] Cable no. 1969 dated May 3, 1956 from State to the U.S. Embassy in Beirut, and Cable no. 1405 dated May 5, 1956 from Embassy, Beirut to Secretary of State.

water in Lake Tiberias are not abandoned but may come if it is advantageous and is carried out under conditions that are completely suitable to them. However, primary storage should be constructed first on Arab lands before storage in the Lake is considered.

4. The plan would be financed by a Jordan Valley Development Authority; run by and for the Arabs. This authority would obtain its funds through capital lent on a no-interest basis by the Arab states, and it would welcome loans from other countries on a similar basis, except that the principle of Arab control would at all times be maintained. Many nations speak much of Arab friendship; the reaction from the world to the financial needs of the Authority would show which countries really have Arab interests and objectives at heart. The Authority would receive income from the sale of electric power and from water fees, which in time would restore the funds expended."

The above text of the Lahoud Plan sounds very much like what Arabs like to hear. In fact it is nothing more than a staged implementation of the Unified Plan produced with an Arab coloring. This proposal by Lahoud paved the way to conceive of projects on the Arab side, and others on the Israeli side, that were compatible with the overall Unified Plan.

3. The East Ghor Canal Project

The implementation of the East Ghor Canal Project in Jordan was rooted in the notion, advanced by Lahoud, of partial and separate implementation of the Jordan Valley (Unified) Plan, with possible integration later as the political environment may permit. The arrangement had the advantage of averting a serious conflict with Israel over the diversion scheme, kept the Jordan Valley Plan alive, and would carry with it substantial benefits to the Arab side. It would, however, permit the initiation of the process to resettle the Palestinian refugees, a factor enshrined in the involvement of the United Nations and the United States in the Jordan Valley Plan from the beginning. The Lahoud approach carried with it advantages to all sides concerned. It was a positive sum approach. Jordan started tapping that potential, as it was the country most in need to achieve economic development and create jobs for an increasing labor supply expanded by the influx of Palestinian refugees.

To that end, the Jordan Ministry of National Economy asked the Food and Agriculture of the U.N. (FAO), on 18 June 1956, to look into a proposal whereby a "pilot scheme" for the development of the Jordan Valley would be undertaken in preparation for the implementation of the full Jordan Valley

Plan. Mr. S.N. Simansky, an expert working with the FAO, undertook the study of such a proposal, and came up with a plan to draw off some of the normal flow from the Yarmouk River and develop a Pilot Diversion Scheme. Three variants for such a scheme were considered:

a) A short main pilot canal (about 22 km) to irrigate about 30,000 dunums of land.
b) Longer main pilot canal (about 60 km) to irrigate about 30,000 dunums in a first stage and 100,000 dunums in a second, and,
c) Long main pilot canal (105 km) with about 100,000 dunums of State Domain and some other lands of the driest region. This variant was eventually not recommended.

Simansky concluded that the first and second variants were feasible, and recommended the employment of a specialized engineering consulting firm to do the designs and tender documents, and to supervise the construction. Simansky's study was accompanied with an elaborate analysis of the costs and benefits of each of the above variants, and a proposal for the scope of work for the consultants. He submitted his report on 12 July 1956. The U.S. International Cooperation Agency (ICA) Mission, in Amman, covertly obtained Simansky's report by a Mission technician whom the Embassy wanted protected.[141] The report was carefully studied, in preparation for potential involvement by the United States in implementing its recommendations.

Whether or not the Jordanian idea emanated from any discussion of the Lahoud Plan is unclear, but the second variant of the Simansky report was the one that was picked up in 1958-1959 for detailed study and design under ICA finance, with Harza Engineering Company as the consulting engineering firm. The implementation was carried out in several stages. Simansky's Pilot Diversion Scheme was part of the Master Plan that Baker and Harza had developed for the Jordan Valley between 1953 and 1955, while the Johnston's efforts were under way. He identified the part that could be developed using only base flow of the Yarmouk, before the regulation of its flood flow could be made by a dam at Maqarin or by storage in Lake Tiberias. The Pilot Scheme fitted very well with the Master Plan and was an integral part of it.

4. The Suez Crisis and its Aftermath

The summer of 1956 was eventful. Tensions mounted throughout the summer and severe clashes with Israel ensued along the Armistice lines. Jordan perceived the conduct by Israel of military exercises along the

---

[141] Cable from Nelson, Director of ICA in Amman, to Washington, dated 23 August 1956.

ceasefire lines as a provocation. She invited Iraqi troops to reinforce the Arab Legion (As the Jordan Armed Forces were called) in the event of further hostilities. The Jordanian Armed Forced had been "arabized" with the dismissal of General Glubb and the appointment of Jordanian officers the command posts.

In Egypt, several factors prompted the United States to act in a hostile manner to Egypt. She became wary of Nasser's courtship of the Soviet Union and of his increasingly neutralist foreign policy. She was not happy because of the failure by the Arabs to endorse the Unified Plan and because of the lack of breakthroughs in solving the Arab Israeli conflict. Secretary John Foster Dulles suddenly withdrew, in July, his government's offer to help finance a major water project in Egypt, the Aswan High dam. The dam would regulate the flow of the Nile and would have a long-term storage function. The construction of that dam had adverse environmental impacts that had to be mitigated and, more importantly, needed the consent of co-riparian Sudan, part of whose territory the reservoir would inundate. Moreover, a water sharing agreement had yet to be arrived at between the two countries, a situation analogous to the hardships that the Unified Plan of the Jordan Valley was facing. Unlike the case of the Jordan Valley, a bilateral agreement between Egypt and Sudan was reached in 1958, after a military take-over of power in Sudan that year.

Nasser responded to the withdrawal of western support to the Aswan High dam, in an inflammatory public speech in Alexandria on 26 July 1956, marking the anniversary of the exile of King Farouq, whom Nasser's revolution had deposed in 1952. Nasser announced the nationalization of the International Company of the Suez Canal. He passionately stated that, "at this time that I am making this announcement to you, Egyptian experts are on their way to take control of the Egyptian company, and Egyptian guides are steering navigation in the Canal. The revenue from our canal will bring the hard currency we need to build the Aswan High dam." Nasser was highly applauded by the crowds attending the rally in Manshiyya Square in Alexandria, and by all Arabs across the Middle East and North Africa. Nationalization of the Suez Canal was an outright blow to the British and the French, the primary stockholders of the International Suez Canal Company. Nasser's speech gained him the admiration and support of all Arabs from the Atlantic to the Gulf, and of the non-aligned nations of the world.

In North Africa, Tunisia had gained independence from France only a month earlier, and Morocco was on its way to do so shortly thereafter. The revolution in Algeria against France had been escalating for about two years, with active support and fueling by Nasser. The spirit of Arabism characterized the Middle East and North Africa, with Nasser as its uncontested champion. Syria had been overtly allying herself with Nasser, since the downfall of Shishakly in 1954. Nasser had his informational machines wage a propaganda campaign against Britain's influence in the Arabian Peninsula and Sudan, and against France's presence in North Africa. Nasser's nationalization of the Suez Canal Company provoked Britain and France and stretched their patience beyond limits. Israel, on the other hand, wasted no time to benefit from the opportunity of hostile confrontation between Britain and France on the one side, and Egypt on the other. She had been trying, through the good offices and quiet diplomacy of the United States, without success, to gain the right of transit passage of her ships through the Suez Canal and through the Straits of Tiran at the entrance to the Gulf of Aqaba. Additionally, as cited in the above, Israel wanted delivery of arms and aircraft from the West to match the Soviet military shipments to Egypt. In Washington, the American Jewish Committee was lobbying for Israel to obtain aircraft because, the Committee claimed, she was threatened to be invaded by Egypt, and for her to obtain a loan from the Export-Import (Exim) Bank to develop water resources outside the Jordan basin, especially since Israel "had accepted the Johnston Plan."[142] The American Jewish Committee proposed to the Secretary of State that settlement of the Suez controversy should be in the context of an overall settlement of Near Eastern problems.

In Jordan, general elections were held on 16 October and, for the first time since her independence, Jordan had a Prime Minister from the ranks of the political party that had scored highest in the elections. The cabinet was formed on 27 October 1956, and the new government opted to terminate the Treaty with Britain, under which Jordan received financial, technical, and military assistance. Egypt, Syria, and Saudi Arabia pledged to replace the financial assistance that Britain provided under the Treaty. There was jubilance displayed by the people and, with Nasser's blow to the old colonial powers, the mood on the streets was optimistic for a triumph of Arab nationalism and an honorable solution of the Palestine problem. Syria maintained strong ties with Egypt, and the support for Nasser in Arab ranks was skyrocketing. In Iraq, the popular mood was sympathetic with the nationalist movement that Nasser symbolized and championed, but the Iraqi government was critical of his moves and considered them dangerous to Arab interests.

In late October, 1956, Israel, in collaboration with Britain and France, and under the pretext of retaliating against Palestinian Fedayeen bases in Gaza, mounted an all out attack on Gaza and over-ran it in 24 hours. Her armed forces proceeded into Sinai towards the Suez Canal. Britain, who was made to evacuate the Canal Zone in 1954, and France, under the Suez Canal Treaty, had the right to intervene and protect the Canal and the freedom of navigation through it, if it came under threat of war. As Israeli forces were

[142] Memorandum of Conversation between the Secretary of State and the American Jewish Committee, August 9, 1956.

advancing, and before they engaged in any fighting with the Egyptian forces, Britain and France issued an ultimatum to the "fighting parties" to stay away from the Canal. Before Egypt had time to respond, British and French paratroopers dropped down on Port Said and Port Fuad, on each side of the entrance of the Canal on the Mediterranean, and marine forces landed on the Egyptian coast to occupy the Canal Zone. The task of the Egyptian forces was suddenly compounded: to face the invading Israelis in Sinai and to engage with the invading British and French troops. Egypt's forces evacuated from Sinai to reassemble west of the Canal to defend Egypt's cities against the invading British and French forces. Israel's forces ended up occupying Sinai from Egypt and stopped at the eastern bank of the Canal. The fighting was ended by an ultimatum from the Soviets and from the United States, issued to the aggressors. They finally pulled out under American pressure, and a United Nations Emergency Force (UNEF), was stationed on the borders between Egypt and Israel, and at Sharm El Sheikh at the mouth of the Gulf of Aqaba. Israel gained the right of shipping through the Straits of Tiran and the U.N. forces would guarantee the freedom of Israeli shipping through it. Ten years later, the termination by Egypt of the presence of the UNEF proved to be a critical factor in triggering yet another all out war between Israel and the surrounding Arab states, as shall be shown later.

With Israel's hawkish attitudes and aggressive actions so clearly displayed in the Suez Campaign, there was more rejection by the Arabs of accepting her or dealing with her in any capacity. There was no more talk of the Jordan Valley Plan, and the United States intention to have the Secretary General of the United Nations launch another initiative for the Jordan Valley development died out.

Relations between the Arab states themselves became strained, as a result of the Suez Campaign and the Baghdad Pact. Nasser's relations with the United States took a turn toward more cautious diplomacy than before, in light of the announcement of the Eisenhower Doctrine that was approved by Congress. Under that Doctrine, Congress gave authority to the President to provide assistance to any country in the Middle East that was being threatened by another, dominated by International Communism. Nasser suspected that Egypt was the country intended to be posed as an aggressor, and decided to escalate his campaign against the allies of the West in the region, primarily the partners and sympathizers of the Baghdad Pact. Nasser's influence in the Arab countries after the Suez Campaign grew very fast, and public sentiments were highly in his favor. He capitalized on that and soon became the Arab leader to reckon with on the Arab and international fronts. Earlier, he emerged from the 1955 Bandong Conference of non-aligned nations as a world figure with strong ties to India's Jawaher Lal Nehru and Yugoslavia's Joseph Broz Tito. Nasser was able to move the Arab masses on the streets, arouse opposition, and threaten the influence of foreign powers in the Middle East and North Africa. Nasser's followers were not only ordinary Arab citizens, but included followers in the ranks of the military and other officials in many Arab countries, especially Syria. His influence in Jordan was showing strength. An attempt by some pro-Nasser military officers to overthrow the monarchy was aborted shortly after the attempted coup reportedly went into action in April of 1957. King Hussein rushed to the scene of a rebellion in the Army barracks at Zerka and was able to take control of the explosive situation, assume command, and proceed with sure steps to protect the country against revolutionary attempts. The parliamentary government of Mr. Suleiman Nabulsi was forced to resign, and Hussein Fakhry al Khalidy, the Foreign Minister at the time of Johnston's first visit to the region, was asked to form a Cabinet. It lasted for only nine days before it gave way to another, headed by Ibrahim Hashim, that lasted less than a month in power before it was succeeded by a government headed by Mr. Samir al Rifai. These fast developments in Jordan and the arrest of Nasser's supporters brought about hostile reactions from Egypt and Syria. Several Jordanian suspects defected to Syria, where they were granted political asylum. Jordan's hopes for Arab delivery of financial support, pledged by Egypt, Syria, and Saudi Arabia to replace British aid, soon evaporated, and Jordan was in bad need for financial relief.

In Lebanon, the political scene was not rosy either. There was an attempt to break the tradition observed since independence, whereby the President of the Republic serves a non-renewable tenure of six years. President Camile Chamoun, in office since 1952, intended to have his tenure renewed or extended. This move drew criticism and protest from many foes and critics, especially the political forces with inclinations towards Syria and Egypt. The domestic situation escalated and the political heat transformed into a hot confrontation between militias, organized mostly along ethnic lines. A "civil war" broke out, and the United States Marines landed in Beirut, by invitation of the Lebanese President, to keep law and order. In many ways, Iraq, Jordan and Lebanon appeared to be lining up on one side, and Egypt and Syria on the opposite side, of a new Arab polarization trend. The first 'camp' was looked upon as allies of the West and the unpopular Baghdad Pact, and the second camp's members were viewed as champions of Arab nationalism and representatives of the hope that Arab pride would be redeemed and Palestine liberated from the Zionist occupation. Saudi Arabia and Yemen stood on the sideline, although the Saudis were nervous about the acceleration with which Nasser was gaining command of Arab peoples from the Atlantic to the Gulf.

Developments in Syria accelerated and there was fear of a Turkish military threat to Syria. The Syrian military, already known for its readiness to mount coups, became active and pushed for some kind of unification with Egypt. By mid-February 1958, and under pressure from the military and the public, the Syrian politicians and high-ranking military officers[143] went to Cairo and convinced Nasser to accept unity between Egypt and Syria. The United Arab Republic (UAR) was announced on 22 February 1958, as a united country of of the two states. A public referendum in early March confirmed Nasser as President of the new republic. The majority of Arabs were happy and enthused by the move that was a triumph for the leftist camp. The row deepened with the Arab adversaries who were proponents of alliance with the West. In response to the birth of the UAR, Jordan and Iraq conducted intensive talks, and a union was declared between them under the name "The Arab Union", in which the King of Iraq and the King of Jordan would alternate as heads of the Union, and the prime minister of the Union, operating under the King, would be from the other country. The unifications brought Egypt in with Syria as a riparian party on the Jordan, and brought in Iraq as a partner with Jordan, another riparian basin.

5. Launching the Pilot Scheme in the Jordan Valley

The economic situation in Jordan in 1957 could hardly be less enviable. She had an estimated population of 1.6 million people, and was burdened with Palestinian refugees, unemployment, and a heavy military budget. The country was severely limited in natural resources, technical capacity, and transport facilities. The creation of Israel and the prevailing state of war with her cut Jordan off from her traditional markets in Palestine and from her principal outlet to the sea, Haifa. The country was beset by internal and external political pressures. In ten years since her independence, the state had been dependent on foreign aid for more than half her annual budget requirements and over 60 percent of her total revenues. At the time, the balance of payments deficit was the highest of any Middle Eastern country except for Israel, while her per capita annual income of $100 was among the lowest. Agriculture constituted a sizeable share of her economy and employment, and was mostly dependent on rain fed farming and livestock.

With the row developed between Jordan and the "progressive" Arab states next door, it became more imperative to launch development projects, create jobs, and make up for the political frustration that had befallen the people, exacerbated by agitation induced by the hostile radio broadcasts from Cairo and Damascus. Foreign financial and technical assistance was very much needed, and the United States was the most likely party Jordan could turn to.

The "Pilot Scheme" that was proposed by Simansky was the most likely candidate with which to commence serious American aid and to launch

---

[143] Among the officers were Afif Bizri, the Chief of General Staff, Amin Naffouri, Mustafa Hamdoun, Ahmad Abdul Karim, and others. Some became ministers in the central government of the Union.

a development project that responded to the above needs. The ICA Mission in Amman (successor of the TCA) had done a good piece of homework in studying the scheme. The political and economic environments became fit for an agreement to finance that irrigation scheme and to hire a specialized consulting firm to prepare the designs and tender documents and proceed to construction. Action to improve Jordan's domestic situation was needed fast, and the Pilot Scheme provided the opportunity to move immediately.

The Government of Jordan applied for assistance from the United States to finance a unilateral diversion of the Yarmouk River and to build the Pilot Scheme which was Variant 2 in the Simansky Report. In January 1958, the National Security Council of the United States (NSC) authorized in its resolution no. 5001, Paragraph 40, a "piece-meal approach" to the implementation of the Johnston Plan, a notion credited earlier to the Foreign Minister of Lebanon, Mr. Salim Lahoud. In early February, the Jordanians proposed to the Americans to implement an irrigation project in the Jordan Valley in stages. In its final stage, the project would have an irrigated area of 25,000 acres (105,000 dunums). The water needed for irrigation would be diverted from the Yarmouk River through a small diversion structure on the river at Adasiyya. The structure would extend to midstream[144] and would divert approximately seventy per cent of the annual flow of the Yarmouk to irrigate lands in the Jordan Valley.

"We have avoided financing the project in the past for two principal reasons," said a report dated 10 February 1958 from the Director of the Near East Department, Mr. Lampton Berry, to Acting Secretary of State, Christian Herter, "a) Belief that the unified development of the Jordan waters (the so called Johnston Plan) would provide substantially greater economic and political benefits than would unilateral development by riparian states. b) The risk that Israel would seize upon unilateral diversion by Jordan of the Yarmouk as justification for resuming work on the Jisr Banat Ya'qub diversion and that our position on this matter would be prejudiced by our financing the Yarmouk diversion."

"There is no doubt that this project would provide the greatest favorable economic and political impact of any development project now under consideration in Jordan," continued the report, "implementation of the proposed project need not prevent subsequent unified development of the Jordan waters.... The Jordanians state, and the planned structures are so

---

[144] The territories beyond the midstream of the Yarmouk at the diversion site were part of the Demilitarized Zone between Syria and Israel, as defined in their Armistice Agreement of 20 July 1949. Building a structure across the entire river section may be construed as giving Jordan military advantage over Israel. To avoid such complications, the Jordanians proposed a structure protruding to midstream, her border line with the northern territories (former Mandate Palestine).

designed, that the waters diverted will not exceed the share allotted to them under the Unified Development Plan (377 million cubic meters out of a total flow of about 500 mcm per annum). Since agreement on unified development does not seem feasible in the foreseeable future, we think it best to make a modest beginning on the utilization of water which is now being wasted."

The American analysis of the Jordanian proposal addressed the Israeli and Syrian positions, the ruling of UNTSO, and recommendations for U.S. action. The Israeli aspects were concerned with her usage of the Yarmouk waters that could possibly be jeopardized by the Jordanian diversion upstream, and with her plans to implement her diversion of the Jordan River at Jisr Banat Ya'qub. The Syrian aspects dealt with the Yarmouk diversion and with the Banat Ya'qub diversion as well. The American analysis, contained in the aforementioned report to Acting Secretary Herter, addressed these topics as follows:

"**Israel Aspect**

a) Yarmouk usage: The proposed diversion would not conflict with present downstream usage of Yarmouk water by Israel, which now amounts to 25 mcm per year. During Ambassador Johnston's negotiations on the plan for unified development, Israel claimed a potential usage of 40 mcm per year of Yarmouk water.[145] Should the proposed Jordanian project be constructed, approximately 150 mcm would remain available for downstream usage. If the Syrians take 90 mcm allocated to them, approximately 40 mcm would still remain.
b) Banat Ya'qub: Israel might refrain from attempting to block the Jordanians' project in the hope that this would strengthen its case for proceeding with the Banat Ya'qub diversion. Work on the latter was halted following a decision in 1953 by General Bennike, Chief of Staff of the U.N. Truce Supervision Organization, that the project would give Israel a military advantage in a demilitarized zone. The Israelis have indicated to us that they intend to resume work on this project when they complete their present concentration on development of water resources outside the Jordan basin. We understand that this will be in a period of about three years.

Syrian Position

a) Jisr Banat Ya'qub: If the Israelis were to resume work at Jisr Banat Ya'qub we could expect the Syrians to assume a hostile attitude, with resort to the Security Council as a minimum reaction and use of force as their maximum. They could probably rely on full support from other Arab states.

---

[145] This issue remained as a point of disagreement between the U.S. and Israel, even after the fourth round of Johnston's negotiations.

b) Yarmouk: The Syrians are understood to have agreed to the unilateral diversion of the Yarmouk by Jordan.[146] It has also been reported that they mean to proceed with their own unilateral irrigation projects in the Yarmouk basin with Soviet assistance.

UN Truce Supervision Organization Position

The Acting Chief of Staff of UNTSO, Colonel Leary, has determined that if the Yarmouk structure does not extend beyond midstream, it will not give military advantage to any party and will not come within the preview of the Syrian-Israeli Armistice Agreement.... Col. Leary also raised the question of the seventy years concession granted to the Palestine Electric Corporation by the High Commissioner for Palestine in 1926, which gave the Corporation exclusive rights for the utilization of the hydroelectric potential of the Yarmouk and Jordan Rivers (the Rutenberg Consession). The proposed diversion would fall within the concession area. We do not believe this should prevent execution of the project since present circumstances preclude utilization of the concession and because the assets of the Corporation in Jordan have been nationalized.[147]

United States Position

a) The Jordanians indicate that they intend to proceed with diversion of the Yarmouk whether or not we finance it. They do not have the capital to do so but may well be able to obtain enough to make a start from friendly Arab sources. If the project is to be implemented in any event, it is to our interest to gain what credit is to be had from participation in it.
b) The Israelis indicate that eventually they are going to resume work at Jisr Banat Ya'qub. We have no assurance that the Israelis will indefinitely respond affirmatively to our advice that this work be deferred. Therefore, we are likely to be faced with resumed Israel activity on this project whatever we do about the Yarmouk
c) There is a superficial similarity between the Jisr Banat Ya'qub and the Yarmouk diversion since the former lies within the central Israel-Syria demilitarized zone and the center of the Yarmouk forms the southern boundary of the southern Israeli-Syrian demilitarized zone. However, the

---

[146] The Jordanian Undersecretary of National Economy, Mr. Hazim Nuseibeh, stated to officers of the U.S. Department of State, on several occasions in November 1957, that the Syrian Representative on the Yarmouk Joint Committee had stated that Syria had no objection to Jordanian implementation of this project. Nusseibeh further stated that, for political reasons, Syria could not very well object to a project of this nature, one which would be used to develop Jordan's economy.

[147] The concession itself was cancelled by decision of the Government of Jordan and ratification by the Board of Regents in 1953.

Israeli project is designed ultimately to divert Jordan waters out of their natural basin, which is not the case with the Yarmouk diversion. Furthermore, the proposed Jordan structure would not actually lie within the southern demilitarized zone (like the Israeli diversion structure would). Therefore the two projects are not parallel in nature. Since the Yarmouk diversion would not provide a military advantage to any party, we do not believe that our participation in it should affect our position of support for General Bennike's 1953 decision. We believe these two projects should be treated separately and that our financing of the Yarmouk does not imply our approval of Jisr Banat Ya'qub."

That was the analysis of the situation as expressed in the memorandum submitted to the Acting Secretary of State on 10 February 1958. The memorandum proceeded to make the following recommendations, which the Secretary approved:

"**Recommendations:**

1. That we inform the Jordan Government that we are prepared to assist in financing first year costs of the Yarmouk diversion from Fiscal 1958 funds already allocated for economic development in Jordan, and that we are prepared to assist in financing future year costs subject to the availability of funds and to our determination that such assistance is needed and desirable. This assistance would be based on an explicit undertaking from the Jordanians that they *will not draw from the Yarmouk more than the share allotted to them under the Unified Development Plan.*[148]
2. That we inform the Israelis at an appropriate time that we are assisting Jordan with this project in the interests of area economic and political stability, and that we believe this to be consonant with the Israelis' recommendation that we concentrate on economic development in Jordan. Our interest in Israel's economic development has been amply demonstrated, most recently with regard to Israel's application for an Exim Bank loan and development Loan Fund assistance. We do not believe this affects our position on Jisr Banat Ya'qub, based as it is on the findings of the U.N. Truce Supervision Organization and the Security Council resolution of October 27, 1953."

The above constituted the position of the United States on the Yarmouk diversion. The U.S. contacted the new-born United Arab Republic which had emerged on 22 February 1958, to advise President Nasser of the American intention to support the unilateral Yarmouk diversion. In addition to his role as a leader in the region, Nasser became responsible for Syria who,

---

148 Italics by the author to indicate emphasis and importance of the condition of financing.

as a riparian, has direct interests in the Jordan River basin. On 26 February 1958, a cable from State to the U.S. Ambassador in Cairo read:[149]

"On February 27 you should arrange to have conveyed to Nasser on confidential basis that USG has advised Jordan it prepared assist in financing to amount estimated at approximately $2 million first year costs unilateral Yarmouk river diversion by structure to be erected entirely on Jordan territory approximately 5 kilometers upstream from Adasiyya village and construction East Ghor Canal. Purpose is to irrigate estimated 25,000 acres Jordan Valley with Yarmouk water in accordance expressed desires Jordan Government HKJ expected announce shortly. As proposed diversion structure is simple one extending only to midstream it would not, of course, affect upstream users. We informing UAR in view previously expressed interest Jordan Valley waters and trust they will welcome this as constructive step in development area resources."

Before an agreement could be concluded between the United States and Jordan, the latter had to issue a commitment to the effect that her use of the Yarmouk River would not prejudice the Unified Plan. Such a commitment was issued through a note by the Foreign Minister of Jordan in a memorandum no. 58/14/6719, dated 25 February 1958.[150] The note acknowledged the statements made the day before by the Charge d'Affaires of the Embassy of the United States of America as documented by the following *Aide-memoire*:[151]

"*AIDE-MEMOIRE*[152]

The American Charge d'Affaires called on Foreign Minister Rifa'i this morning and informed him that the United States Government had instructed him to state that it is prepared to assist Jordan in financing the first year costs of the proposed Yarmouk Diversion Project and the East Canal. It is expected that this financing will be accomplished by use of fiscal year 1958 funds already allotted for economic development in Jordan. The Charge said that he had also been instructed to inform the Foreign Minister that this assistance will be extended provided there is an explicit undertaking from the

---

[149] State Department Records, National Archives of the United States, Cable no. 2336, copied to Amman (cable no. 2033) and Damascus (cable no. 1901).

[150] Foregn Minister's note number 58/14/6719 dated 25 February 1958 addressed to the U.S. Charge d' Affairs in Amman and signed by the Foreign Minister, Mr. Samir al Rifai.

[151] See also the memorandum submitted to William M. Rountree, Assistant Secretary for Near Eastern, South Asian, and African Affairs at the Department of State, Washington D.C., by Mr. John O. Bell, Regional Director for Near East and South Asia, dated July 1, 1958. State Department Records, National Archives of the United States, Washington D.C.

[152] Italics by the author to indicate emphasis and importance.

Hashemite Kingdom of Jordan that it will not draw from the Yarmouk River more water than the share allotted it under the Unified Development Plan.

Embassy of the United States of America,
Amman, February 24, 1958."

The explicit commitment of the Hashemite Kingdom of Jordan was issued the following day and signed by the Foreign Minister, Mr. Samir al Rifai. With this important condition met, as well as, other conditions pertaining to the technical aspects of the project as well, a Project Agreement was entered into on 31 May 1958 by the International Cooperation Administration of the United States (ICA) and the Development Board of Jordan.[153] The Project Agreement was made under the General Agreement on Technical and Economic Assistance concluded between the United States and Jordan in July of 1957. In that Project Agreement, the Jordanian Government undertook to take "appropriate, timely action to provide for only economic, resident-operated farm irrigation units to receive water under the East Ghor Canal Scheme which will provide a reasonable standard of living, and which will enable such resident operator (a) to pay a pro-rata share of the annual costs of operation and maintenance of the East Ghor Canal Scheme and (b) to repay a pro-rata share, over a reasonable period of years, of an appropriate share of the construction costs incurred in the construction of the Scheme." By February 1959, a law was passed that provided for the establishment of an East Ghor Canal Authority to be in charge of implementing the Project and of its operation and maintenance. A land reform provision was contained in that law that provided for the redistribution of farmland and its re-allocation to landowners, in accordance with a sliding scale, and to resident land-less farmers. The law was the first of its kind since the creation of the Hashemite Kingdom.

On 8 August 1958, when Jordan was passing through a very crucial phase of her existence as a state, ground was broken for the Scheme and the Prime Minister, Hazza' al Majali attended the ceremony. Three weeks earlier, the "Arab Union" with Iraq had collapsed in the wake of an Iraqi military coup, and neighboring Syria and Iraq closed their borders with Jordan. Crucial supplies were airlifted to Jordan from Cyprus. Nonetheless work on the East Ghor Canal, implemented by Force Account,[154] began on the excavation for a trapezoidal section canal that would pass 10 cubic meters per second. The

[153] The Development Board was the predecessor of the National Planning Council (1972-1984). The Council was superceded in November 1984 by the Ministry of Planning.

[154] Force Account method is the method of construction by which the owner takes on the actual construction work with his own forces. Other methods of implementation include outsourcing by awarding the works to contractors.

working force was composed of 851 Jordanian laborers recruited from 114 Jordanian villages from all over the country. That was a positive change for the country. On 16 July 1959, the Development Board of Jordan contracted the services of Harza Engineering Company for the engineering supervision of the canal construction and for the designing of the distribution and drainage systems for the farms that would benefit from the Scheme. Also in mid July, bids were received from five companies to do the lining of some 22.8 kilometers of the canal, drill the diversion tunnel, and construct the diversion facility at the Yarmouk. The Italian firm, Impresit, was awarded that contract.

6. Israeli Response to the Yarmouk Diversion Project

Israel was informed of the United States intentions to support the East Ghor Canal Project shortly after Jordan issued the commitment not to draw more water than allotted to her under the Unified Plan. There was a verbal and bitter Israeli complaint about the way the Yarmouk diversion project was handled.[155] They wanted to know why there could not be a private understanding with Israel in advance. The United States' officials ruled that out, for fear of word spreading, that the Jordanians had a secret deal with Israel. On 2 April 1958, the Israeli Ambassador in Washington forwarded a note to the State Department, in which Israel expressed worries about the East Ghor Canal Project causing appreciable harm to Israel. The note requested assurances from the United States that ways and means be found to safeguard her established rights as to the Yarmouk waters, as defined in the Unified (Johnston) Plan. They mentioned that Israel's share in the Yarmouk waters as per that plan was 40 mcm/year.[156] An attachment to the Israeli note stated that the maximum summer Israeli diversion was 2.5 cubic meters per second (cms),[157] and that the Syrian summer diversion from the river would be 3 cms. The total diversion by these two riparian parties would add up to 5.5 cms which would leave less than 2 cms for Jordan's use in the summer. The note expressed concern over a possible increase in water salinity, and over work in the Demilitarized Zone across the Yarmouk River.

---

[155] Memorandum by James Ludlow.

[156] This number is the Israeli figure contained in their Memorandum of Understanding of 5 July 1955, as opposed to the figure of 25 mcm/year contained in the Unified Plan circulated in January 1956 to the parties. This point remained issue of contention between Israel and the United States (and the other riparian parties) until finally settled in the bilateral negotiations between Jordan and Israel in the morning of 18 October 1994, and was stipulated in the Water Annex to the peace Treaty.

[157] This Israeli claim was repeated throughout the years 1979-1985, in the UNTSO meetings to settle differences between Jordan and Israel over diversions from the Yarmouk.

The United States' response, conveyed in a note, dated 1 August 1958, was firm and to the point. It stated that the U.S. believed that the East Ghor Canal Project was well within the provisions of a future unified development of the Jordan river basin and that the diversion planned from the Yarmouk did not exceed 70 percent of the flow in the dry months, which would be about 5.1 cms, leaving adequate flow for Israeli use downstream. There would be no work anticipated in the Demilitarized Zone between Syria and Israel across the River.

The Israelis seized this opportunity to request a similar treatment, i.e., for the United States to help finance their Jordan River projects. In particular, they were hoping for American endorsement of their Tiberias-Beit Shean Project that would supply the Beit Shean Ghors with water from Lake Tiberias, and of the transfer of Jordan River water outside the basin through the National Water Carrier. On 4 September 1958, the political correspondent of Haaretz published a report in which he mentioned that Israel would submit to the Government of the United States an application for assistance in the implementation of an irrigation scheme using the waters of the Jordan. "Jordan," the report said, "is already starting the implementation of her scheme for exploitation of the waters of the Yarmouk with American economic support."

On the same day, the Jerusalem Post carried an article by their Diplomatic Correspondent that said, "It was recalled that Israel has always been ready to agree to an over-all regional scheme for the use of water resources, and that it has been Arab intransigence that has blocked the many proposals put forth over the years. If Jordan does begin a water project on her own, this would tend to give strength to Israel's claim that, barring the existence of a regional scheme, she has the right to go ahead unilaterally on the development of resources available to her."

Indeed, Israel submitted applications to the United States for assistance in the financing of the remaining portions of the National Water Carrier, which would pump water from Lake Tiberias to the Negev, and of the Tiberias-Beit Shean Project. The Israeli note to the State Department was dated 28 January 1959. Several exchanges of notes took place between the Israeli Embassy in Washington and the State Department. The second note from the United States to Israel was on 7 May 1959, expressing reservations towards Israel's position in regards to the Unified Plan. The Israelis followed up with a third note on 17 July 1959, in which they objected to the U.S. position, assured that implementation of Israel's plan would not affect her attitudes toward the Johnston Plan, and expressed readiness to participate in the Johnston Plan, whenever the other governments concerned agreed to accept. The note suggested that the Beit Shean project was similar to the East Ghor Canal Project, and that it was meant to protect Israel from the effects of the East Ghor Project. A third note was issued by the U.S. State Department, on 19 November 1959, responding to the third Israeli note. Additionally, there were meetings of experts in an attempt to reconcile the differences between the two parties as to the understanding of the Johnston Plan. The Israelis insisted on 40 mcm allocation from the Yarmouk waters to them, as opposed to Johnston's 25 mcm, and they ignored the need to store the Yarmouk floods in Tiberias. In their third note of 17 July, they ignored the allocations to the Hashemite Kingdom from Lake Tiberias, which were meant for the West Bank in Jordan. The U.S. maintained that a unified plan of the basin best served the interests of all riparian parties, and in the congressional presentations for Fiscal Year 1961, the State Department reported that "The United States continues to believe that the unified development plan worked out by Ambassador Johnston is the best solution yet offered for this complex and explosive problem. We hope that the progress made by Ambassador Johnston on the technical level can be preserved. It is clear, however, that a solution to the Jordan water problem must in the first instance be the responsibility of the riparian states themselves." The U.S. also maintained that the fundamental premise of the Johnston Plan was that "all irrigable lands within the basin are to be allotted adequate irrigation water before water is taken from the basin for use outside."[158]

A turn of events took place as a result of technical meetings between Aharon Wiener of Israel's Tahal and Wayne Criddle, pursuant to the U.S. note of 19 November. The meetings were held on 20 and 21 November 1959 and Mr. Wiener assured that Israel's Jordan Project would not be incompatible with the Johnston Plan. Based on this assurance, the U.S. State Department issued a note on 7 January 1960 that it would have no objection to consideration of a loan to the Israeli project, on condition that documented assurance was provided by Israel that the project is consistent with the Johnston Plan. An injunction was ordered against publicity of this position of the United States. The loan would be in the amount of $15 million and would be channeled through the Israel Bank of Agriculture, which was proposed for use as a "cut-out", i.e. the U.S. funds would appear as funds extended to agricultural development and not overtly given to finance the National Water Carrier. Clearly, the United States feared that her support of the Israeli project would jeopardize her role as an intermediary and would upset the Arab states, particularly the United Arab Republic (UAR).

---

[158] A report for the International Cooperation Administration submitted by K.F Vernon and Wayne D. Criddle, August 12, 1959. The report reviewed the Israeli Jordan Project and pointed out several discrepancies between the Project and the Johnston Plan. The review indicated that the Israeli project would be capable of transferring 425 mcm of water from Tiberias as opposed to 231 mcm envisaged in the Johnston allocation. It also indicated that Jordan would not have enough water in May and June for 76,000 and 80,000 dunums respectively.

Egypt had been duly contacted, in October 1959, through the U.S. Ambassador in Cairo, and was tipped to the possibility that the United States might approve a loan for Israel's water project. This loan they contended, would be the right thing to do, if Egypt were to receive a loan from the World Bank, with U.S. support, to improve the Suez Canal. The Americans also wanted to feel out President Nasser, to see if his approval (as President of the UAR) could be obtained for the Syrian part of the Jordan Valley Development. The Israelis widely publicized their project, and it was not a secret that they were going to proceed with the implementation of the Jordan River diversion.

As Israel was actively pursuing the implementation of her Jordan water projects, she exerted very serious efforts to have her Memorandum of Understanding of 5 July 1955, given to Ambassador Johnston, recognized as the acceptable version of the Unified Development Plan of the Jordan River basin. The United States stood firm against the Israeli efforts, and stood by the version that Ambassador Johnston had concluded with the Arab Technical Committee and with the Israelis, in October 1955. The State Department was eager to do everything possible to contain the situation. The Israelis did not spare any effort to bring the United States to their side, and to have the U.S. government communicate a U.S. supportive position to the Arabs. President Kennedy was said to have pledged to David Ben Gurion, the Prime Minister of Israel, that the U.S. would inform the Arabs of her position on Israel's withdrawals from Lake Tiberias.

The Arab reaction to Israel's widely publicized diversion project is detailed in the following chapter. The reaction enabled the resurrection of joint Arab action and the birth of yet another institution within the Arab league: the Arab Summit. The United States, however, followed a line of preventive diplomacy aimed at preserving the integrity of the Unified Plan, as the best practical plan for the development of the basin, and at controlling the political situation so as to avoid an outbreak of hostilities in the region. In a meeting at the State Department,[159] in April 1963, between American and Syrian diplomats, Mr. Crawford of the State Department said to his Syrian guest: "You were not willing to see the Johnston Plan into effect; what then are your ideas? We seek to impose no specific plan, but believe that the work of the past year and a half constitutes a solid foundation on which to continue this examination."

Mr. Crawford informed his guest: "Israel is expected to begin taking some water from Lake Tiberias by late Fall this year or early in 1964. We have supported the concept of unified development, but have equally come to

---

[159]Memorandum of Conversation dated April 11, 1963, in a meeting between Dr. Jawdat Mufti, Minister, Embassy of the Syrian Arab Republic, Washington D.C. and Mr. William R. Crawford, Jr., Officer in Charge, Lebanon-Israel Affairs.

believe that in the absence of international agreement on such unified development, the riparian states, Arab or Israeli, should not be prevented from proceeding with national plans for water usage provided these do not impinge on the primary rights of in-basin users (who are mainly the Arabs below Lake Tiberias), and do not conflict with the general allocation patterns of the 1955 Unified Plan. We are not trying to impose this as an ideal, but do think that it laid down an equitable distribution and serves as a good model until something better comes along. We have always followed Israel's plans closely to satisfy ourselves that these were consonant with the Unified Plan, and we intend to take a further close and careful look in coming months." The U.S. State Department then summoned Mr. Wayne D. Criddle, conferred with him on 14 June 1963, and dispatched him to the region on a close follow-up mission.

Criddle's visit came at a time when both Jordan and Israel had legitimate concerns about the impending developments in the basin. The Joint Jordanian-Syrian Committee on the Yarmouk[160] had, in its meeting of 11-13 April 1963, approved a draft agreement to be concluded with an engineering consulting firm to prepare the blueprints of the Great Yarmouk Project and supervise its construction. They authorized the Jordan Development Board[161] to sign the agreement with the companies that offered the best bids, and to put the agreement into execution. The contract was awarded to a Yugoslav firm, Energoproject, and included the preparation of the designs of a dam at Maqarin and the extension of the irrigation system in the Jordan Valley of the Hashemite Kingdom of Jordan. The project was essentially the same as outlined in the Master Plan of Baker and Harza in July 1955.

7. Israeli - American Differences

The United States held the position that the Unified Development Plan for the Jordan River basin provided the best practical arrangement that existed thus far for the economic development of the basin. Israel went along, except that she had a Unified Plan version of her own. Her version of the plan was reflected in the Memorandum of Understanding, dated 5 July 1955 (see

---

[160]Comprised Messrs. Akram Zurayq, Secretary General of Public Works, Abdul Hadi Nihlawi, Secretary General of the Ministry of Finance; and Fawzi Al-Khabbaz, Assistant Secretary General of the Ministry of Public Works for the Syrian side; and Messrs. Ahmad Al Tarawneh, Head of the Board of the East Ghor Canal Authority; Omar Abdallah Dokhgan, Assistant Director General of the Central Water Authority; and Sweilem Haddad; Head of the Engineering Department at the East Ghor Canal Authority for the Jordan side.

[161] The predecessor of the National Planning Council which, in turn, was succeeded by the Minstry of Planning.

Appendix 2), which the Israeli side handed over to Ambassador Johnston for negotiations with the Arab side. Ambassador Johnston went on record as saying that the Memorandum expressed the Israeli point of view, and that it would have been the basis for the Unified Plan had the Arab parties accepted it. He returned to Israel, in October of 1955, and informed the Israeli side of the latest agreement he had with the Arab Technical Committee, in which Israel's share in the Yarmouk was set at 25 mcm/yr and not 40 mcm/yr as stated in the Israeli memorandum. Prime Minister Moshe Sharett, Finance Minister Levi Eshkol, and the Israeli team accepted the final agreement Johnston had worked out with the Arab side and postponed the ratification of that acceptance by the Israeli Cabinet until after the Arab League accepted the modified Unified Plan. Three points of difference clouded the common understanding of the Unified Plan by the two countries:

1. Israel's insistence that her share from the Yarmouk waters to irrigate the Yarmouk Triangle was 40 mcm per annum as opposed to the 25 mcm that Johnston managed to have the Arabs agree to in lieu of their proposed share for Israel of 17 mcm for that same triangle;
2. Israel's further insistence that the share of the Hashemite Kingdom of Jordan from Lake Tiberias would be 100 mcm including 30 mcm of saline water as opposed to the upper limit of 15 mcm of saline water that Johnston worked out with the Arab side; and,
3. Israel's refusal to accept international supervision on the distribution of water, as stipulated in the Unified Plan that Johnston worked out with the Arab side.

All three issues affected Jordan, of all the Arab riparian parties. In essence, Israel, with her position as outlined above, had not been in full agreement with the final modified version of the Unified Plan, a position that the United States viewed as unconstructive and, coupled with Israel's plans to divert the Jordan, potentially explosive. The United States requested that Israel avoid mentioning the "Johnston Plan" *per se* and to refer instead to the Plan for Unified Development of the Jordan River System,[162] and that Israel issue a declaration regarding her own position. Such declaration by Israel was set as a condition for U.S. support of the Israeli diversion project. The United States proposed the following draft for that purpose on 10 October 1963:

"Israel will keep its withdrawals within the limits established by the Plan for the Unified Development of the Jordan River System which was formulated (worked up) by impartial experts in consultation with all riparian states in the years 1953-55."

---

[162]Communication between William R. Crawford, Officer in charge of Arab-Israeli Affairs at the State Department, and Mr. Mordechai Gazit, Embassy of Israel in Washington D.C., dated 18 October 1963.

It is clear that the United States was fair in her dealing with the Israeli and Jordanian sides, with respect to her support of water projects in the basin. She had imposed a similar condition on Jordan before she seriously entertained Jordan's request for assistance to build the first stage of the East Ghor Canal Project (the Yarmouk diversion) in Jordan, and Jordan abided the following day, as stated above.

The United States hoped that Israel, on the occasion of her making such a public statement about the Jordan waters, would also make reference to international supervision, a matter that Israel had been contesting. The divergence of positions between the U.S. and Israel was the subject of contacts between officials of the two countries in the field (Israel) and in Washington. In one such meeting, Deputy Assistant Secretary of State, John D. Jernegan, summed up United States' understanding of, and stand on, Israel's position:[163]

"We understand Israel believes the divergence issues in our understandings will not be of practical importance for some years to come, and that when they do become important, Israel will negotiate them with Jordan. Meanwhile, Israel prefers that we let 'sleeping dogs lie'."

"The United States," continued the Deputy Assistant Secretary, "does not agree with this position. We must be the moral guarantor who assures the Arab riparian states that Israel is acting within the limits of the Unified Plan and vice-versa. If Israel is not in agreement with part of the Plan, we should not give these assurances. We note the Israeli argued that there are really two alternative plans. However, the Arabs are basing their thinking on the version they have, which reflected Ambassador Johnston's views on these two points. The Israeli version was never agreed to by Ambassador Johnston and he so informed the Israelis on several occasions. For example, on July 16, 1955, he told Ambassador Eban that under no circumstances could he indicate his approval of the plan outlined in the Israeli Draft Memorandum of Understanding or of the Israeli position as reflected in that Memorandum. He said he could not make any expression which would indicate that he regarded the Israeli position as a satisfactory basis for final agreement."

"Conceivably," Mr. Jernegan went on, "Israel could go its own way, unsupported by the United States on these two points. But the Arabs could then say that Israel was not abiding by the Plan and hence the Arabs were not bound by it. They would endeavor to drive a wedge between the U.S. and Israel. It would be preferable for us to have Israel's agreement with the Plan

---

[163]Memorandum of Conversation dated December 6, 1963 on Jordan Waters in a meeting held at the State Department and attended, in addition to Mr. Jernegan on the American side, by Willian Crawford, Jr., and Lucien L. Kinsolving; and from the Israeli side by Ambassador Avraham Harman, Mordechai Gazit, Minister, Israeli Embassy, and Shaul Bar-Haim, Counselor, Israeli Embassy.

on these points so that it can have our support in all respects. The matter is taking on added importance now in view of the increasing Arab interest demonstrated by the meeting of Arab military chiefs in Cairo[164]. This may not of itself produce serious consequences, but it could be an ominous harbinger. We want to be on the most solid ground to deal with Arab reactions."

"The third point on which we have not reached agreement is the necessity for some voluntary public statement by Israel indicating its willingness to accept some form of international observation if the Arabs will do the same. We are not discussing a mechanism. If it were to come to that, the mechanism is provided for in the Plan itself. In fact, we doubt that this mechanism is ever likely to be established. If the Arabs were to raise the issue, we would refer them to the Plan and they would shy away from it. We would like to be able to point to some Israeli statement to reassure the Arabs, who do not trust us alone as the guarantor because they consider us a friend of Israel. Such a statement need not be highlighted; it just has to be there for reference."

Israel's Ambassador to Washington, Mr. Avraham Harman, said the proposal put by the United States on 10 October 1963, as discussed in full afterwards by Mr. Crawford with Mr. Bar-Haim was that Israel should hold her summer withdrawals of the Yarmouk to 25 mcm but that she could take the balance of her 40 mcm entitlement during the winter.[165] "Was it now being suggested that Israel would owe some of this water to Jordan?" He asked.

Mr. Crawford said there had evidently been a misunderstanding. "On October 10, we had very carefully said that Israel's entitlement was recognized as 25 mcm. Until such time as Jordan was able to make use of the Yarmouk waters other than this amount, Israel should hold its summer withdrawals to 25 MCMs. As regards the winter we could leave an inference but specify nothing. When Jordan states she is ready to make use of all water over 25 MCMs, she would be entitled to receive this water. We would not support a retroactive claim by Jordan."

There followed, in the reference meeting, a brief discussion of the interests of Israeli users along the southern Jordan, who, Ambassador Harman said, had an entitlement to Yarmouk water. Mr. Crawford said that under the Johnston Plan, the needs of these users were met by the Beit Shean pipeline, which Israel had completed. This pipeline was the parallel of the East Ghor

---

[164] Details of Arab reactions are in the next Chapter.

[165] The Israel Embassy subsequently telephoned to acknowledge a misunderstanding on this point, arguing that the U.S. had recognized only a 25 mcm allocation and had left the matter of temporary winter withdrawals to inference.

Canal on the Jordan side and had the same purpose of protecting farmers who would otherwise be hurt when the old course of the Jordan became too saline for use. Ambassador Jernegan stated that nothing had been included in the Unified Plan regarding carrying water from the Yarmouk to the west bank of the Jordan in Israel.

Referring again to the Adasiyya allocation (Yarmouk Triangle), Mr. Gazit said that, in the notes Israel received from the U.S. at the time the East Ghor Canal project in Jordan was initiated (1958-1960), there was no mention of Israel's share being limited to 25 MCMs. "Thus," said Gazit, "the U.S. notes supported Israel's contention that there had already been US-Israel agreement that 40 MCMs was the proper allocation."

Mr. Jernegan recalled that even the 25 MCMs was a substantial concession by Ambassador Johnston. Israel's usage at the time was between 12 and 15 MCMs, at most. The two sides did not appear ready to reconcile their differences regarding the Israeli share from the Yarmouk.

About the same time, more political hardships were caused to the United States by Israel. The Government of Israel made statements in the Knesset disavowing paragraph 11 of U.N. Resolution number 194, of 1948, which called for the compensation and/or repatriation of Palestinian refugees. The Israeli Knesset took action to confirm those statements. Such developments, Jernegan represented, posed serious problems for continuation of U.S. talks and for their efforts to find a solution.

8. Contacts with the World Bank

While working on the Israelis to have them conform to the Unified Plan, the United States officials concurrently moved to assist Jordan in implementing the East Ghor Canal Project and to solicit financial support for Jordan from the International Bank for Reconstruction and Development (IBRD) to build the dam on the Yarmouk River. A meeting was held at the State Department with officials of IBRD on 12 December 1963 for the purpose, and the IBRD decided to look into the potentials for their involvement. They instructed an IBRD mission that happened to be in the Near East to proceed to Jordan with the aim of sounding the possibility of IBRD's involvement in the construction of the dam at Maqarin. The concurrence of the States that shared water rights on the River was an essential condition for the Bank's involvement. However, because Jordan, a country of low income economy, was eligible only for credits from the International Development Association (IDA) of the World Bank, and not for loans from IBRD, the Banks involvement would not have been more than a few million dollars, "so little as to be a drop in the bucket for a project that might end by costing $150 - $180 million."[166]

9. Criddle's Follow Up Mission- Relaxation of U.S. Pressure

The State Department requested, and the riparian parties agreed, to receive Mr. Wayne Criddle who was on a field mission to the region aimed at following up on the activities of the riparian states in the basin. Two projects were being implemented at the time -- the Jordanian East Ghor Canal project and the Israeli diversion project, with the diversion point shifted to Tabgha on the northwest shore of Lake Tiberias. Part of the Israeli diversion was the Tiberias-Beit Shean pipeline. Criddle had a meeting with States Department officials on 14 June before his departure. Mr. Criddle reported: "It appears that to date [first week in July 1963] there has been no fundamental divergence from the Plan in any of the countries visited."

It is noteworthy to review what Mr. Criddle had to report about the activities in the Hashemite Kingdom. He wrote:[167]

"Jordan has not exceeded her total allocation, nor would her construction to date indicate plans to exceed such allocation. In fact, some of the planning sheets would indicate a misunderstanding adverse to Jordan as to the quantity of water that can be taken under the Plan. Such sheets indicate that 30 percent of all ranges of flow of the Yarmouk at Adasiyya must be by-passed. This is not so. It was understood and confirmed by the Israeli engineers this week that 30 percent of the <u>low</u> flow must be by-passed during the irrigation season, or about 2 cms (cubic meters per second). However, the total annual quantity to go to Israel was between 25 and 40 mcm depending on which document is used as a base."[168]

In his trip report, Criddle was softening on Israel. Several points are worth analyzing. He threw doubt as to the right document of the Unified Plan to take as a base. He proceeded to acknowledge the reasonableness of the Israeli insistence that only 70 mcm per year of Tiberias water was due to the

---

[166]Mr. Stoakes M. Talbert, South Asian Division, World Bank, commenting to William Crawford, Officer in Charge, Arab-Israel Affairs, in a meeting between them at the State Department on 12 December 1963. State Department Records, U.S., National Archives, Washington D.C.

[167]"Jordan River Plans & Development As of June, 1963," a report prepared by Wayne D. Criddle for the Near East Division, State Department, a personal copy was made available by Criddle to the author in June, 1984.

[168] This confirmed understanding was not readily accessible by the Jordanian officials later on, and was to cause friction and confrontation in 1979 as water demands increased in the Jordan Valley and water shortages became more frequent.

Hashemite Kingdom. "It might be pointed out," he reported,[169] "that guaranteed annual releases of 70 mcm of normal Lake Tiberias waters on demand might be equivalent to a higher figure if Jordan had to take unregulated flows. Thus, the "effective" release from the upper Jordan River for use by Jordan might be considered somewhat greater per mcm released than originally envisaged under the Unified Plan." Such views were not consistent with the provisions of the Unified Plan that Ambassador Johnston had worked out and had distributed to the Parties. The Plan was on United States' public record and Criddle had a copy of it.[170] The history of the Johnston's negotiations and the measure of technical agreement reached was made clear in the President's 1955 report to Congress on "US Participation in the UN", pages 70-71.[171]

Additionally, Mr. Criddle acknowledged that "Israel has structures, which, if used to capacities at all times would undoubtedly exceed the 1955 proposed allocation. However, it is not feasible to use a system at 100 percent capacity continuously throughout the year, nor would the seasonal demands for water require it." Such a statement, by an engineering authority like Dr. Criddle, does not reflect absolute accuracy. Whereas the statement is true of all systems, especially those requiring periodic maintenance and on which demands have certain peaks, Criddle's report did not indicate the efficiency of the system operation that was required to keep extractions from Lake Tiberias within the limits of the Unified Plan. Moreover, the report ignored a primary objective of Israel's National Water Carrier, which was to pump water from the Lake for artificial recharge of the coastal aquifer, an undertaking that was not dependent on crop water requirements or seasonal demands, thus increasing the extraction rates from the Lake and thus increasing the total water quality extracted. In the same report, Criddle confirmed the feasibility of Israel taking Yarmouk water in winter by pumping it through the Beit Shean pipeline for storage in Lake Tiberias or for direct use in the Beit Shean valley. This idea was implemented by Israel later, and was done in addition to her abstraction of about 25 mcm of Yarmouk water during the summer season, which spanned five months of each year.

---

[169] "Jordan Valley Programs," by Wayne D. Criddle. A report on a field trip (11 pages) communicated to the State department as an attachment to a cable from the U.S. Embassy in Tel Aviv as Enclosure No. 1, Records of the Department of State, U.S. National Archives, Washington D.C.

[170] The first full version of the Unified Plan given to the author by Dr. Wayne Criddle on June, 1984.

[171] Memorandum of Conversation between Mr. Rodger Davies, Deputy Director NEA/NE Department of the U.S. State Department and Mr. Shaul Bar-Haim, Counselor, Embassy of Israel, February 18, 1963.

Criddle's report can be viewed as paving the way to a relaxation by the United States of her pressure on Israel to have her abide by the Unified Plan, and as giving technical justification to potential violations by Israel of that plan that Criddle himself helped formulate.

In a cable to the U.S. Ambassador to Israel, dated 12 April 1963, the Department of State instructed the Ambassador to assure Israel that the U.S. had the problem of the Jordan waters and preventive diplomacy in connection therewith very much in mind. "We have already had certain background discussions with Arab leaders and will take further appropriate action as critical period of initial withdrawal[172] approaches." The Department viewed Criddle's June visit as another important aspect of their groundwork.

"In the meanwhile," the cable continued, "it essential Israel keep us fully and accurately informed of diversion plans and timing. Also, as we have earlier informally suggested to Israel Embassy here, Israel can help by an occasional low-key repeat low key reference in unostentatious public statements (supported by remarks in UN corridor and appropriate few world capitals, and by articles, etc) to: (1) its belief in desirability unified equitable development of Jordan waters in manner benefiting all riparians, (2) its willingness discuss unified development with other riparians at any time, (3) its intention hold its withdrawals to level which will protect traditional usages and rights of Jewish and Arab in-basin users even though international agreement on unified development has not yet been reached, and (4) its willingness at any time to accept international observation of its Jordan water usage provided Arabs do likewise. What is needed as critical phase approaches is some sugar with this spice that will put Israel's action in best possible international light and thereby facilitate our support of it."

Israel's response came indirectly, through policy statements in three documents. The first was a radio interview of Dr. Yacov Vardi,[173] Assistant to the General Manager of the Israel Water Planning Authority, TAHAL; the second was a document on the national water plan; and the third was a pamphlet entitled "Water Development," No. 11 in the "Israel Today" series published by *Israel Digest*. These documents mentioned Israel's conforming to three of the four themes that the U.S. wanted. On 10 July 1963, Prime Minister Levi Eshkol, in a press conference, conformed to all but the fourth theme above. The fourth theme pertaining to international supervision remained unresolved. Israel's view was that such a commitment by Israel would shift the matter to the U.N. Security Council in which the U.S.S.R. would support the Arabs, and the western countries would try to seek a

---

[172] Withdrawal from Lake Tiberias through the Israeli National Water Carrier.

[173] Dr. Vardi was to play an important role during subsequent Jordanian-Israeli frictions over the Yarmouk between 1979 and 1994.

compromise. That compromise, Israel feared, would call on the delay of the diversion of the Jordan waters, a delay Israel refused to entertain.

The United States continued to follow a course of preventive diplomacy, as Israel actively was implementing her National Water Carrier and approaching the date of test-pumping of water from Lake Tiberias. The Kennedy Administration was more sympathetic with Israel's water needs and plans, which explains the relaxation of United States' pressure on Israel regarding the Unified Plan. A report by the political correspondent of the Israeli newspaper, *Lamerhav*, organ of the Coalition Ahdut Avoda, wrote, on 12 November 1963 that, "at a certain stage of the Israel-US dialogue on this question [pumping of water from Lake Tiberias], to which Israel attached great importance, an explicit promise was received from the White House itself to the effect that the United States Government recognizes Israel's needs in the sphere of development of water resources and will stand by Israel in this matter." In fact, the U.S. Government made it clear that it would not oppose, nor would it delay, the Israeli diversion.[174] Israel still insisted on the allocation figures: 70 mcm from Tiberias to Jordan and 40 mcm from the Yarmouk to Israel's Yarmouk (Adasiyya) Triangle.[175] On international supervision, Israel opposed any United Nations' involvement and proposed to have two experts appointed, one by Jordan and the other by Israel, and the two would select a third as a neutral water engineer.

10. Water in Arab Political Rivalries

The Arab countries were busy in the 1960's with their inter-fights: the Arab Ba'th party, ruling in Syria and Iraq since early 1963, was feuding with Nasser over Arab unity and competing for leadership of the Arabs; Lebanon, with the former Chief of Staff Fuad Shihab as President, leaned towards Nasser without unduly upsetting the delicate relations with Syria. Nasser

---

[174] U.S. communication to this effect was made by Mr. Roger P. Davies, Director of the Near East at the State Department on his visit to the region in October 1963. When the issue was brought to publicity through a story in *Newsweek*, it caused a stir in the Arab countries. Foreign Minister Philip Takla of Lebanon told the Foreign Affairs Committee of the Lebanese Parliament on 14 November that "Lebanese Government has received no communication to this effect". Takla was technically correct because the U.S. did not send written demarche saying that the U.S. supported Israeli diversion; but the story of *Newsweek* also was correct because Davies had communicated this stance to the Arabs in his conversation with the concerned officials. See cable No. 455, November 15, 1963 from Ambassador Meyer in Beirut to the Secretary of State, State Department Records, U.S. National Archives.

[175] Cable from U.S. Ambassador in Tel Aviv (Barbour) to Secretary of State dated December 20, 1963, No. 671, with copies to U.S. Embassy in Amman.

simultaneously campaigned against Jordan and Saudi Arabia over the bloody confrontation in Yemen.

The Arab Ba'th Party held its Sixth National Conference in Damascus during the period 5-23 October 1963. The Manifesto issued at the conclusion of the conference included resolution 18, the warning that any move by Israel to divert Jordan River waters would "constitute a direct threat to the Arab homeland... a military danger of the first order which must be resisted by force." All Arab countries, but particularly Syria and the UAR (Egypt), were called on to "face this danger."

The Baghdad press for sometime prior to the Manifesto had taken the same line, and on 4 November 1963 the Ba'thist Weekly, *Sawt al-Jamahir*, took it up again and developed it at length. According to the paper editorial, Israel intended to reclaim the Negev desert in order to house three million Jews there. This would plan constitute an intolerable military danger for the Arab countries. The paper said all Arabs must "discard our wrangling and disputes and consider the Palestine issue seriously." It proposed specifically that the Arab states face Israel by:

"1. Preventing the changing of the course of the River Jordan by every means and under any circumstances,
2. Continuing to deprive it of access to the Suez Canal,
3. Closing "Israel's vital gateway - Elat"; Jordan and Saudi Arabia particularly were called on to prevent Israel from using the Gulf of Aqaba,
4. Forming an armed Palestinian Liberation Front,
5. Strengthening the economic boycott of Israel and "*ignoring*"[176] Johnston's American plan for the distribution of the Jordan waters."

It is noteworthy that the editorial of the Ba'th paper did not reject the Johnston American Plan, but called for ignoring it.

The American preventive diplomacy approach played down the impact of the Jordan waters diversion on Israeli population growth. "There is a general feeling, at least among Israel's neighbors," noted one American report,[177] "that withdrawal of 320 mcm of Jordan basin water allocated to Israel under the Unified (Johnston) Plan would radically increase the country's water consumption, and vastly enhance its ability to absorb immigrants. Corollary and equally erroneous suppositions are that if Israel's diversion project were blocked, immigration would have to be slowed down and the Negev could not be settled".

"In fact," the report continued, "there is only a tenuous relationship between immigration and water, since future development of Israel is

---

[176] Italics inserted by author.
[177] Cable from U.S. Embassy in Tel Aviv dated 23 October, 1963 to Department of State.

expected to be industrial - using small amounts of water - rather than agricultural, where vast amounts of water are required. To put it otherwise, the future population level of Israel will be determined by factors other than the water obtained under the Unified Plan...... Israel fresh water potential, including the Johnston allocation, is estimated at 1,800 mcm's. Of this, about 1,200 mcm's are now [1963] being utilized. Thus, the full implementation of the Tiberias withdrawal scheme would increase Israel's now available water supply by roughly 27%. However, other water resources will have been developed before the withdrawal of Jordan's basin water reaches 320 mcm's. That the Johnston's allocation represents approximately 18% of the country's water potential may render the project somewhat less frightening to Israel's neighbors."

The United States estimate on timing of the pumping commencement was mid 1964.[178] Full use of the pumped water necessitated blending with fresher water to control salinity, and therefore the full capacity of the project would not be reached before the diversion of the saline springs[179] was completed, and control of salinity levels in Lake Tiberias waters became easier.

The above did not allay Arab fears, which were based on political rather than technical grounds. Much of the high pitch had to do with rivalries among the Arab ranks, and with the competition for championship of the central Arab cause, the Palestine problem. In a conversation on 29 October, in Beirut, between the U.S. Ambassador, Mr. Meyer, and the Ambassador of the UAR, Mr. Abdul Hamid Ghaleb, the latter registered real concern over the above Ba'thist resolution concerning Israeli diversion of Jordan waters. Ghaleb considered it "cheeky" for Syrians and Iraqis (each ruled by the Ba'th Party) to determine to take military action and call on UAR to join; clearly putting Nasser on spot. Ghaleb noted the story carried in the Lebanese press, reporting that the Israeli Government had notified great powers it would consider Arab interference in water diversion as aggression and would respond militarily. Meyer disclaimed knowledge of such Israeli *démarche*, and replied with the hope that Israel's water development program would

---

[178] Based on Criddle's June 1963 report above, Appendix B- Timing of Construction, "the following schedule is the earliest dates of withdrawal from the lake through the new pumping stations:
1963 (Late Fall) Some testing of lines, no appreciable quantity will be pumped.
1964 (Beginning about March 1) 150 mcm 250 mcm 320 mcm.

[179] These are saline springs discharging into Lake Tiberias from the west. Some emerge below the water surface in the Lake. Israel had a project to redirect these springs away from the Lake and discharge them into the Lower Jordan. This project was completed and was an item on the water negotiations agenda between Jordan and Israel, during the Peace Process.

proceed according to schedule. Meyer concluded from the conversation that "Ghaleb obviously hoped some means can be found for keeping UAR out of hostilities which Syrians were apt to provoke."[180]

The early appraisal of the U.S. State Department[181] was that the Arabs would most likely "blow off steam" in the U.N. framework and use the U.N. as an excuse for non-recourse to promised military action. The Department viewed "that the Arab riparians' progress on their own water development projects utilizing equitable shares of the Jordan system would make Israel's project less a focus for their attention and frustration. It is for this reason [but obviously not this alone] that we have generously supported Jordan's East Ghor project." The U.S. supported the Jordanian-Syrian joint effort to have a dam built at Maqarin, albeit at a lesser storage capacity than the total (475 mcm) envisaged in the contract awarded to the Yugoslav firm, Energoproject. Jordan was eager to secure Arab involvement in financing this undertaking to, avoid any notion by others that she alone was going along with the Unified Plan.

Many other issues emanating from the Israeli diversion of the Jordan waters were particular to the Hashemite Kingdom. Most important was the salinity problem of the lower Jordan. While some lands irrigated by the river could have their irrigation source switched to the Yarmouk via the on going project of the East Ghor Canal, others were beyond the command of the new project and would have to be treated differently.[182] Many of the farmers so dependent on the river and away from the reach of the Canal were on the West Bank of the Kingdom. Criddle addressed those problems in his trip report, and proposed that Jordan be given her water allocation from Lake

---

[180] See cable no. 381, dated October 30, 1963, 1 p.m. from Meyer in Beirut to Secretary of State. Department of State Records, U.S. National Archives, Washington D.C.

[181] See cable from State to the U.S. Mission at the U.N., no. 3247 dated 24 June 1963.

[182] The areas in the Hashemite Kingdom that pumped water from the Jordan River for irrigation were, according to Criddle's June 1963 Report as follows:

West Ghor Lands	
Opposite Stage 1 of the East Ghor Canal 14,700 dunums	72 pumps
South of Stage 1 14,500 dunums	104 pumps
East Ghor Lands	
Opposite Stage 1 19,500 dunums	118 pumps
South of Stage 1 19,800 dunums	109 pumps
68,500 dunums	403 pumps

Tiberias by releasing it in the lower Jordan, storing it in the Rutenberg pool, and subsequently pumping it to the East Ghor Canal. The Canal can then be extended to reach those areas in which Jordan River water would become too saline for irrigation. Causes of salinity in the lower Jordan were the Jordanian progressive diversion of the Yarmouk waters and the impending Israeli diversion of the Jordan waters.

In political terms, Jordan's action, in the view of the State Department, could be cast as insistence that there be no interruption of the waters from Tiberias traditionally used by Arab in-basin users. Therefore, there would be minimum political vulnerability and, in fact, advantage in Jordan demonstrating that it would brook no interference with this traditional usage. This reasonable option of releasing Tiberias water to Jordan as recommended by Criddle, was never implemented.

In view of the direct adverse impact Jordan would sustain from the diversion of the Jordan waters by Israel, "its political reaction is likely to be more pronounced and shrill than that of the UAR," said a cable from the American Embassy in Amman. "However," continued the cable, "except during the most acute crises caused by other events, it appears unlikely that Jordan would actually consider the start of pumping from Tiberias to be a casus belli against Israel. This should be true despite public statements already made by Jordanians in official positions that war would be the answer. This relatively optimistic assessment can, of course, prove inaccurate if the Government of Jordan is under particularly strong internal political pressures at the time the pumping, either experimental or operational, commences. If, however, one of the other Arab states such as Syria, inadvertently or otherwise, involved itself militarily with Israel on this issue, Jordan would almost certainly have to come to that Arab state's assistance."[183]

11. The Adasiyya Diversion Weir

At the time the frenzy of Jordan River diversion was on, the U.S. Embassy in Amman recommended to the Department of State, for political reasons, not to sound supportive of the Maqarin Dam, nor to be involved in its financing. Their reasoning was that important elements in the Arab world, in view of the U.S. association with Israel, were likely to interpret such an action as "payment" or "blood money" for Arab restraint in the face of what they consider a direct Israeli provocation (prompted by her diversion scheme). Another important related structure, the diversion weir across the Yarmouk at the Adasiyya off-take to the East Ghor Canal, had not been constructed either. In his field report, Criddle stated that the diversion weir would be essential for

---

[183] Cable from the U.S. Embassy in Amman to the State Department, no. A-151, dated September 25, 1963. State Department Records, U.S. National Archives, Washington D.C.

Jordan's project. In addition to the diversion weir, he was in favor of building a "lip" across the Yarmouk downstream, so that the water level could be raised for the Israeli pumps to abstract water for the Yarmouk Triangle. The Israelis had piled some rocks across the river to the midstream and relied on such a temporary arrangement to assist in pumping Yarmouk water. The Israelis raised the matter in the Mixed Armistice Commission (MAC), in 1962, and the Jordanians had approved to have Israel build a dike to midstream. As a result, the Israelis constructed a rock dike across the river contrary to what the Jordanians approved. The Jordanians viewed that as an expression of bad faith and refrained from considering its reconstruction in subsequent dry seasons. The Jordanians refused a bank-to-bank Israeli dike on the grounds that such a dike would give Israel a military advantage. Actually, no vehicles could possibly be driven on such a narrow rock dike, and it would be washed away by the first flood, to be rebuilt later before the following dry season. The more likely reason behind Jordn's refusal was the fear of being accused by other Arabs of cooperation with Israel.

When the Jordanians wanted in 1963 to build a "lip" across the Yarmouk to facilitate water diversion to their canal from the river, the Israelis demanded that they be consulted because the opposite bank of the river was part of the southern demilitarized zone between Israel and Syria, although it was under defacto Syrian control. The Israelis were hoping for a trade-off whereby the Jordanians would acquiescence to the construction of a lip across the Yarmouk, at the location of their pumps downstream, in return for a no-objection by Israel to have the Jordanians build a lip/diversion structure across the Yarmouk at their diversion point. Again, such a structure was not built, and was to be a reason for subsequent confrontation a decade and a half later.

12. Summary of the Jordan Waters' Problem

An attachment to a cable from the U.S. Embassy in Beirut to the Department of State, towards the end of 1963, provided an appraisal summary of the Jordan waters. It addressed three topics in relation to that issue as quoted below:

"A. PEACEFUL RESOLUTION OF PROBLEM IS POSSIBLE - E.G., UNIFIED DEVELOPMENT PLAN.

1. Agreement was reached with Arab technical experts in 1955.
2. First priority: provide adequate water ior all irrigable areas of Jordan Valley.
3. Adequate allocations for Syria (132) and Lebanon (35 mcm)
4. Each riparian can use its allocation wherever it wishes.
5. All but 10% of plan can be implemented without co-operating with Israel.

6. The 10% represents 100 mcm, which Israel must deliver to Jordan.
7. An international water-master is envisaged.
8. U.S. is until now holding Israel to Unified Plan allocations.
9. Key Arab need is storage dam on Yarmouk.
10. Arabs can publicly say they are protecting Arab interest.

B. PROBABLE RESULTS OF REFERING PROBLEM TO UNITED NATIONS
   1. Unlike 1953 affair, this is not a violation of Demilitarized Zone.[184]
   2. World community will undoubtedly recommend direct negotiations.
   3. Arab case will be poor one since equitable solution, which was agreed upon by Arab technical experts, is available.

C. PROBABLE RESULTS OF ARAB RESORT TO MILITARY ACTION
   1. Israel's military capability is adequate to repel attacks.
   2. In any case, United Nations and the United States will intervene:
      a. Militarily to stop Arab aggression,
      b. Politically along lines of B above.
   3. Israelis will probably occupy West Bank of Jordan River.
      a. UN and U.S. will probably compel Israel to withdraw,
      b. But only after Israel gains concessions ala Sinai.
   4. As happened with 1947 United Nations resolution on partition, Israel will no longer feel bound by Unified Development Plan allocations.
   5. Thus once again Arab extremists will cause Arabs to lose ground.

The appraisal was proven mostly accurate after the June war of 1967 in which Israel dealt a devastating blow to the surrounding Arab countries and occupied the West Bank, Sinai, and the Golan. Israel became no longer bound by the allocation under the Unified Plan. She even unilaterally annulled the Armistice Agreements she had signed in 1949 with Egypt, Lebanon, Jordan and Syria. Israel withdrew from Sinai only after a Peace Accord was signed with Egypt in 1979, as shall be seen later.

---

[184] Refers to Israel's off-take structure at Tabgha on the northwest shore of Lake Tiberias, inside Israel, as opposed to the previous intake at Jisr Banat Ya'qub that was abandoned.

CHAPTER V

Spinning Out of Control

The efforts of Ambassador Johnston culminated in an agreement with the Arab Technical Committee (Egypt, Syria, Lebanon, and Jordan) on the basics of the Unified Development of the Jordan Valley. The Committee submitted its recommendation to the Council of the Arab League to approve the Plan. Ambassador Johnston had the support of most of the Arab leaders involved.[185] When the time for decision came, however, the Palestinian question and related inter-Arab politics, exacerbated by frequent changes in governments (especially Syria and Jordan) and overbidding for championship of Arabism, all helped to promote indecisiveness in the ranks of the Council of the Arab League. The Council, instead of accepting the recommendations of the Technical Committee, decided to refer the matter back to it for further study. The Council, after intervention by the Syrian Prime Minister, favored an arrangement whereby the Arab riparian states could utilize the waters of the basin without a need to include Israel in a "Unified Plan". The lack of a positive decision on the part of the Arab League Council provided Israel with the justification to proceed unilaterally with her diversion project.

1. The Arab Response to the Israeli Project

In early 1956, the Prime Minister of Israel, David Ben Gurion, announced that Israel would resume work on her diversion project in the demilitarized zone with Syria upon the approval of the Chairman of the UNTSO. Syria was quick to remind the Chairman of UNTSO, in a memorandum dated 22 February 1956, of the decision, and still in effect, of the Security Council taken on 27 October 1953, in which it confirmed the suspension of the diversion works until the Security Council resolved the issue. The Security Council, Syria contended, and not the Chairman of UNTSO, was therefore the competent party to decide on the resumption of the works.

The political developments in the Arab East took a distinct turn after the Suez Campaign. Jordan's King Hussein, through his personal courage and initiative, aborted an attempted military coup in April of 1957. Jordan's relations with Egypt and Syria deteriorated thereafter. In February of 1958, the United Arab Republic emerged through a union between Egypt and Syria

---

[185] Such support was accorded during meetings of Ambassador Johnston with the leaders of the countries concerned. Support was reported given by President Jamal Abdul Nasser of Egypt, King Hussein of Jordan, President Camille Chamoun of Lebanon, and the Prime Minsiter Fadhil Al Jamali, of Iraq.

and was soon followed by the Arab Union between Iraq and Jordan. Civil disturbances in Lebanon peaked as President Camille Chamoun, approaching the end of his term in office, had plans to extend his tenure for another term. Developments there accelerated and became rather bloody. The U.S. Marines landed in Beirut to maintain law and order. On 14 July 1958, in the early hours of the morning, a brigade of the Iraqi armed forces was passing through Baghdad on its way to help stabilize the situation in Lebanon, turned their guns towards the Royal Palace, Qasr al Rihab, killed King Faysal II, his uncle Prince Abdul Ilah Ben Ali, and all the members of the Iraqi Hashemite Royal family. The officers mounted a military coup that transformed Iraq into a republic headed by their chief, Brigadier Abdul Karim Qassim. Only those members of the Royal family who were abroad escaped the massacre. The Jordanian Prime Minister of the infant Arab Union, Ibrahim Hashim, the Defense Minister, Suleiman Toukan, and other military and civilian Jordanian officials, who were on an official visit to Iraq, were also massacred. The five months old Arab Union between Iraq and Jordan collapsed. The new ruler announced Iraq's withdrawal from the Baghdad Pact, and the revolutionaries made friendly overtures to Nasser. Syria and Iraq closed their borders with Jordan, whose imports came primarily from Beirut through the borders with Syria. Britain provided badly needed assistance to Jordan by airlifting crucial supplies to her from Cyprus. The situation in Jordan became very tense; a new government was formed headed by Hazza' al Majali, and a state of emergency was declared and Martial Law put into effect. The country was run with an iron-fist policy that had state security as top priority. In Lebanon, attempts to renew the tenure of President Chamoun were frustrated and, by October 1958, Lebanon's Parliament elected the Army Chief of Staff, General Fuad Shihab, as the new president. The domestic scene there quieted down. The Iraqi revolutionary regime under Abdul Karim Qassim soon turned against Jamal Abdul Nasser of the UAR and jailed the second man in the regime, the one who had mounted the coup, Colonel Abdul Salam Arif, and who was close to Nasser. The governments of the Arab East were basically at odds with each other. Many had withdrawn their ambassadors or severed diplomatic relations with many others!

In response to the Israeli-wide publicity for the Israeli Jordan Water Project, and in the above environment of Arab disarray, the United Arab Republic (UAR), of which, Syria formed the Northern Province, suggested that a meeting be held between the Ambassadors to the Arab League of Jordan, Lebanon and the UAR to look into Israel's insistence to divert the Jordan River. The UAR initiative was made through a cable, dated 6 December 1959, addressed to the Ambassadors of Jordan and Lebanon in Cairo. Jordan responded positively the following day, on 7 December. Jordan's relations with the UAR had been aggravated since April 1957, and were further strained to the limits in the wake of the Iraqi coup on 14 July

1958. Hazza' al Majali, the Jordanian Prime Minister, publicly dared Nasser to prevent the Israelis from diverting the Jordan waters, part of which was Syria's. A propaganda war dominated the Middle East scene: the UAR against Jordan; Iraq against Jordan; UAR against Iraq; UAR against Saudi Arabia; UAR, Jordan, Saudi Arabia, and Iraq against Israel, and so on. The situation was more like an Arab cold war.

In this atmosphere of reciprocal inter-Arab animosity, and in response to the invitation by the UAR, the Arab League Ambassadors[186] of Jordan, Lebanon and the UAR met during the period 19-24 December, 1959 and issued a communique recommending that the governments of the three countries:

1. Wage an information and diplomatic campaign to mobilize public opinion and enlighten it as to Israel's intentions;
2. Request that the Council of the Arab League be convened soon to treat the case of the Jordan River diversion, with all the consequences thereto pertaining, and to take whatever necessary measures it sees fit; and;
3. Submit to the Council the studies that each government has of the technical, legal and political dimensions to help the members during their deliberations.

The Secretary General of the Arab League, in accordance with Recommendation 2 above, invited the member countries, through a memorandum dated 23 January 1960, to send delegates who should be experts in dams and reservoirs and others eloquent in the provisions of the Armistice Agreements[187] to discuss the Jordan River diversion, and to submit their studies to the League Council. A month later, the Council of the Arab League met and issued resolution 1640, dated 29 February 1960, recommending that the waters of the Jordan basin be utilized for the benefit of the concerned countries and the Palestinian Arabs. It called on the Arab states, jointly and severally, to accelerate the implementation of the necessary works. It also recommended that a special commission be set up for the purpose of coordination and follow up.

---

[186] Ambassador Mohammad Shurayqi Pasha, Jordan's Ambassador to Egypt (and the League) represented Jordan, Ambassador Dr. Najib Sadaqa represented Lebanon, and Ambassador Ahmad Abdul Magid, Director of the Arab Directorate at the Egyptian Foreign Ministry, represented the UAR.

[187] Pursuant to the cease fire in the war of 1948, four such agreements were concluded between Israel and the following countries: Egypt at Rhodes on 24 February 1949; Lebanon at Ras En Naqura on 23 March 1949; Jordan at Rhodes on 3 April 1949; and Syria at Hill 232, near Mahanayim, on 20 July 1949.

In its meeting on 28 August 1960 during the regular session of the League, the Council issued resolution no. 1696, calling for a meeting of the above commission to follow up on the works and for a meeting of the Permanent Military Committee (PMC) of the Arab League. Representatives of the new member states who were not parties to the Joint Arab Defense Treaty, concluded in 1951, were to participate in that meeting. The member states increased after 1951 with the independence of Sudan, Libya, Tunisia, and Morocco. The PMC was authorized to prepare a plan for a comprehensive confrontation with Israel, taking account of all possibilities. The Secretary General made contacts with the concerned Arab states and, by his decision of 14 September 1960, formed the Technical Committee for the Jordan Waters from representatives of the UAR, Jordan and Lebanon.[188] Through another decision, no. 124 in 1960, the Secretary General called on the Committee to meet and start its work.

The Committee met during the period 5-10 November 1960, deliberated the issue, and submitted its report in accordance with the above resolutions and decisions. The report of the Technical Committee reviewed the Israeli diversion project and contained the following proposals that formed the essence of an Arab Plan:

"A) In the Republic of Lebanon:

Diversion of the Hasbani Waters

The Hasbani has an estimated annual discharge of 157 mcm. The majority of it can be utilized as follows:

1. Construction of three storage dams on the river course:
   - The Yabisa dam near Rashayya with a storage capacity of 20 mcm and a height of 40 meters,
   - Wadi Hosh Daniyeh dam with a capacity of 25 mcm and a height of 50 meters, and,
   - Suq El Khan dam with a capacity of 60 mcm and a height of 70 meters.
2. Construction of a 4750 meters tunnel to connect the Hasbani with the Litani with a diversion capacity of 75 mcm and a potential to generate 15 thousand kilowatt.hr per year.
3. A pumping station on the Wazzani to irrigate 1200 hectars in Lebanon.

[188] Engineer Sweilem Haddad represented Jordan on this committee, Dr. Mohammad Ahmad Salim (Egypt) and Dr. Subhi Mazloum (Syria) represented the UAR, and Engineer Mohammad Al Ghaziri and Joseph Amasian represented Lebanon. The representatives of the UAR were the same that worked on the Committee during the Johnston shuttle diplomacy.

The proposals were based on a preliminary study and warranted conducting detailed studies to better define the framework of the project.

B) In the Northern Province[189] of the UAR

<u>Diversion of the Banyas</u>

The Banyas springs outflow from a point close to the southern borders of the Syrian Province at elevation 340 m.a.s.l. Some of its waters are used to irrigate about 600 hectars on the right [western] bank of the river, and the remainder flows to the occupied territories. The river discharge is about five cubic meters per second [cms] in winter and between 1.0 and 1.5 cms in the summer.

Syrian lands on the right bank could be irrigated from the Wazzani springs in Lebanon with Lebanese Government approval. The idea is to divert the entirety of Banyas southwards via a canal extending parallel to the Armistice line for a distance of about 40 km to the Hasbah area overlooking plains that connect with the Boteiha area. The canal should extend southwards to convey excess water to one of the Yarmouk tributaries. In case the Wazzani springs are used to irrigate Lebanese lands only, part of the Banyas waters would be conveyed to irrigate the Syrian lands on the right bank lying between the Lebanese borders and the Armistice line. The diversion canal should have a capacity to carry a discharge of 5 cms to carry the Banyas spring flow plus whatever spare water from the Wazzani springs and the flow of any cross wadis on the way.

<u>C) In the Hashemite Kingdom</u>

The on-going East Ghor Project, implemented in three stages, is meant to irrigate 12 thousand hectars in the East Ghor. It represents the first step towards the implementation of the Great Yarmouk Project to irrigate 52 thousand hectars in the Hashemite Kingdom. The on-going project does not interfere with the demilitarized zone (between Syria and Israel) and all of its structures are located in Jordan. Substantial benefits will accrue to the Kingdom upon completion of the on-going project by the end of 1963."

"The above various projects," wrote the Committee in its 10 November 1960 report, "aim in general at utilization of the Arab tributaries to the Jordan River in Arab lands to prevent their utilization in the occupied part of Palestine." However, since these projects are long term, expensive, and warrant technical studies and research that could take a long time contrary to

---

[189] "Al Iqleem Al Shamali" (the Northern Province) was the official name to Syria as part of the UAR. Egypt was referred to as "Al Iqleem Al Janubi" (the Southern Province).

the current status that requires an immediate answer to the attempt by Israel to divert the Jordan River, the Committee recommends that the following urgent measures be implemented immediately:

"In the Republic of Lebanon:

Essentially listed the above measures outlined for Lebanon to divert the Hasbani and the Wazzani springs, except that one dam at Suq El Khan was to be implemented among the immediate measures.

"In the Northern Province of the UAR

The above measures outlined for the UAR to divert the Banyas.

"In the Hashemite Kingdom of Jordan

Speed up the implementation of the East Ghor Canal Project to divert the Yarmouk and prevent the flow of water to the existing pump stations in the Occupied Area[190], whether they are in the Yarmouk Triangle or in the western bank of the Jordan River."

The Committee estimated the flow that can be diverted in accordance with their plan as follows:

Hasbani (not including Wazzani)		110 mcm
Wazzani		30 mcm
Banyas		110 mcm
	Total	250 mcm[191]

The Committee further recommended that the concerned Arab governments be requested to:

"1) Implement the immediate projects as recommended in response to the Israeli aggression on the Jordan River waters and its diversion of these waters outside the basin.

2) Simultaneously resume and complete the studies and research necessary to utilize Arab waters inside Arab territories.

3) Determine the extent to which each country can utilize its waters inside its territories, with due consideration given to the needs of Jordan as a result of the damages that will result from the Israeli aggression.

4) Supply the Technical Committee with the results of the studies and research referred to under (2) above, so that it will be enabled to

---

190 This terminology was used to refer to the territories that comprised Israel.

191 These estimates must be over-optimistic. The flood flows of the rivers could not all be diverted. The Committee admits that its figures are preliminary and that more detailed studies are warranted.

coordinate between the various projects, and present the necessary recommendations regarding them."

The delegate of Jordan expressed reservation regarding the above recommendations; the reason was that these individual projects did not emphasize the compensation of the Hashemite Kingdom for the water that she would lose as a result of the Israeli diversion of the Jordan River. He presented a "Unified Arab Project" to utilize, in an integrated way, the waters of Hasbani, Banyas, and the Yarmouk for the benefit of Arab countries.

The Jordanian Proposal

The Jordanian proposal for a "Unified Arab Project"contained the measures specified above for Lebanon and the Northern Province of the UAR. It called for the construction of a dam on the Hasbani as proposed in the Main/TVA Plan of 1953, and for the diversion of the excess water in a canal towards the Banyas. No portion of the Hasbani would be diverted towards the Litani in Lebanon. The Banyas itself would be diverted by a diversion dam at elevation 300 m.a.s.l. via a southeasterly canal with a carrying capacity of 7 cms, to accommodate the additional water diverted to it from the Hasbani. The canal would discharge into one of the tributaries of Wadi Raqqad and thence flow towards the Yarmouk and be regulated by a dam at Wadi Khalid. The Yarmouk itself would be regulated by two dams: one of 118 meters height at Maqarin, and another of 93 meters height at Wadi Khalid. A diversion dam would be built across the river at Adasiyya to divert the waters to the East Ghor Canal and irrigate the Jordan Valley in the Hashemite Kingdom. The Yarmouk arrangements were essentially based on the works of Baker and Harza (1953-1955). See Figure V.1 for a schematic of the Unified Arab Project.

The Jordanian proposal would eliminate the need to store water in Tiberias for Jordan's use, and would dispense with the reverse conveyance of water from Tiberias to Jordan. It would also eliminate the need for international monitoring and controls as proposed in the Unified Plan. There would be no project component outside Arab lands, thus avoiding political complications that would emerge if project components were located inside Israel. The proposal claimed that it would secure to Jordan the same amounts of water that the Unified (Johnston) Plan would bring to it, or a bit more. It estimated that the excess water that could be diverted to Jordan from the Hasbani and Banyas amounted to 140 mcm.[192] The proposal would also deny Israel any waters from the Hasbani and Banyas,[193] and would thus abort the Israeli diversion project. Even if the Israeli project were implemented, it

[192] Estimates were preliminary and were on the optimistic side.

[193] Flood flows would continue to flow in the Banyas, down to Lake Tiberias. The technical feasibility of a dam on the Hasbani was never established.

Figure (V.1)- The Unified Arab Project

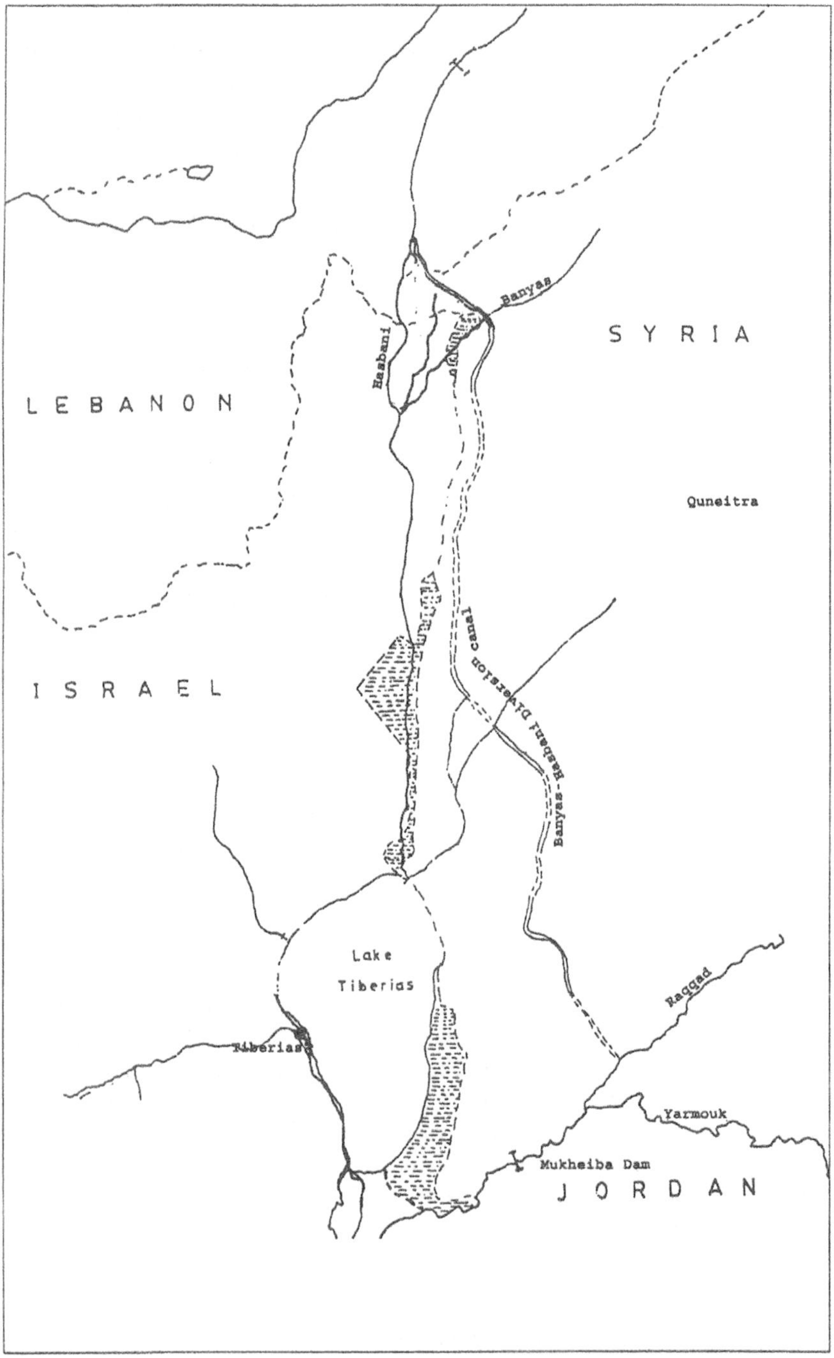

would not have any effect on the Arab Project, which was located in its totality upstream of any Israeli intake point.

Compared to the proposals advanced by the delegates from the UAR and Lebanon as outlined above, the Jordanian project, in essence, added the dams on the Yarmouk, and was not in favor of any pumping of the Wazzani springs, nor of the diversion of any part of the Hasbani to the Litani; instead, the excess Hasbani water was to flow towards the Yarmouk along with the excess Banyas waters for Jordan's benefit.

The Political Committee of the Arab League approved, in its meeting in Baghdad between 30 January and 4 February 1961, the reports of the Technical Committee and of the Permanent Military Committee (PMC) and asked that the Consultative Commission of the Arab Defense Council, along with the Chiefs of Staff of the new League members, to meet and deliberate on the military position and possible military action to prevent the Israelis from diverting the Jordan River. The Consultative Commission (composed of the Arab Chiefs of Staff) met in the Arab League Headquarters in Cairo on 22 April 1961 to discuss the matter, and prepared recommendations for action by the Joint Arab Defense Council. The Defense Council met on 10 June 1961 and decided to implement the utilization of the Jordan waters by the concerned Arab countries. The Defense Council authorized the PMC to implement those decisions. The Secretary General of the Arab League advised the Jordan Government that the UAR and Lebanon had expressed readiness to implement the scheme for the diversion of the Banyas and the Hasbani; that the technical preparations were completed, but the two governments were looking for the material support.

In honor of the decision of the Council of the Arab League, Jordan proceeded, and diverted the flow of the Yarmouk, through a carefully designed drop inlet, into a 1 km tunnel and on to the East Ghor Canal. The diversion structure protruding to the mid stream of the Yarmouk was never built because of Israeli objections. Thus, the East Ghor Canal Project originally conceived along the lines of the "Lahoud Plan" in partial and staged implementation of the Unified Plan, ended up to be a project demanded by the Arab League Council and was the only component of the Arab Plan that was implemented.

2. The Arab Cold War Heats Up

In July of 1961, allocations were provided in the budget of the Northern Province of the UAR to implement the Banyas diversion. However, in September 1961, a group of military officers headed by General Abdul Karim Nahlawi staged a coup in Damascus and announced the separation of Syria from the UAR. Nasser's immediate response was to send military forces to subdue the rebellion, but he soon changed his mind and accepted the facts on the ground. Conservative Syrian politicians[194] ascended to power, and a wide rift developed between the two former partners to the UAR, Syria and Egypt. The priorities of the new regime in Syria were revised and the Banyas diversion plan was shelved before any implementation commenced.

Jordan, feeling the forthcoming danger of the Israeli project, tried to have the Arab joint effort resurrected after the setbacks in the aftermath of the disintegration of the UAR. Jordan's Foreign Minister cabled the Secretary General of the Arab League and invited him to Jordan to initiate preliminary talks on the Jordan waters, which could be expanded to include all the member states. The cable was followed by a Jordanian memorandum in which Jordan asked the Secretary General to explicitly solicit the opinions of the member states regarding the Israeli diversion project, and asked that collective Arab efforts be exerted to prevent it.

The Assistant Secretary General of the League, Sayyed Nawfal, paid a visit to Jordan and discussed the matter with the Jordan Government. He proceeded to visit the other two concerned Arab riparian states, Syria and Lebanon, and discussed with high officials the topics he had discussed with the Jordan Government. Finally he met with the Foreign Minister of the UAR (composed of only Egypt then). In his letter to the Jordan Government, which formed his response to her memorandum, the Assistant Secretary General wrote:

"The Governments of the Syrian Arab Republic and of Lebanon assured their full preparedness to implement their shares in diverting the rivers of Banyas and the Hasbani, and to secure the necessary finances and allocate them in their respective budgets. As for the technical preparedness, they expressed that it has been completed, and what remains is the floating of the works in a tender after the Joint Arab Defense Council so decrees.

The opinion was shared among all [parties], that the subject of the Jordan River diversion is considered fully dealt with technically and militarily by the Arab League. What we need is to commence the implementation of the decisions of the Joint Defense Council taken in June of 1961. The implementation of these decisions is dependent on the formation of the Joint Arab Command, and for the governments of the member states to prepare the suitable atmosphere to cooperate in this field, like the atmosphere that prevailed last June."

The response of the Assistant Secretary General spelled out indirectly the differences that befell the concerned Arab states (Egypt, Jordan, Syria and Lebanon), as well as other member states, for reasons explained in the above.

The Israeli diversion project put the Arabs in a serious dilemma. They had not opted to accept the "Unified Plan" at the political level in 1955, and they followed a path aimed at the utilization of the Jordan tributaries

---

[194] Leading conservatives included Ma'arouf Dawalibi and Ma'moun Kuzbari.

without any contact or cooperation with Israel. This path provided Israel with the justification to do the same in response, i.e., utilize the Jordan waters away from any contacts with or approval by the Arabs. However, through conducting diplomatic dialogues with the United States, Israel was able to gain the political support of the important world capitals. It even convinced the United States to provide a loan for the implementation of her project.

In their own response to the Israeli project, the Arabs had to act collectively and to rely on the provision of Arab military protection for their diversion works. In view of the complications that prevailed in their ranks, the riparian Arab states (Lebanon, Syria, and Jordan) had no trust of Egypt and would not place on their territories any forces under Egyptian command to provide the protection. Not that the three Arab riparian states were in alliance with each other for they, in turn, had their own political and ideological differences. Subsequent developments would aggravate this situation.

In Yemen, a military coup toppled the long rule of the Hamid Eddin family in September 1962. The ruler, Imam el Badr, escaped and took refuge next door in Saudi Arabia, from where, he directed the forces of the Royalists who remained loyal to him. The leader of the coup, Colonel Abdallah el Sallal, asked Nasser for military support to face the Royalists who were supported by Saudi Arabia. Soon, Jordan joined the line-up against the revolutionary regime and supported the Royalists. The internal strife in Yemen would commit a sizeable Egyptian military force and civil servants there, and caused friction within Egypt between the proponents of intervention in Yemen and those against it.

In Iraq, the Ba'th party seized power on the 8 February (14 Ramadan) 1963, through a coup that ended the rule of Qassim and had him executed by a firing squad, and appointed Abdul Salam Arif in his place. In Syria, the Ba'th party seized power exactly a month later, on 8 March 1963, through a military coup that toppled the "secessionist" elements responsible for disintegrating the union with Egypt in 1961. The Ba'th regimes in both countries concluded a union agreement with Egypt's Nasser on 17 April 1963, only to abort in July. A military and "popular" revolt in Syria led by a Nasserite Colonel, Jassim Alwan, took to the streets of Damascus in mid-day on 18 June 1963. The reaction of the Ba'thist government was in kind and there were death causalities, something that rarely happened in the Syrian tradition of coups. Nasser broke with the Ba'th regimes on 22 July, in his speech on the eve of the eleventh anniversary of the 1952 Egyptian revolution. The Iraqi Ba'thists got into internal competition for power that made it easy for the non-Ba'thist President, Abdul Salam Arif, to engineer a peaceful takeover of power on 22 November 1963, and the Ba'thists in Iraq went underground, only to resurface in July 1968 when they seized power again.

By the end of 1963, more Arab states were simultaneously at each other's throats than ever before. Syria was feuding with Egypt after the collapse of the Unity Accord in July and, since Arif's take-over in November, with Iraq. Egypt and Saudi Arabia were locked in a struggle for the future of Yemen, where 40,000 Egyptian troops had failed to win a final victory for the revolution there. Syria deemed it politically correct to be unfriendly to Jordan, who was classified as belonging to the Arab "reactionary camp", as opposed to the Arab "progressive camp" to which Syria belonged, and was among the first states to recognize Syria's secession from the UAR in 1961.[195] She also deemed it ideologically fashionable to be unfriendly to Morocco, and was trading complaints with Lebanon over border incidents. Algeria, newly independent from France, had come to blows with Morocco over a border dispute and had another dispute with Tunisia. Tunisia and Morocco had been cool to each other ever since Tunisia had recognized the independence of Mauritania. Egypt was hostile to Jordan as well as Saudi Arabia, as a matter of ideological principle and because of the dispute over Yemen, and took the side of revolutionary Algeria against her more conservative neighbors. Of the thirteen member states of the Arab League, only three were on satisfactory terms with everyone: Kuwait (toward whom Qassim's successors in Baghdad had relented), Sudan, and the Kingdom of Libya.[196] It was hard at the time to keep track of which Arab country severed diplomatic relations with which other Arab country; and was not easy to guess at who took care of the interests of one feuding party in the country of the other.

The worsening in relations was reflected in the various ties between these countries, and university students were affected subtstantially. They drifted with the ideological currents that supported one Arab country against another, and waves of students were deported from one Arab country to another because of such allegiance. In short, not only the ruling class was affected by the feuds, but also every aspect of bilateral relations was adversely affected. Families of university students were worried about their sons and daughters, and it was not unusual for governments to apprehend people without charges or to deport them.[197]

---

[195] In Nasser's speech to the nation on 30 September, he attacked the four states that had already extended recognition to the new regime in Syria: Jordan, Iran, Taiwan and Guatemala.

[196] Kerr, Malcolm, "The Arab Cold War (1958-1967)," New York: Oxford University Press, 1967, p. 127.

[197] On July 23, 1963, the Egyptian Secret Service apprehended over two dozen Jordanian and Iraqi graduate students, in Cairo and Alexandria, and had them deported by passenger ship, Esperia, from Alexandria to Beirut on 26 July. Staff of the Iraqi Embassy in Beirut was at the port to receive them.

The majority of the inter-Arab quarrels pitted revolutionary against conservative or moderate regimes: Egypt, Algeria, Iraq, the Republic of Yemen, and Syria, variously, against the others. But of all disputes, the most bitter and least soluble was between the rival revolutionary regimes of Damascus and Cairo. The Arab League, under such circumstances, was paralyzed. Israel, on the other part, was getting stronger with support from the West, and was moving fast to complete implementation of her diversion project and commence pumping Jordan waters from Lake Tiberias.

3. Arab Efforts Resurrected: The Arab Summit

At the end of 1963, at the climax of Arab quarrels, a turn of events took place. In mid-December, the Cairo press was exchanging the usual insults with Damascus, Amman and Riyadh, and Damascus radio was exchanging similar insults with Amman and Riyadh. On 17 December, the Cairo weekly, *Rose al-Yusif*, published an article containing two principal points. The first was that "the United Arab Republic will not let itself be pushed into a battle with Israel before the attainment of unity among all the Arab countries." The second point, which seemed to contradict the first, was that the UAR "knows how and when it will eliminate Israel, and she knows herself to be capable of shouldering this burden by herself."

The reaction from Damascus was furious. "*Rose al-Yusif*, the nationalized Egyptian Magazine," wrote the al Ba'th Newspaper, "had committed a great crime which even the rotten reactionaries, the stepchildren of imperialism never dared to commit."[198] Nasser was compared to marshal Pétain who had surrendered France to the Nazi invader in 1940, and whom, it was pointedly noted, the French people had punished without regard for the glory he had previously earned at Verdun,[199] an implicit reference to the glory Nasser had brought to the Arabs at Suez.

Nasser responded in a public speech at the city of Port Said. After routine castigation of Ba'thist "immorality" and "arrogance", he turned to the subject of Palestine.

"In order to confront Israel, which put a challenge to us last week, and whose Chief of Staff stood up and said, '*We shall divert the water against the will of the Arabs, and let the Arabs do what they can*'[200], a meeting between Arab Kings and Heads of State must take place as soon as possible, regardless of the strives and conflicts between them. Those with whom we are in strife, we are prepared to meet; those with whom we have quarrel, we are ready, for the sake of Palestine, to sit with.... We will sit and talk seriously at the

---

[198] "*Al-Ba'th*" newspaper, Damascus, Syria, 19 December 1963.
[199] Ibid, 23 December 1963. Text quoted from Kerr's, the Arab Cold War 1958-1967'
[200] Italics inserted by author.

meeting, and it will be no shame if we come out and say that we cannot today use force. We will tell you the truth, we will tell you every word that was said, that we cannot use force today because our circumstances do not allow us; be patient with us, the battle for Palestine can continue, and *the battle of the Jordan is part of the battle of Palestine.*[201] Or we may say that we will be able, if they divert the waters of the Jordan, to stop this diversion by force. But we will not say one thing behind closed doors and another thing outside... For I would lead you to disaster if I were to proclaim that I would fight at a time when I was unable to do so. I would not lead my country to disaster and would not gamble with its destiny.

Let us forget all the stupidities and irritations which we have seen in the past few years; also the disputes that took place, the words that were spoken, and the treachery and so on."[202]

The next day the Syrian Revolutionary Command Council issued a reply. It complained that Abd al-Nasser's speech had been "dominated by vituperation" and, in return, reminded him of some of his latest sins, but it did add that co-operation in the face of Israeli diversion plans was a national duty and that, therefore, it would accept the invitation. It added its hope that propaganda campaigns could be terminated.[203] This latter request was quickly fulfilled: within a few days the Cairo press and radio, scarcely pausing for breath, switched to describing the personal virtues and accomplishments of the men they had for so long been reviling.[204]

The Arab Kings and Heads of State met in Cairo, in the first Arab Summit Conference, between 13 and 17 January 1964. Arab Summits since then became a tradition, broken only at times of severe inter-Arab quarrels. The catalyst that brought the Arabs together was the Israeli project of diversion of the Jordan waters. This project was the immediate reason to summon the Summit. Egypt had other reasons that made the Summit a convenient political vehicle. The results of the propaganda war with the "reactionary" regimes in Jordan and Saudi Arabia, and with the "revolutionary" regime in Syria, yielded more headaches for all than benefits to any. Cairo started to realize that positive influence over other governments' policies might be obtained more easily by conciliation and tolerance than by threats and confrontation. Then, there was also the continuing stalemate in Yemen and its very negative impact on an ailing Egyptian economy. One of the major results of the Cairo Summit was the resumption of Saudi-Egyptian cordiality and the initiation of a serious effort to negotiate a compromise

---

[201] Italics inserted by author.

[202] "Nasser's Speeches," text reproduced from Kerr's, "The Arab Cold War 1958-1967", pp. 132.

[203] Text of statement in "*al-Ba'th daily*", Damascus, 25 December 1963.

[204] Kerr, Malcolm, "The Arab Cold War 1958-1967," pp. 132.

settlement in Yemen. Cordial relations were resumed with King Hussein of Jordan. However, relations between Egypt and Syria remained cool and diplomatic relations, non-existant since the secession, did not resume despite Syria's willingness to have them re-established. Relations between Syria and Iraq maintained their prevailing level of hostility.

The Summit decided on several important matters: first, a plan to divert the upper tributaries of the Jordan River would be implemented, essentially along the recommendations submitted to the Council of the Arab League by the Arab Technical Committee in 1960. To that end, a special fund would be set up at the League to supervise the plan, and a budget of $17.5 million would be allocated for its execution. An institution would be formed and charged with designing and supervising the construction of the needed works in Jordan, Syria and Lebanon. The new organization would be called the Jordan River Tributaries Regional Corporation (JRTRC). It was expected that the diversion would be operational within eighteen months.

Second, a joint Arab military command would be established under Egyptian leadership to defend the diversion works from possible aggression, presumably Israeli strikes. The Arab states would contribute a sum of $42 million to set up the unified command and reinforce the armies of Jordan, Lebanon and Syria. Third, the Arab leaders pledged to end inter-Arab differences and consolidate their relations. Moreover, they agreed to organize their political and economic affairs with other States in the international community on the basis of the attitude those States adopted to the Jordan waters issue and to the "Palestine Question" in general.[205]

Syria was disappointed with the decisions taken at the Summit. The regime doubted that implementation of the resolution would foil the "Zionist Project," which posed "a threat to the very existence of the Arabs as a nation." Syria, instead, wanted a decisive war with Israel.[206] To this end, she called upon the Arab governments to act immediately to establish and support a joint military command, which is "the effective means of destroying the plot to rob the Arab waters. It is the means capable of liberating the usurped homeland from the colonialist Zionist occupation." Syria, moreover, reserved for herself the freedom of action. Her objectives went beyond protection of the diversion plans to the liberation of Palestine from Zionist occupation. Syria's rulers at the time, one may recall, were the Arab Ba'th party.

---

[205] BBC, "Israeli Press Comments on the Arab Summit Conference," no. 1459, 22 January 1964, pp. A/1-2; Miriam Lowi, "Water and Power," Cambridge University Press, 1993.

[206] BBC, "Damascus Press Comment on the Cairo Conference," no. 1459, 22 January 1964, p. A/5; Ibid, p.122.

4. Appraisals and Reactions

The reaction of the Israeli press varied with regard to the political persuasion. For the most part, there was consensus on three points: first, the fact that the Summit had taken place and that it passed several resolutions unanimously constituted a victory for Nasser. Second, the Arabs were united in their hatred of Israel. The resolutions indicated that they were simply waiting for the day when they would be ready to fight Israel. Hence, the decision not to go to war was not an expression of the desire for peace.[207] Third, Syria posed the greatest concern, since it was not convinced, like the other Arab States, that it could not face Israel in a military confrontation. Israel was wary of Syrian moves that could drag other Arab States into a war with Israel.[208]

The United States saw the mere convening of the Summit as a success for Nasser.[209] It helped him mend fences with the Arab states with which he had been quarreling. He backtracked from his "unity of aim" theory of foreign relations to an earlier conceptualization of the Arab League, which became once again the instrument for cooperative action among Arabs. The Egyptian position was strengthened by clever isolation of the Syrian Ba'th rulers. Nasser's policy of containment of Israel, rather than military attack upon her, had explicit endorsement of all Arab leaders with the exception of Syria. The Summit improved Egypt's relations with Tunisia and Morocco and won deeper commitments of North African states to the Palestinian cause.

The American appraisal did not assume that recovery of Egyptian leadership would last indefinitely. It pointed to the ability of Arabs to quarrel, reconcile, and quarrel again. "Nonetheless," continued the appraisal, "role of Egypt revealed by conference has obvious implications for United States policy in the area."

The lack of positive reaction in the West to the Summit hurt the Arabs' feelings. A routine speech by the U.S. Deputy Undersecretary of State, soon after the close of the Cairo meeting, failed to mention it and, in an address to the Friends of the Weizmann Institute a few days later, President Johnson also ignored it and went on to suggest U.S.-Israeli cooperation to use nuclear energy to desalinate water. The mention of nuclear energy, water, and Israel in the same speech further stimulated Arab criticism of the United States.

The Israeli Prime Minister and Defense Minister, Levi Eshkol, announced that his country would resist "unilateral and illegal steps" by the

---

[207] See Lowi, p. 123.

[208] Ibid.

[209] See cable from the U.S. Ambassador to the Secretary of State, no. 1643, dated January 21, 1964, Fiche no. 274 D, date of issue 10/01/1976, Doc. no. 00301, National Archives and Records Administration, Washington D.C.

Arab states, and would "act to safeguard her vital rights. Israel would pump from Lake Tiberias within the limits of quantities allotted to it under the Johnston Plan... Furthermore, the Arab riparian states were each drawing water from the Jordan system, but they wanted to prevent Israel from drawing its own share." He concluded by saying that "the Arabs' argument is not about water but is an argument which denies Israel's right to exist."[210]

This conclusion by the Israeli Prime Minister, and the articles in leading Egyptian papers, in addition to the broadcasts from Damascus radio, prove that the Jordan River waters were not the actual cause of confrontation between Israel and the Arab states. It was rather caused by the forcible creation of the Zionist entity and the resentment by the Arabs of that State. The diagnoses by the Israeli and the Arab sides of the causes of that confrontation were alike, each referring the cause to the denial of Israel's right to exist.

In May, 1964 five months after the first Arab Summit, Israel began to test-pump the National Water Carrier. The Government announced that it was carrying out military maneuvers around the sites of the project. In response, the Assistant Secretary General of the Arab League, Sayyid Nawfal, announced: "victory is in the hands of the Arabs, the sole owners of the usurped right. They are bound to carry out their duty, speed up the implementation of their legitimate defensive plan until they achieve its goal." Syria's reaction was not so fervent. She urged the Arab states, once again, to abandon their idleness and implement the resolutions to resist Israel's plans.[211] Again, these statements go beyond the cause embedded in the Jordan waters issue and refer to usurped rights. These rights were the rights of the Palestinians in their homeland. No water from the Jordan had been usurped yet, because the National Water Carrier was not yet operational.

5. The Second Arab Summit

A second Arab summit conference was held in September of the same year in Alexandria, Egypt. Its purpose was to chart the next phase of the "struggle". Several important resolutions were made: first, work on the Arab diversion scheme was to begin at the end of the month; contractors were given orders to proceed on those projects that were ready for execution. Second, and in accordance with the diversion scheme, a dam would be constructed at

---

[210] BBC, "Eshkol's Statement on the Cairo Conference," no. 1459, 22 January 1964, pp. A/1-2. The last comment was corroborated by the Egyptian paper *Al-Ahram*, which wrote that: "The battle against the diversion of the Jordan river is one against Israel as a hostile political entity forcibly implanted in the midst of the Arab land." The comment also draws on numerous broadcasts by Damascus Radio, commenting to the same effect.

[211] See Miriam Lowi's "Water and Power," pp. 123.

Mukheiba on the Yarmouk across the Syrian-Jordanian border, and would be linked by a tunnel to the Wadi al Arab in Jordan[212] and thence to the East Ghor Canal. This tunnel would replace a diversion dam to be built across the Yarmouk at Adasiyya, as was originally envisaged in the East Ghor Canal Project. The Mukheiba dam would have a storage capacity of 200 mcm and would impound water of the Yarmouk as well as the waters diverted to it via a 70 km canal from the Hasbani in Lebanon and the Banyas in Syria. Third, aggression against any one Arab country would be considered as aggression against all. Hence, Arab leaders agreed that, as the work on the diversion scheme proceeded, their armed forces should remain in a state of alert, ready to repel attacks. Fourth, a Palestine Liberation Organization (PLO) was established to represent the Palestine Arab people and to defend their interests. And a military wing, the Palestine Liberation Army (PLA), was to be formed among the refugees and would operate under the aegis of the Joint (Unified) Arab Command.[213]

The Summit made the linkage between immediate concerns over Israel's National Water Carrier and the Palestinian cause. The fact that a Palestine Liberation Organization and a Palestine Liberation Army were created in the same meeting was indicative of that linkage. Hence, water, although the immediate cause of bringing the disarrayed Arab countries together in their first Summit, was not the core cause of the confrontation. The cause, again, lay in the protracted conflict over Palestine between Jews and Arabs.

Official communiqués on the Summit were laudatory. Abdul Khaliq Hassuneh, the Arab League Secretary General (Egypt) announced that the Arab States had unanimously defined the "national cause" as that of "liberating Palestine from Zionist imperialism."[214] To achieve their goal, they had committed themselves to pursuing a joint plan of action.

But despite public affirmations of consensus among the member States, serious differences emerged in Alexandria. Military preparations and collective action were issues of contention between Egypt on the one hand, and Jordan, Lebanon, and Syria on the other. The latter member States categorically refused to have Egyptian-led troops on their soil. They saw in

---

[212] Diversion from the dam at Mukheiba would leave the springs of Himmeh and Mukheiba downstream flowing in the river westward toward Israel. Thus an average of about 80 mcm would be made available to Israel instead of the 25 allocated to her from the Yarmouk under the Unified Plan.

[213] Ibid., pp. 123-124.

[214] Water, as such, was not the primary target of the Arab leaders. The confrontation was based on the need to liberate Palestine. Note how the PLO and the PLA were established in this Summit. These would play a vital role in war and peace in the Middle East.

that a hidden desire on the part of Nasser to interfere in their domestic affairs. The Arab diversion works thus dragged on until the concerned States received adequate military guarantees.

The other major source of contention was Syria's avowed determination to fight Israel. The Syrian Ba'th leadership was demanding that a definite date be set for serious action to "recover the usurped homeland". This was truly a curious position, given that Syria insisted that the diversion works within her territories required the military protection of other Arab States. Nonetheless, the regime was pressing for war. The media services asserted that, "the Arab masses today feel that they are capable of fighting the battle and winning victory." The Syrian leadership was clearly making every effort to drag Egypt into a hot confrontation with Israel in which the Syrian forces would also participate. Diversion works inside Syria were started on schedule in early October 1964, and only by the end of December 1964 did the diversion works on the Hasbani in Lebanon to the Banyas in Syria begin.

6. U.S. Interventions with Israel

The above and other developments in the region were taking place while the arms race was intensifying. Israel, on her part, had embarked on a campaign to persuade the United States to provide her, on grant basis, with five hundred tanks to match, they argued, the UAR's T-54 tanks obtained from the USSR, who also supplied Syria and Iraq with weapons. The build-up of armaments was most certainly leading the Arabs and Israelis to a clash. Israel was keeping a close watch on the military build-up on the Arab side, and was eager for an opportunity of whatever nature to destroy the military power of her Arab adversaries. The U.S. had maintained the policy not to be a major arms supplier to the Middle East and breaking away from that policy would upset the Arabs even more. At stake for the U.S. were the interests of U.S. oil companies in Arab countries. Those companies might come under pressure to provide better shares of revenue to the host countries. At stake also was the weight of the Arab countries in the United Nations, at a time when China was attempting to get into the world body and the U.S. was blocking it. The votes of the Arab states were needed to reinforce the U.S. position in the U.N. Moreover, a period of sharply deteriorating U.S.-Arab relations would give the Soviets opportunities that they would almost certainly use to improve their position in the Middle East. The Israelis would probably help the USSR in this respect, since the dominant policy makers in Tel Aviv would welcome the opportunity to use the Cold War as a device to assure themselves of closer U.S.-Israeli ties, particularly in the military field.[215]

Jordan, too, had approached the U.S. to purchase her arms and supersonic planes for the Arab build-up against Israel's water diversion scheme, claiming that otherwise they will have to buy Soviet items being touted by the UAR (Egypt). A memorandum to the President[216] dated 21 January, 1965 spelled out the strategy of the United States toward the subject, "if we sell supersonics to Jordan for an announced Arab buildup against Israel's diversion scheme," the memorandum explained, "we'll come under acute Israeli and U.S. Zionist pressure to sell arms directly to Israel too... If we get into overt arms sales to Israel, then all Arabs may turn against us, and Jordan may have to accept Soviet and UAR arms anyway.

We owe King Hussein an answer by the end of January. Our proposed strategy is to come through with some carefully phased ground equipment sales, and to tell Hussein he can't have either U.S. or Soviet fighters without jeopardizing U.S. aid. But Department of Defense and now State's Arab experts fear that Hussein won't play. So they propose as a fallback an offer of one U.S. squadron by 1968-1969 (which they think Hussein will buy) if he is adamant. Others of us oppose going this far now. We should try our hardest to turn Hussein aside before deciding on any fallback because of the acute risk of a violent Israeli reaction."

The memorandum revealed the way the U.S. had been promoting supplies of arms to Israel. She undertook the modernization of German tanks, on the provision that the Germans supply Israel with M-48 tanks. The memorandum continued:

"A new complication is that the Israeli/German/U.S. tank deal is now leaking (though our commitment to modernize the German M-48s is still not out). This could be played up by the Arabs to force Hussein to take Soviet arms. We told the Israelis we might have to call the deal off if it leaked prematurely, and may want to threaten this as pressure on them to keep their lips buttoned (they may have been the leakers).

"Another important part of this problem package is our policy toward the UAR. We face a growing crisis in 1965 on Jordan Waters, and if Nasser wants he can make it highly painful to us. The one issue on which he can always rally the Arab brethren is Israel. So you might want to send Luke

---

[215] Central Intelligence Agency, Office of National Estimates, "Special Memorandum No. 6-64- Storm Warnings Up For US-Arab Relations," dated 25 February 1964. Fiche No. 278 C, date of issue 07/01/1981, Document No. 00042, NARA, Washington, D.C.

[216] "Memorandum for the President," dated January 21, 1965 by R.W. Komer, Deputy Special Assistant to the President for National Security Affairs, NARA, Fiche No. 98, date of issue 03/01/1994, Document No. 1104.

Battle back with some quiet word that we'd rather talk than fight if only Nasser would show a little understanding of our problem too."

It was clear that the diversion project, and the Arab reaction to it brought several other issues into play simultaneously. Above all, the question of direct arms supplies by the United States to the Middle East, and the question of competition for influence in the moderate Arab States, primarily Lebanon and Jordan. The U.S. was trying her best to keep the USSR out of these two countries, even if that would cost her an overt supply of arms to them. The issue of U.S. aid in the region came also into play; it interacted with the issue of armament and the timing thereof.

In late February 1965, Governor W. Averell Harriman accompanied by Robert W. Kromer, Deputy Assistant to the President for National Security Affairs, made a trip to Israel upon request of President Lyndon B. Johnson. They met with Israel's Prime Minister, Levi Eshkol, in the presence of the Foreign Minister, Golda Meir, the Deputy Defense Minister, Shimon Peres, and the Chief of the General Staff, Yitzhak Rabin. Harriman struggled to obtain Israel's consent to the U.S. selling tanks to Jordan and making it public. Israel contended that the arms balance in the region had tipped in favor of the UAR, after the latter concluded deals with the USSR. Israel asked for the U.S. to supply her with tanks and fighter aircraft to maintain the balance, a factor the U.S. always was eager to maintain. "However," continued the Israeli negotiator, "if in the global view of the United States arms to Jordan seemed the best course, the U.S. could go ahead. But this should not be done before agreement was reached on direct U.S. supply of arms to Israel."

Israel asked that there would be no tanks deployed in the West Bank of the Hashemite Kingdom. "Shortly after the Jordan arms deal becomes known," said the Israeli Prime Minister, "President Johnson would make a public statement reaffirming U.S. support of Israel on the Unified Plan and on maintaining the arms balance." The Government of Israel would agree to enter discussions with the U.S. on ways to prevent Arab diversion *outside the Unified Plan*."[217]

"The Arabs planned to build a 70 km canal to Mukheiba (for the diversion of the headwaters), but Jordan already had its share of the waters[218]. You get your friend Jordan to agree not to cooperate in this diversion," said Prime Minister Eshkol. He maintained that there was an

---

[217] Italics inserted by author. Note how Israel portrays herself as an advocate of the Unified Plan.

[218] Eshkol was referring to the East Ghor canal Project. That project alone would not secure Jordan's share. Regulation of the Yarmouk was still missing, and a transfer of 100 mcm of water from Israel to the canal still needed to be decided on and guaranteed.

important matter of principle involved. "If the Arabs deprived Israel of one cubic meter of water, and got away with it, they would then take more of Israel's share," he added.
The Prime Minister then shifted to the issue of navigation of Israeli vessels, which were allowed through the straits of Tiran to the Gulf of Aqaba after the Suez Campaign of 1956, but not yet admissible through the Suez Canal. "What would happen if Nasser told UNEF[219] to go home?" said the Prime Minister, "This means war to us."

It is important to note here that the presence of UNEF was crucial to Israel, more crucial than the Arab diversion scheme. In the latter, the Prime Minister expressed fear that if the Arabs took some of Israel's water and were allowed to get away with it, they would then take more. He did not say that such an act by the Arabs would mean war. But it meant war to Israel if UNEF departed at the request of Egypt. This important note is crucial to keep in mind when reading the assessment of the eruption of the 1967 war, is made below. The Arab diversion works had commenced in 1964, and no war was waged, but a few days after Nasser demanded that UNEF be evacuated, in May of 1967 war broke out. The U.S. position, expressed by Harriman, may explain why.

Governor Harriman alluded to the Jordan waters issue in his talks with the Israelis. "The Arab diversion," he said, "was still a questionable matter. Nobody knew how far it would go. In any case, the U.S. would not change its position; we had accepted the Johnston Plan. The U.S. simply could not accept Israel's military intervention against the Arab projects. The problem should be resolved by other means like using the U.N. When the problem became more acute we would discuss with Israel how it might best be handled. But now we wanted a clear understanding that Israel would not use military force."

The Israelis asked the U.S. mission to secure, in a binding undertaking from Jordan, that the Mukheiba Dam would not serve Arab diversion plans involving Jordan River headwaters. The dam, conceived under the new contract with the Yugoslav firm, Energoproject, was underway for construction by an Egyptian contractor and financed by the fund established as a result of the Cairo Summit. The dam, 200 mcm in storage capacity, would replace the Khalid dam envisaged in the Harza studies and would regulate the Yarmouk, in addition to the Maqarin dam. The Mukheiba dam was so located as to be able to regulate the flow diverted to the Yarmouk from the Banyas and the Hasbani under the Arab diversion plan. It would also do away with the need to store Yarmouk floods in Lake Tiberias.

---

[219] United Nations Emergency Force, which was stationed in the Sinai between Egypt and Israel as an arrangement to keep peace after the Suez Campaign. Israel gained shipping rights through the Straits of Tiran under protection of UNEF.

The Israeli Chief of the General Staff, General Yitzhak Rabin, gave the American mission a military briefing in which he said that the Israel Defense Forces (IDF) considered the air and armored forces its key elements, for what was important was not to contain an enemy attack but to destroy the enemy on its own ground.[220] General Rabin said that the UAR armed forces outnumbered the IDF in heavy bombers, fighter-bombers, and armor, without counting the capabilities of Syria, Jordan, Lebanon, and Iraq. He noted Israel's need for long-range bombers, fighters, air defense systems, and for armor to match the UAR capabilities. Governor Harriman told the Israelis that the President was contemplating a fundamental change in U.S. policy--a willingness to reconsider the question of direct supply of arms to Israel. This shift would alter a policy that had exisited unchanged for seventeen years.

A secret memorandum of understanding was concluded between the two sides on 3 March 1965. Israel agreed to the sales of U.S. arms to Jordan, and the U.S. undertook to examine with Israel periodically the balance of power between her and her Arab neighbors, and to directly supply her with the arms needed to maintain a balance of power. On the Arab diversion scheme, the memorandum stated: "the United States agrees to review with Israel the dangers posed by the Arab diversion scheme and the [Arab] Unified Command build-up. The United States believes that the Unified Plan provides the equitable standard for determining the allocations of the Jordan waters and will use its best efforts to see that the shares of the Jordan waters allocated to Israel and the Arabs conform to that standard."

It is able that the U.S. stood by the "allocation"and "shares" stipulated in the Unified Plan as equitable; she did not insist on the way those shares were delivered, although such delivery was stipulated also in the Unified Plan. The memorandum further committed Israel not to use force over the water diversion scheme. "Israel recognizes," the memo continued, "that any resort to force to assure water supply could create a serious danger to peace and therefore agrees to seek earnestly for all peaceful means to achieve such protection."

As a part of the over-all agreement, Prime Minister Eshkol was to send a letter to President Johnson committing Israel as follows:[221]
"My Government will support US efforts in the international field to prevent proliferation of national nuclear weapons and will agree to accept IAEA

---

[220] "Memorandum of Conversation," Harriman Mission to Israel, February 25-26, 1965. Fiche Nos. 204, 205; date of issue 04/01/78; Doc Nos. 587, 589. NARA, Washington, D.C.

[221] Telegram to Harriman and Komer dated 2 March 1965, Fiche No.126, date of issue 05/01/1998; Document No. 1464, Department of State declassified documents; NARA, Washington D.C.

[International Atomic Energy Agency] controls on all Israeli nuclear facilities if the UAR similarly accepts such controls.

"Israel should not expect the United States to support or condone any Israeli use of force in connection with Arab diversion of Jordan waters".

The United States strategy was clearly not to provide impetus for any expansion of communist influence in the Middle East. She saw that "certainly the growth of Communist influence in this part of the world would not be to Israel's advantage. There had been an up-cycle in Arab unity. There were pressures on Lebanon. There was the unfortunate expression of support by Nasser and Ben Bella of Algeria for the rebels in the Congo. The Unified Arab Command had created a very worrisome situation.

"This was very much a matter of our concern as well as Israel's. But Arab unity movements tended to be cyclical. Above all, it was extremely important not to allow the Near East to become polarized along East-West lines. It would be most unfortunate if Israel became a point of East-West confrontation as Berlin had. We had been successful in maintaining special relationships with some Arab countries, with Saudi Arabia, Lebanon and Jordan."[222]

The U.S. policy was to refrain from condoning an armed conflict over the Jordan waters and to have the allocation specified in the Unified Plan as the standard for determining the allocations of Israel and the Arabs. Interestingly, the U.S. statements as elaborated in the Harriman Mission, did not express concern as to how the water shares were to be conveyed to each riparian, and lumped the allocations of Arab States together. This attitude meant that, at least as far as the U.S. was concerned, the Arab diversion scheme could proceed provided that the water thus diverted (including the Yarmouk diversion) would not exceed the amounts allocated to the Arabs under the Unified Plan. Should the diversions infringe on Israel's allocations, Israel was to undertake all peaceful means, and the U.S. would come to her aid, to protect her shares. This U.S. stand is important to keep in mind as the assessment of the causes of the June 1967 war is made below.

7. Clashes across the Cease Fire Lines

Within six weeks of the commencement of work on the Arab diversion project, both Syria and Israel reported a border clash between their armed forces. Each side accused the other of initiating the confrontation. Whatever the truth of the matter, the clash was the first in a series of military responses to rival water projects. The Prime Minister of Israel announced at a press conference: "water is a question of life for Israel. Hence, any attempt to

---

[222] Ibid.

prevent Israel from using the part of the Jordan River waters due to it in accordance with the Unified Plan would be considered a violation, not simply of the country's borders, but of its right to exist." The linkage made by the Prime Minister was not the customary linkage between water and life, but was the linkage between the current immediate cause of clashes and the real cause of confrontation, the right of Israel to exist. He further emphasized Israel's adherence to the Unified Plan. Cairo radio scoffed at Eshkol's statement and took the line of usual rhetoric and challenge to Israel.

Clashes along the Syrian-Israeli Armistice line continued during the spring and summer months of 1965. They prompted the ruling Ba'th party to accuse the Arabs of laxness in the face of the "Israeli challenge," and re-awakened the Arab debate between those in favor of and those against immediate war. Nasser was not enthused to the idea of immediate war. "How can we talk about attacking when we are not able to defend ourselves?" he remarked publicly. He accused the Ba'th party, his rival, of having "reduced the liberation of Palestine to claptrap and bravado."

Other serious Arab disagreements emerged. President Habib Bourguiba of Tunisia made a serious statement, in the spring of 1965 on his visit to Jordan, the first ever to come publicly from an Arab head of State. He urged the Arabs to sit down and negotiate peace with Israel. His statement aroused anger across Arab lands and ranks, and the President was accused of committing treason. Bourguiba was isolated, and a vindictive slander campaign began between Egypt and Tunisia, and between Syria and Tunisia as well. The environment was set for more inter-Arab disagreements. Chairman of the newly established PLO, Ahmad Shuqayri, accused Jordan of doing nothing for the Palestinian cause, and that the activities of the PLO were barred in Jordan. A rift developed between Jordan, the major host of Palestinian refugees, and the PLO. In the meantime, the military clashes across the Armistice line on the West Bank continued, increasing in number and intensity. Israel blamed Jordan for the entry of *Fatah* infiltrators , a Palestinian guerilla group established on 1 January 1965, into Israeli territories to commit acts of sabotage, some against the National Water Carrier.

By mid summer of 1965, there were rumors that work had stopped at the diversion site in Syria. Beirut newspaper reported that the Lebanese government had halted the diversion works in her territory as well, because the funds had run out.[223] Work on the Mukheiba dam in Jordan had not started until August of that year. One month later, representatives of Arab Kings and Presidents met for a follow-up of the Summit decisions. Cairo radio was optimistic: "we can conclude that the rifts and contradictions which have

---

[223] According to *Le Monde* (5 June 1965), the Lebanese Government suspended the work after very little progress because it was not assured of adequate air cover.

appeared in the Arab world recently have not been able to hamper the river diversion plans or smother the spirit of Arab summits." Contrary to this optimism, it was confirmed by quasi-official sources that the diversion works had ceased. Mohammad Hassanein Haykal, the chief editor of the influential Egyptian newspaper, *Al Ahram*, and a close confidant of Nasser, announced that the Syrian regime, demonstrating a lack of "psychological readiness," had abandoned the project after it had had to face Israel "in small local skirmishes."[224] Despite the interruptions, work on the diversion in Syria resumed, albeit only for a short period of time before they were smothered by Israeli attacks on the diversion works. The air forces of the two countries got involved and, in April of 1966, the work was suspended again. Levi Eshkol asserted that, "with remarkable firmness and ability, the Israel Defense Forces foiled the beginning of the water diversion scheme in the neighboring countries..."[225] Work on the Mukheiba dam in Jordan proceeded unhindered until it was interrupted by the Six Day War of June 1967.

8. Arab Disarray Prior to the June Explosion

8.1 Saudi-Egyptian Relations

Beginning in September 1961, when Syria broke away from the UAR, Egypt turned away from cooperation with the Arab moderate regimes and followed a revolutionary path in the conduct of her Arab relations. She seized on the Yemeni revolution of September 1962 to make a comeback, on the basis of a revolutionary leadership. Yemen represented more to Egypt than just an opportunity to come out of her isolation. It was a foothold in the Arabian Peninsula, neighboring oil rich Saudi Arabia and the British-protected, South Arabian Federation, which encompassed South Yemen and the Gulf protectorates. From that foothold, Egypt could be a major player in the affairs of the Peninsula and in Arab affairs in general. Nasser backed the new regime in Yemen with some 40,000 troops to help fight the royalist forces backed by Saudi Arabia and, to a lesser extent, by Jordan. Soon there was a stalemate, and the future of Yemen was hanging in the air.

Repeated efforts were made to negotiate a solution: first by American and then by United Nations' mediation. The Alexandria Summit of September 1964 brought Crown Prince Faysal of Saudi Arabia together with Nasser, and they agreed to bring the opposing Yemeni factions together to negotiate a settlement on neutral grounds. A conference for the parties was organized in the town of Erkwit, in the Sudan, but did not bring about any change.

---

[224] BBC, "Haykal on Arab Strategy," no. 1988, 18 October 1965, p. A/2, cited by M. Kerr in "the Arab Cold War."

[225] BBC, "Eshkol's Speech on Memorial Day," no. 2145, 26 April 1966, ibid.

Crown Prince Faysal Ibn Saud, also Prime Minister, assumed the throne from his brother, King Saud, in a bloodless coup in November 1964.[226] To break the stalemate, and to assure his leadership, Nasser made a surprise visit to Jeddah, Saudi Arabia on 24 August 1965, abruptly calling off a barrage of threats to invade Saudi Arabia, where the support to the rival royalists came from. The two leaders came quickly to an agreement whereby the adversaries in Yemen would meet in November, at Harad, a village near the Saudi frontier with Yemen, to arrange for the creation of a provisional regime. The outcome of this conference was no more impressive than the one at Erkwit, and the stalemate continued. By March 1966, encouraged by a British announcement that they would evacuate their military base in Aden by 1968, Nasser announced that his troops were prepared to stay in Yemen indefinitely. The Jeddah agreement was dead, and the threats to attack Saudi Arabia were renewed. Nasser was probably more upset because of an initiative that King Faysal came up with and campaigned for through visits to Islamic states. The initiative called for the establishment of an Islamic Conference that would align Islamic states. Faysal visited the Shah of Iran, for whom Nasser harbored no particular love or admiration. He then visited Jordan's King Hussein in January 1966 and received his vigorous support for the idea. Later, Faysal visited Turkey, Sudan, Pakistan, Morocco, Tunisia, and Nasser's revolutionary friends, the Moslem president of Guinea, Ahmad Sikotori, and of Mali, Modibo Kita. Nasser saw in Faysal's initiative an attempt to revive the Western influence that had faded with the collapse of the Baghdad Pact and the demise of the Eisenhower Doctrine.

8.2 Egyptian-Jordanian Relations

Since the first Arab Summit in Cairo in January 1964, King Hussein worked hard to improve relations with the revolutionary regimes. He made frequent trips to Cairo, recognized the revolutionary regime in Yemen, and took a leading role in the establishment of the Palestine Liberation Organization, despite the potential danger its establishment might pose on the Kingdom. He raised no objection to the establishment of a Unified Arab Command under an Egyptian general to defend the diversion works of the Jordan River tributaries. The King was the first leader to translate into action the decision of the second Arab Summit to set up a regional authority for the Arab diversion project, and he ordered legislation for the establishment of the

[226] The author was a witness to the siege of Saud's Nasiriyyah Palace in Riyadh on a Thursday evening. He was on a routine visit to Nasiriyyah as an engineer for the Royal Guards Brigade, and was denied the usual access. He was ordered to turn back on foot leaving his 1963 Opel car behind by the back entrance gate. The following day, Faysal was pronounced King right after the Friday noon prayer. King Saud was kept at Nasirriyah for a few days before he was allowed to depart to Greece.

Jordanian branch of that authority. In 1965, a law establishing the Jordan River Tributaries Regional Corporation was enacted, and the organization was swiftly created under the chairmanship of the Prime Minister. The King issued a general amnesty allowing those Jordanians, self-exiled for political reasons, to return. Some of them were placed in high government positions.

All of these improvements were outweighed, in Nasser's eyes, by the King's support of King Faysal's initiative for an Islamic Conference, to be convened in Mecca. Nasser used the recent verbal assault that Shuqairi of the PLO waged at the King, and renewed the campaign in the Egyptian press against Jordan. He did so in public speeches as well. Another reason for Nasser's attacks on the King was the Syrian pressure from the left. Syria favored, as a matter of principle, attacks on "reactionary" regimes, and Jordan was a favored target. In addition to her reactionary character, the King's regime twarted a few attempts, organized by the Ba'th party and its sympathizers, to overthrow him. The more general Syrian objective was to end the "detente" that the Arab Summit brought about and force Nasser into an alliance with the revolutionary regimes, just as he was campaigning before the Summit.

Nasser's attack on Jordan started through the Voice of Palestine transmitting from Egypt. The PLO was allowed to say whatever it liked against the King. On 14 June 1966, when the King publicly ruled out further cooperation with the PLO, Shuqairi and the Syrian leaders announced that "the liberation of Jordan from Hussein was a necessary precondition to the liberation of Palestine." The Egyptian press and radio waited until 1 September 1966 before they started their criticism of the King, and Nasser himself waited to do so until 23 December, when he declared that Hussein, like Faysal and Bourguiba, was "ready to sell the Arab nation in the same manner as Abdulla [Hussein's grandfather] sold it in 1948." Nasser was assaulting the King at the same time that Israel was intensifying her raids across the Armistice line with Jordan. In December 1966, when Nasser was making these verbal attacks, the Israelis attacked the village of Samu' in the southern sector of the West Bank, and the King ordered the Jordanian Royal Air Force to take to the air in support of the ground troops that rushed to defend the town. Two Jordanian Hawker Hunters were downed, and a fighter pilot was killed as he was parachuting from his downed plane.[227]

Two months later, Nasser's public attacks against the King of Jordan became personal and mean, and Jordan protested by withdrawing her Ambassador from Cairo in February 1967.

---

[227] The pilot was Muwaffag Badr Salti of the Royal Jordanian Airforce, a high school friend of the author. An airbase east of Amman was named after him, in honor of his combat record.

8.3 Egyptian-Syrian Alliance

The Ba'th rulers in Damascus finally gave in to one of their strong generals, Amin al Hafiz, who served as Minister of the Interior at the beginning, and then took over the premiership from Salah al Bitar. He eventually became president of the Syrian Arab Republic. The National Command of the party elected in April 1965 a new Secretary General, Dr. Munif al Razzaz, a Jordanian medical doctor of Syrian extraction. Al Razzaz administered party affairs while Michael Aflaq, the party's quiet thinker and ideologist, retained moral status as the party's elder statesman, and Bitar served intermittently as Prime Minister. As long as these men remained on the scene, Syria's relations with Egypt continued to be colored with indecisiveness and lack of assurance. The Syrian rhetoric against Israel's intransigence did not diminish, but the action on the ground was very passive. Syrian forces and diversion works came under Israeli fire on 13 May 1965, and the Syrian military refrained from the use of force. Syria publicly announced that her army did not retaliate. It was believed the reason for that was two fold:[228] "a) Syria's own relative weakness and present disarray among Arab states, and, b) fear that at this time Syrian counter-fire might be used by Israelis as excuse for a major attack against Syrian military positions and diversion sites."

Eventually, these veterans of the Ba'th Party were challenged by younger, more decisive men, both military and civilian, who shared none of their ambivalence. The prominent military leader among the challengers was Salah Jadid, an Alawite who seized the post of Chief of the General Staff in December 1964. The civilians were young professionals, the most prominent of whom were Dr. Nur Eddin al Atassi, Dr. Ibrahim Makhus, and Dr. Yusif Zu'ayyin, all of whom were in their mid-thirties. Zu'ayyin became Prime Minister, Makhus became Minister of Foreign Affairs, and Atassi became a member of the Presidential Council under Amin al Hafiz. The newcomers were Marxist-socialist radicals who were more militant towards Israel and the moderate Arab regimes than their counterparts in the other faction of the Ba'th party. The competition for power between the two factions soon led to a military coup, staged on 23 February 1966, by the socialist radicals. Hafiz was arrested and so were Bitar, Razzaz and others, and the radicals assumed power in Syria. Atassi replaced Hafiz as President. This was the ninth military coup in Syria since her independence in 1946.

The new rulers in Syria were not interested in Arab unity for its own sake; They were more eager than their predecessors to see an end to peaceful co-existence with the 'reactionaries'. They did not plead for Nasser's favor,

---

[228] Cable of U.S. Embassy in Damascus to the Secretary of State, dated May 14, 1965. Fiche no. 161, date of issue 07/01/1997; Document no. 1990, NARA, Washington, D.C.

but were bent on forcing him to the left into an alliance on their own terms. During this time (April 1966), the United States moved to effect a change in her policy on arms supply to the Middle East, and gave Nasser advance notice of her intent to sell advanced aircraft to Israel.[229]

Nasser responded cautiously to the Syrian desires. He received Makhus, Syria's Foreign Minister, and agreed to exchange trade missions.[230] Soon afterward, the Foreign Minister of Egypt, Mahmoud Riyadh, who served as Egypt's ambassador to Syria before the union, and who chaired the Arab political-military committee at the time of the Johnston negotiations, paid a visit to Damascus, the first ever by a member of the Cabinet of the UAR since the secession in 1961. But the Syrian radical leaders were more interested in action on the ground than they were in the exchange of visits. They wanted to abort the emerging tradition of Summit conference and cancel the forthcoming conference scheduled to meet in Algiers in September. Nasser gave them what they wanted when he announced at a public rally that he would not deal with the 'reactionaries' (meaning moderate Arab regimes) until they mended their ways. The era of the Summit was terminated. The Syrian leaders pushed still further. They openly sponsored a wave of Palestinian guerilla raids into Israel and started to engage their army in skirmishes with Israel along the Armistice line. These developments disturbed Nasser at a time when the Summit conference was no longer operative. He could have had the Summit restrain the Syrians and talk them out of that kind of 'recklessness'. He had no one close to the Syrians who could do that service for him and for the Syrians themselves. Nasser's options were obviously reduced to one: to try and restrain them himself. He invited Dr. Yusif Zu'ayyin, the Prime Minister of Syria, to Cairo, and on 7 November 1966 signed a treaty of mutual defense. By this treaty, Nasser would bind Syria to advance consultation with Egypt before any future skirmishes or acts of violence were triggered with the Israelis.

8.4 Egyptian-Iraqi Relations

President Abdul Salam Arif had maintained cordial relations with Nasser ever since the Iraqi leader succeeded in ousting the Ba'th party from the power structure in November 1963. He, too, was under attack by the Syrian Ba'thists. Under Arif's leadership, Iraq failed to fill a role that was naturally cut for her in the region. Located alongside Syria and three other 'reactionary' Arab states (Kuwait, Saudi Arabia, and Jordan), populous, with a relatively well-equipped army and considerable oil income, Iraq could have

---

[229] Cable from State to U.S. Embassy in Cairo, dated 14 April 1966, Fiche no. 68, date of issue 01/01/1984; Document no. 1137, NARA, Washington, D.C.

[230] Egypt had not exchanged diplomatic missions with Syria after the secession in September, 1961.

played a decisive role in Arab politics, at a time when Nasser was caught between the Syrian leftists and the rightists of Saudi Arabia and Jordan.

Ever since the collapse of the monarchy in Iraq, the country had been suffering from instability, economic stagnation, and a continuous headache caused by the rebellion of the Kurdish community in the northeast. The rebellion complicated Iraq's relations with her two powerful, and one time allied, neighbors, Turkey and Iran, into which the Kurdish population extended. In May 1964, Arif had agreed to work toward unification with Egypt within two years. A ministry was created in Iraq for the purpose of preparing for the unification. A series of Presidential decrees nationalizing private assets were set in motion to bring the Iraqi economy closer to the structure of the Egyptian economy. An Arab Socialist Union, ASU, was established along the lines of Egypt's ASU. As time passed, however, the pledge for unification was forgotten, mainly because of the failure to maintain domestic political stability within Iraq. On 29 June 1966, the Kurds came to terms with the Prime Minister of Iraq, Dr. Abdul Rahman Bazzaz, and the Kurdish uprising ended, at least for a while.

Another trouble came to the Iraqis from the young radical rulers of Syria, who seized power in February 1966. They demanded higher fees from the Iraq Petroleum Company (IPC) for the transit of Iraqi oil through Syria to the port of Banyas on the Mediterranean. When the British owned IPC refused to abide by the Syrian demand for drastically higher fees, Syria closed the pipeline. Iraq's royalties from the oil exported through that route were $250 million per annum. The Iraqis could not protest the Syrian measures because they were aimed at 'exploiting' the foreign oil company. A build-up of political and financial pressure on the Iraqi government could ultimately force it to nationalize the oil company and get in trouble with Britain, or run the risk of being overthrown by more radical elements. Either solution would have pleased the Syrians, who had never forgiven the Iraqi leaders for double-crossing the Ba'th three years before. Fortunately for the government of Iraq, the Syrians and IPC came to a compromise agreement on 2 March 1967.

Iraq missed the chance to influence Middle Eastern political trends between 1963 and 1967. Iraq could have been a stronger support for Nasser in trying to restrain Syrian radicalism. The camp of the 'revolutionaries' led by Nasser suffered a severe blow by the downfall of Ahmad Ben Bella as President of Algeria in June 1965. But the Iraqi domestic stability issues, coupled with the death of Abdul Salam Arif in a helicopter crash in April of 1966, and the watchful calculating eyes of all her neighbors, made an Iraqi-Egyptian alliance, let alone unification, rather difficult.

By the beginning of 1967, the Arab states were in such disarray that a gathering of the Arab League Council, meeting in Cairo in mid-March, could find nothing better to agree on than a denunciation of racial discrimination in the United States.[231] The institutions created by the Summit conferences: the Jordan River Tributaries Regional Corporation, the Unified Military Command, the Palestine Liberation Organization, and Palestine Liberation Army, were virtually defunct for lack of funds and lack of co-operation. Jordan and Saudi Arabia were now boycotting meetings of the League's Defense Council, and demands had been raised for their expulsion from League membership. There was clearly a sharp ideological cleavage between two distinct Arab groupings: the Damascus-Cairo-Sana'a alliance, and the Riyadh-Amman alliance. Many members of the League, including the Kingdom of Libya, Sudan, Lebanon, and, to an extent, Iraq, were openly distressed at that polarization, believing that the cleavage could only cause them harm. They preferred to keep good relations with everyone, and disliked being pressed to take sides. They were all, more or less, vulnerable to ideological denunciation by the Syrians and subversion by the Egyptians. Iraq was recognized as a 'liberated' state by Egypt, but not by her ally, Syria; Lebanon was the center of odious free enterprise and a constitutional government; and the regimes of Libya, Sudan and Kuwait were distinctly conservative.

The cleavage lines between Arab alliances ran parallel to their international affiliations. The West, led by the United States, was closer to the Riyadh-Amman alliance, but was keeping reasonably good contact with Nasser, especially after the success of the Summits; and the 'revolutionary' regimes were much closer to the USSR. From her they obtained military and technical assistance, as did the 'reactionary' alliance from the United States and Western Europe. There was a real danger that the superpowers, with whom the Arab alliances were thus aligned, could get into conflict with each other as a result of the inter-Arab confrontation, or as a result of Arab clash with Israel.

9. The June 1967 Explosion

In the above state of Arab disarray, the Jordanian Government announced in January 1967 that the work on the Mukheiba dam was progressing. The contract was given to an Egyptian firm, The Arab Contractors--Othman Ahmad Othman, and the supervising consulting engineers were the Yugoslav firm, Energoproject. Reviewing consultants were appointed to assist the Jordan River Tributaries Regional Corporation (JRTRC) in reviewing the designs and assessing the Consultants' proposals

---

[231]Malcolm Kerr, "The Arab Cold War 1958-1967, Oxford University Press, New York.

on technical matters.[232] Soon after, the Board of the JRTRC, in its meeting on 2 February 1967, made it known that the Arab diversion scheme was not progressing as scheduled. Work had started on various parts of the project and then came to a standstill. Other parts had not begun at all. Moreover, financial obligations had not been met.[233] This situation, one would conclude, was the result of Arab quarrels and their disarray.

In April, tensions erupted along the Syrian-Israeli Armistice line. Each party blamed the other for starting the hostilities. These tensions had nothing to do with the diversion works that were, for all practical purposes, announced stopped. They were rather a manifestation of the moods and strategies of the rulers in Syria, and the hawkish attitudes of the rulers in Israel. Syria fired at Israelis farming in the DMZ and shelled Israeli settlements. Israel fired at Syrian military positions. Both sides called upon their air force, and over a dozen fighters from each side engaged in an air battle on 10 April 1967. Six Syrian MIG fighters were downed, and the tensions escalated. The Israeli Prime Minister announced that by retaliating, Israel was exercising her right and duty of self-defense against repeated attempts to violate her sovereignty. He warned the Syrian leaders that, "the miscalculation they were making regarding the extent of patience was very dangerous for them."[234] There was no mention of the water diversion in the context of this important air battle.

Egypt did not rush to help her ally, Syria, in her military engagement with the common enemy. The Egyptian abstention provided Jordan with ammunition to attack Egypt for letting down their Syrian brothers. They accused Egypt and the Unified Arab Command of having abandoned Syria. Amman found it a golden opportunity to taunt the Egyptian leader, as he had taunted the King and his grandfather four months earlier. In explanation, Nasser announced that despite the alliance with Syria, he was not in a position to rush to Syria's defense, since the latter continued to refuse to have Egyptian planes stationed on her soil. Throughout April, there were repeated complaints to the Syrian-Israeli Mixed Armistice Commission of violations of the Armistice Agreement. Israel accused Syria of firing at Israeli settlements and Syria accused Israel of crossing the Armistice line. None of these engagements was attributed to water diversion works. At the end of April, a Palestinian guerilla operation of Fatah targeted a water pump station in northern Israel and destroyed it. Eshkol repeated his stern warning: "Israel views with extreme gravity the recent acts of sabotage in her territory and, if

---

[232] Consultants were Energoproject of Belgrade, Yugoslavia; the Reviewing Consultants were Coyne et Bellier of France; and the Contractor was the Arab Contractors -- Othman Ahmad Othman of Egypt.

[233] See Miriam Lowi, "Water and Power," pp. 131.

[234] Ibid.

there is no way out, we will be forced to take suitable retaliatory measures against the sabotage nests..."[235]

Despite the Israeli warnings, the situation on the front remained volatile. By mid-May, the Syrian government announced that the enemy was massing troops along her side of the Armistice line. Nasser responded by announcing that Syria would not be alone in her confrontation with Israel, and declared a state of emergency for the Egyptian armed forces. Cairo radio, somewhat less sensitive to the gravity of the situation several hundred miles away, and sounding more like its Syrian counterpart, made the following statement on what proved to be a decisive day, the 18th of May 1967:

"We challenge you, Eshkol, to try all your weapons; put them to the test; they will spell Israel's death and annihilation."[236]

On that day, and in an attempt to deter Israel from attacking Syria, Nasser demanded the immediate evacuation of UNEF from the Sinai Peninisula. This request was made in response to Syrian and, separately, Jordanian critics who wondered how Nasser would engage his forces in combat with Israel with the presence of the United Nations Emergency Force acting as a buffer between the two countries. In an associated move a few days later, and in response to Jordanian taunts and challenges, Nasser announced the closure of the Straits of Tiran to Israeli shipping. Israel responded by announcing that any interference with the freedom of shipping in the Gulf of Eilat (Aqaba) and in the Straits constituted an act of aggression against Israel; and Nasser's aggression "endangers Israel's existence." The Israeli interpretation of the evacuation of UNEF as a threat to Israel's existence was made clear to Governor Harriman as he came on mission to Israel in 1966 (see above).

With the events fast developing, Jordan came to grips with the situation and realized that matters were getting very serious. King Hussein knew that he could not afford to stay out of an all-out confrontation with Israel and survive the aftermath, whatever it might be. He knew only too well that Israel's forces were superior to those of the UAR and Syria combined, but could do nothing to stop the deterioration towards a decisive confrontation. By the end of May, he made a surprise visit to Egypt, made up with Nasser, and signed a defense pact by which Jordan gave permission for Iraqi and Saudi troops to enter her territory. A military Unified Command was established under an Egyptian general and, as far as issues related to the confrontation with Israel, the Jordanian troops were put under the Unified Command; so were the other troops that reported for duty from other Arab

---

[235] Foreign Broadcast Information Service of the U.S., 12 May 1967, p. H-1, ref. cited by Lowi.

[236] BBC, "Cairo Radio on UAR support for Syria," no. 2468, 18 May 1967, p. A/6. Ref. cited in Lowi's "Water and Power," p. 132.

countries. King Hussein and Ahmad Shuqairi made up as well, and the King brought back to Amman with him from Cairo the Chairman of the PLO, as a sign of a new era of solidarity. The Israeli reaction was that it would intervene if the status quo were violated by the entry of troops into Jordan: "Along with Nasser's aggression in the south, this step will leave no alternative for Israel..."[237]

9.1 American Feelers- A Presidential Envoy

The United States and the USSR were watching the situation in the Middle East very closely. The U.S. was eager to contain the deterioration in the area, and was careful not to be a direct party in any hostilities that might erupt. President Johnson sent an envoy[238] to President Nasser to talk with him over the issues that had induced tension between Israelis and Arabs. It is worthwhile to read his report to the President and the Secretary of State. The report details the views of Nasser, his confidence of the outcome of any war between the Arabs and Israel, and his preparedness to retaliate instantly after Israel made the first move. He intended not to start a war, but would respond when Israel made the first strike. Nasser was worried that the Syrian's might do something out of anger at Nasser for making up with Hussein, and assured the American envoy, Robert Anderson, that if hostilities between Syria and Israel broke out, he would have to act. On the closure of the Straits of Tiran to Israeli shipping, Nasser held the view that the straits were in Egyptian territorial waters and that he was returning back to the conditions that prevailed in 1956, before the tripartite aggression (Israel, Britain and France) resulted in opening the straits to Israeli shipping. Nasser responded to the envoy's question on whether or not he would, in principle, accept the right of Israel to exist, by saying that a lasting peace in the Middle East can not be attained without the return of Palestinian refugees to their homeland. Nasser expected that Israel would strike first and would target Cairo as a starter. He expressed desire to have the friendship of America and proposed to dispatch his deputy, Zakaria Mohyeddin, to America to explain the situation. The text of Mr. Robert Anderson's cable to the President of the United States and the Secretary of State is shown in Appendix 5.

On a separate attempt, Israel was reported to have made contacts with King Hussein and assured him that they would not start hostilities against

---

[237] BBC, "Israeli Comment on the Crisis," no. 2474, 25 May 1967, p. A/18; no. 2475, 26 May 1967, p. A/16. cited in Lowi's.

[238] Robert Anderson met with Nasser on 1 June 1967 in an attempt to diffuse the tension by involving the UN or the International Court of Justice. See cable from U.S. Embassy in Lisbon where Anderson stopped, addressed to "Eyes only for President and Secretary of State from Robert Anderson", Fiche no. 68, date of issue 01/01/1984; Document no. 1144.

Jordan if he kept out of the potential clash between them and both Syria and Egypt.[239] Messages to this effect were also reported to have been conveyed by the Chairman of the Jordanian-Israeli Mixed Armistice Commission, General Odd Bull.[240] The King realized he could not win either way. He knew he would lose the West Bank if he joined the upcoming clash with Israel, and that he would most probably lose his life and the Kingdom if he sat on the sideline.

9.2 Israel Strikes From the Air

In fact, Zakaria Mohyeddin, Egypt's Vice President, started out on his trip to the United States on Sunday 4 June as desired by Nasser and as expressed in the Anderson report. Before he took the flight to Washington from Paris, Israel had started the war with a devastating air strike against the Egyptian Air force. As Nasser had predicted, Israel's first target was Egypt and Cairo in particular. With the loss of most of the military aircraft on the ground, the Egyptian ground forces in Sinai and elsewhere were left without air cover, and they were over-run by the advancing Israeli armor. By 8 June the Israelis were on the eastern bank of the Suez Canal and had occupied the whole of the Sinai Peninsula, including Sharm el Sheikh. From their new positions, the Israelis controlled the navigation through the Straits of Tiran into the Gulf of Aqaba. By the same day, they had occupied the West Bank of the Hashemite Kingdom, including the old walled city of Jerusalem. On that front, the Jordanian Armed Forces, under command of the Egyptian General of the Unified Arab Command, started the hostilities on the morning of 5 June by opening fire across the Armistice Line through Jerusalem and occupying the building of the United Nations on Mount Scopus. It did not take the Israelis long before they occupied the West Bank. They then turned in force to the Golan Heights of Syria, and deferred the issuance of a Security Council Resolution ordering a cease-fire until they were well in command of the Golan and had occupied its territories up to the city of Quneitra, the provincial capital.

In six days, the Israeli forces were able to conquer three Arab armies, occupy the Golan in Syria, the West Bank in Jordan, and the Sinai Peninsula in the UAR (Egypt). Waves of displaced Palestinians left their towns and villages in the West Bank and crossed the Jordan into the East Bank of the Kingdom, adding to the hundreds of thousands of Palestinian refugees already there. The Israelis sat on the occupied territories and demanded face-to-face

---

[239] Moshe Zak, "Hussein Makes Peace," The Begin-Sadat Institute for Strategic Studies, Bar Ilan University, Israel. Translated into Arabic by Dar al Jaleel, P.O.Box 8972, Amman, Jordan, 2000, pp. 202.

[240] Odd Bull, "War and Peace in the Middle East," pp. 113, reported by Moshe Zak, pp. 224, Ibid.

negotiations with each of the surrounding Arab states, separately, to negotiate the terms of peace. On 22 November 1967, the Security Council issued Resolution 242 and specified the steps to be taken to maintain peace in the area.

9.3 The Soviet Policy

The Soviet policy on the Middle Eastern War of 1967 as expressed by a medium-level Soviet[241] and reported to the White House by U.S. intelligence was as follows:

"1. The Soviet told [the U.S. source] there had been 'miscalculations' by the Soviets and by the Arabs. The Soviets overestimated the Arabs' ability to employ their substantial military strength against the Israelis while the Arabs overrated their own strength and underrated the Israeli military capability and determination to win. When [the U.S. source] asked if that meant that the Soviets had encouraged the Arabs in their hostile attitude toward Israel, the Soviet replied affirmatively, stating that the USSR had wanted to create another trouble spot for the United States in addition to that already existing in Vietnam. The Soviet aim was to create a situation in which the U.S. would become seriously involved, economically, politically and possibly even militarily and in which the U.S. would suffer serious political reverses as a result of its siding against the Arabs. This grand design, which envisaged a long war in the Middle East, misfired because the Arabs failed completely and the Israeli blitzkrieg was so decisive. Faced with this situation the Soviets had no alternative but to back down as quickly and gracefully as possible so as not to appear the villains of the conflict.

2. The Soviet thought that Nasser 'must go' and that he would be 'most probably' assassinated in the near future by his own disillusioned people. He said that Nasser's charge that U.S. and British aircraft had aided the Israeli forces was a desperate attempt to save face in the Arab world after suffering a humiliating military defeat and that no one, certainly not the USSR, believed the charge. In a final comment the Soviet said the war has shown that the Arabs are incapable of unity even when their vital interests are at stake."

9.4 Was it a Water War?

It is clear from the above that the deterioration of conditions that led to the explosion in the morning of June 5, 1967 were not directly related to

---

[241] Cable to White House situation room from the Central Intelligence Agency dated June 8, 1967; Fiche no. 280, date of issue 07/01/1981; Document no. 00047, NARA, Washington, D.C.

water or the diversion projects on either side of the Jordan. Admittedly, the conditions on the Armistice lines between Israel and Jordan and Israel and Syria were disturbing throughout a whole year, from the spring of 1966 to the summer of 1967. The majority of the disturbances were in relation to infiltrations into Israel from the Jordanian territories by newly established Palestinian guerilla organizations, whose recruits were mainly Palestinian refugees. While Jordan could not maintain tight enough control on the Armistice line to avoid untimely clashes between her armed forces and the Israeli IDF, the incidents on the Syrian front were condoned by the officials, if not promoted by the radical wing of the ruling Ba'th in Damascus. The escalations, started by the radical socialist Marxist faction of the Ba'th party, eventually led events to spin out of control when Egypt was pulled in. The rivalry between the 'revolutionary' and the 'reactionary' alliances in Arab ranks contributed to the escalations that led to the explosion. With regard to the Arab diversion works, it was shown in the above that the United States exercised her leverage on Israel in order that wide scale hostilities would not break out because of them, and that a joint appraisal of the situation would be made before any action was taken. Israel's Prime Minister blamed Syrian attempts to violate Israel's sovereignty, not the diversion works, for the outbreak of the April hostilities that started the quick escalation that led to the war. Finally, the water diversion works, with the exception of the Mukheiba dam, had been stalled for months before the hostilities broke out.

The claim that the June war of 1967 was a 'water war' is not substantiated. One observes that water was never brought up in any of the resolutions that came out of the U.N. Security Council as a result of that war. Neither resolution 237 of June 1967, nor resolution 242 of November 1967, addressed the water conflict over the Jordan River basin. As a matter of fact, a U.S. assessment of the Israeli objectives focused primarily on her future security. A memorandum prepared by the Central Intelligence Agency (CIA) was submitted the day after the war broke out, and discussed by the Control Group at the State Department,[242] summed up the Israeli objectives as:

"1. Immediate and primary GOI [Government of Israel] war aims is destruction of the center of power of the radical Arab Socialist movement, i.e., "the Nasser regime". Beyond this the GOI will attempt to see the emergence of a balance to the Arab Socialist-Moscow influence in the Middle East. Prior to the war, Israeli's hopes were based on a loose coalition of Iran, Turkey, the Kurds and the moderate Arabs. After the war, the future of the moderate Arabs

---

[242] Talking Paper Prepared for a Meeting of the Control Group at State by Richard Helm, the Director of the CIA; dated 6 June 1967. Fiche no. 416 A, date of issue 10/01/1981; Document no. 00036; NARA, Washington, D.C.

became a great question mark if not an element to be written off. Hussein's trip to Cairo marked Jordan, in Israeli eyes, for elimination.

2. Maximum destruction of Soviet weapons in the hands of the Arabs. When this goal is achieved, the GOI assumes, Turkey, Iran and Israel will represent an overwhelming military power in the area.

3. Israel will attempt to destroy the Syrian regime and to eliminate both Syria and Jordan as modern states. Occupation of the West Bank is probably a high priority task once control of the tactical air is achieved.

4. Israel will oppose:
a) An early cease- fire,
b) A return to anything resembling the status quo, and,
c) A successful Soviet salvage operation vis-a-vis Nasser.
They would prefer to see the U.S. remain "neutral," remaining aloof from any cooperation with the Soviets. The Israelis would like to see the U.S. Government active laying the groundwork in consultation with Iran, Turkey and Western Europe for a revised U.S. foreign policy that will provide Israel minimum security. Also they would favor the U.S. Government attempting to preserve Faysal as a moderate leader around which the Arab world could be rebuilt."

Again, nothing about water had been mentioned as part of the Israeli objectives, at least in as far as the U.S. strategists could see.

The above meeting of the Control Group discussed a more detailed memorandum presented by the CIA.[243] The memorandum outlined Israeli objectives in general and vis-a-vis each of the surrounding Arab states, the Arab objectives, and the objectives of the USSR. Water was not among the objectives of any of these parties. It looked like the water issue evaporated altogether, and was not directly linked to the causes, effects or objectives of the war. The memorandum read:

"June 6, 1967

MEMORANDUM

SUBJECT: Objectives of the Middle East Combatants and the USSR

---

[243] "Objectives of the Middle East Combatants and the USSR," a paper presented to the Control Group Meeting at State, on 6 June 1967. Fiche no. 416 B, date of issue 10/01/1981; document no. 00037, NARA, Washington, D.C.

Israeli Objectives

1. When Israel began its offensive, its objectives were probably limited to giving the Arab side --especially Nasser-- as severe and humiliating a beating as possible. As the momentum of the Israeli attack developed, with the UAR's air force shattered and with the resistance in Jordan collapsing, the Israelis probably have begun to see a major Arab defeat looming, which would give Israel the opportunity to dictate terms of a semi-permanent peace. To the degree they succeed in humbling the Arabs, they will so dictate a settlement. At least they will aim for the following objectives:

A. General

1) Compelling its Arab neighbors to recognize that Israel will not accept any major Arab political victory over it, e.g., closing Straits of Tiran; that Israel has a right to exist, and that Arab physical harassment has very dangerous consequences.

2) In the above context, destroying as much of the military force of its Arab neighbors as possible.

3) Extracting maximum political advantage by retaining conquered Arab territory until demands satisfied.

B. UAR

1) Humiliating Nasser by inflicting a major military defeat, which, hopefully, would destroy him.

2) Attain use of the Strait of Tiran; possibly by holding territory to insure such use.

3) Right of Suez Canal transit.

4) Gaza represents a real problem in view of its large refugee problem; at a minimum Israel will seek to insure its demilitarization,

C. Jordan

1) Israeli objectives probably developed in the course of the fighting. It seems unlikely that Israel intended attacking

Jordan or destroying Husayn two days ago.[244] Israeli objectives probably are:

a) Seizure of strategic territory on West Bank, or perhaps demilitarization of the West Bank.
b) Holding the old City of Jerusalem.

D. Syria

1) Israel will probably try to seize strategic heights and destroy Syrian military power. In any final settlement, Israel would seek at least to demilitarize the heights.

E. Lebanon

1) Israel content to leave Lebanon alone, except in unlikely case of Lebanese provocation.

2. Summing up, the degree to which the Israelis expect to satisfy their objectives depends on two factors:

A. How far the Israeli forces advance, and,
B. Whether or not Nasser is toppled as a consequence of the war.[245]

Arab Objectives

3. When the crisis began, the Arabs hoped for a political victory, short of war, achieved by political and diplomatic means.[246] Egypt, as well as Syria and Jordan, probably expected to be able to delay any Israeli assault sufficiently to permit recourse to the UN. Given the extent of the defeat suffered to date by Jordan and Egypt, the principal aim of the Arab belligerents is now to salvage what they can. Jordan has already asked for a truce.

---

[244] In the first memorandum, it was mentioned that Israel had listed Husayn to be written-off after his Cairo visit.

[245] In a pre-emptive move, Nasser announced his resignation on 10 June, and turned over the post to the Vice President, Zakari Mohieddin. Massive popular rallies all over Egypt reinstated him.

[246] This Diplomatic approach might have been the objective of the responsible governments, but the Arab peoples, under the influence of sustained propaganda, were sure of a military victory. The outcome of the war was nothing less than a disaster that generated many shock waves across the Arab world.

4. Egypt would probably like to see a cessation of hostilities, retrieving losses through diplomatic means. If it believes the situation to be desperate, Egypt might flail out in a variety of extreme actions, e.g., use of lethal gas or physically blocking the Suez Canal.[247] In any event, Egypt will almost certainly initiate action against the US and UK interests in the Arab world.

5. Jordan is now virtually at the mercy of the Israelis. At the least Husayn will try to retain the throne and will continue to seek US budget support without the former visible US presence.

6. Syria aims will be to minimize Israeli danger and take advantage of Husayn's weakened position.

7. Iraq and Saudi Arabia will want to get out of the war without the overthrow of their regimes. They will seek to minimize damage to the pipelines and oil installations,

8. Algeria will continue actively to support Nasser.

9. Other regimes, i.e., Libya, Lebanon, Kuwait, Morocco, and Tunisia are hangers on. They seek to be involved only to the extent necessary to show solidarity with the Arab cause."

That was the appraisal by the U.S. C.I.A of the objectives of the combatants on the second day of the war. By 12 June, Israel had over-run the Golan Heights and occupied strategic heights there and on Jebal al Sheikh (Mount Hermon).

Some of the Arab regimes were overthrown shortly after that stunning defeat. Iraq's regime of the Arif brothers (Abdul Rahman Arif took over after the death of his brother, Abdul Salam, in a helicopter crash in April 1966) was overthrown on 17 July 1968, and the Ba'th once again took control. Libya's Monarch was overthrown in September 1969 and a young Colonel, Muammar al Ghaddafi, took over. Syria's rulers were overthrown by their own military supporters in October 1970 and Asad took control. Sudan's regime was overthrown in a military coup and Ja'far Numeiri took over. The royalists in Yemen lost their campaign and the republicans took control in 1970. Aden (South Yemen) became an independent republic run by Marxists in 1968, and the Arab protectorates formed independent states in the Gulf, and joined with each other to form the United Arab Emirates (UAE).

---

[247] Egypt actually exercised the second option, physically blocking the Suez Canal with sunken ships.

The second part of the CIA appraisal dealt with the USSR:

"**<u>Soviet Objectives</u>**

10. As the Arab-Israeli conflict rages, Moscow's immediate objectives are:
    a. To contain the conflict within the limits of its present locale and without the intervention of outside (i.e., Western) powers;

    b. To effect a cessation of hostilities with the minimum possible damage to their standing in the Arab world.

        1) The Soviets are anxious to forestall a disastrous Arab defeat that would open Moscow to the onus in Arab capitals of not having done what it could to defend the Arab cause.

        2) The Soviets will avoid giving the impression that they are selling out the Arabs by continuing their public support for the Arab side and by continuing to condemn Israel as aggressor, inspired and backed by the imperialist US and UK.

        3) Whatever form the negotiations take following the end of hostilities the Soviets are likely to weigh in very heavily on the Arab side to repair the damage already done to their standing. They will seek to forestall any situation in which the Arabs are abandoned to the dictates of Israel.

11. Implicit throughout the above is Moscow's desire to avoid a Soviet military confrontation with the West. The USSR has no intention of intervening militarily, and can be expected to do whatever is necessary to avert a major flare-up with Western intervention while appearing to support the Arabs."

The Security Council issued its famous resolution 242 on 22 November 1967, calling for the rights of all States in the region to live within recognized and secure boundaries, for the withdrawal from territories occupied in the recent conflict, and for the settlement of the problem of Palestinian refugees. The efforts by a United Nations' special envoy, Gunar Jarring, to have the resolution implemented were not successful.

The occupation of the Arab lands continued. Arab resistance turned to popular and para-military Palestinian organizations. Fatah picked up popularity, and other guerilla organizations emerged. It is not the intent here to review the developments after the June war as much as it is essential to look into the causes of that war. It is important to pin down the causes of that war, and to underline the fact that the war had its own motives and causes as the foregoing presentation clarified, and that it was not, as some analysts and journalists analyzed in retrospect, a 'water war'. The only ongoing works of the Arab diversion project at the time the war broke out were limited to the activities on the Mukheiba dam that straddled Jordanian-Syrian territories near Mukheiba. Israel did not object to its construction.

The Syrian territory across the Yarmouk at the Mukheiba damsite was occupied by Israel as part of her occupation of the Golan, and the works there were suspended. Jordan settled her financial accounts with the Special Fund of the Arab League in 1973, and did so with the Egyptian contractor. The partially completed works comprised the township, the excavation for the foundation of the dam, and the supply of construction equipment and spare parts.

It is important to note here that the Jordan Valley of the Hashemite Kingdom became an important theater for the paramilitary operations mounted by Palestinian organizations in their resistance of Israeli occupation. The Valley transformed into a front with guerilla operations waged against Israel from it, and Israeli shelling from across the river in retaliation. On 22 March 1968 Israeli forces crossed the Jordan River eastward in a military operation against the refugee camp of Karama which had been made into a base for guerilla warfare against Israel in the Jordan Valley. The Israeli forces engaged in an early morning battle with the fighters of Fatah before the Jordan Armed Forces engaged with the Israeli invading forces and dealt them their first beating. Several other encounters ensued, and the Jordan Valley was all but depopulated of civilians. Sixty percent of its houses were destroyed, and its population fled eastward to take refuge in the towns and cities on the Jordanian Plateau. The water structures of the East Ghor Canal became a target only on one serious occasion. The Canal was shelled from across the river, and a short section was knocked out. It was hurriedly repaired with earth dikes that lasted in service for about eighteen years, before it was properly repaired in 1986.[248]

The war was therefore not a water war, and water structures, with the exception of the one above incident, were not targets for the adversaries during the war nor during the paramilitary campaigns after it.

---

[248] The author organized the proper repair of the canal when he was President of the JVA.

CHAPTER VI

Water for the Jordan Valley

1. The Aftermath of the 1967 War

The occupation of the West Bank dealt a severe blow to the economy of the Hashemite Kingdom of Jordan, and burdened it with a new wave of displaced Palestinians, who crossed the Jordan River seeking refuge. Some of those had been Palestinian refugees living in camps on the West Bank since 1948. New camps were hurriedly set up in Jordan for the newcomers. The contribution of the West Bank to the economy of the country suddenly ceased, and shifted to be negative as the Kingdom decided to keep open bridges with the occupied West Bank, with people and goods freely moving in one direction: eastward, and with special Israeli permits, people could travel westward.

The confrontation with the Israeli occupation was waged from the Jordan Valley. The Valley had been booming with agricultural development, as a result of the implementation of the East Ghor Canal Project and the diversion of Yarmouk water for irrigation. By June 1967, about 114,000 had been brought under irrigation through staged implementation of the Project under grant financing from the United States. The Valley population approached about 70,000 people and formed a cohesive rural community. Shortly after the war ended, the PLO paramilitary forces moved in that border zone and its various factions maintained bases there from which they mounted raids against Israeli occupation in the West Bank. The Israeli forces retaliated with fire from their new positions on the mountains of the occupied West Bank, and from old positions facing the north Jordan Valley. The confrontation lasted until 1971 and the results were devastating to the Jordan Valley and to Jordan. Agricultural production came to an all time low, exacerbating the economic hardships caused by the loss of the West Bank to Israeli occupation. About sixty percent of the houses in the Valley were destroyed and it was all but depopulated. Most of the people took to the Plateau seeking security and safety, with about 5,000 people remaining to care for perennial crops. The irrigation infrastructure came only once under deliberate serious attack by the Israelis, and other hits were only of minor importance.

The PLO factions established headquarters in Amman and were present in almost all of the cities of Jordan. They soon became involved in the affairs of these cities, and defied government authority. The factions were not all united and had their internal rivalries. The domestic situation in Jordan deteriorated and was finally exacerbated by the skyjacking of international airlines to a desert mud flat, northeast of Amman, where they were made to land. That operation brought the Jordanian authorities and the PLO to a serious confrontation. King Hussein formed a military government headed by a Jordanian officer of Palestinian origin, Mohammad Daoud, and the Jordan Army assaulted the capital, Amman, on 16 September 1970 to dislodge the PLO fighters who had taken control of the city in 1969. Syrian forces crossed the borders to come to the aid of the PLO but were soon ordered home after they clashed with the Jordan Army in the north. The Arab League intervened, an Arab Summit was convened and a ceasefire was arranged. Finally, in April of 1971, the paramilitary forces of the PLO evacuated Amman and the other major cities to the mountains in the northwest of the country. Another armed clash ensued, and the PLO forces were routed out of Jordan by the Jordanian army. They found a convenient environment in Lebanon from where they resumed their attacks against Israel.[249] Their presence there was a primary factor in the eruption of the civil war in Lebanon (1975-1989).

As soon as the confrontation with the PLO came to an end, in the summer of 1971, Jordan initiated efforts to enforce law and re-establish order.

In Egypt, Nasser had passed away on 28 September 1970, and was succeeded by Anwar Sadat, who soon entered into rivalry with his comrades and won the bout with them. In Syria, the radical leftist regime was toppled in October of 1970 and a more accommodating leader, General Hafez Assad, took control. In Iraq, the Ba'th party ascended to power through a coup in July 1968, and a familiar face, General Ahmad Hasan el Bakr, was named president. The strong man, however, was his deputy, Saddam Hussein, who replaced him a decade later.

In Jordan, with the internal and regional political pressures easing off, the government started out on the development trail to achieve economic recovery and growth. In November 1971, the strong Prime Minister of Jordan, Wasfi Tell was assassinated in Cairo, where he was attending a meeting of the Arab Joint Defense Council. The King gave his youngest brother, El Hasan Ben Talal, then Crown Prince, a mandate to lead the efforts of development planning. He led teams of Jordanian experts and specialists to achieve that goal and formulated a "Three Year Development Plan, 1973-1975". The Jordan Valley received special attention, and El Hasan headed a special team to formulate a plan for the rehabilitation and further development

---

[249] For a detailed account of this topic see "Palestine, from Jerusalem to Munich," a book by Abou Dawoud (in French), translated into Arabic and published in series by *Al Hayat* newspaper, Kensington Centre, 66 Hammersmith Road, London, W14 8 YT.

of the East Jordan Valley.[250] A separate plan (1973-1975) was formulated for the Rehabilitation and Social and Economic Development of the Jordan Valley - East Bank. Both plans were presented to His Majesty King Hussein in November 1972, and he gave the green light for their implementation. The young prince El Hasan, then 25 years of age, displayed unique leadership qualities and led the efforts for the implementation of the plans, including securing financial support for their projects from friendly countries and specialized development institutions. The energy needed to overcome the static inertia that had befallen the country was enormous, and the young prince met the challenge with equivalent vigor.

2. Rehabilitation and Development of the Jordan Valley

The Jordan Valley Development Plan was the first attempt by the Jordan Government to pursue an integrated social and economic development approach. The backbone of the Valley development was irrigated agriculture. The Plan, however, focused on economic and social endeavors to develop the rural Jordan Valley and optimize its contribution to the Gross Domestic Product. The Valley was to provide job opportunities capable of supporting increasing population and by halting the phenomenon of migration to urban areas, while simultaneously, attracting urban population to move to the Valley and take up economic opportunities in agricultural and related enterprises. The infrastructure needed to support agriculture and a rural community was to be built and maintained. A new autonomous government organization, the Jordan Valley Commission, was established to attend to the implementation of the Plan, and to continue the process of further planning for economic and social development. The Commission was to rely first on the existing government organizations to perform its tasks, but could perform them independently provided that the projects would be turned over to the concerned government agency once they were completed. The concerned government agency would care for the operation and maintenance of the respective project. That was a first attempt in government administration to assure integration of planning and implementation. The Commission worked efficiently, with minimum manpower that was mostly borrowed from other government ministries.

The development effort accelerated with generous assistance from Arab countries and from friendly western countries. The oil boom that accompanied the 1973 Ramadan/Yom Kippur War brought about unexpected growth in the wealth of the oil rich countries, and those, in turn granted

---

[250] The head of the technical team was Omar Abdallah Dokhgan, then Minster of Agriculture in the Cabinet headed by Wasfi Tell. Dokhgan was chosen to be President of the newly established Jordan Valley Commission in February 1973. The author was appointed as his deputy in May 1973.

Jordan financial assistance for military support and economic and social development, and their labor markets absorbed a lot of qualified Jordanians. Unemployment disappeared from Jordan's labor market, and the country welcomed the unskilled labor that came to it from Egypt, Syria, and a number of Asian countries.

3. Water Supply From the Yarmouk

In 1977, the Jordan Valley Commission was transformed into the Jordan Valley Authority (JVA). It differed from the Jordan Valley Commission in that it was allowed to retain water projects and care for their operation and maintenance. That task had been the responsibility of an existing government organization, the Natural Resources Authority, NRA. The JVA acquired other government departments concerned with water in the Jordan Valley. One such organization was a division within NRA in charge of management of the East Ghor Canal Project and irrigation infrastructure in the Jordan Valley, and another was the Jordan River and Tributaries Regional Corporation (JRTRC) that was established to implement the Arab Project after the September 1964 Summit, and a third was the Jordan Valley section of the Domestic Water Supply Corporation. The geographic domain of the JVA was extended southwards to include the Dead Sea and the part of the Rift Valley running south of the Dead Sea to the city limits of the port of Aqaba on the Red Sea. The projects in the sectors other than water projects were to be implemented by the JVA and handed over to their respective ministries for operation and maintenance.

Water for irrigation came primarily from the Yarmouk River via the East Ghor Canal, a project that was conceived under the various plans for the utilization of the Yarmouk waters and was part of the staged implementation of the Unified Plan, as outlined by the "Lahoud Plan". The flow diverted from the Yarmouk to the East Ghor Canal in 1966 was sufficient to irrigate an area of about 12,000 ha extending between the Yarmouk and the Zarqa rivers. The design of the project indicated that an area of 3,000 ha needed an average flow of 1 cubic meter per second (cms), making the total flow needed for the entire irrigated area about 4 cms. The average river discharge in the dry months (April through October) was in the order of 6 cms, and Jordan was entitled to 70% of the river's summer flow at the diversion point. The balance, or 30%, was to be made available for Israeli uses downstream. This arrangement was worked out by the United States' responsible offices when the grant was advanced to Jordan to build the East Ghor Canal Project.[251]

[251] See the June 1963 report by Wayne D. Criddle, prepared for the Near East Division of the Department of State and cited in Chapter 4 above; copy declassified and available from the U.S. National Archives, State Department Records, Washington, D.C.

Other sources of supply were the side wadis on which agricultural activities also depended.

It is important to note that the river flow at the Adasiyya tunnel is sensitive to Syrian abstractions upstream. Syria had historic uses of the spring waters in her territories that fed into the Yarmouk. Her uses were irrigation, during the dry months of April through October. The monthly average of Syrian abstractions was 3.8 mcm which left a residual flow in the Yarmouk downstream at the tunnel ranging between a 13.7 mcm low for the month of August, and an 18.5 mcm high for the month of April. Of these flows, Jordan was entitled, as stated above, to 70%. The side wadis inside Jordan contributed between 8.4 mcm (0.25 cms) during the month of September and 10.5 mcm (0.31 cms) during April.[252] Thus the water available to Jordan could irrigate an area of 10,400 ha in August, and 12,400 ha in January. However, the water balance for a specified cropping pattern with a specific crop intensity allowed the irrigation of 12,400 ha.

The Jordan Valley Rehabilitation and Development Plan included the construction of a dam on the Zarqa River to increase water availability in dry months and enable the expansion of irrigated agriculture. Construction of the dam, 56 mcm in capacity, started in 1971, and was completed in 1977. A total average yield of about 65 mcm (about 2 cms) would accrue from the Zarqa River as a result of regulating its flow by the King Talal Dam, and about 6,000 ha would be irrigated using its waters. The Plan based its assessment of water resources on the average contribution of the Yarmouk and the side Wadis throughout the history of operation of the East Ghor Canal Project. However, drastic changes had taken place and continued to take place as the Plan was being implemented and thereafter.

These developments were critical to the sustainability of irrigated agriculture in the Jordan Valley. On the one hand, Syria had intensified her efforts to utilize the waters of the springs and tributaries to the Yarmouk inside her own territories and, on the other, the flow of the side wadis inside Jordan were being diminished by abstraction of water from their aquifers for municipal and agricultural purposes. The Yarmouk and the side wadis are the primary water sources for the Jordan Valley. Their exploitation by Syria upstream and by Jordanians respectively reduced the amount of water available for use in the Jordan Valley and impacted water quality. The water shortage was felt there especially after 1979, when the irrigated areas expanded under the Plan became ready for cultivation.

Water from the Yarmouk was diverted to Jordan without any structure protruding into the river course. Although such a structure was

---

252 "Water Available to the East Ghor Canal Irrigation Project," a memorandum submitted by Fahd S. Natur, Gene T. Thompson, N. Simansky and James Metzger, 1962, Files of the East Ghor canal Authority, Amman, Jordan.

envisaged under the unilateral diversion scheme financed by the United States, its construction was vetoed by the Israelis on the grounds that it might prejudice her share of the Yarmouk water and give military advantage to her adversaries in the southern DMZ between her and Syria. Instead, a simple drop inlet was dug-in on the Jordanian bank of the Yarmouk to "induce" diversion into a short channel, from which the flow entered a 1 km long, concrete-lined tunnel, emerging into daylight at the beginning of the concrete-lined East Ghor Canal. Across the river from the diversion drop inlet is the Himmeh strip, a part of the southern DMZ that separated the Syrian and the Israeli forces as per the Armistice Agreement of 20 July 1949. However, the Himmeh strip across the river was not inhabited and was under de-facto Syrian police jurisdiction. Until June 1967, the Jordanian maintenance crew could easily access the entire Yarmouk river bed, perform work there at any time they wanted, and keep it always clean of deposits brought about by the flood seasons.

As a result of the June 1967 war, not only the West Bank fell to Israeli occupation in 1967, but so also did the Golan Heights of Syria and the Sinai of Egypt. All three DMZ's between Syria and Israel, north, middle and south, fell to the Israelis, as their forces advanced to occupy the Golan. A new fact on the ground dictated itself and affected the operation of the Jordanian water intake from the Yarmouk to the East Ghor Canal. The midstream of the Yarmouk between its confluence with the Jordan River to the west and the confluence of Wadi Raqqad with it to the east became the cease-fire line between the Israelis and the Jordanians (Figure VI-1). Across the river from Jordan, Israeli military positions were dug-in, and Israeli roaming patrols surveyed the area on a continuous basis. The Jordanian operators of the East Ghor Canal and the maintenance crew were not allowed to approach the river, let alone cross it to clear the riverbed of flood debris.

Prior to the June war, there was enough water for Jordan's needs, and more to bypass the tunnel and flow downstream for Israeli uses. The irrigation water requirements in the Jordan Valley diminished after the war because of a lull in agricultural activities and a reduced cropping intensity. This lull was caused by the paramilitary activities of Palestinian guerilla organizations in the Jordan Valley, as outlined above. In fact, there was more water than the brave farmers who stayed behind could possibly use. Regular farming activity was resumed in 1973, after law and order were re-established and after the Government announced her serious plans for rehabilitation and further development of the Jordan Valley. More water was needed from the Yarmouk, but despite increasing Syrian uses upstream, Yarmouk water there was enough for the irrigation of the lands north of the Zarqa River with a total area of about 11,500 ha.

Figure (VI.1)- Cease Fire Line on the Yarmouk after 1967

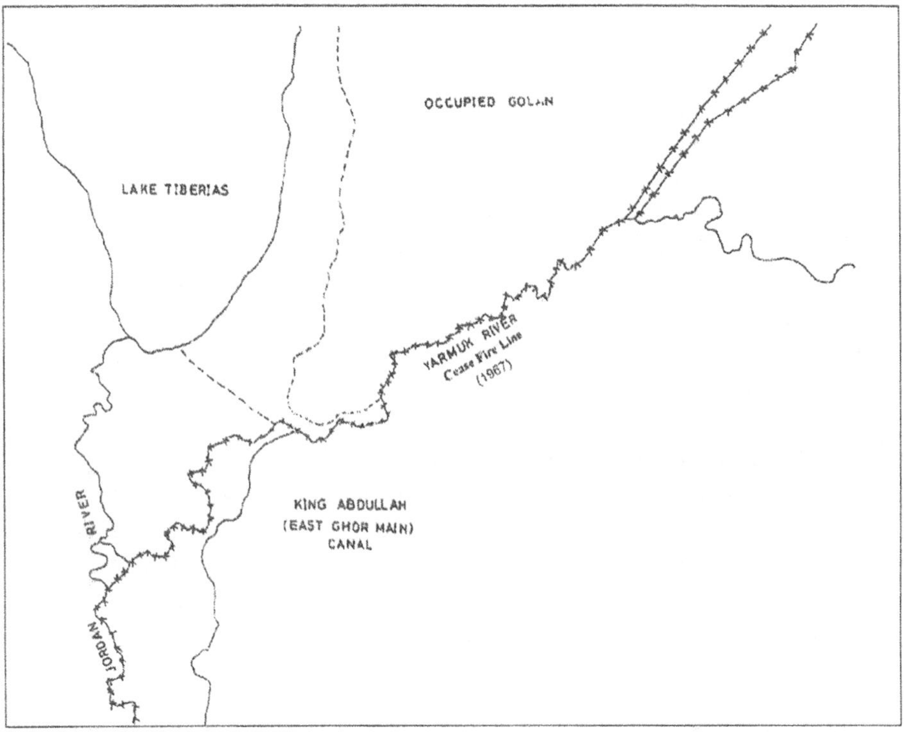

The Jordan Valley Development Plan called for the expansion of irrigated agriculture in two stages. Stage I would extend the irrigated area to about 22,000 ha, and Stage II would increase it to 36,000 ha. To achieve these goals, higher irrigation efficiency had to be achieved to reduce water losses and waste; furthermore, dams to regulate the flow of the Yarmouk and the side wadis were to be built to increase the supply of water during the dry months. The Plan included the construction of the Maqarin Dam on the Yarmouk to acquire Jordan's share as stipulated in the Unified Plan of 1955 and in the Jordanian-Syrian Agreement of 1953. Jordan approached the World Bank to solicit its support for the project in 1973, but the Bank turned down Jordan's request.[253] Jordan moved to build a much smaller dam[254] on Wadi al Arab in 1979 to support the expansion of irrigated agriculture in the northern part of the Valley.

4. The Ramadan/Yom Kippur War of 1973

The political situation in the area was transforming. President Sadat of Egypt, having won his battle against his domestic opponents, moved to terminate the services of the Soviet military experts working with the Egyptian Armed Forces since the 1967 war. Egypt had been engaged in a war of attrition against the occupying Israeli forces on the opposite bank of the Suez Canal. Sadat pledged a war of liberation when the time was ripe. On 6 October 1973 (10th of Ramadan 1393 H),[255] the Egyptian and Syrian armed forces mounted a surprise attack across their respective fronts with the Israeli occupation forces. The Egyptians were able to cross the Suez Canal, over-run the Israeli Bar-Lev fortification line, route the Israeli forces there, and advance towards the passes to liberate Sinai. The Syrians mounted a simultaneous attack, advanced on the Golan, recovered Jabal al Sheikh (Mount Hermon), including the springs of Banyas, and were able to take Israeli soldiers as captives. Jordan and Iraq sent troops to support the Syrians in their combat on the Golan. However, before the Egyptians had time to

---

[253] A Jordanian delegation visited the Bank in the spring of 1974, led by Crown Prince El Hasan and included Omar A. Dokhgan, President of the Jordan Valley Commission, Dr. Khalil Salem, President of the National Planning Council, Dr. Hanna Odeh, Secretary General of the National Planning Council, and others.

[254] Wadi al Arab dam was initially designed to have a live storage capacity of 7.5 mcm. It received the support of the Government of Japan who extended a loan through its Overseas Economic Cooperation Fund (OECF), for the purpose of design and construction.

[255] "H" stands for Hijrah, the point in time when Prophet Mohammed moved his base from Mekka to Yathrib (Madeenah); it marks the beginning of the Islamic Calendar (lunar).

enjoy their initial victory over the Israeli forces, the Israelis managed to cross the Suez Canal westwards with armor, under General Ariel Sharon, and besieged the Egyptian Third Army in south Sinai. They were also able to turn to the Golan, route the Syrian army, and advance past the cease-fire line of 1967 towards Sa'sa' on the way to Damascus. The United Nations Security Council issued its resolution no. 338 of 1973 calling for a cease-fire and for the implementation of its resolution no. 242 of November 1967. The Arab oil producers declared an oil embargo in support of the Arab and Palestinian cause. The embargo sent oil prices up to unprecedented levels.

The United States moved quickly to bring the situation under control. A Peace Conference was convened, under U.N. auspices, in Geneva for a short while in 1974, but was adjourned without positive results. The United States became active in a spree of shuttle diplomacy to arrange for the disengagement of forces in the Sinai and on the Golan front. The Secretary of State, Henry Kissenger, exerted strenuous efforts and achieved disengagement agreements on the Egyptian and Syrian front in January and May of 1974 respectively. He was unable to achieve a similar agreement on the Jordanian front. Prime Minister Yitzhak Rabin refused to entertain a Jordanian request to disengage on the Jordan River.[256] Soon after the disengagements were achieved, President Sadat, having established secret contacts with Moshe Dayan of Israel in Morocco, decided to make a historic trip to Israel to break the barriers on the road to peace. The trip was made in 1977 and was followed by peace negotiations between Egypt and Israel under the auspices of President Jimmy Carter. The negotiations culminated in the Camp David Accord in 1978 and subsequent agreements between Egypt and Israel in 1979. The Arab states were in dismay, because of Egypt's unilateral action to make peace with Israel, and Iraq called for an Arab Summit, which convened in Baghdad in 1978. Decisions were taken to reinforce the eastern front against Israel, provide support for Jordan, Syria and the PLO, and to move the Arab League Headquarters from Cairo to Tunis. Egypt was expelled from the League and the Egyptian Secretary General was replaced with a Tunisian diplomat, al Shadhily al Qulaybi. After an era of strong Soviet influence in Egypt, the United States made a successful comeback and Sadat pursued a policy to reform the Egyptian economy and reverse the socialist trend that Nasser had followed since 1959. The Soviet ties were maintained with the

---

[256] In a panel on 18 October 1999 at Tel Aviv University which the author attended, a former Israeli intelligence chief panelist mentioned that Kissenger told Rabin that he had to brace himself with talks with Arafat if he declined the Jordanian request of disengagement. The panel was part of a conference activity organized by the Rabin Center for Israel Studies to commemorate the fourth anniversary of Rabin's assassination and the fifth anniversary of the Jordan- Israel Peace Treaty.

Arab leftist regimes in Syria, Iraq, and South Yemen in the Middle East, and with Libya and Algeria in North Africa.

5. Resumption of the Development Effort

It was in this complex political atmosphere that Jordan was trying hard to jump-start her development efforts in 1973..She received financial support from a score of friendly countries and specialized development lending institutions.[257] The Jordan Valley projects aimed at economic and social development and included the following:

a. Water resources development projects that called for the construction of dams on the side Wadis, the development of groundwater resources, and the expansion of irrigated areas. The irrigation networks were converted to pressure pipe networks, in lieu of surface canals, to improve conveyance and distribution efficiency. Donors to these projects were the Kuwait Fund, the Abu Dhabi Fund, the IDA (World Bank), USAID, KFW (Germany), OECF (Japan), the Saudi Fund, and the Government of Italy.
b. Village Development projects, which included the development of existing villages, the creation of new population centers, an attractive program for house mortgage, and the construction of village streets, schools, health centers, government buildings, community centers, and public utilities of electricity, telecommunications, and domestic water supply. Donors to these projects were USAID and KFW.
c. Valley infrastructure, including highways, farm to market roads, high tension power lines, microwave telecommunications links, agricultural marketing centers, agricultural research and extension facilities, and others. Donors were USAID, KFW, the Netherlands, the Kuwait Fund, the United Kingdom, and IDA.

The development effort that was kicked off in 1973 gained momentum and peaked in the middle half of the 1980's. The first batch of projects started to come on line in 1977, when the main Valley highway was completed and the electricity grid was finished. By 1979, a total of 8,300 ha were added to the area of irrigated agriculture. The irrigation conveyance and distribution

[257] For the Jordan Valley, the United States headed the list, followed by the Federal Republic of Germany, the Kuwait Fund for Arab Economic Development, the Saudi Fund for Development, the Abu Dhabi Fund for Economic and Social Development, the International Development Association of the World Bank Group, the Arab Fund for Economic and Social Development, the Governments of Japan, the United Kingdom, the Netherlands, and Italy.

networks adopted for the new areas were pressure pipe networks to reduce water losses to a minimum and improve water use efficiency. The new systems were installed on 1,000 additional hectars of existing irrigated areas in lieu of the traditional surface irrigation canals. As a result of the expansion in irrigated area, the need for water to irrigate both existing and new areas by 1980 had increased by about 80%, and continued to grow in subsequent years as cropping intensities increased. In the meantime, the availability of water from the Yarmouk receded, and the contribution of the various side wadis decreased in quantity and in water quality. The King Talal Dam on the Zarqa River was put into operation in 1978, but was filled and spilled in November of 1979 and provided much needed water thereafter.

6. Water Contribution From the Yarmouk and Side Wadis

The Yarmouk was the primary contributor of water to the Jordan Valley, averaging about 130 mcm per year with minimum contribution of 4.5 cms in August of 1963 (the rest of the Yarmouk flow, about 2 cms, bypassed the diversion point for Israeli uses downstream). The side wadis inside Jordan contributed much less. However, both contributions started to diminish in the 1970's because of abstractions by other users upstream. Cities and towns in the Jordanian uplands witnessed sudden increases in population, as a result of people pouring into them from the West Bank. The water supply to these cities came primarily from tube wells drilled in aquifers that also fed the flow of the side wadis. In effect, water was diverted from the side wadis to municipal uses in towns and cities. Amman, the capital city, hosted a higher percentage of Palestinian refugees and displaced persons from the Palestinian territories. Its local supplies came from the aquifer that supplied the Zarqa River with its base flow. The aquifer was over-pumped to meet the increasing water demands of the capital and as the water level of the aquifer receded, springs dryed up. Amman's historic perennial freshwater stream dried up at the source and transformed into a floodwater drain running only seasonally. A similar impact was experienced in the catchments of the other side wadis. Those impacts became more pronounced after the June 1967 war, when more Palestinians were displaced from their homeland and crossed the river into the East Bank of the Kingdom.

In the case of the Yarmouk, which provided the bulk of irrigation water to the Jordan Valley, Syria had accelerated her efforts to utilize the water from springs feeding the river, and to build dams on the Yarmouk tributaries located inside her own territories. The Israelis, on the other hand, started a water development program for their settlements in the Golan that included the construction of small dams and the drilling of wells for groundwater utilization. They even installed pumps on the Yarmouk and pumped water up the escarpment of the Golan for irrigation purposes. More water was used by Israel from the occupied Himmeh spring for fisheries, including alligator farms, and for banana farming. The Israelis also built small dams on tertiary side wadis in the Golan. All these upstream developments took place without serving any notification to Jordan. As a matter of fact, the Syrian dams, the construction of which started in 1967, came to the attention of the Jordanians by sheer coincidence in 1976.

The decline in irrigation water flow prompted the Jordan Valley Authority to invest in improving the irrigation efficiency by replacing the concrete lined surface canals with pressure pipe networks. By 1994, all the networks in the Jordan Valley (with the exception of the East Ghor Main Canal itself) were converted to pressure pipe networks, using mostly ductile iron pipes. The irrigation distribution efficiency[258] increased from about 55% to about 95%, with an overall irrigation efficiency as high as 80%. However, the water thus saved would not last long for Jordanian uses, because of the expansion in Syrian abstraction from the Yarmouk upstream. Despite close ties between the two countries that were resumed in 1973, Jordan could not persuade Syria to stop at the ceiling of her share of the Yarmouk waters, as specified in their bilateral agreement of June 4, 1953. Instead, Syria expanded summer and winter irrigation from the Yarmouk sources to the detriment of agriculture in the Jordan Valley. An expansion in the irrigated area of some 6,000 ha in the southern part of the Jordan Valley, completed in 1988, remained fallow for lack of irrigation water. The additional flow that the designers of the project counted on did not accrue to Jordan. While the water saving was achieved through the use of the advanced networks of pressure pipes, Syria took advantage of the water that Jordan had saved through more diversions from the Yarmouk.

The lack of maintenance of the Jordanian diversion point on the Yarmouk exacerbated the difficulties of reduced water diversion. Sediments were deposited in the riverbed at the intake area every winter, calling for periodic dredging and cleaning before the next rainy season. Jordanian workers had done that without difficulty, before the June War of 1967. They were not able to continue with this maintenance work after the Israeli occupation of the opposite bank of the Yarmouk (the Himmeh strip) and the Golan. The sediments accumulated during each flood and retarded the water diversion into the East Ghor Canal.

When the water needs of the Jordan Valley were low, during the turbulent years between 1967 and 1973, the reduction of diversions from the Yarmouk did not pose a serious problem. With the passage of time, the sediments deposited in the Yarmouk riverbed developed into a small, long,

---

[258] This measure is to be distinguished from the on-farm efficiency that improved also. An efficiency of 85% was achieved with the use of drip irrigation methods. The overall efficiency became about 80% in the dry months.

narrow island or a "sand bar". The Yarmouk summer flow, progressively reduced by increasing Syrian abstractions, continued around the little island; a portion was diverted into the drop inlet of the East Ghor Main Canal, and the remainder kept flowing past the Jordanian drop inlet to where the Israelis could pump it from four consecutive locations on the northern Israeli bank of the river. The Israeli pumps, installed by Jewish settlers before the creation of Israel to replace older, surface canals, irrigated portions of the Yarmouk Triangle in Israel. The irrigation of the Triangle was the basis of its water rights on the Yarmouk (defined in the Unified Plan at 25 mcm per year). Because of the absence of a diversion structure to regulate the division of water at the Jordanian drop inlet, the amounts diverted to Jordan were very sensitive not only to Syrian abstractions, but also to the physical condition of the diversion point and the intrusion of the "sand bar". The Syrian abstractions impacted the total River flow at the diversion point, and the sand bar impacted adversely the diversion efficiency to Jordan, and positively impacted the remaining flow that passed by the tunnel towards Israel.

7. The Maqarin Dam

The Jordan Valley Commission realized that the full development of the Jordan Valley required the regulation of the Yarmouk River flow. The plans of the 1950's for such regulations called for the construction of a dam at Maqarin or at Wadi Khalid, and for the diversion of the residual unused flow to Lake Tiberias. The Unified Plan, as agreed between the Arab Technical Committee and Ambassador Johnston, included the dam at Maqarin and the storage in Lake Tiberias, or any other economic location, as determined by a neutral engineering board. The Arab Plan, devised after the Alexandria Summit in 1964, proposed the construction of a dam on the Yarmouk at Mukheiba that could help regulate it and to which the excess flow from the Hasbani and Banyas would be diverted. The diversion of water from Mukheiba dam to the East Ghor Canal would be via a tunnel under the mountain to Wadi Arab. The Arab Plan was stalled in 1966 for lack of funds, and was discontinued altogether in 1967 by the occupation of the Golan by Israel. The location at Wadi Khalid has the advantage over Maqarin of impounding the flow from Wadi Raqqad, a major tributary to the Yarmouk originating at the foothills of Mount Hermon in Syria. However, that site became the end of the zone along the Yarmouk occupied by Israel, in June 1967, and was not accessible by Jordanians for site investigations, let alone construction. The Maqarin site is further east, and is located between Jordan and Syria, a few kilometers away from the new cease-fire line between Syria and Israel (Figure VI.2).

Figure (VI.2)- Maqarin Site

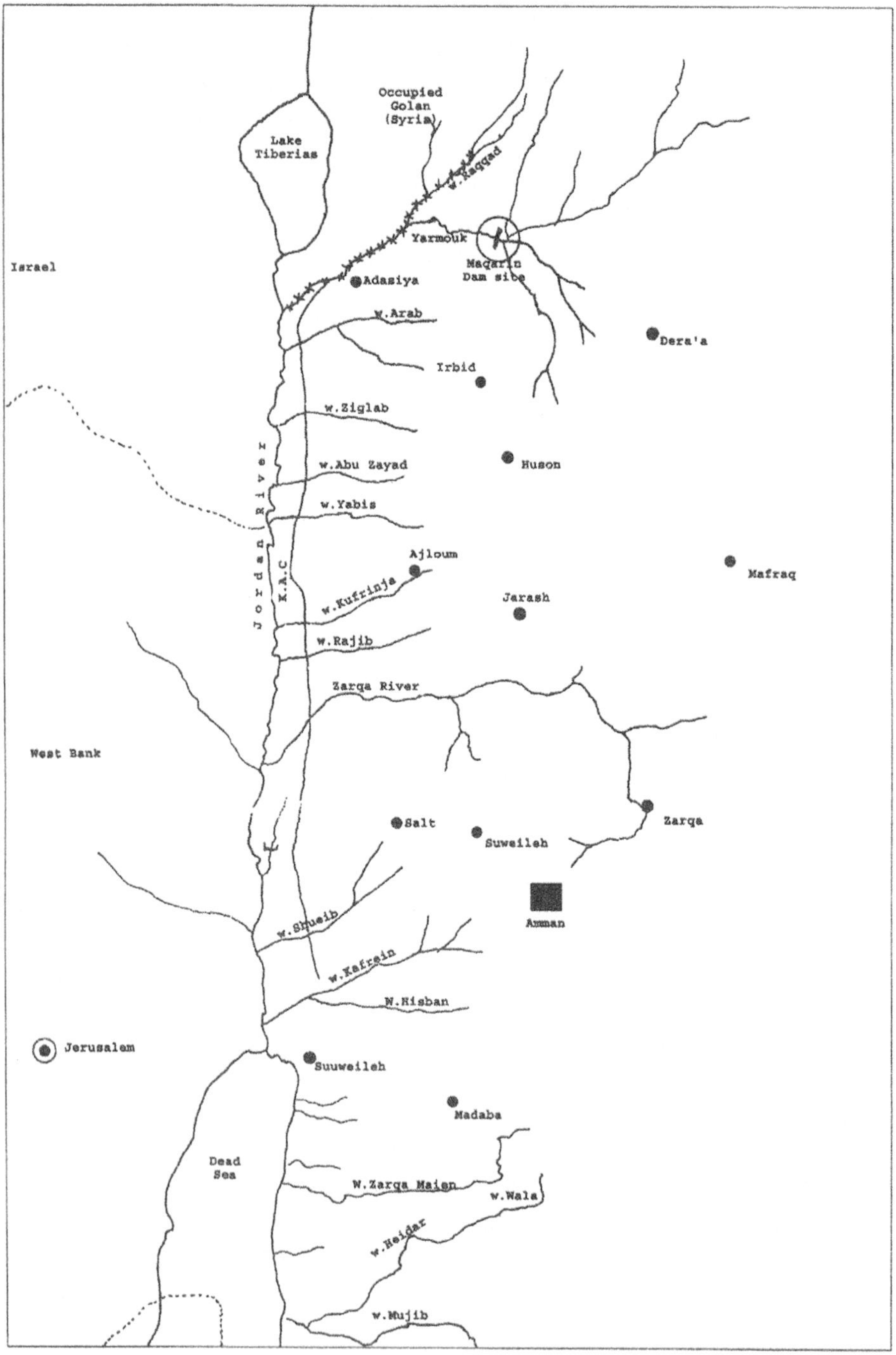

7.1 Contacts with the World Bank

The support for the development of the Jordan Valley was wide, both inside Jordan and in the ranks of donor countries. It was felt that a developed Valley would reduce the risks of military confrontation and the risk of people abandoning it, if hostilities with Israel erupted again. Additionally, a densely populated Valley would make it unattractive for Israel to invade and occupy. Capitalizing on the support the Valley development was receiving, and on the attention paid to the region in the wake of the war of October 1973, the Jordan Valley Commission intensified its efforts to secure support for the construction of a dam on the Yarmouk at Maqarin, and contacts were resumed with the World Bank to win support for the construction of that dam.

Mr. Robert McNamara, the President of the World Bank, paid a visit to Jordan in November 1974, accompanied by the high-ranking officials at the Bank in charge of the Middle East and North Africa. The visiting officials were briefed on plans for economic and social development, including the plans for the development of the Jordan Valley. A major project for the Valley development was the construction of the Maqarin Dam and the extension of the East Ghor Main Canal for its final stretch of 14.5 km. The McNamara visit culminated in a development dialogue between Jordan and the World Bank.[259] Agreement was reached to have Jordan prepare a report on its development priorities and projects, their justification and projected benefits and schedule of implementation. A select team of the Jordan Valley Commission labored through November and a good part of December 1974 to prepare the report requested by the World Bank. The report was presented in Washington, D.C. in January 1975 and was discussed with the specialized group there. It was well received, and the Maqarin dam project was revived.

7.2 The Maqarin Studies

The JVC approached USAID in 1975 for support of the construction of the Maqarin dam. The response was positive and $1.0 million was allocated to study the feasibility of the project. This was a promising sign and gave the Jordanians hope that the dam would be built, because of the influence the United States had in the area, and over Israel, in particular. Involvement of the United States would encourage the World Bank and would help overcome Israeli objections, a factor that was key to World Bank involvement. Through a process of competitive bidding among five pre-

---

[259] The Bank's team included Mr. Benjink, Vice President for Europe, Middle East and North Africa; Mr. Bart, Director of Middle East Department, and others. The Jordanian team was headed by Omar Abdallah Dokhgan, President of JVC, and included Dr. Hanna Odeh, Secretary General of the National Planning Council and a member of the JVC Board; Dr. Ahmad Mango, Economic Advisor to the Crown Prince, and the author, then Deputy Director of the JVC.

qualified U.S. consulting firms, Harza Engineering Company, the veteran of Jordan Valley development planning and the co-author of the Jordan Valley Master Plan in 1955, was selected to do the study. The proposed dam-site would make it straddle territories in Jordan and Syria, and its reservoir would impound lands in both countries. The bilateral agreement of June 1953 formed the basis of co-operation between the two countries in respect to the development and utilization of the Yarmouk.

Jordan's relations with Syria had been sour throughout the 1960's, but took a turn to the better in 1970 after the "military" faction of the ruling Ba'th Party, headed by General Hafez al Assad, toppled the radical Marxist faction. The relations between the two countries became cordial. In the October 1973 war with Israel, Jordan sent in armored brigades to help liberate the Golan Heights. King Hussein paid a visit to Syria and the ties with Syria were further strengthened. An Arab Summit was convened in Rabat in 1974 and decreed, against the will of King Hussein, that the Palestine Liberation Organization (PLO) is the sole legitimate representative of the Palestinian people, thus shifting the burden of liberation of the occupied West Bank to the PLO. The King's eventual acquiescence helped improve Jordan's relations with many Arab countries, including Syria. The political environment became suited for bilateral co-operation.

Harza Engineering Company started work on the feasibility study. A field visit to the Yarmouk catchment was necessary to evaluate the gauging stations that were once installed on the tributaries of the Yarmouk in the 1950's when Harza was working on the Master Plan. The Deputy Director of the JVC accompanied two Harza specialists to Syria and made arrangements with his Syrian counterparts at the Ministry of Irrigation to make the field visit. The team headed towards the Golan from Damascus and got as far as the disengagement line with Israel at Quneitra. They traveled down Wadi Raqqad, and over to the east to inspect the other tributaries and the sites of the springs that contributed to the Yarmouk flow. The Jordanian official was surprised to see a number of earth-fill dams that Syria had constructed on the Yarmouk tributaries of Wadi Zaydi, Wadi Allan, Wadi Dahab, Wadi Hreir and Wadi Raqqad. The projects that Syria had on the springs were, as far as the visiting Jordanian could judge, consistent with the provisions of the 1953 agreement between Jordan and Syria, but the dams were not. The agreement allotted to Syria the flow of the springs above elevation 250 meters above sea level. The total discharge of these springs was about 90 mcm per year, equal to the share of Syria in the Yarmouk under the Unified Plan.[260] The flood waters coming to the river from the Syrian part of the catchment were allotted

[260] The share of Syria, as requested by the Syrian members of the Arab Technical Committee, was 90 mcm per year, as opposed to the 45 mcm proposed by the Main/TVA Plan that Johnston carried to the area in 1953. See Chapter II for details.

to Jordan and were to be impounded, along with the runoff from the Jordanian part of the catchment, by the dam at Maqarin. The 1953 Agreement stipulated that the dam was to have a yield of 10 cms. Any impoundment of flood water by dams built on the tributaries upstream will have a direct effect on the yield of the Maqarin dam and, consequently, on its economic feasibility. Jordan's attempts to obtain information about Syrian projects in the Yarmouk catchment were not successful, and the officials in Damascus did not release any information except to say that the projects were minor and were meant to supply domestic water to some villages. The concerned Jordanian officials had to seek information about the Syrian projects through all available means, including the satellite imagery taken to prepare maps for the disengagement agreements between Syria and Israel. The lack of Syrian data posed a problem in the attempt to update the hydrology of the Yarmouk. Jordan, nonetheless, proceeded with the feasibility studies of the Maqarin dam and the consultants had to make certain assumptions as to the amounts of water that were depleted by Syria.

The project raised another complication, this one with Israel, related to the need to construct a diversion weir across the Yarmouk at the Adasiyya diversion point downstream. The design diversion efficiency of the drop inlet i.e., the percentage of the river flow that could be diverted to Jordan, was about 70% at best. That efficiency was progressively diminishing due to the existence of the sand bar and its progressive growth. The efficiency of diversion affects the total benefits to Jordan from the construction of the dam, and thereby affects the economic feasibility of the dam construction. A diversion structure had to be built to assure the needed diversion efficiency, something Jordan had been unable to achieve since 1962. Moreover, Israel's share of the River had to be agreed upon and guaranteed.

7.3 Political Complications

The complication of the diversion weir stemmed from the fact that, unlike the Maqarin dam that would straddle Jordanian and Syrian territories, the diversion weir would span the Jordanian bank on the south with the Israeli occupied Himmeh strip on the northern bank. Naturally, when the diversion to Jordan has a low efficiency, like the case was with the drop inlet, the flow that bypasses the diversion point to go to Israel increases. Israel, therefore, had a material interest in keeping the Jordanian diversion efficiency low, and a political interest as well, because it would prompt Jordan to seek an understanding with Israel on water division and diversion.

The economic feasibility of the Maqarin dam is sensitive to the Syrian water depletions upstream, and to the existence of a diversion weir downstream. Any increase in Syrian abstraction would negatively impact the economic feasibility because of decreased availability of water to be impounded by the dam. Similarly, the absence of the diversion weir reduces the diversion of water flows to Jordan and will have an adverse effect on feasibility. Moreover, the construction of the Maqarin dam cannot proceed without Syrian approval, a riparian on the Yarmouk, and the construction of the diversion weir cannot proceed without approval of Israel, the downstream riparian party on the Yarmouk who, in addition, a de-facto control of the northern bank of the river at Adasiyya. Additionally, approval of the project by the riparian parties, and the settlement of the water rights of each, are necessary conditions precedent to the construction of the water structures on the Yarmouk, especially if foreign finance is needed. Thus Jordan, the middle riparian party on the Yarmouk, sits at the mercy of Syrian activities in the catchment upstream, and needs Syrian approval of the dam construction. Simultaneously, she faced a dilemma of having to settle water rights with Israel and of making arrangements for the construction of the diversion weir across the Yarmouk at Adasiyya. Neither Syria, nor Jordan, nor any other Arab State recognized the State of Israel, and all Arab States had been in a state of war with her since she was created in 1948.

8. Reverting to Shuttle Diplomacy

The United States had been at the center of activities pertaining to the development of the Jordan River System in Jordan and in Israel. Her mediation role through shuttle diplomacy was described in the preceding chapters. She had sponsored the Unified Development Plan for the Jordan Valley. As such, the United States made it a condition upon the concerned parties to abide by the Plan, when any of them aimed at developing any part of the Jordan system with United States financial support. That was how Jordan and Israel, separately, qualified for financial support to build the water projects in their respective territories. Before the June 1967 war, the United States followed closely the activities of both parties, and played an intermediary role to assure that the works of one party did not conflict with the works of the other. As stated earlier, the division of Yarmouk water at Adasiyya between Jordan and Israel was made through an understanding with the United States, at the time of her financing of the East Ghor Canal.

The financer of the studies of the Maqarin dam was USAID. They contributed U.S. $ 1.0 million for the feasibility study in FY 1975, and $5 million towards preparing the final designs and tender documents of the project in FY 1976. The United States further pledged to finance the construction of the dam, and Congress appropriated U.S. $ 150 million in FY 1979 toward that goal. Jordan, therefore, approached the United States for possible resumption of her intermediary role with Israel over the Yarmouk issues. The preparedness of the United States to assume that role was encouraging, and hopes were high to thresh out differences with Israel, through the United States' good offices, over the construction of the Maqarin and the associated diversion weir. On the Syrian side, Jordan undertook to maintain direct contacts with Syria at all levels, to secure her consent to the construction of the dam, and, at the same time, talk her out of building more dams on the Yarmouk tributaries so that the feasibility of the Maqarin dam would not be further compromised. The Jordanian-Syrian Joint Committee on the Yarmouk established under a special article of the 1953 Agreement was reactivated.[261]

To compensate for the reduction of flow at Maqarin and to enhance the feasibility of the Maqarin dam, Jordan proposed to divert the flow of Wadi Raqqad in Syria eastwards to Wadi Allan and thence to the Maqarin reservoir, instead of into the Yarmouk downstream of the dam site.

In response to Jordan's request, the U.S. Department of State appointed a special envoy, Ambassador Philip Habib, to explore the possibilities of working out an understanding between Jordan and Israel over the Yarmouk. He started his mission in November 1977 and was aided by officials of the Agency for International Development in Washington, D.C.[262] By that time, a comprehensive study was completed for the National Planning Council of Jordan on the needs of North Jordan for municipal and industrial water.[263] The conclusions came as a surprise to officials. The needs were more than many of them had expected, and the sources to supply these needs were limited. The deficit in municipal water supplies to North Jordan was estimated to run at about 160 mcm by the year 2000. Consequently, the purpose of the Maqarin dam was modified in 1978 to supply not only irrigation water for the Jordan Valley, but also to supply municipal and industrial water for use in North Jordan[264]. The reallocation of the dam water enhanced the feasibility of the dam, and the urgency for its construction increased.

The efforts of Mr. Philip Habib extended between November 1977 and November 1981, the year when Jordanian-Syrian relations went sour again, on account of the Iraq-Iran war. Jordan stood by Iraq in that war, whereas Syria supported Iran. Throughout Mr. Habib's mission, the region was going through drastic political and strategic transformation. The Civil

---

[261] Omar Abdallah Dokhgan chaired the Jordan side and the members were Dr. Hanna Odeh, President of the National Planning Council and a Board member of the JVA, and Mr. Dhafer Budeiri, Vice President of JVA for Dams. The Syrian side was chaired by Mr. Lutfi El Khas of the Mega Projects Corporation, which merged into the Ministry of Irrigation in the early 1980's.

[262] Mr. Selig (Sy) Taubenblatt, Deputy Director of the Near East Department at AID, and later its Director, played a central role in the talks.

[263] The study was conducted by Howard Humphry, a British consulting firm.

[264] North Jordan was defined as the area between Wadi Wala in the south and the Yarmouk River in the north, bounded by the sea level contour to the west, and the Hidjaz railroad to the East. Ninety-one percent of Jordan's population lived in that quadrant at the time.

War in Lebanon, triggered in April 1975, was escalating. As earlier stated, President Anwar Sadat of Egypt made his historic visit to Israel in 1977 and made peace with Israel, to the dismay of all the other Arab countries. In 1979, the power in Iraq was transferred from its president, Ahmad Hasan al Bakr, to his strong deputy, Saddam Hussein. A row between Iraq and Syria soon followed. Relations between Iraq and Jordan, frozen since the Iraqi military coup of 1958, started to thaw after the Baghdad Summit in 1978. In the Iraq-Iran war, Jordan, the Gulf States and Yemen sided with Iraq and supported her war efforts against Iran, while Syria sided with Iran. In the meantime, the Lebanon Civil War intensified, and Syria led military contingents of Arab States, authorized by the Arab League, to interfere in Lebanon and keep law and order there. The PLO was an active party in that war, and had its own show against Israel in South Lebanon. Israel, on her part, started to mount military incursions into South Lebanon in response to PLO shelling of Israeli settlements. She occupied a "security zone" there in 1978, and established the South Lebanon Army under Colonel Saa'd Haddad, a Lebanese military officer. Despite all these developments, Jordan had no time to lose. She continued her development efforts with vigor and achieved high rates of economic growth. Most notable of her achievements were: a) the first stage of the development of the Jordan Valley; b) the construction of elaborate networks of infrastructure projects in transportation, communications, electricity, and domestic water; c) expansion of phosphate mining; and, d) implementation of the long awaited project of extracting potash from the Dead Sea.

While the region was boiling with the above developments, Mr. Habib continued his mission. He shuttled between Israel and Jordan, with stops at times in Lebanon to appraise the volatile situation there, and to conduct some talks with the warring factions as well.[265] Mr. Habib had talks with the Jordanian officials aimed at narrowing the differences between the positions of Jordan and Israel over the Yarmouk.

9. A Source for Municipal Water

A landmark in municipal water supply to North Jordan was laid that year (1978). A project had been conceived in 1976 by the Amman Water and Sewerage Authority (AWSA), in cooperation with the Jordan River and Tributaries Regional Corporation (JRTRC) and the National Planning Council, to supply Amman with water from the King Talal dam that was being built for the JRTRC. The project to supply Amman with municipal water was to be financed by USAID and the Saudi Fund for Development. With concerns rising regarding the possible pollution of the King Talal Dam

[265] Ambassador Habib was an American of Lebanese extraction.

from municipal and industrial wastewaters,[266] the National Planning Council recruited experts from the United States to look into the possible pollution hazard. Officials from the Jordan Valley Authority (that acquired the JRTRC in 1977) participated in the meetings. Attendants also were representatives of the consulting firm and of the USAID. The JVA participant contended that the dam reservoir would be so polluted that safe treatment of municipal water would be subject to serious hazards. Although the reservation came late, as the contract for construction was about to be awarded, the President of the National Planning Council entertained it and so apprised the Prime Minister. The project was abandoned and the Prime Minister asked the JVA to come up with a workable alternative to supply Amman with much needed water.

The JVA President, Omar Abdallah Dokhgan, assisted by his deputy, got busy working on a new proposal: to supply municipal water to North Jordan from the Jordan Valley. The central structure for the purpose would be the Maqarin Dam on the Yarmouk River. A ministerial committee discussed the proposal, but a decision on such a shift in water strategy had to be blessed by the King. A meeting was held at the Royal Court,[267] where the situation was reviewed, and a decision was made to cancel the King Talal to Amman project, and to have JVA work on the supply of municipal water to North Jordan from the Jordan Valley. The new project would take more time than Amman could wait, and depended for its full capacity of 90 mcm per year on the construction of the Maqarin dam. In the meantime, AWSA was authorized in that meeting to implement a project to transfer an average of 12 mcm per year of water to Amman from the groundwater aquifer at Azraq, 72 km to the northeast.

By the time Habib was intensifying his efforts, Jordan had the final designs and tender documents of the Maqarin dam completed. Harza's work on the dam and the site investigations were financed by USAID with an add-on amount of $10 million disbursed from the Congressional allocation of $150 million earmarked for dam construction. The cost of the associated irrigation designs was covered by a loan from the Kreditanstalt fuer Wiederaufbau (KFW), of the Federal Republic of Germany. Concurrently,

---

[266] The major cities of Amman, Zarqa, Ruseifa and Jerash are located in the catchment of the Dam.

[267] HM King Hussein chaired the meeting that was attended by Crown Prince El Hasan; the Prime Minister, Mr. Mudar Badran; the Chief of the Royal Court, Sharif Abdul Hamid Sharaf; the Minister of Municipal Affairs, Ibrahim Ayoub; the Mayor of Amman, Ma'an Abu Nuwwar; the President of the National Planning Council, Dr. Hanna Odeh; the President of the Jordan Valley Authority, Omar Dokhgan; the Vice President of the Natural Resources Authority, Ahmad M. Dokhgan; the Director General of AWSA, Tahseen Sabbagh; the Director General of the Natural Resources Authority, Yousif Nimry; and the author, who was Senior Vice President of JVA.

the Jordan Valley Authority commissioned two U.S. consulting firms[268] to prepare a plan for the supply of domestic water to North Jordan from Jordan Valley water resources, including the Maqarin dam. The contribution to domestic water supply from the Jordan Valley was to be 120 mcm per year, of which 30 mcm would be pumped from the Maqarin dam directly to Irbid district, and 90 mcm would be pumped from the East Ghor Main Canal at Deir Alla to Amman. The water balance in the Jordan Valley took into consideration that the municipal wastewater generated from the urban areas in the Jordan River catchment would be treated and left to flow back into the side wadis to the Jordan Valley for re-use in irrigation. The wastewater treatment and reuse was another major modification of the water strategy.

Jordan called for three successive conferences of the potential donors for the grand plan of the Maqarin dam and its contributions to agriculture and domestic water supply. The first conference was convened in London, in June 1977, upon the completion of the feasibility study, and the conferees recommended that Jordan proceed with the preparation of final designs and tender documents; the second was convened in Amman, in April 1978, in which the conferees asked the World Bank to conduct the appraisal of the project; and the third conference was convened in London, in October 1979, to agree on a schedule of implementation and a financing plan. The implementation schedule called for clearing agreements with Syria over the diversion of Wadi Raqqad and the use of her territories to build the Maqarin dam, and a target date was set to have a draft agreement reached with Syria by 18 February 1980. The conferees also required that an arrangement be worked out to have the diversion weir at Adasiyya built. No mention was made of Israel or the United States' role.

Mr. Habib was supplied with the designs of the Maqarin dam and the diversion weir at Adasiyya. USAID, the financier of the Harza studies, received copies of the consultants' reports and final designs, and these were made available to Mr. Habib. Harza was also authorized to respond directly to any technical queries that Mr. Habib and his assistants might have. The JVA presumed that Mr. Habib would share with the Israelis the technical features of the Maqarin dam and the diversion weir. The Israeli response to the technical designs was communicated to the American side and, through them, to the Jordanians.

Concurrently, the JVA supplied their Syrian counterparts with copies of the Harza studies to make whatever comments they saw necessary. Contacts were maintained to speed up the Syrian review of the project and to have them issue the statement of approval. By 1979, no such written

---

[268] These were Stanley Consultants of Muscotine, Iowa, and Boyle Engineering Co. of Newport Beach, California. The USAID covered the foreign exchange cost of their work.

statement was issued, although the Syrian verbal response to the repeated Jordanian request was positive. In anticipation of a successful resolution of issues directly with Syria, and of issues with Israel through the U.S. envoy, Mr. Philip Habib, Jordan proceeded with her implementation plan as agreed in the London conference. She intensified contacts with Syria to secure her agreement to the dam and the diversion of Wadi Raqqad. In a side meeting between King Hussein and President Asad in Mexico, on the occasion of the conference of the non-aligned nations in 1980, King Hussein urged President Asad to issue Syrian approval of the Maqarin dam project. President Asad was quoted as responding favorably and asked that the technical teams complete their recommendations. The joint Jordanian-Syrian Yarmouk Committee had met several times, completed the work, and awaited the political decision.

10. A Mission to Damascus

On a nice Tuesday morning in July 1980, the Prime Minister of Jordan, Mr. Mudar Badran, asked the Senior Vice President of the JVA, Dr. Munther Haddadin, to prepare a draft letter to the Prime Minister of Syria urging him to speed up the process of the project approval. Mr. Badran reviewed the draft letter and introduced paragraphs to "tighten the screws". Mr. Badran asked the President of the National Planning Council, Dr. Hanna Odeh, and the Senior Vice President of the JVA to travel to Damascus that same afternoon and deliver the letter to the Syrian Prime Minister, Mr. Mohammad Ali el Halabi.

Ambassador Hussein Hamami, Jordan's Ambassador to Syria, joined the Jordanian officials upon arrival in Damascus. They first stopped by the office of Dr. Lutfi al Khas, Director of the Mega Projects Corporation, discussed the matter with him and proceeded with him to the Prime Ministry in the Mazra'a district, to meet with the Syrian Prime Minister. The Prime Minister, dressed in safari field clothes, received the Jordanian delegation warmly. Dr. Odeh handed him the letter, and Mr. al Halabi read it carefully. He frowned as he passed over the strong-worded paragraphs that Mr. Badran used.

"What made Mudar Pasha so mad at us? The whole matter is not worth his anger at all. We are brothers and we will so remain," said al Halabi. "Excellency," answered the polite and diplomatic Dr. Odeh, "we really are in a hurry to get your approval of the project. We have lined up the financial support from many international and Arab donors,[269] and Syria's endorsement of the project is a must before the dam can be built. Our joint technical

---

[269] Funds had been sucured from USAID, the Kuwaiti Fund, and the KFW. Other donors pledged support.

committee reviewed the project components and no reservations were expressed. The Syrian comments have all been incorporated in the project design."

"Oh, let me tell you my frank opinion," said the Prime Minister. "You are pursuing a project that is not feasible; the Americans tell you that it is feasible but they are fooling you. If you want to increase agricultural production, I have a better idea. Why don't you bring your farmers, the future beneficiaries of the project, here to Syria; we will form a joint company and your farmers can get busy in irrigated agriculture in the Euphrates basin, here in Syria. This is more feasible than the expensive project you want to build with borrowed money. Besides, we want to make sure that Jordan will not face (because of its debts) a fate similar to what Egypt once faced as a result of her indebtedness incurred by building the Suez Canal."

The Euphrates project in Syria was facing difficulties with soil types and drainage. Syrian officials, including three Prime Ministers, had visited the Jordan Valley and were impressed with what Jordan had achieved. They thought that maybe a partnership of the type described above would help transfer agricultural technology from the Jordan Valley to the Euphrates basin and from there to the rest of Syria. Besides, there would not be pressure to deliver Jordan's water share in the Yarmouk, and Syria would benefit from that.

Dr. Odeh tried his best to convince the Prime Minister of the advantages of the Maqarin dam and the hopes Jordan placed on its completion to supply the urban areas with domestic water, and its other economic benefits.

The Senior Vice President of the JVA, mindful of the comments made by the Prime Minister and of his insinuations, assured al Halabi that Jordan was in no need for guardians, and contended that the dam was proven feasible and that competent Jordanian professionals had reviewed the Consultant's studies. He compared the Maqarin project with the earth dams that Syria had been building in the Yarmouk catchment and explained the adverse impacts those had on the feasibility of a dam at Maqarin. The dialogue became a bit heated but the diplomatic skills of the Jordanian Ambassador saved the day.

Prime Minister al Halabi assured his visitors of his support for the dam and said that the letter of approval had been forwarded to President Asad for signature. The letter was not signed, however, and Jordan did not ease her pressure.

King Hussein paid a visit to Syria shortly after the visit of the delegation. The issue of the dam was brought up but the Syrian answer was the same: Let the technicians revisit the project details and come up with a joint recommendation!

That was the last visit the King paid to Syria. The Iraq-Iran war erupted late that summer, and the Jordanian-Syrian relations went sour again, because each supported one of the warring parties. Syria never issued an approval of the dam, and the Jordanian efforts were stalled.

11. Habib's Last Visit

On the Israeli front, Mr. Habib continued his intermediary role. At the outset, Israel inflated her demands for a share in the Yarmouk. The Israelis fell back on the figure contained in their Memorandum of Understanding with Ambassador Johnston, dated 5 July 1955, i.e, 40 mcm per year. They further inflated a figure they said was due to the West Bank from the Yarmouk and tried their best to take advantage of Jordan's need for a diversion weir, coming up with proposals that Jordan could not accept. Such proposals included: forming a joint committee to manage the operation and maintenance of the weir; bringing a foreign contractor, not a Jordanian contractor, to build it; conceding to Israel the use of the Yarmouk excess flow not diverted for use by Jordan; agreeing to release Israel's share in the Yarmouk during the summer months only; guarantees that understandings will be honored; adjustment of water rights in light of groundwater availability;[270] and other similar issues aiming at creating a need for frequent bilateral contacts between Israeli and Jordanian personnel. Mr. Habib and his assistants worked diligently to weed out issues that were premature and others that were politically motivated, and did their best to conduct their mission as neutral intermediaries.

It appeared, however, as though the neutrality of the U.S. envoy became clouded after the Camp David Accord, and the refusal of Jordan to play the role the Accord carved for her in her absence. On one of his trips to Jordan, in June 1980, Habib met with the Prime Minister and the concerned Jordanian officials at the Prime Ministry.[271] Mr Habib asked for Jordan's position in writing, and the President of JVA, Mr. Dokhgan, prepared it the following day. It conformed to what the Arab League Technical Committee had approved in September of 1955; i.e., Israel's share would be 25 mcm per year, and the remainder would be the share of the Hashemite Kingdom of

---

[270] These matters were previously raised with Ambassador Johnston and were settled. The Israelis revisited these topics hoping that their Jordanian counterparts would not know what went on in Johnston's time. In many ways they were right, but the Jordanian response was not far from the Arab response during Johnston's negotiations.

[271] The Prime Minster was the late Abdul Hamid Sharaf, and the officials were the Minister of Foreign Affairs, Mr. Marwan al Qassim; the President of the JVA, Mr. Omar Abdallah Dokhgan; and the President of the National Planning Council, Dr. Hanna Odeh.

Jordan. The Jordan position was reviewed and finalized by the Jordanian team without any material changes to what Dokhgan prepared. A meeting was convened at the house of the Foreign Minister, Mr. Marwan al Qassim, in the evening,[272] and Jordan's position was discussed. When the Jordanians remarked that their position was basically in line with the Unified/Johnston Plan, Ambassador Habib remarked that the Plan was too old and not quite workable anymore. That remark shocked the Jordanians.

Habib asked, on Israel's behalf, about the quantity of water Jordan planed to allocate to the West Bank from the Yarmouk, and the Jordanians answered that they would not entertain any notion that could imply Israel's right to speak on behalf of the West Bank. The debate went on and on, and it became clear that Ambassador Habib was not about to start the talks from where they had ended back in the 1960's before the June War. The Unified Plan, apparently, was no longer the yardstick with which the United States was measuring the performance of the riparian parties in the Jordan basin, as she actually had been doing until the June war of 1967.

Habib stressed that Israel's demands were for 40 mcm from the Yarmouk, and that the 25 mcm stipulated in the Unified Plan were only her summer share. Dokhgan stressed that if that was Israel's position, Jordan would likewise fall back on the original Arab figure for an Israeli share of 17 mcm per year for the Yarmouk Triangle from the Yarmouk, and that the rest of her needs should be satisfied from Lake Tiberias.

The Foreign Minister employed his diplomatic skills to conceal dismay similar to that visible on the faces of his colleagues, and kept the conversation going. He expressed, in unambiguous terms, that Jordan refused to have Israel speak on behalf of the West Bank, which was at that time occupied Jordanian territory.

Between the Syrian abstention and the Israeli attempts to exploit her physical control of the West Bank along with Jordan's need for more water, the Maqarin project faced more hardships than could be reasonably managed. Several years later, Mr. Selig A. Taubenblatt,[273] who accompanied Habib, outlined in his statement to the U.S. House Foreign Affairs Committee on 11 March 1987, the reason behind the failure to proceed with the Maqarin dam:

"Although progress was made during the late 1970's and early 1980's between Jordan and Israel, eventually Jordan's inability to reach agreement

---

[272] Attendants from the Jordan side were the Foreign Minister, the President of the National Planning Council, the President of the Jordan Valley Authority, and the author, then Vice President of the Jordan Valley Authority. From the American side were Mr. Philip Habib and Ambassador Nicholi Veliotes, U.S. Ambassador to Jordan.

[273] Director of Projects for USAID 1975-1981 covering Near East, North Africa, and Europe. He was in charge of the Maqarin dam project and was Chairman of the State/A.I.D. Steering Committee on Jordan River basin water rights.

with Syria became an immediate cause of indefinite postponement of the Maqarin dam project. Since the end of 1980, the project has been in abeyance."[274] The progress referred to in Taubenblatt's statement concerned Habib's success in weeding out some of the unreasonable Israeli demands, but did not include anything close to Israeli approval of having the diversion dam built. In the absence of the agreements Jordan had to work out with Syria and Israel, it was impossible to conclude a financial plan and to obtain financial support for the project from international donors. As the Jordanian-Syrian relations became more strained,[275] the project lost practically all its chances to see the light in the foreseeable future. Contacts between the Jordanians and their Syrian counterparts were discontinued and not resumed until the summer of 1986. The matter could not be discontinued with Israel, primarily because of the need to clean up the Yarmouk riverbed at the diversion site at Adasiyya, and to remove the sand bar that was growing in size each winter season. This pressing need kept strings attached to the Israeli dimension for some years to come.

The shelving of the Maqarin dam project dictated a modification of the domestic water supply to urban North Jordan. The amount that could be contributed by the Jordan Valley water resources without Maqarin could not exceed 35 mcm per year. To increase that amount, the JVA decided to modify the design of the Wadi al Arab dam, then at its early stages of construction with financial support from Japan, to increase its live storage capacity from 7 mcm to 17 mcm. The added storage capacity, which was above the natural flow of Wadi Arab, would be filled by pumping from the King Abdallah Canal[276] (KAC), diverted from the Yarmouk winter flow. The domestic water supply project to Amman and environs was designed to carry 45 mcm per year, half the planned flow of 90 mcm per year. The latter could be achieved only after the Maqarin dam was built. So, in addition to the KAC supplying water for irrigation for an expanded area in the Jordan Valley, it would also need enough flow to cater for the new need of storage in the Wadi al Arab dam and the requirements of domestic water supply to Amman, when its project became operational in 1986. The diversion efficiency from the Yarmouk River became even more important and demanded more attention.

---

[274] Letter addressed by Mr. Taubenblatt to the author with enclosure of the statement, March 17, 1987.

[275] In addition to the row created by the Iraq-Iran war, Syria accused Jordan of plotting to cause unrest in Syria by covertly supporting the Moslem Brotherhood movement inside Syria and training its members on the techniques of urban warfare. The Moslem Brotherhood movement was crushed in Syria in 1982, in an all out Syrian military assault on a claimed uprising in the Syrian city of Hama.

[276] The King Abdallah Canal is the name that was given to the East Ghor Canal in 1987.

CHAPTER VII

The Sand Bar

1. Importance of the Yarmouk

The King Abdallah Canal (KAC), formerly called the East Ghor Canal, irrigated in 1967 about 114,000 dunums in the East Jordan Valley with water diverted mainly from the Yarmouk. About 8,600 more dunums received water only during the winter season. About 18% of the area irrigated by the KAC was planted with citrus trees, and the cropping intensity was about 115%; i.e., there was demand for irrigation water during the dry months to support spring and summer crops. Up until 1979, all the areas served by the KAC had surface irrigation methods, and the distribution of irrigation water was done through a network of concrete-lined surface canals. On-farm irrigation was done through surface methods that required higher water duty per unit area of land. The overall irrigation efficiency was therefore lower than could be achieved with updated modern irrigation networks and farming methods. The average water duty as stipulated in the Unified Plan was about 1,460 m^3 per dunum at the source of water. All irrigation infrastructure, up until 1967, was built with grant assistance from the United States that totaled US $12 million.

To expand irrigated agriculture, the King Talal Dam was built with financial support from the Kuwait Fund and the Abu Dhabi Fund, and began operation in 1978. Modern networks, in the form of pressure pipe networks, were installed in the expansion projects. Farmers in the new lands were receptive to new, on-farm irrigation techniques such as drip methods. However, the investment Jordan made in improvement of irrigation efficiencies to save water for expansion helped only in sustaining existing agriculture. The water saving was mostly used to offset the effects of reduced flows into the KAC from the Yarmouk. The reductions in flow resulted from increasing Syrian abstraction upstream and the obstruction to efficient diversion from the Yarmouk caused by the "sand bar".

Prior to the occupation by Israel of the Golan Height in June 1967, Jordan had no problem in accessing the diversion point on the Yarmouk to maintain it whenever she needed to. The cleaning process was normally done after the flood season was over, readying the riverbed conditions for efficient diversion during the following dry months, mid-April through mid-October. Year after year since 1967, the deposits accumulated in the riverbed at the diversion point and formed a sand bar along the riverbed, shifted slightly off midstream towards the Jordanian bank of the river (Figure VII.1). Soon, wild plants started to grow on it, including reed, bushes, and others. By 1979, the tip of the sand bar was protruding in the cross section of the drop inlet, thus reducing the amount of flow that could be diverted to Jordan. Any flow that was not thus diverted would keep flowing downstream to where the Israelis could pump it for use into the Yarmouk Triangle, and, further downstream, for storage of Yarmouk winter flow in Lake Tiberias (Figure VII.2). The Israelis, therefore, had a vested interest in the growth of the sand bar. The immediate interest was the gain of more water, especially during the dry months, and the corollary interest lay in pressing the Jordanians to make arrangements with them to have the sand bar dredged and removed. Jordan's need for water was increasing, as outlined above, at a time when the Yarmouk contribution was receding due to increasing abstractions by Syria upstream, and with reduced diversion efficiency into the KAC because of the sand bar.

Irrigation water shortages became frequent and prompted the JVA to ration distribution of water. Priority was given to perennial crops, because of the capital investment made in developing orchards. The situation was exacerbated in the dry months, as water shortages became more pronounced. Major water projects were planned to expand the irrigated area further, with assistance from friendly countries. Evidence of lack of water would deal a blow to Jordan's efforts to secure financing. In her approach to Syria, Jordan was careful not to rock the boat for fear that Syria might impart more delays on the implementation of the Maqarin dam. As far as Jordan was concerned about Israel, Ambassador Philip Habib had been active to work out an arrangement for the construction of a diversion weir in the same location on the Yarmouk where the drop-inlet diversion is. Jordan was optimistic that such an arrangement could be found, and did not want to rock that boat either.

2. Contacts to Clean the Sand Bar

Jordan started to feel the pinch of the reduced diversion from the Yarmouk in the summer of 1979. It was then that the new irrigation projects were put into operation. A total of 83,000 dunums of new lands were added to the irrigated area, and the surface canals of 10,000 more dunums were replaced with pressure pipe networks. The King Talal Dam was also completed and would add to the water stock available for use.

The 1978-1979 water year was a bad drought year. Although the effect of the drought on groundwater and springs flows is usually delayed, the effect on the Yarmouk was very visible. Several factors combined to have Jordan feel the pinch in May of 1979: a) the drought in the previous rainy season; b) increased Syrian uses of the spring water upstream, some within their recognized rights under the Jordan-Syria Agreement of 1953, and others over and above those rights; c) the sudden increase in Jordan's water needs as a result of operations in expanded irrigated areas; and, d) the decrease in diversion efficiency from the Yarmouk, caused by the presence of the sand bar.

Figure (VII.1)- The Sand Bar

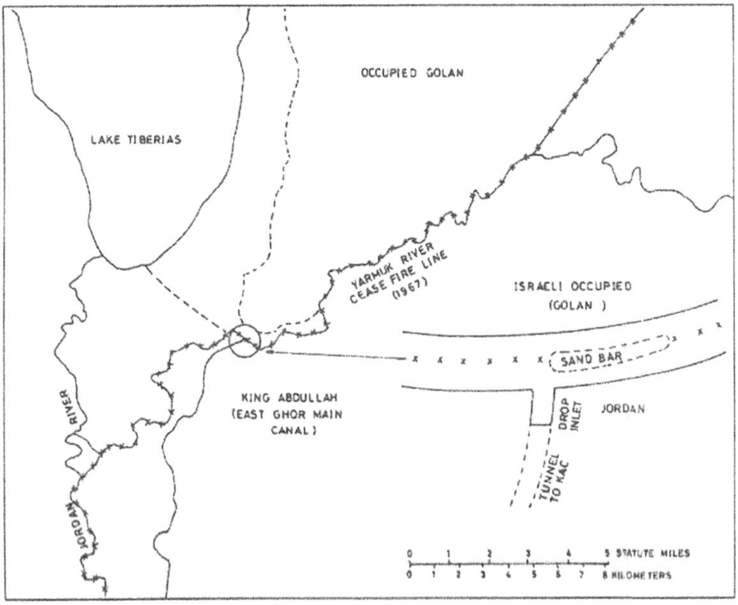

Figure (VII.2)- Israeli Pumping to Lake Tiberias

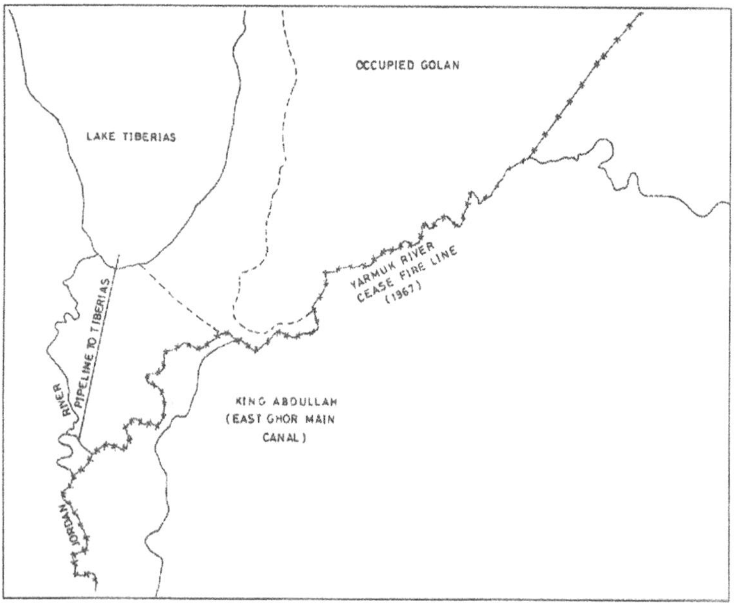

The management of this situation posed a challenge to the JVA, who had taken over the operational responsibilities of the KAC from the Natural Resources Authority only two years earlier. Several measures were taken to manage the drought situation. Contacts were maintained with Syria to brief them on the critical water situation and ask them to save as much of the water as was practically possible so as to spare some water for Jordan. The Syrian response was less than what Jordan hoped for. Another measure was to ration the irrigation water in the Jordan Valley and to advise farmers of the critical water shortage. That measure worked well, with some adverse economic and social impacts; usually, orchards that command priority for water are owned by better-to-do farmers than farmers of seasonal crops. The third measure was to make an attempt to clean up the sand bar in the bed of the Yarmouk River at the diversion point, and thus increase diversions from the limited flow of the Yarmouk. That flow, which had to be shared with the Israelis, was in the order of only 3.2 cms in the summer of 1979 compared to its average of 6.5 cms in the same month in 1963. The reduction in Yarmouk flow was due to the combined effect of increased Syrian uses and the drought of the previous rainy season.

At that time, the United States envoy, Philip Habib was at work to moderate between Jordan and Israel over the construction of the Maqarin dam. The President of the JVA thought that American diplomacy might be helpful in arranging for a way to clean up the sand bar located on the cease-fire line; especially since the Jordan Valley Project was heavily financed by USAID. He was confronted with a basic question: "what was the agreed arrangements, worked out by the United States in 1958, to share the flow of the Yarmouk with Israel?" There were hardly any records available to the JVA on that matter.

Several answers were given to that question. Some JVA employees working in the operation of the KAC gave what they thought was the formula for sharing; other current and former officials gave a different formula. They all agreed that the formula was deposited by the USAID with the Development Board, the predecessor of the National Planning Council of the Government of Jordan. The offices of both, as well as the officials in both, had since changed more than once. The Development Board moved its offices twice, and the USAID moved its offices at least twice. The concerned former officials were not around to be consulted. The attempts by the President of JVA to obtain records of that formula from USAID never succeeded, and there was no record found at the National Planning Council. The JVA had to play it by ear! However, there was a consensus that there had never been a need to apply a sharing formula throughout the previous life of the project (operational since 1962), because there had always been enough water to satisfy both Jordan's needs and the needs of the Yarmouk Triangle in Israel.

During the second half of May 1979, JVA maintained almost daily contacts with USAID in Amman. The presentation by JVA was that the KAC project was initiated and financed by USAID and therefore the Agency should have a stake in its normal operation, and in the protection of investments made thereunder. On the other hand, the State Department and USAID were then active in their pursuit of Israel's consent to build the diversion weir across the river at the diversion point, as part of the Maqarin dam project. The contacts resulted in expression of willingness by the Israelis to discuss the problem of the sand bar and to agree on the detailed steps for a solution.

Arrangements were made by the American Embassies in Amman and Tel Aviv to have the Jordanians and the Israelis hold a meeting between their technical people on both sides, to resolve the problem under the auspices of UNTSO. Two such meetings were held at the site, in the presence of the competent security authorities of both sides. The Jordanian technical team was headed by a competent engineer very familiar with the KAC project since the time of its initiation. There were no facilities on the site at the diversion point for the teams to meet. The talks were conducted between the two heads as they stood in midstream, which was about two feet deep. The Jordanian delegate explained the problem of decreased diversion efficiency and asked for the removal of the sand bar. The starting point was to measure the flow, and find out how much of it was going to each side. Each team measured the flow separately. It turned out that, of the 3.2 cms river flow, about 1.8 cms was flowing to Jordan and 1.4 cms to Israel. Jordan asked that the riverbed conditions be mitigated so that the sharing conformed to the agreed formula. The Israeli delegate was open to suggestions, but said that the removal of the entire sand bar was out of question. Instead, its protrusion in the cross section of the drop inlet could be shaved to increase diversion to Jordan. Before they departed for the day, the heads of the two teams agreed when next to meet.

The Israelis liked the idea of meetings with their Jordanian counterparts and dragged their feet as the teams were looking for a satisfactory resolution. They asked for the details of the Jordanian operation intended for cleaning of the protrusion of the sand bar, the size and type of the equipment to be used, the operators and their numbers, and all other details. The President of JVA kept the Prime Minister of Jordan informed through verbal reporting only.[277] Finally, agreement was reached to allow a Jordanian bulldozer to enter the site and do the cleaning of the tip of the sand bar in a pre-arranged way. Both teams would attend the cleaning operation and make sure it went according to the agreed plan.

---

[277] Written reports were the responsibility of the military escorts.

3. The First Cleaning Operation

On 8 July 1979, the JVA President left for the United States on a mission in connection with the Maqarin dam designs, and designated his Senior Vice President to be Acting President in his absence.[278] The head of the Jordanian technical team briefed the Acting President on the agreement and informed him that the cleaning operation would start on 12 July, and asked for instructions.

The bulldozer operator, a veteran worker of the KAC project, had so deep an allegiance to the project that he risked his life more than once rushing to the aid of maintenance workers during the turbulent years between 1967 and 1971. The Acting President had a meeting with the operator and urged him to do a good job in recovering Jordan's water share.

Because the cleaning day was critical for the Jordan Valley, and to be prepared for solving any complications that could arise, the Acting President of JVA decided to go down to the site and stay close to the scene. He stayed with the Army soldiers in their tent, some 150 meters away from where the teams met. The Jordanian delegate could consult with him concerning any problem the delegate could not handle alone. The delegate actually came to him twice, one was to know what to do to reconcile differences in flow measurements, and the other to specify what flow rate the Israelis should be entitled to. The flow of the river went down to 2.8 cms that day as measured by the Jordanian team, and 3.1 cms as measured by the Israeli team.

The Acting President instructed the delegate to emphasize that the topic to be addressed was only to solve a transient problem and avoid misunderstanding. Water was short for both parties and their aim should be to minimize damages to each side. The talks should not address any other topic.

Lacking a sharing formula, the Jordanians debated the matter with the Israelis vigorously and played it by ear and by their wits. They were able to get the Israelis to accept, after much debate, a share of 0.75 cms, applicable only throughout that dry season of 1979. Out of an average flow of 3 cms, that share constituted 25% of the total flow. The Israelis were cooperative and meant well. The bulldozer then went into the water to do the cleaning, and several Israeli personnel were standing on the Israeli occupied bank of the river watching the operation. As the bulldozer moved forward, an Israeli official would shout giving the operator directions in broken Arabic. The operator acted like he was deaf and kept doing his job. There was much shouting and counter-shouting as the Acting President watched from inside the Army tent. The Israeli senior engineer got in the water to come close to

[278] The author was the Senior Vice President.

the bulldozer and the Jordanian delegate did the same, and they debated the matter while standing in the water. As the Israeli engineer accepted the Jordanian argument, he would watch the bulldozer move in a way that did not produce the agreed result. The Israelis on the bank would shout in Hebrew, and their engineer in the water would shout back at them in an apparent internal argument. The Israeli team had representatives of the farmers in the Yarmouk Triangle (called the Jordan Valley by Israelis), officials of TAHAL, of whom the engineer in the water was one, and of the Israeli Defense Forces. The Jordanian team was composed of a senior engineer, an assistant, and two technicians to measure the flow. A high rank officer of the Jordan Army accompanied them. A United Nations' Blue Beret heading the UNTSO office in Amman managed the meetings and helped avoid verbal clashes. The shouting continued, and the TAHAL engineer did a very good job to control chaos in their ranks. He moved with the bulldozer and once fell flat on his back in the water.[279] Work on the cleaning continued until sunset, and there was an achievement in shaving the tip of the sand bar and cleaning in its vicinity. The length of the sand bar was about seventy meters and the width was about twenty meters, and it was covered with wild plants.

The work was resumed the following day, 13 July, with the same arrangement. More shouting was heard but more cleaning was done, and the flow to the KAC was visibly improved. It turned out that the operator of the bulldozer was over excited and he quietly defied instructions and worked in accordance with his own intuition to bring more water to Jordan, something that soon brought accompanying headaches.

4. A Call After Midnight

Two days after the cleaning was done, the Acting JVA President received a call from the USAID in Amman, on 15 July, informing him that the Israelis were complaining that their share of the flow had diminished, and that they demand that it be restored promptly. He abided, but dragged his feet in implementation. The Israelis had dragged their feet in responding to the Jordanian request for cleaning (it took three weeks before the meeting took place), and the Acting President thought that the Israelis should not mind a similar response!

In the early hours of 19 July, the Acting JVA President received an untimely phone call, past midnight. The caller was the duty officer in the Operations Room of the Jordan Armed Forces. After apologetic statements for calling at 02:30 hrs., the officer read to the Acting President an urgent cable

---

[279] The scene became suddenly funny, but the Acting President held respect for that Israeli senior engineer, who later became known to him as the famous Dr. Yacov Vardi, a renowned Israeli water expert.

addressed to him from the Minister of Defense who was, *ex-officio*, the Prime Minister.

The cable told of Israeli soldiers who crossed the cease fire line on the Yarmouk at the diversion point, overwhelmed three Jordanian soldiers in their tent, cut-off their communication line, and re-arranged some rocks in the river bed. The result was a re-adjustment of the flow with substantially more water going to the Israeli side. The Defense Minister instructed the Acting President to restore Jordan's water. "The commander of the 2nd Division," said the cable, "will be waiting for you at the site at 06:00 hrs. He has orders to obey your instructions to accomplish the mission."

The Acting President had time to get ready and he drove from Amman towards the Jordan Valley, then up the main highway towards the KAC intake tunnel on the Yarmouk. As he approached the site on a curving road, the rising sunrays beamed into his eyes. He reacted by dropping the sunshades and was stunned to see what was in sight--across the Yarmouk, there were Israeli military deployments of infantry, personnel carriers, tanks, artillery and machine guns. These were the first he saw as he drove up around that curve. Soon after that, he could see the Jordanian side with similar military deployments, and soldiers taking position in their trenches, armed with machine guns. Artillery, tanks, and other weapons were on the site. Only about 200 meters separated the two armies. They could actually throw stones at each other and, with concentration, make hits. The Acting President never imagined the situation could be allowed to escalate that seriously.

As he disembarked from his car, he could see the Israelis across the river with a video camera taking shots of him. An officer in battle uniform and a helmet received him and escorted him down to the Army tent by the bank of the Yarmouk. The tent was there for use by three guards to watch over the water. The commander of the military Division in charge of defense of Northwest Jordan was inside the tent alone, at 06:00 hours.

The Brigadier General briefed the JVA Acting President on what happened the day before. A mechanized Israeli force overwhelmed the Jordanian guards, cut off their communication line and managed to adjust the water flow such that Israel got more water at Jordan's expense. Jordan responded with mobilization of force and the Israelis reacted with similar mobilization. "My orders," said the Commander, "are to follow your instructions to recover Jordan's water."

The Acting President asked to inspect the diversion point, which was at the bottom of a cliff where the Israeli soldiers were standing. The Commander agreed on condition that he lead the way. He took his military rank off his shoulders, and walked towards the diversion point on the river, ahead of the civilian Acting President who followed two steps behind. The Israeli soldiers were overlooking from a cliff above, and they could, if they wanted, deal the Jordanian officials a fatal blow by dropping a couple of rocks on their heads. The Israelis watched carefully and their video cameras were taping.

After inspecting the site carefully, the Acting President realized that the Israelis had effected a unilateral adjustment of the flow under protection from their military. He weighed the odds and decided to diffuse the military tension. He believed that constructive solutions would come, not from dangerous confrontation, but from dialogue and understanding. He instructed the Commander to have the military evacuate the site and provide a chance for dialogue. The Commander declined to carry out the instruction on the ground that he had orders to implement "active" and not "passive" instructions; and evacuation was passive. The Acting President insisted that the instruction he gave was the only instruction he could give, and that it was as active as he could be under the circumstances. The Commander would not budge.

As they walked back to the tent, the Acting President stared at the entrance gates of the intake tunnel, lifted half way up to let water flow into the tunnel to the KAC. He dreaded the possibility of that water structure being hit and damaged by a shell from the guns across the river, 200 meters away. The gates could drop, twist and block the entrance to the tunnel leaving the KAC, the main conveyor of irrigation water, dry. It would take weeks to fix it, but the crops would have wilted. The Acting President was determined to avoid any such possibility and did not mind denying the military the opportunity to do the job they had been trained to do, and open fire. He stood unarmed between the two armies and was determined to solve the problem peacefully.

To break the stalemate with the Commander, the Acting President asked to be taken to the Rutenberg pool, where he could see if there was any Yarmouk water flowing unused past the Israeli pumps to the Jordan River. The Commander escorted him to that destination and the Yarmouk bed was dry. Obviously, the Acting President concluded, the Israelis could beneficially use more water than even the adjusted flow brought them; but so was also the case of the Jordanians. He believed that one should live and let live, and that no party should attempt to solve his water shortage problem at the expense of the other.

The dialogue continued with the Commander. He insisted that his orders did not include a pull-out from the site, but rather to recover the water the Israelis had taken away. The Acting President assured him that the Jordanian side was not all that innocent; that the bulldozer operator had shaved a lot more debris from the river bed than was agreed, and that resulted in more water flowing to Jordan than she actually was entitled to. It was a zero-sum game, where a gain by one party was the loss of the other. But the Commander persisted, and the Acting President did not go along with his desires to make a clash.

The Acting President drove back to Amman and headed to the offices of the Mixed Armistice Commission (MAC), located on Jebel Weibdeh across the street from the old U.S. Embassy building. The MAC was set up under the Armistice Agreement that Jordan and Israel concluded in April of 1949. From there, he called the Prime Minister, who was eager to hear from the civilian official his assessment of the situation, and asked if he was able to have the water recovered. The following conversation went on:

"Sir," said the Acting President, "there is a workable peaceful way to recover our water. I just can not propose it while the two armies are facing each other with tanks and guns."

"How do you propose to respond when your enemy encroaches on your territory, insults two lightly armed soldiers, cuts off their communications and diverts your water to his side?" Asked the Prime Minister angrily.

"I deplore their action, but I am more concerned about recovering the water without firing one shot."

"If you are not willing to support the military action," said the Prime Minister, "and you are not willing to do it unilaterally with your own men, I will call upon the students of the Yarmouk University to have them do the job. If it is a contest of who can impose facts on the ground I will have them cast cement bags across the river, block the flow to the Israelis and have all the water diverted to us."

"Sir, such an action will trigger an Israeli response that could harm these college kids, and damage our water structures," said the Acting President.

"I will expose them in the United Nations, I will show the world what aggressors the Israelis are, and will face their aggression with equal force if need be."

"Sir, it is better done the quiet peaceful way. Please, as Minister of Defense, pull back our forces from the site. I came to know that we mobilized first and the Israelis mobilized in response to our move. Please have the army withdraw, and I expect the Israelis to react in kind," pleaded the Acting President.

"What!" shouted the Prime Minister, "you want me to withdraw before recovering our water? Do you back down in times of confrontation? I knew it, you are *khawweef*"[280]

"Sir, *khawweef* is an under-statement. The fact is that my knees buckled; they could not carry me and they shake with fear from the potential clash. But let me remind you, sir, of one fact: if you do not know when to be afraid, you will never know how to scare others. I am now scared for one

---

[280] The Arabic word for frightened, very scared...very fearful.. usually said to a coward.

simple reason. It is that the trees in the Jordan Valley cannot walk to the water; I have to bring water to them. A shot from one of the guns I saw this morning will knock off our water supply facilities, and heavy losses will be incurred by the people who have invested in agriculture and by those whose livelihood depends on it. I feel responsible. This is the reason why my knees are shaking with fear."

The Acting President knew that the Prime Minister was trying to invoke the guts he knew he had. He continued:

"Sir, how long have you been Prime Minister? Three years now? Have I ever asked you for any favor?"

"No, I do not recall," came the answer.

"Now, for the first time I try. Please, I beg you to pull the army out, and give me a grace period during which I can get to work and recover our water peacefully."

"How long of a period?" Asked the Prime Minister.

"72 hours," suggested the Acting President.

"That is too long."

It sounded to the Acting President that the principle was accepted, but that he had to negotiate the grace period.

"Okay, sir, 48 hours."

"Fine, I will have the Army pull back today, and you have 48 hours to recover the water."

"Thank you sir, thank you. I will never forget this favor. I consider it a personal favor. I will start action right now."

After the conversation ended, the Acting President looked at the Austrian Blue Beret officer in charge of the MAC office, thanked him for use of the phone, and asked him if he could arrange for a meeting of the MAC. It was the mechanism Jordan had to use to deal with the Israelis in a specified manner to prevent eruption of hostilities along the cease-fire line. MAC's presence there was part of UNTSO and it was their job to prevent a military clash. The officer said he would get on it promptly.

That afternoon, the Acting President received a call from the Deputy Charge of Mission (DCM), of the U.S. Embassy. It was a coincidence that the concerned bosses in the region were all out of the countries in which they carried their responsibilities. The President of the JVA was in the United States, and his deputy cared for the shop in his absence; the United States' Ambassador to Jordan, Mr. Nicholi Veliotes, was on his annual leave, and so was Mr. Samuel Lewis, the American Ambassador to Israel. Their deputies ran the respective embassies; and the party at the State Department who handled the crisis was Mr. William Draper, Deputy Assistant Secretary of State. It was a show of deputies!

The DCM at the U.S. Embassy in Amman, Mr. David Zweifel, was more worried than the situation warranted. He knew that the two armies were almost looking each other in the eye, and was afraid that hostilities would break out while he was in charge of the diplomatic mission. The Acting President calmed him down and told him that the matter had just been brought under control. He could not imagine any interest for either side in starting hostilities. The mood in the region was that of watch-and-see. People were waiting to see what would come out of the Egyptian-Israeli peace deal, and none expected any hostilities to break out.

That evening, the Prime Minister called for a meeting at his office. The Acting President was invited, along with the brass of the Army and other ministers. The Chief of the Royal Court, the late Abdul Hamid Sharaf, was there too. The Jordanian mood was uncompromising. The Acting President explained to the meeting how he intended to resolve the conflict through a meeting of the MAC under U.N. supervision. The Army would have to delegate an officer to accompany a senior delegate from JVA to try and resolve the issue peacefully, in accordance with the 1949 Armistice Agreement between Jordan and Israel.[281] He appreciated the willingness of the government to go as far as using military force to recover the water, but he harbored fears that the losses would be substantial at a time when Jordanians were starting to cultivate the fruits of Jordan Valley development. Lucky for him, the President of the Jordan Valley Authority returned home two days later, and rolled up his sleeves to shoulder his responsibilities. He was good at that.

The MAC met twice in two days. The Israelis would not entertain the idea of moving a bulldozer into the water again. The meeting culminated in an agreement to use sand bags[282] to control the water level which, in turn, determined the rates of flow of water to either side. The Israelis refused to allow the Jordanians to remove the sand bar. The likely reason was that this obstacle in the way of flow to Jordan would force the Jordanians to meet more frequently with the Israelis to adjust the flow every time the river bed conditions, or the flow rate, changed. However, all the meetings were held under the auspices of the United Nations in accordance with the Armistice Agreement of 3 April 1949, and in the presence of an officer from UNTSO, an officer from the Jordan Armed Forces, and an officer from the Israeli Defense Forces.

Jordan had high expectations to have the Maqarin dam built, including the diversion weir at Adasiyya. JVA officials thought that it would

---

[281] "General Armistice Agreement (with annexes)." Signed at Rhodes, on 3 April 1949, Article VIII. The registration at the United Nations took place on 6 October 1949.

[282] These were sacks filled with sand that weighed about 50 kg each. They would be placed in the rivver bed to raise the water level and control the flow going to each side at the diversion point.

be a matter of a short period of time before the diversion weir would be built, through Habib's efforts, and the problems of diversion resolved without the need to hold meetings of the MAC. Habib's efforts to remove the obstacles in the way of building the dam stalled in 1980, and thus the problem at the diversion site persisted. The sand bar grew bigger and bigger as time passed, and the water requirements were increasing to meet irrigation demands and the supply requirements of municipal water to Amman. As the hopes to build Maqarin diminished, JVA officials decided to raise the Wadi al Arab Dam currently under construction, to make more water available for use in Amman in the summer. Water to fill the increased capacity of the dam, about 10 mcm, would be pumped from the KAC, which, in turn, got its water from the Yarmouk. Concurrently, they decided to have the King Talal Dam raised to increase its live storage capacity from 52 mcm to 82 mcm with finance from the Kuwait Fund. Financing of the Amman water project from the East Ghor Canal had been secured from the USAID, the Kuwait Fund, and the Saudi Fund. The contracts were awarded and signed in the spring of 1982, and construction commenced in May of that year. The diversion of water from the Yarmouk would soon supplement municipal water availability in Amman, in addition to irrigation water availability.

Thus, Jordanian, dependence on the Yarmouk increased, but so did the Syrian abstraction of water from its watershed upstream, at the expense of water availability to Jordan. The diversion of water to Jordan had to be adjusted through the sand bags more frequently than anticipated, to secure a fair share for Jordan, and water shortages became more frequent despite the investments made in more efficient irrigation networks. On the Syrian front, no progress was ever made to move on with the implementation of the Maqarin dam, and the plans in that regard had to be suspended.[283] Mr. Philip Habib made his last visit to Jordan in connection with the project in November 1981, and his mission ended without the success he was hoping for. That was bad news to Jordan where need for water to supply North Jordan with domestic water was intensifying.

In June 1982 the Senior Vice President of the JVA succeeded Mr. Dokhgan as President and Chairman of the Board of the JVA He assumed his new responsibilities amid growing shortage of water in the Jordan Valley, especially in the summer. The nightmare of military confrontation gave him chills every time he thought about the possibility of them recurring. He could not stand the idea of Jordanian water structures being forever vulnerable to Israeli shelling without a Jordanian "answer back". Finally, he embarked on a

---

[283] The JVA was following the schedule of events agreed upon in the donors' conference. A pre-qualification of contractors was made, and letters were signed to advise the contractors of their status. Fourteen contractors were prequalified out of twenty-six international contractors who responded to the announcement.

solution that would provide Jordan with a deterrent. He pinpointed on a map the sensitive Israeli water structures that were within the range of several Jordanian artillery batteries. He arranged for a meeting with the Army Chief of General Staff, explained the situation to him, and requested that those targets be clearly marked on the military assault maps with a recommendation that these be assaulted only as a last resort and only after the Israelis score damaging hits on Jordanian water structures. Whether or not the military heeded the advice was never tested, but that was the end of the water-instigated military confrontations with Israel.

5. The Pleasant Surprise

When the domain of the Jordan Valley Authority was expanded southward in May 1977, to include the Dead Sea and Wadi Araba, the visionary President of the Jordan Valley Authority, Mr. Omar Abdallah Dokhgan, initiated a comprehensive study for groundwater investigation in the Jordan Rift Valley. The project received the support of USAID, and drilling was started in Wadi Araba in 1978 using the government rigs of the Natural Resources Authority. The drilling depths were modest but deep enough to investigate the shallow renewable aquifers.

Soon, JVA officials started to notice a reaction from the Israelis to the west in Wadi Araba. They mobilized drilling rigs directly across the Armistice Line from the Jordanian rigs, obviously to investigate groundwater on their side. The JVA staff engineers were very careful to keep the drilling logs and data obtained from their wells classified the best they could. Modest discoveries were made and their assessment of groundwater availability was in the order of 18 mcm per year. The water quality was marginal at best. With that conclusion, they proceeded to drill about two dozen wells to introduce agricultural development in that arid area. Additionally, they awarded a contract to investigate the groundwater resources in the Jordan Valley between the Yarmouk and the Dead Sea.

The first surprise came in March 1982, when the contractor was drilling in Wadi el Arab, the tributary immediately to the south of the Yarmouk and close to the town of North Shuna. At a drilling depth of about 550 meters, the groundwater gushed out under pressure. It was hot, with a temperature of 57° Celsius (135° deg. Fahrenheit). The salinity was below 600 parts per million (ppm) of total dissolved solids, and that was very good news to the JVA. When the United States' Ambassador, Mr. Richard Viets, visited the well (the contract was financed by USAID) and saw the water flowing out at about 750 cubic meters per hour, he turned to JVA Vice President and said, "two more wells like this and you will qualify for the Nobel Peace Prize!" The Ambassador's comment reflected the importance water had in the conflict between Israel and Jordan, so much so that adding water to the known stock would relieve tensions and help remove one obstacle on the way to some arrangement for peace with Israel.

The bigger surprise came about a month later, in April 1982. The contractor's rig was drilling an investigation well at upper Mukheiba in the Yarmouk river gorge, a couple hundred meters away from the Israelis, who patrolled the territories across the river. JVA's groundwater technical expert, Dr. Omar Joudeh, assigned a design depth of 850 meters to that well, where he hoped a deeper aquifer, below the one feeding the springs there, would be reached. The rig drilled the specified depth but there was no sign of water. He asked for authorization to drill 100 meters more, and the authorization was made with the approval of the financer, USAID. The drilling reached a depth of 950 meters but no sign of an aquifer. Dr Joudeh made another similar request, and it was granted with the approval of the financer. The drilling reached a depth of 1,050 meters, but no sign of water. It was embarrassing for the JVA President to ask USAID for a third increment of another 100 meters, but he did. To that request, the USAID officer in charge, Mr. Thomas Pearson, thought that JVA was wasting their time and USAID's money, and he advised that the drilling site be abandoned. The JVA President then turned to the Minister of Finance and asked for an advance of money from the following year's budget to go deeper in drilling at the same site. The Minister, Mr. Salem Masa'deh, approved the request. The rig continued drilling. When the drilling got to a depth of 1,146 meters, four meters shy of the 100 meters approved by the Minister, the water came up under artesian pressure to a depth of 6 meters below ground surface; i.e, the pressure head drove the water up a column of 1140 meters! The freshwater aquifer Dr. Joudeh was looking for was finally discovered. The President issued his instructions to keep the drilling logs away from unauthorized eyes, and away from the financier who refused to pay for the last 100 meters! He made it clear to USAID that it was Jordan's money, not USAID's, that finally made the discovery.

With that discovery of groundwater, hopes for a source of water supply to Irbid, the second largest city in Jordan, increased. More work was needed to establish the properties and extent of the new-found aquifer, and to study its dynamics. Promising sites for production wells were chosen at the intersection of two faults at the bottom of Wadi Shag el Barid in the Yarmouk gorge, a few kilometers upstream from the site of the above well. The location was about one kilometer upstream from the foundation trench of the Mukheiba dam, whose construction was interrupted by the June 1967 war. An Army observation post was close by. The drilling rig was moved to that promising site, and it started drilling.

On 6 June 1982, the Israelis started a military operation in Lebanon, Peace for Galilee (*Salam el Jaleel*), aimed, as they claimed, at cleaning South Lebanon of the PLO bases. The Israeli Minister of Defense, General Ariel Sharon, in the Likud government headed by Menachim Begin, was the advocate and champion of that operation. The Israeli tanks and troops crossed the Lebanese borders on 6 June, and advanced northward with sporadic resistance offered by the PLO troops. By 22 June, the Israeli Army was besieging the castle of Shgeif, in South Lebanon, and there was fierce resistance.

The Director of the JVA President's office, Ms. Raja' Kamel Dawoodi, received a phone call from Dr. Omar Joudeh who had gone out to the field. He informed her of the need to talk to the JVA President. When she informed him the President was in the conference room chairing an important meeting, Joudeh insisted that he be interrupted to answer the call. Daoudi came into the hall where the President was chairing a tender opening meeting, and related Joudeh's message, who was calling, she said, from Wadi Araba.

"But there are no phones in Wadi Araba for him to call!" Remarked the President.

"I don't know," responded a confused Daoudi.

"Could he be calling from Wadi al Arab?"

"Yes, it is very likely!"

"There is a hell of a difference between Wadi Araba south of the Dead Sea, and Wadi al Arab next to the Yarmouk," said the President while he walked to his office to answer the phone call.

Joudeh was calling from Wadi al Arab, where he could find the closest JVA phone. He was shaken up, and for good reason. The drilling rig at Mukheiba was in danger of over-turning. Water was gushing up the unprotected drillhole sooner than he had expected, and the gushing water was eroding the soil as it headed for the Yarmouk nearby, clearly posing the danger of undermining the drilling rig. He was afraid of the responsibility of losing an expensive drilling rig. The President asked his deputy to continue the meeting and headed for the Yarmouk gorge, about two and a half hours away. On the way up the Jordan Valley, he could see heavy traffic on the parallel road across the Jordan River, inside Israel. The traffic was mostly military vehicles travelling northward, in what looked like logistical support for the advancing Israeli forces inside Lebanon.

The scene at Shag el Barid was something very unfamiliar. Water was gushing out from the hole under the heavy rotary drilling rig and had, during a couple of hours, dug up a deep trench as it opened its way down the steep slope to the Yarmouk river. It gave a sound similar to that of a waterfall, as the water gushed down with high velocity toward the river. The erosion of the soil was progressing toward the rig; the contractor who owned it was standing close to it, and the President stood next to him, along with other engineers. The workers on the rig were trying hard to pull the drilling equipment out from the hole and salvage whatever they could.

While the Jordanians were managing their emergency situation on the drilling site, an Israeli helicopter appeared in the Yarmouk gorge flying over the northern occupied bank. It was flying low and the Jordanian group, standing on a slope about 80 meters above the Yarmouk, could see the pilot and his co-pilot at their own altitude. The helicopter flew westward and disappeared behind the Golan western slopes. The Jordanians thought it might be carrying wounded soldiers from the siege of the Shgeif castle in Lebanon to hospitals in Israel.

The soil erosion got to the rig and the foreman in charge shouted at his men to abandon it; one of them was so attached to the equipment that he barely had time to jump, before the rig fell westward. Its mast appeared as though it would fall on the President and his company, standing close by. They ran back for safety and, when they stood in a safer place, they saw the Israeli helicopter making another pass over the occupied bank of the river, slower in speed this time, and they could see the co-pilot with a video camera, taking shots of what was happening on the Jordanian side. The chopper then flew westward and disappeared again behind the Golan western slopes. So, it couldn't possibly be carrying wounded soldiers as it was first thought, but was on a surveillance mission to observe and document what was happening with the rig and the groundwater investigation. The rig fell as Joudeh had feared, sustaining heavy damage, and the gushing water could not be easily arrested nor controlled for the following six months.

What actually happened was beyond expectation. With water struck at a depth of 1,146 meters, in the well about two kilometers away, Dr. Joudeh expected the water level at the new site to be at about 850 meters depth. He approved the drilling plan so that the drilling would proceed down the hole unprotected, without casing for a depth of about 500 meters, after which they would install the casing. The water surprised them under pressure at a depth of 354 meters, and the result was as described above. The drilled hole had obviously intercepted a big cleavage crack that was connected to the aquifer underneath, and the water gushed up under artesian pressure in an un-cased hole that was expanded in diameter by the action of the gushing water.

The Prime Minister in Amman was briefed. He had just finished breaking his fast that Ramadan evening, when he received the call from the JVA President. He received the news with obvious pleasure. The problem of supplying Irbid with water was obviously on its way to a good solution without the construction of the Maqarin dam, which the Syrians had not yet approved.

A few days later, Prince El Hasan piloted a helicopter of the Royal Airforce to the well site. The Jordanian airspace available for the maneuvering of the helicopter before landing in the Yarmouk gorge was very limited, and the Crown Prince almost crossed to the occupied airspace before landing his helicopter. The Prince disembarked, inspected the well and the damaged rig, and sat down for tea with the guard and the workers before he took off in the helicopter. His trip was spontaneous and he just piloted the helicopter with his military escort and headed towards the well. The people in the nearby village, Mukheiba, thought that an Israeli helicopter had landed near the new well, especially as there was no sign on the ground of the Prince coming to visit. Such sign was usually in the form of a small contingent of the Royal Guards Brigade deployed on the way to the well before the Prince arrived. Soon after that, the rumor spread in Amman that an Israeli helicopter had landed near the well uncontested, Israeli soldiers disembarked, and took samples of the well water before they flew back out of the area. Several newspaper correspondents, Jordanian and foreign, converged on the President of the JVA to confirm or deny the report. His answer was:

"Why would the Israelis fly a helicopter, trespass against Jordanian territories, and run the risk of a serious military setback if that chopper were shot down, just for the sake of taking a water sample? They could do that without flying a chopper or facing any material risk because the well water is gushing down a slope of about 200 meters to join the Yarmouk river, which is the cease fire line. It is less dangerous to sneak a few meters across the cease fire line and scoop a bucket of that water than the risk inherent in landing a helicopter that would be very vulnerable to Jordanian fire."

The helicopter of the Royal Airforce suddenly became Israelis in the eyes of the suspicious!

6. The Israeli Reaction

JVA's high concern, however, was not the sampling of the water by the Israelis. The more serious concern was that the aquifer that was discovered might extend across the cease-fire line into Israeli occupied territories, where they could drill and tap it. The JVA President spent days with long hours of desk reviews and inspection of geological maps assisting Dr. Omar Joudeh to answer the question: "is there a risk of the Israelis tapping the same aquifer?" The conclusion of their work was that the aquifer was predominantly Jordanian, but that a pocket or two could be extending across the Yarmouk into the occupied territory. They located the domains of the pockets and kept their fingers crossed that the Israelis would not guess at them. They had no doubt that the Israelis would mobilize a rig and drill across the river from the Jordanian site. They had been in that habit since the Jordanian program of groundwater investigation was started in the Jordan Rift Valley. The Israelis waited until the Jordanians moved their rigs before they moved theirs on the other side of the cease-fire lines.

JVA's concern was high over the real potential of the Israelis tapping the new aquifer from one of the pockets across the Yarmouk. The Army post at Shag el Barid was instructed to keep close watch over Israeli civilian activity across the river, and to report the movement of any Israeli rigs in that area. Two weeks later, the Army post reported that a rig was being mobilized, and the JVA President immediately drove to see for himself where they would be positioning it for drilling. He used the facilities of the Army observation post and walked to a cliff closer to the river, from where he could see the rig in better detail. The location where they were going to drill was away from either pocket, but the concern persisted. The JVA President worries settled only after he watched the Israelis test their well, and he could estimate a yield of about 600 cubic meters per hour. That was confirmed in an article published in the Jerusalem Post about the well. It reported a pumped yield of 550 cubic meters per hour, less than one tenth of the yield from the Jordanian well, and theirs was not a flowing type, an indication that their aquifer was different from Jordan's.

Soon after the Jordanian well was discovered, officials of USAID approached JVA and related Israeli concerns and worries. The Israeli concern, as related by the American officials, was that the aquifer the Jordanians discovered could be historically feeding Lake Tiberias. If that was true, the Jordanians would then be intercepting water that had been flowing into Tiberias for millennia, and therefore they should not be abstracting from that aquifer. Another hypothesis advanced by the Israelis was one that JVA themselves entertained. The flow of water at that high rate from the well could impact the nearby springs, one of them was used for tourism on the Jordanian side before it discharged into the Yarmouk downstream of the new well, and another spring was basic to Israeli tourism and fisheries in the occupied Himmeh, across the River from the Jordanian springs.

The JVA officials requested that the Israelis establish the grounds for their fears by taking measurements of the spring flows on their side as the Jordanians had started to do on theirs. On the Israeli concern of interception of flow that otherwise would be going to Lake Tiberias, the Jordanians asked for scientific evidence to support such fears.

Soon, and by virtue of these findings, the JVA went back to drill in Wadi al Arab, upstream from the dam that was under construction, and water came flowing out from a depth of 500 meters. A well field in Wadi al Arab would be more convenient to supply Irbid with water, and so JVA shifted the analysis accordingly. Simultaneously, the uncontrolled flow from the well at Shag el Barid had to be diverted away from the Yarmouk to keep it separate from the water sharing arrangements at the Adasiyya diversion downstream.

A plan was hurriedly put together to divert the flow of the well, and the Prime Minister had it discussed and approved in a mini Cabinet meeting. It took the JVA eight months to design and build a diversion canal with all the appurtenant works, and the water was diverted to the KAC inside Jordanian territories, in March 1983. The Israelis were watching the canal work from their side every day, and did not feel happy about it. They filed unreasonable complaints with the MAC. When Harold Saunders, an American diplomat who worked with Secretary Kissinger, visited the Shag el Barid well field, his son who was accompanying him wondered why the Jordanians did not want to share the water of the well field with the Israelis. Saunders, the father, answered his son's question and said that the reason was the same that the Americans would not share the Texas oil fields with their neighbor, Mexico.

7. Off-Canal Storage Found

The newly discovered aquifer was a rescue for Jordan, at least for a while. The JVA concern was focused on Syrian activities, especially since Jordan's relations with Syria went cold in 1981. The JVA realized that Syria's written approval of the Maqarin construction would not be coming anytime soon. The Yarmouk floods and part of the winter flow was passing the Adasiyya diversion toward the Jordan River and the Dead Sea. The Israelis had installed pumping plants on the Yarmouk, before its confluence with the Jordan, and were pumping a good part of the winter flow that was not diverted to Jordan's KAC. New means had to be found to store Yarmouk floods and make water available to the Jordan Valley. One such means was the raising of the Wadi al Arab Dam and pumping water to it from the KAC; that was done, but it provided only modest amounts (10 mcm per year).

Accidentally, and while the JVA President was taking a morning walk in the Middle Jordan Valley with his 7 years old son, in October of 1982, they came to a *Katar*[284] cliff where they looked over a depression that could possibly be used as a storage reservoir. The depression was shown on the maps as "Wadi Mallaha." It was wider in reality than the line on the map indicated. The whole wadi lay in an area of the Jordan Valley whose land was classified as "Class 6", unfit for irrigation, and it was therefore never the subject of surveying and detailed mapping like the rest of the Jordan Valley had been. The President thought that only the MacDonald's report of 1951 could possibly contain topographic maps covering that part of the Valley. After three days of search in the stores of the Department of Lands and Surveys, the MacDonald report was found, and a contour map of the entire Jordan Valley was included. Using that map, JVA discreetly worked on the "topographic" feasibility of Wadi Mallaha, and found that a dam could be built across the wadi to create a reservoir of about 55 mcm capacity; it could be raised further with additional work on the perimeter of the reservoir, which would increase that capacity to 75 mcm. The water could be diverted from the Yarmouk floods down the KAC and dropped into Wadi Mallaha for storage. JVA further studied the possibility of pumping into the reservoir from the close-by Jordan River, and that possibility was real. It was concluded, therefore, that the JVA could make up, at least partially, for the stalled

---

[284] *Katar* is the name given to salty clay soils that are eroded, believed to be marine deposits from old times when the Dead Sea submerged the Jordan valley.

Maqarin dam by building a dam at Wadi Mallaha, and provide summer water to irrigate the intended extension of the Canal.

The next step was to employ a reputable consulting firm. The Kuwait Fund for Arab Economic Development expressed preparedness to finance a feasibility study, and an announcement for pre-qualification of consultants was placed in the papers. Two days later, the Ambassador of the United States, Mr. Richard Viets, asked for a meeting with the JVA President.

The Ambassador was concerned that the construction of a dam in Wadi Mallaha to store Yarmouk floods would contradict the Unified /Johnston Plan.[285] The President and the Ambassador reviewed the provisions of the Unified Plan and concluded that the dam was not in contradiction with it. The Plan called for the storage of the Yarmouk floods, past Maqarin, in Lake Tiberias or in any other economic location as a neutral engineering board may validate. It was agreed that the study of feasibility of that site would fall within the investigation for possible other economic locations where Yarmouk floods could be stored. Following are excerpts of the dialogue that took place in that meeting:

"Why is it that the rules should apply on one side of the river and not on the other side?" Wondered the JVA President.

"You know, Excellency," responded the Ambassador, "we have been trying our best to have all the parties conform. We do not want troubles to erupt because of water. The region is getting ready for a meaningful dialogue pursuant to the Reagan Plan.[286]"

"I am after arrangements to remove the sand bar. It does not look like the Israelis are responding."

"We will keep trying. Let us hope for the best," said Ambassador Viets.

8. The Picnic Table

The issue of the sand bar persisted. The JVA President raised it with every United States official who visited Jordan with whom he had the chance to meet. The Jordan Valley visit had been on the itinerary of visiting dignitaries, and the subject was brought up with visiting Congressmen, Senators and State Department officials. The importance of improved diversion efficiency from the Yarmouk became more important with time. For one reason, Syria started to implement a project in the Yarmouk

---

[285] Later, during the peace negotiations in November 1992, the Israelis demanded that Jordan stop work on the construction of the Karama dam (the dam on Wadi Mallaha).

[286] President Reagan announced the intent of the United States to renew contacts with the parties to look for a peaceful settlement in the Middle East. His initiative came in September 1982, as a result of the massacre at Sabra and Shatella refugee camps in Lebanon, in the wake of the Israeli invasion of Beirut.

catchment aiming at winter irrigation of field crops with an area of about 11,000 ha; and, for another, the JVA was planning on implementing the irrigation component of the Maqarin dam project by extending the KAC 14.5 more kilometers and planting the area primarily with one winter crop and a limited area of perennial crops. For that purpose, the King Talal dam would be raised, and the diversion efficiency from the Yarmouk should be improved. The Israelis, on the other side, would not respond to Jordanian requests to clean up the sand bar, despite the many interventions of United States officials. For the Israelis, the issue provided an opportunity to meet with Jordanian technical staff under the auspices of MAC, and to regulate the water sharing. The meetings at the Yarmouk diversion point became more frequent. Tall trees existed on the Jordanian bank with nice shade suitable for holding the meetings on-site. A pre-cast concrete roof panel, transported from the Middle Jordan Valley where JVA was conducting experiments on low cost housing at the village of Damya, served as a meeting table. Sand bags used for the control of the water level in the river served as seats around the table. When some news leaked about the MAC meetings, excited journalists in the United States referred to it as the "picnic table." They claimed that Jordanians were negotiating substantive water issues with the Israelis around that picnic table. The fact was that the United Nations' mechanism was used to avoid an outbreak of hostilities over disagreements and misunderstandings over water diversion from a point on the cease-fire line. Nothing but the transient water diversion arrangements and the sand bar issue were ever discussed between the two sides. The other substantive water matters were cared for by Mr. Philip Habib and his mission.

The river flow would change from week to week as a result of Syrian abstraction upstream, and the diversion to Jordan (and consequently to Israel) would change accordingly. The meetings would discuss the updated circumstances and agree on a sharing arrangement. Sand bags would be used to control the water level in the stream and thereby control the rates of diversion. The Prime Minister of Jordan was briefed on the outcome of each meeting, and so were the military, each through their respective official channel.

9. A Daring Operation on the Yarmouk

The winter season of 1983-1984 brought little rain, and citrus trees in the Jordan Valley needed irrigation in the winter months, an unusual phenomenon, especially in the north. Diversions from the Yarmouk, needed to meet increased irrigation demand, continued to be impeded by the sand bar. The President of the JVA understood what the price for its removal would be: a higher-level Jordanian meeting with the Israelis, in which the Israelis would have the advantage because of Jordan's pressing needs. Another important factor that increased the need to clean up the diversion point and remove the sand bar was a new project that the JVA announced--the extension of the KAC for a distance of 14.5 Km to add 60,000 dunums to the irrigated area. The JVA President had negotiated the terms of a loan from the Federal Republic of Germany to help finance that project. More water would be needed to serve this expansion. Thirdly, improved diversion efficiency became mandatory if the dam on Wadi Mallaha, (the Karama dam), were to be built soon. Such a Jordanian pressing need, it was realized, would handicap the Jordanian official who would undertake to attend the meeting of the MAC to resolve the issue of the sand bar. Something had to be done to remove that constraint on Jordan. The JVA had to make the removal of the sand bar less important that it had been. There was no way to make its removal totally un-necessary.

In early February of that 1984 drought season, farmers and people noticed how little the water diverted from the Yarmouk to Jordan was. The irrigation demand was high, and the water supply in that winter month was at an all time low. Farmers were complaining and their hopes for relief were very low. The JVA officials were so annoyed that they had to do something to relieve the farmers. The President of the JVA decided to conduct an operation that would bring some relief and improve JVA's negotiation position with regard to the sand bar. The operation he thought of had a high risk for safety and security. But he decided to take that risk anyway. He benefited from a military circular to the border military units ordering them to facilitate his access to the waters of the Yarmouk and the Jordan rivers. The President used to get to the banks of both rivers accompanied by an armed soldier, and the units would facilitate his missions.

He approached the diversion point one afternoon in early February 1994, in the company of a soldier armed with an M16 rifle. They walked upstream to inspect the river bank, which was crowded with bushes and wild plants. He wanted to survey the bank and decide on the point of approach for a daring operation that would help remove some of the diversion constraint. When the JVA President noticed an Israeli patrol across the river, he asked the soldier accompanying him to keep cool, and not to return fire if the Israelis opened fire. The soldier protested and said he would not allow the Israelis to shoot freely, but would use his M16 very efficiently to return the fire!

The JVA president worked out an arrangement with the soldiers on the site for the conduct of a "maintenance operation" a few days later. He further decided to attract the attention of the Israeli Army patrols, operating across the river, away from the diversion point and the sand bar. About three kilometers upstream, the newly built Mukheiba Canal underwent a minor slide and it needed to be repaired. An urgent operation was planned by which work would be conducted in three shifts over 24 hours per day for as long as it took to finish a daring operation on the sand bar downstream from those repairs. The repair work was noisy and was equipped with a generator to produce electricity for work at night. The site of repair was close to the Yarmouk bank, but some 100 meters above it. The noise and night activity attracted the attention of the Israeli patrols across the river, and they parked for hours watching the Jordanians work diligently to repair a broken canal. That reduced the focus of Israeli patrols on the diversion point down stream. Additionally, for about a week, the JVA president asked a fisherman who had military clearance to fish in the Yarmouk to watch for patrols on the other side, and he benefited from the fisherman's reports.

The President worked out an arrangement with the Director of Operation and Maintenance in Deir Alla in the Valley to mobilize a dozen men for the operation. They were to be equipped with the hand tools and ropes needed to access the river stream. In Amman, he contacted the Prime Minister, and went to see him instantly. Following are excerpts of their conversation:

"Sir, you know the situation in the Jordan Valley. I have never been exposed to such pressure from the farmers to bring them more water. I have decided to do something about it."

"What is it that you want to do?" Asked the Prime Minister.

"I want it to be secret because it carries high risk. I am afraid you would talk me out of it if you knew its details. Just trust me, and if you hear that I have been taken captive to some Israeli detention camp in the Negev, please disclaim me. You do not know anything about the operation." The Prime Minister smiled, and gave no comment. He did not know what the JVA President's operation was, nor did he know anything about the timing.

The timing of the operation was set at 0600 hrs on the morning of 8 February 1984.

The President met with the twelve-man crew under a tree close to Adasiyya at 0600 hours, in the morning of that day, and briefed them on the operation. The crew was to approach the river with the hand tools and dig a channel through the sand bar and across the cease-fire line. The aim was to divert more water to the KAC. Their foreman, Mohammad Zuhair Mashriqi, was excited and very enthused. He said he knew what the President was after when he was told to mobilize with hand tools and ropes. The group was ready and was enthused to see the President leading them. The President explained to the crew that the operation was not without a high safety risk, and that was why he wanted to lead them personally. He noted that the Israeli patrols surely recognized the car he used--a purple Caprice 1979 with a red license plate and a number 105. Having briefed the workers, the President took with him three of the crew, Mashriqi and two others, and told the rest he would be back for them in a short while.

The President drove past the Army tent by the diversion up toward the site where the Mukheiba canal was being repaired. He disembarked and noticed the Israeli patrol parked directly across the river at that early hour of the morning. After a short inspection of the repair work, he left his Caprice, mounted a Toyota field car from the repair site and asked Mashriqi to join him with his two other workers. He drove back, dropped the workers at the Army tent, and asked the soldiers to have them escorted to the river bank, where they should wait under a huge eucalyptus tree and not be seen by the Israeli patrol. He then drove to get the rest of the workers using the field car in two shuttles. Finally, he joined the workers on the river bank, where the water was about 3 meters below the bank and flowing with appreciable velocity.

The President insisted to be the first to be lowered in and, over-ruling protests by the workers, he was lowered into the water. The rest of the crew followed, one by one. Only one man stood guard on the bank and held the rope to help lift or lower workers and tools.

"Give me that axe," the President asked Mashriqi.

"In the name of God, the Merciful and Most compassionate," the President said in a loud voice as he hit the first blow on the sand bar. He continued with more strokes before Mashriqi stopped him.

"Sir, if you do not leave right now, we will not be able to do our job. You are actually standing in our way. Please go back up, attend to your own job and let us do ours as quickly as we can. You are in danger here and the country needs you more than it needs us. Please leave."

The President gave Mashriqi a communication radio and agreed with him on a code they made up on the spot for the occasion, and was lifted up by the guard.

The sand bar was really more like an island, not a small bar. Its width that had to be dug was over 20 meters, and its average thickness was over one and a half meters. The crew had quite a task at their hands. The President kept communicating with the crew throughout the operation using the special code. By sunset, the crew would come out of the river, and would return to their task before sunrise the following morning. It took them two weeks to dig a sizable channel across the sand bar, and water ran down the diversion channel into the tunnel and on to the KAC. They had a good reason to celebrate.

The Israelis detected a drop in the flow going to their pumps only late the next day. They inspected the site from their side and immediately established a permanent observation post on the cliff overlooking the diversion site. They decided to keep it under continuous watch. The President reported the results to the Prime Minister in a phone conversation:

"Sir," said JVA President, "mission accomplished. It took us a bit longer than I expected. But we did it and water is now flowing into the KAC twice as much as it used to during this winter season. Congratulations."

The Prime Minister was pleased. The President returned to the Jordan Valley the following week to watch and enjoy the fruits of the operation. He was relaxing on the back porch of his house in the Valley when a farmer, Mohammad Musa el Ya'qub, also called Abu Ali, came to see him.

"We all are so happy you brought the water into the canal," said Abu Ali.

"This is our water and we have every right to it. So, we brought it home."

"Have you heard what people say?"

"What do people say?" Asked the President.

"They say that Dr. Munther brought the water from the Yarmouk because he and the American Ambassador in Israel are married to sisters," said Abu Ali.

"What for heaven's sake!" Exclaimed the President.

"I know your wife is American, but is her sister married to the American Ambassador in Israel?"

"No, Abu Ali, no. Don't believe a word of what some people are saying to spread a rumor."

No Government official was aware of what the operation was, nor of the timing of its implementation. Soon after it was accomplished, an official of the United States Embassy in Amman, Acting for the Director of USAID, who was on vacation, asked to see the JVA President urgently. He explained that they had received a complaint from Israel about some action that JVA had unilaterally taken on the Yarmouk. He added that the Israelis complained to the U.S. Embassy that the JVA had sent a bulldozer into the river, and managed to change the ratio of water division.

"A bulldozer," the President noted to his guest, "does not fly in the air. It is driven on the ground and it leaves tracks, be it a rubber tire track or a chain track depending on the type of bulldozer. You and I are going there to check the validity of the Israeli complaint. If we find tracks, we as JVA are to blame, but if there are no tracks, you know who to blame for such allegations."

They drove to the diversion site, walked down to the river. There were, of course, no tracks of any machine to be seen. The Israeli allegations were obviously not founded, and the American official said he would file a report to Washington. Later, it turned out that the official was acting on basis of a rumor his Ambassador had heard in Jordanian circles. The visit was over, and the President thought he made his point clear and that he had fooled the American official like he had the Israeli patrols, but not quite, as will be explained below.

10. A Scary Presentation

The joy expressed by farmers was a source of satisfaction to the JVA officials. This whole feeling of triumph was clouded by rumors spreading in

Amman, after they were spread in the Jordan Valley, that the success in bringing more water to the KAC was credited to an alleged connection the JVA President had with the Ambassador of the United States in Israel. That rumor was spreading quickly, so much so that colleagues of the President's wife at the school where she taught heard it and believed it. It became the talk of Amman and was circulated at dinner parties. People further reckoned that the American Ambassador to Israel (then Samuel Lewis) was of the Jewish faith and concluded, therefore, that the JVA President's wife must be of the Jewish faith too. There was little logic and ability by the President to deny these rumors and to assert that his wife came from South Dakota, a descendant of a German-Russian Protestant family, with no connection or acquaintance with the American Ambassador to Israel! Actually, his wife had never seen a person of the Jewish faith until she was 20 years of age and attending Augustana College in Sioux Falls, S.D., where she heard that the shop keeper some two blocks away from the campus was of the Jewish faith. She made a special trip to that shop to see for herself if that person was different than the rest of the people around her! When the JVA President jokingly asked his wife if that person was different, she seriously answered that he was not.

"Okay, sweet heart," continued the President, "brace yourself for surprises in this town. It is widely rumored that you are an American Jewess."

"You must be kidding."

"I could be, but they are not. It is the habit of some competitors here to hit below the belt. I think my achievements in the Jordan Valley development are driving some of them crazy. These achievements are just too much for one man to make in such a few years. Pay no attention if you hear these rumors."

"This is mean; who spreads all these rumors?"

"I wish I knew; just ignore them. They are insinuating that I could have connection with the enemy, Israel, through one whom they claim is my brother in law!"

A week or so later, a senior official of USAID came to see the President in the afternoon.

The President asked him: "Did you know of the Israeli allegations against us about the Yarmouk? Have you been told about the complaint that they filed with your embassy about us moving a bulldozer into the Yarmouk? Have you been told that?"

"Listen, Munther, I will tell you something that I would later swear I did not should it leak out," said the old friend.

"Please do, and, mind you, you know I do not keep secrets when it comes to our national interests."

"What we had was not an Israeli complaint," said the USAID official, "but a very high-ranking official in your government, and don't ask me who, told our Ambassador about you moving a bulldozer into the Yarmouk."

The President was stunned. He could not collect his thoughts to carry on an intelligent conversation with his visitor, and found difficulty in responding to the important issues he was raising about the Jordan Valley development.

"I am sure glad you came to visit," said the President, "and we shall have ample time to compare notes and follow up on what we all have to do. We have to maintain our jobs and make a living."

As soon as he saw the official out, the President returned to his desk and called for an appointment with the Prime Minister. He drove up to the Prime Ministry and waited until the Prime Minister had time to see him.

"Excellency," said the President, "I have just been visited by an official of USAID, and he told me that the Israeli complaint I told you about was not actually an Israeli complaint after all. He told me that a very high-ranking official in our government told the American Ambassador that I had moved a bulldozer into the Yarmouk. If this is true, it is very serious."

"How did that Jordanian official know?"

"I did not ask, and the USAID official did not volunteer the information like the high-ranking official in our government has done. Excellency, this is very serious. I will never ever venture to conduct such operations in the future, because next time, with such leaks, we may encounter mines in the water. Excellency, the foreman who played the major role, Mohammad Zuhair Mashriqi, has a wife and fourteen kids who depend on his support. What would I tell them if a mine exploded in his face and he kissed us good bye?"

"Are you sure of the authenticity of what that official was telling you?" Asked the Prime Minster.

"Sir, I am telling you what he told me, and that was scary, very scary."

"Who do you suspect made that leak to the Ambassador?" Asked the Prime Minister.

"This, I think, is your job to find. But please, sir, remember that no one knew of the nature of the operation except you and me. You even did not know of the operation or of its details or timing. On my part, I have told no one about it."

"What if the Americans want to have us suspect each other and initiate a string of distrust between us?" Asked the Prime Minister.

"If that was their objective," I responded, "I am telling you, Excellency, that they have succeeded. I have no trust to tell any one any such or similar plans in the future. Believe me, sir, my own wife and kids know nothing about what I have done, let alone my brothers and extended family."

"I suggest that you ignore the matter and just proceed as you have always done, dedicated to your job, and doing what it takes to make a success out of it."

"I assure you, sir, I would not repeat what I have done even if the Jordan Valley went dry. Human life is to me more important that the bragging of officials or their acquaintances in dinner parties. I will remain as dedicated to the job as you have always known me, but please help me find the leaker."

The President left the office of the Prime Minister, not knowing what to think. He was sure his back was exposed, and that his own position was so vulnerable. He did not mind so much the rumors that claimed a new inlaw that he never had met, but he did care for the safety of those men who carried on their dangerous mission to bring more water to the Jordan Valley in service of the country. He was depressed.

More reasons for concern came a few days later. Fortunately for him, a rainstorm had come in late February that caused a high flood in the Yarmouk. A USAID staff member happened to be passing by the Yarmouk and witnessed the flood and informed the JVA President how high it was. The Director of USAID, Mr. Walter Bollinger, who had come back from vacation, came to visit the President. He had in his possession what looked like large maps. After they exchanged greetings, Bollinger asked to move to the meeting table where he could spread out the maps he had brought. When he did, it became clear that they were satellite photos. The first was a large photo of the Rutenberg pool and Lake Tiberias. The surface of the lake was so blue and still that one could recognize any disturbance in the surface of the Lake.

Bollinger pointed to a white spot in the photo appearing in the water surface at the southern end of the Lake, and asked what it was. The President, being familiar with what the Israelis had been doing on the Yarmouk, answered that the white spot was a bubble formed by the water gushing out of a pipe under the the surface of the Lake, and that the pipe was a conduit in which Israel pumped water from the Yarmouk to the Lake. He could trace the pipe back to the edge of the Rutenberg pool on the Yarmouk.

"Good," said Mr.Bollinger, "now let us look at the other photo," and he spread another photo on the meeting table. In the second photo one could clearly see the diversion point on the Yarmouk, before the Jordanian operation was conducted. The photo was so clear that the stones in the riverbed could be counted. Both men identified the photo very easily. Then Bollinger spread out a third photo, taken two weeks later, that showed the new channel the Jordanians had dug-out across the sand bar! The channel was clearly identifiable. The President had cold feet.

"What is this?" asked Bollinger pointing at the new channel.

An idea sprung up to the President's mind, and he said, "You must have heard about the high flood that occurred in the Yarmouk a while back, as a result of the rain storm. I even first knew about it from your assistant, Dick

Johnson. Ask him how violent the flood was... See, it even toppled these trees," and he pointed to some trees on the Jordanian bank at the diversion point that were truly toppled by the flood, "this channel must be the works of the flood."

"The work of the flood?" Wondered Bollinger with a smile, "I just want you to know that we know." And the President's feet became even colder.

"Where would I go?" He wondered to himself. "On one side I have Jordanian high ranking officials who donated sensitive information to the American Ambassador and, on the other, I have satellites that provide photos more clear than the pictures I took at 5 meters distance away from the diversion point? Additionally, the Syrians have been squeezing us by increasing their abstraction upstream, and the Israelis have been squeezing us downstream by denying us the opportunity to improve our water intake." He felt squeezed from every direction with no where to go for a safe path, and remembered the words of the famous Arab poet, Abu Firas Al Hamdani:

"Walakin itha humma alqada'u ala mr'ien
falaysa lahu barrun walaysa lahu bahru."
(When fate converges on someone, he would have
no land [to ride away], nor a sea [to sail away])

11. American Interest and Follow-Up

On Saturday, 5 May 1984, the Ambassador of the United States, Mr. Richard Viets, accompanied by Mr. Walter Bollinger, the USAID Director, and his assistant, Mr. Jerry Brown, came to see the President of JVA. The conversation focused on the Yarmouk and the problems the JVA was facing because of Israeli behavior. The Israelis were demanding, in the MAC meetings at the picnic table, a summer flow of 2.3 cms[287] at a time when the entire flow did not exceed 3.7 cms. They were obviously being impossible, or they were after other objectives. Additionally, the Israelis proposed to have a diversion weir built across the Yarmouk *upstream* from the diversion point, and not downstream, as all the Yarmouk plans proposed. They suggested that from that upstream diversion, a concrete canal would take-off to divert flood water to their side. The carrying capacity of the canal was proposed at 25 cms.

The Ambassador proposed to the President to meet with the Israelis himself, but the President politely declined. He assured the Ambassador that the JVA delegate to the MAC meetings was capable of handling all the

---

[287] The 2.3 cms would have been the Israeli share (30%) of a flow of 7.7 cms, which was the flow in May 1962 when the KAC was first operated; i.e.,the Israelis did not want their share influenced by Syrian abstraction or by drought.

technical matters, and there was nothing the President could do that the delegate could not.

Their conversation went on, and the Ambassador made a sudden suggestion:

"If you do not like to go to the MAC meeting, would you entertain the idea of meeting with the Israelis in a third country where you would stay on one hotel floor and they would stay at another floor. We will act as mediators as you meet to resolve the standing issues!"

"Excellency," said the President, "the issues that prohibit us from dealing directly with the Israelis are well known to you and to the rest of the world. I am not prepared to forget about those issues and meet with the Israelis for the sake of a few cubic meters of water. Any such meeting will have to come at a time when the Arabs and Israel negotiate a peaceful settlement. I am sorry, I will not entertain such an idea."

"What, then, do you plan to do in connection with the sand bar?" Asked the Ambassador.

"I consider the sand bar a piece of property lying in Jordanian and other Arab territories. If the Israeli consent is not forthcoming by next Thanksgiving Day, I will mount a D9 bulldozer myself and will clean the sand bar come hell or high water."

"Give us some time," said the Ambassador, "I am leaving for Washington D.C. on the 8th, and I hope we will somehow find a solution."

When the Ambassador left, the JVA President immediately called the Prime Minister and briefed him on what the Ambassador proposed, and told him of his rejection of the American proposal.

"Please, Excellency," said the President; "this is something I am sure you will not entertain either."

"Don't worry, I support your position fully. This is our policy," assured the Prime Minister.

Soon after that meeting, the President traveled to the United States at the invitation of USAID, to conduct talks with AID and other agencies over the Jordan Valley and the Yarmouk issues. The session at State was attended by officials from different agencies, and the talks centered around the development of the Jordan Valley. The President pointed to the role that the United States had been playing in the Jordan River basin, and the contributions she had made towards its development. He was asked what he thought of the Israeli proposal to have a diversion weir built across the Yarmouk upstream from the diversion point and with a diversion canal to their side with a capacity of 25 cms.

"The location they propose is all new. Technically, it will not have the same diversion efficiency as the weir proposed in the Master Plan of the Jordan Valley, designed by Baker and Harza in 1955. Politically, their proposal calls for the construction of a canal in territories occupied in the war of 1967, it would have to traverse the southern DMZ between Syria and Israel. Additionally, its carrying capacity, as proposed by the Israelis, is ten times the maximum rate of diversion that their share in the Yarmouk would justify. Supportive as we are of having a diversion weir built, I am afraid the Israeli attitude would lead us toward more political and technical complications."

"Are you sure," said one of the participants, "that there is no error in a decimal point in the Israeli figure of 25 cms?" And the participants laughed.

The topics also covered the Syrian diversions upstream. The Americans sounded like they were more knowledgeable of the Syrian plans than the Jordanians were. They have the advantage of satellite imagery, a facility not available to the Jordanians. They informed the President of more dams being built on the Yarmouk tributaries inside Syria with negative impacts on the feasibility of the Maqarin dam that they would finance. The talks covered the cropping pattern in the Yarmouk Triangle and the Israeli pumping from the Yarmouk in winter for direct use and for storage in Lake Tiberias (Figure VII.2).

"I became aware of the Israeli intentions in 1977, but was able to cite physical evidence of their pumping only last week," remarked the JVA President during the meeting, "I saw high tension lines installed near the outlet of the Yarmouk from the Rutenberg pool, and I noted a pump house, and the beginning of a lined canal." He further asserted that Jordan contested these Israeli actions and considered them an infringement on Jordan's water rights.

The participants asked about the off-canal storage in Wadi Mallaha that was being studied by the consultants to the JVA, and the JVA plans regarding that.

"Please remember that the residual flow of the Yarmouk, after deducting Syria's share and Israel's share, is the share of the Hashemite Kingdom of Jordan. We intend to make use of our share as soon as we possibly can. The Mallaha dam [later renamed Karama dam] provides us with a valuable opportunity, and I hope you will be as supportive of it as you have always been of the Unified Plan."

The meeting ended after three hours of substantive debate, throughout which the President informed the American officials about where JVA stood and the fact that the Jordanians, unlike the Israelis or the Syrians, had never committed one violation of the Unified Plan that the United States considered the most workable plan for the sharing of the waters of the Jordan River basin. He then paid a courtesy visit to Mr. Robert Pelletreau, Deputy Assistant Secretary of State, and raised with him the issue of the sand bar.

"I hope there will be a political change in Israel after the elections," said Mr. Pelletreau. It did not sound like the American interventions would be strong enough to effect a change in the Israeli attitude. However, the JVA officials were not as squeezed after the February operation as they had been before it. That did not mean they could ignore the need to remove the sand bar, however, because increased diversion efficiency was a standing priority for the JVA, in order to:

a) Compensate for the increased abstractions by Syria upstream;
b) Make more water available to the new canal extension, and off canal storage; and,
c) Minimize the flow that would by-pass the diversion point and be subject to Israeli pumping and uses.

The political developments in the region took a new turn. An Arab Summit in Fez, Morocco, in 1982, approved an initiative by Prince Fahd Ben Abdul Aziz of Saudi Arabia to talk peace with Israel in accordance with U.N. Resolutions 242 and 338. That decision by the Summit overturned a previous resolution by the Arab Summit at Khartoum, Sudan, in the wake of the defeat of 1967. In Khartoum, the Arab leaders opposed peace talks with Israel or even to recognize her right to exist. After the Fez Summit, the Palestinian National Council met in Amman in November and narrowed the gaps between the factions of the PLO, who had different views about the Reagan initiative and the settlement of the Middle East problem, in accordance with Resolutions 242 and 338. On the other hand, King Hussein and his government worked diligently to have Jordan and the PLO unify their ranks and participate together, as a joint party, in a process for the settlement of the dispute with Israel, but those efforts fell through. Shortly after that, Secretary of State George Schultz intensified his efforts to get a Middle East Peace Process started, and he made several shuttle trips between the concerned Arab States and Israel. While in Jordan, his staff listened to a briefing about the sand bar but, like other interventions from United States' officials, the efforts led to nowhere.

The issue of the sand bar and the frequent presentations to visiting United States officials turned the relations sour between the President of JVA and his boss, the Prime Minister. The management of the Yarmouk issues triggered a collision between them. The Prime Minister did not want anything to do with the MAC meetings conducted under the supervision of the United Nations. He did not object to them but never cared to respond to the official correspondence concerning the subject. There was no way JVA could adjust the diversions from the Yarmouk in the presence of the sand bar without going through the Armistice mechanism, the MAC. The President continued to instruct the JVA delegate on what to say and defined the goals for him before the delegate went to the meetings in the company of the Jordan Army officer and the officer of UNTSO. That way the JVA kept water coming during the dry months despite the sand bar, but had practically no control on water diversion in winter when the stream level was high.

12. Removal of the Sand Bar

The problem of the sand bar became popular among visiting American officials, including the Secretary of State, George Schultz, who was shuttling in the Middle East in an attempt to get a Peace Process started. The new Prime Minister, Zaid al Rifai, thought that the problem should be solved once and for all. He conferred with the President of the JVA in late August, and concluded that the President himself should go to the MAC meetings and have the issue resolved.

Arrangements were made through the good offices of the United States and UNTSO. As a counterpart to the JVA President, the Israelis would send the Director of the Office of their Prime Minister to the meeting. The meeting took place on 10 September 1985, at the Israeli terminal of the King Hussein Bridge (Allenby) crossing point. The Jordanian party was led by the JVA President, who was accompanied by the JVA delegate to the MAC meetings and by a colonel from the Jordan Armed Forces. The Israeli side included, in addition to their Army colonel: Mr. Amnon Neubach, the Director of the Prime Minister's Office; Mr. Meir Ben Meir, the Water Commissioner; Dr. Yacov Vardi, the technical expert; Mr. Noah Kinarti, the representative of the kibutzes in the Yarmouk Triangle; and a lawyer. The UNTSO officer presided and opened the meeting with the usual formalities. The discussions started between the Jordanian delegate and the Israeli Water Commissioner, along the lines they had been discussing in earlier meetings. Finally, after a protracted debate, the JVA President made a proposal whereby the sand bar would be removed and the riverbed at the diversion location would be leveled to an elevation of 206.85 meters below see level. Downstream of that, the riverbed also would be cleaned and dredged to channel the flow going to Israel in a narrower cross section in order to minimize water losses due to evaporation and seepage. Jordan would remove the sand bar using a D9 bulldozer, and Israel would clean the downstream using a D7 bulldozer.

The Israeli team asked for a break to study the proposal; it took them over half an hour before they came back and announced that they accepted the proposal. A time schedule for cleaning was agreed to, and the entire understanding was a gentlemen's agreement. No maps or statements were signed.

The sand bar, a menace to Jordan for over six years, was finally removed under the auspices of UNTSO (Figure VII.3). The riverbed was cleaned and shaved to elevation -206.85 m and sand bags continued to be used to control the water level and adjust the sharing ratios. The function of the sand bags would, in a normal engineering solution, be done by building a diversion structure across the Yarmouk, something that Jordan (and the

Arabs) had not been able to do since the KAC project was conceived and implemented. The diversion structure would become a primary topic for negotiations, when the bilateral negotiations with Israel started after the Madrid Conference, as shall be revealed later.

The MAC meetings, including the meetings at the picnic table, addressed only transient operational problems that dealt with water diversion. No long-term topics were ever entertained. The first direct dialogue, so to speak, between an Israeli delegate and a Jordanian delegate, took place without any prior planning, in a Washington, D.C. seminar. The Center for Strategic and International Studies, then an affiliate of Georgetown University, held a seminar on Middle East waters in November 1986, and invited the concerned countries to it. A half-day activity was assigned to the Jordan River basin. Mr. Selig Taubenblatt, the former AID official and aide to Ambassador Philip Habib, made a presentation of the Johnston Plan. The Israeli delegate, Water Commissioner Mr. Meir Ben Meir, responded and stated that the Johnston Plan was too old and not practically applicable any more. He further criticized the unilateral actions by Syria on the Yarmouk basin. Next was the intervention by the Jordanian delegate, JVA President Munther Haddadin. He indicated Jordan's willingness to have the American-sponsored Plan implemented, and contested Israel's description of the Unified Plan as too old.

"If thirty years is too old," remarked the JVA President, "what do we consider 3000 years of time that has elapsed since a Jewish state was established in Palestine, on basis of which the Jews are claiming the right to it? Is archaic a proper description of that claim?" He also alluded to the unilateral actions by Israel on the Jordan River, and pointed out that Jordan was the injured party as a result of all unilateral actions in the river basin.

13. The Maqarin Project Revived

Relations with Syria took a turn to the better with the formation of the new cabinet in Jordan in April 1985, headed by Mr. Zaid al Rifari. The King sent a public message to President Asad of Syria, in which he expressed surprise at the unfriendly moves that some official Jordanian circles had made against Syria in previous years, and urged that a new page be opened in the bilateral relations between the two brotherly countries.[288] This message was a prelude to improved Jordanian-Syrian ties, which had been strained since the outbreak of the Iran-Iraq war in 1981. That war was still raging, at the time. Jordan and Syria maintained their respective support to the warring parties. The Civil War in Lebanon was still claiming lives, except that Syria alone was involved in that complicated situation, after the peace-keeping forces of

[288] The message was made public and broadcast over Radio Jordan in October 1985.

the other Arab states evacuated. The Lebanese President-Elect, Bashir Jumayyel, was assassinated, before he took office in September 1982, and his brother, Amin Jumayyel, was elected President. The Israeli invasion complicated matters further, and Menachem Begin, the Israeli Prime Minister came under sharp criticism for allowing the Israeli forces to advance into the Lebanese "swamp." He resigned and was succeeded by

Figure (VII.3): Cleaning of the Sand Bar (Picture)

Yitzhak Shamir as leader of the Likud. Elections in Israel produced a Keneset that was fit for the formation of a coalition government between Likud and Labor. Yitzhak Shamir and Shimon Peres alternated as Prime Minister.

Syria was hoping to get Jordan on her side in connection with the Iraq-Iran war, and in connection with the Lebanese situation. The efforts of the United States to convene a conference for Middle East peace continued. Syria rightly thought that a coordinated stand with Jordan would benefit both. Syria had broken ties with the PLO Chairman, Yassir Arafat, after the Israeli invasion of Lebanon.

In Lebanon, the civil war was escalating and there were Syrian troops attempting to pacify the fighting factions. The Lebanese Prime Minister, the veteran politician Rashid Karami, was assassinated in a helicopter crash when a bomb brought down his military helicopter.

Under the improved ties with Syria, JVA's contacts with their Syrian counterparts over the Yarmouk were renewed, but no sign of a breakthrough was in sight. On one of the visits to Damascus to induce the Syrians to approve the Maqarin dam project, the dialogue was resumed, though with little hope for a happy ending.

"How would you treat the Syrian lands that would be inundated by the reservoir? And how would we delineate the borders on the water surface of the reservoir?" Asked the head of the Syrian side, Mr. Barakat Hadeed, Deputy Minister of Irrigation.

"We will compensate the owners of the Syrian lands just as we compensate the owners of the Jordanian lands that would be inundated, if not a bit better. The borders would be the same, and there would not be a problem. The side slopes of the reservoir are so steep that people can not climb up from the water to either side without being noticed." Answered the head of the Jordanian delegation, Munther Haddadin.

"What do you think of this idea: you give us land to the east of the reservoir in return for the Syrian land that would be inundated?" Asked Mr. Hadeed.

"Please tell me what the difference is between Jordanian and Syrian sovereignty over any peace of territory? Are not we each an integral part of one Arab nation?" answered the head of the Jordanian delegation, with an explicit reference to the slogan of the Ba'th Party ruling in Syria.

The Syrian request was dropped, and the debate focused on the height of the dam. The Syrian side insisted that the dam height should not exceed 70 meters, compared to the height, recommended by the Joint Jordanian-Syrian Committee and endorsed by the Arab Technical Committee, in 1955 of 147 meters.

"We do not want to be blamed for approving a high dam that can not be filled," said Mr. Hadeed.

"If the dam can not be filled, it is because of the many earth dams you have been building in the catchment. Those dams hold water that is allocated for the Maqarin reservoir under our bilateral Agreement of 1953. Anyway, we expect that their storage capacities would diminish with time, and flood waters would spill over to be stored at Maqarin where they belong," commented the head of the Jordanian delegation.

"We gather that Yarmouk water bypasses your diversion point and that Israel pumps it to irrigate occupied lands. Why don't you stop that water from flowing towards Israel?" Asked Mr. Hadeed.

"Because," responded the JVA President, "the land has the right to irrigation regardless of who is holding it. It had a right to irrigation before Israel was even established. You and I consider that the subject land is Arab and the occupation in 1948, as far as we are concerned, did not change its status. Neither of our countries recognizes Israel. To stop the water from flowing to that land is a serious matter. We are not stopping the water from flowing to Israel because of that and also for the same reason you have not stopped the Banyas or the Hasbani on your borders with Lebanon from flowing to Israel." The Jordanian responses were not very comforting to the Syrians, and neither was the Syrian attitude comforting to the Jordanians. Soon after that, the JVA President was told that the Syrian Prime Minister, Abdul Ra'uf al Kasm, wanted to see him, and that the Minister of Irrigation, Mr. Abdul Rahman al Madani, would accompany him there. The President of JVA asked that Jordan's Ambassador to Syria, Mr. Nayef al Hadeed, should attend the meeting, and so it was.

"I am told," said the Syrian Prime Minister, "that you reject the dam height proposed by your Syrian colleagues, and that you insist on the dam of 150 meters height. Would you please tell me why?"

"Because, Excellency, it was the Syrians who insisted back in 1955 that the Maqarin dam be as high as 147 meters, despite proposals from the United States delegation to limit the dam height to 126 meters. The Syrian motive for a higher dam was to minimize the spills that obviously become available to Israel. Syria, then, was determined to deny Israel the chance to get more water from the Yarmouk. Now, if the Syrians today have changed their mind, please tell us that clearly so that we give the proper answer."

The Prime Minister defended the Syrian activity in the catchment and justified it on the grounds that those dams that the Jordanians were protesting were meant to hold water that would otherwise go to Israel. The President assured him that the Maqarin dam would do a much better job towards that objective, and he asked that Syrian approval of the Maqarin dam be expedited. The conversation with the Prime Minister ended with a pledge that he would raise the matter with President Asad and try to expedite the approval that had been withheld for so long.

The Joint High Committee of the two countries, chaired by the two Prime Ministers, was supposed to hold its meeting on 12 June 1987. JVA President consulted with his boss, the Jordanian Prime Minister, on 10 June, and they agreed to hold a special meeting of the Joint High Committee in the Jordan Valley to discuss the Maqarin dam. The President indicated that he had been playing the role of a technical professional and a diplomatic representative of Jordan throughout his talks with Syria, since 1975. He asked that, in the presence of the Prime Minister, he be relieved of the diplomatic role, because he wanted to speak out his views on the stalling of the Maqarin dam, and the Prime Minister concurred.

The members of the Joint High Committee flew by helicopters on Friday, 12 June 1987, to the Maqarin site. Before the helicopters took off, the Jordanian Prime Minister made a suggestion to his Syrian counterpart.

"I think that the name of the dam has been behind its misfortunes and delays of implementation. Maqarin is an odd word, what does it mean anyway?"

The JVA President explained the meaning of the Arabic word and how it denoted the point where several branches (tributaries) meet. The Jordanian Prime Minister was not as much interested in the linguistics as he was in some means to jump-start the project.

"Why don't we give it a different, nicer name?" He asked.

"Like what?" asked the Syrian prime Minister, Dr. Abdul Ra'ouf al Kasm.

"Like the Wehda (unity) dam," said Rifai

"Sounds okay to me," said al Kasm.

"What do you think, Dr. Munther?" Asked the Jordanian prime Minister, Zeid Rifai.

The JVA President viewed the name, Maqarin dam, as having a heritage value, and did not like the idea of changing it to whatever name. He thought the original name carried with it a sense of history of the entire Yarmouk development.

"Sir, [unity] with Syria has a lifetime of three years!" He said implying that the agreement would last for three years only. The answer was rooted, as he meant it, in the unity between Egypt and Syria in February 1958, which lasted only until September 1961. Neither Prime Minister liked the answer, but it set the tone for JVA's presentations later that day. Three helicopters carried the delegates to the Maqarin dam site. As they overlooked it, they could see some terracing work on the Syrian escarpment and pumps installed below elevation 250 m to irrigate vegetables by the river below.

"You know, Excellency," said the JVA President, as they all viewed the site below, "if I asked my government for a budget to do such projects like we see on the Syrian bank of the Yarmouk, I would most likely be fired because it is such a waste of money!" His comments throughout the field visit were critical and cynical. When they got to the Guest House at Deir Alla in the Jordan Valley, it was past lunch time.

"Sir," He whispered in Rifai's ear, "please let me order lunch served and have our guests eat before I make my formal presentation, not after. For I am afraid they would not have an appetite to eat after they hear what I have to say."

After lunch, over which Rifai told funny stories about his experience in learning how to fly small airplanes, the party adjourned to the presentation hall. There, the JVA President started his presentation using phrases of Arab solidarity normally used by the information campaigns of the Ba'th Party that had been ruling in Syria since 1963. He told the audience of the extortion that Israel practiced against Jordan over the Yarmouk diversion, and the trespassing that she had been making against Jordan's rightful share of water. Similarly, he presented what he knew about the Syrian practices on the Yarmouk water sources inside her own territories and how, by virtue of the 1953 Agreement between Jordan and Syria, those practices infringed on Jordan's rightful water shares also.

"So, Excellency, you see how Jordan is simultaneously squeezed by Syria upstream and Israel downstream. Our hopes have been that Syria would enhance Jordan's steadfastness against the aggression of the enemy, and not join in making our life more miserable and our confrontation with the enemy more vulnerable!" He said.

There was tension and a visible uneasiness displayed by the Syrian Prime Minister. Rifai's diplomatic skills came on line immediately and he steered the discussions to a satisfactory conclusion. Al Kasm promised that Syria would look seriously into the project and give an answer very soon. As the party walked out of the hall, the Syrian Minister of Agriculture, Mr. Mohammad Ghabbash, whispered flattering comments in the ear of the JVA President.

At the airport the following day, Dr. Al Kasm made a statement to the press upon departure. "We will look with serious favor to the construction of the Maqarin dam, now renamed Wehda dam, on the Yarmouk River on the borders between the two countries. This vital and national project is of interest to us. We hope to see it built as soon as possible." Mr. Rifai and all Jordanians were very happy with the Syrian response.

The King made quick visits to Kuwait and Syria in early July. The Prime Minister accompanied him. Upon their return from Syria, the Prime Minister called the JVA President, Munther Haddadin, to his office on the first Thursday in July 1987.

"Congratulations," said the Prime Minister, "President Asad thinks very highly of you." The Prime Minister then quoted comments President Asad had told His Majesty: "Where did you get this Dr. Haddadin who had turned the hat of Prime Minister al Kasm up side down?" Asad told His

Majesty, "the Prime Minister and I now approve the construction of the Maqarin dam, renamed Wehda dam," Asad added.

"So," said Mr. Rifai to JVA President, "we did it! But I had to shuttle between the Ministry of Irrigation and the Presidential Palace. It was quite a job. Congratulations!"

"You could have saved yourself the trouble by ordering me to join," commented the JVA President, "I am actually good at doing such chores."

"Well, the Syrians placed three conditions, and His Majesty accepted them, and I want you to execute," said the Prime Minister.

"What are they?" Asked the JVA President.

"First, the dam should not be more than 70 meters high; the second is that they should build the dam with our finance; and the third is to replace the Agreement of 1953 with a new one so that we recognize their right to the dams they have built in the catchment."

"The first condition," responded the JVA President, "is negotiable. As far as the second condition goes, we have partners. We are seeking international financing for the dam and we have to go by the rules agreed to under the financing agreements, but I pledge to treat the Syrian contractors as though they were Jordanian contractors."

"The third condition, Excellency," he continued, "is extremely troublesome, and it has to pass over my dead body!"

"But His Majesty agreed," said the Prime Minister.

"When His Majesty is briefed about the risks and dangers involved in such a development, I am sure he would not agree."

"Can you tell me what these are?" asked the Prime Minister, as he rolled back his chair further away from his desk, and stared at his sub-ordinate in concentration.

"This would be a very valuable present to the Americans, and a golden opportunity for the Israelis to legally upset the formula by which we have been working. I mean the Unified Plan and the previous commitments the United States had made in connection with it. The current Agreement with Syria is compatible with the Unified Plan, not that their deeds are conforming to it or to their agreement with us. The replacement of our agreement with the Syrians in the manner they want will throw away the provisions of the Unified Plan, and a new regime of water sharing would have to be worked out. Obviously, under any new update of the Agreement, Syria would be allocated more water than allotted to it under the Unified Plan and our 1953 Agreement. It would be Israel's golden opportunity to ask for equal treatment and have us agree to the pumping of the Yarmouk winter water that she has been doing over and above her share under the Plan. This happens at a time when Israel occupies the West Bank, and the Arabs are in no shape to stop Israel at a reasonable level of a water share.

"I am afraid, Excellency, we will be breaking with our own hands the very jug that keeps our water. I also know that Israel had tried to speak on behalf of the West Bank for its water rights, something that we did not accept. Any disturbance of the current arrangements at this time would be a valuable service to Israel, and would surely exempt the United States from any previous commitments she had made regarding the Unified Plan, not the least of which was the Jordan Valley development program since the late 1950's.

"Besides, Excellency," continued the JVA President, "the Syrians have consistently violated the provisions of the 1953 Agreement. What guarantees do we have that they would not violate any new agreement?"

The Prime Minister agreed to the difficult status of the water rights, especially when the status of the occupied West Bank was considered.

"Water issues will be a primary obstacle in the way of any Israeli withdrawal from the West Bank," said the Prime Minister.

"Well," the Prime Minister concluded, "in this case, you should go to Syria and salvage whatever you can."

Arrangements were made to hold an urgent meeting of the Yarmouk Joint Committee in Damascus on Saturday, 4 July 1987. The Jordanians spent 30 hours there, 18 of which were spent in negotiations. They were able to set the dam height at 95 meters to be raised to the ultimate height of 150 meters in the future. The Syrians were informed that they would be welcome to compete along with Jordanian construction firms and joint venture with foreign firms in the bidding for the dam construction. Such competition would be done in accordance with the conditions of finance, most likely those of USAID. They were also advised that Jordan could not entertain any motions to disturb the existing Agreement of 1953.

The JVA President briefed the Prime Minister on Tuesday, 7 July, and his resignation was accepted on the 8 July 1987. He had submitted five resignations in the previous sixteen months for personal reasons, and the last understanding with the Prime Minister was that the resignation would be accepted after the visit of the Syrian Prime Minister and the finalization of the Maqarin issues with them.

Despite the strong opposition by the JVA President to any change in the Yarmouk Agreement with Syria,[289] a new Agreement was hurriedly drafted after his exit, processed through government and through the Lower House of Parliament, followed by the Upper House of Parliament in a record

---

[289] The President made his opposition to the Prime Minister in July 1987. He qualified his position by the need to keep the Yarmouk-sharing stable and conforming to the Unified Plan, as the existing Agreement did. Any change would give Israel a golden opportunity to impose changes on her own share, as well. Indeed the Israelis raised the subject of this Agreement when the bilateral peace talks got underway as shall be noted later

time of two months. A Royal Decree issued it into law in 1987. The Syrians got what they wanted in that new Agreement, and she approved the dam project after the 1953 Agreement was replaced by this new 1987 Agreement. What remained was the consent of Israel to have the associated diversion dam built.

The efforts of the United States were therefore invited to pave the way to the construction of the Wehda Dam. Mr. Richard Armitage was assigned the task of an intermediary to obtain the Israeli consent. His efforts were met with the same old positions of Israel, and the United States did not use any pressure to persuade them, nor did her team even refer to the Unified Plan. The Jordanians were left to respond to the demands of the Israelis as carried by the intermediary. Armitage's efforts were still going on when the Peace Process was launched in Madrid. When the bilateral negotiations started in Washington, D.C., Jordan's Prime Minister asked Mr. Armitage to suspend his efforts and give a chance to the direct negotiations in Washington. The Prime Minister thanked Armitage and asked that the door be left open to return to the diplomatic efforts of the United States, should the peace negotiations collapse in the future.

CHAPTER VIII

The Middle East Peace Process

1. Back to Shuttle Diplomacy

The end of the 1980's witnessed a promising turn of events in the Middle East. The Iraq-Iran war ended; an Arab Cooperation Council was established encompassing Egypt, Iraq, Jordan and Yemen, which had just been re-united; the civil strife in Lebanon was coming to a close; a national conference was convened, in Tai'f, Saudi Arabia, for the Lebanese ethnic and political factions, and a charter was passed stipulating some constitutional changes, and national reconciliation started. In North Africa, another Arab grouping was established, the Maghareb Union, encompassing Libya, Tunisia, Algeria, Morocco and Mauritania. More drastic developments were taking place in Eastern Europe. The Berlin Wall came down in 1989. The communist alliance of the Warsaw Pact collapsed and the communist regimes were brought down one after the other in Poland, Romania, Czekoslovakia, Hungary and East Germany. The two parts of Germany soon re-united, and the Soviet Union started to crack.

The promising trends in the Middle East were soon shattered by the surprise Iraqi invasion of Kuwait, on 2 August 1990. Afterwards things were never the same in Arab ranks, and the Arab States split up into basically two camps: one that deplored the Iraqi invasion and supporting foreign intervention to free Kuwait, and another that opposed the invasion but stood against the intervention of foreign powers. They called for the employment of the "Arab Order" through the Arab League, to have Iraq pull out of Kuwait. The new born Arab Cooperation Council all but collapsed when Egypt aligned herself with the first camp, while Jordan and Yemen led the second camp. A coalition led by the United States, was quickly assembled, and Iraq was driven out of Kuwait by force of arms. Thirty-three countries participated in the coalition. Allied forces bombarded Iraq with bombs and long-range missiles, before they invaded her with ground troops in operation "Desert Storm" and evicted the Iraqi forces from Kuwait. As soon as the Iraqis signed and accepted the terms of a cease fire, President Bush announced[290] that, in light of the drastic changes that had taken place in the world, he would actively pursue the establishment of a New World Order, and work to bring peace to the Middle East, which had suffered decades from the agonies of war.

The Secretary of State in the Bush Administration, Mr. James Baker, took to the road in a new attempt to work out a formula by which the parties

[290] In his State of the Union Address before Congress, 1991.

to the Arab-Israeli conflict would enter into negotiations to settle their disputes. He picked up where his predecessor, Mr. George Schultz, had left off in 1988. Mr. Schultz's efforts were complicated by the *Intifada* that had erupted on 8 December 1987, in the occupied Palestinian territories, and were made more difficult by King Hussein's decision to disengage with the West Bank. By that decision, all administrative and legal ties with the West Bank, heretofore part of the Hashemite Kingdom, were severed. Only the affairs of Islamic shrines were retained, to avoid creating an administrative vacuum in which the Israeli Ministry of Religious Affairs could take them over. The disengagement decision, taken on Saturday, 31 July 1988, ruled out any Jordanian representation of Palestinian issues, and opened the way for an independent Palestinian representation. The only competent party to represent the Palestinians was the PLO, which was not so recognized by the United States and her allies, including Israel. It was actually referred to as a terrorist organization and contacts with it were banned.

The expert hands of Secretary Baker, working tirelessly with the leaders of the concerned Arab parties, and Israel, and in cooperation with the Soviet Union, put together the unique formula for a peace process in the Middle East. He held talks with "independent" Palestinian individuals in the occupied territories who spoke on behalf of their fellow Palestinians under occupation. Most notable of those representatives were Faysal al Husseini, Hanan Ashrawi, Sari Nuseibeh, and others. The Arab leaders had been calling for an international conference, with the participation of the five permanent members of the United Nations Security Council, to resolve the Arab-Israeli conflict and bring peace to the Middle East in accordance with U.N. resolutions 242 and 338. The Israelis, on the other side, demanded bilateral negotiations to settle the dispute and implement the above resolutions. In such negotiations, Israel would face each Arab party separately to negotiate peace with him. However, the interpretation by Arabs and Israelis of resolution 242 was not the same, especially with regards to withdrawal from the territories Israel occupied in the 1967 war.

The formula that Mr. Baker devised was that the process would start with an international conference under the chairmanship of the process sponsors, the United States and the Soviet Union, from which two tracks of negotiations would ensue—a multilateral conference with wide international participation and separate bilateral negotiations. The multilateral conference would reinforce the work of the bilateral negotiations, but would not replace them. The political dispute would be resolved in the bilaterals, and the multilateral conference would focus on the measures needed to reinforce the resolutions of conflicts pursued in the bilaterals. The agenda of the multilateral conference would look into topics for regional and international cooperation, such as Water, Environment, Refugees, Regional Economic Development, and Arms Control and Regional Security.

The dilemma that faced Baker's efforts was the representation of the Palestinians. They were not a State that could be invited to an international peace conference, and there was no legitimate representative except the PLO, which neither Israel nor the United States recognized. It is inconceivable, however, to convene an international conference to resolve the Middle East conflict without participation of the representatives of the Palestinians, the party at the core of the conflict. Finally, King Hussein provided the way out. The Palestinians would participate under the State umbrella of Jordan, and a joint Jordanian-Palestinian delegation would attend the international conference. The parties were invited and an international conference was convened in Madrid on 31 October 1991. It lasted for two days. The Palestinians made an impressive show on the media and appeared to the world for the first time as the people they really are: worthy of respect, support and independence. The occasion gave the Jordanians world exposure, and it became known that Jordan was not Palestine, as some Zionist circles were claiming.

2. Jordan Gets Ready

By the summer of 1991 it appeared very likely that the efforts of Mr. Baker to convene a peace conference would finally succeed. Jordan started preparations for it in July, and the King placed on the shoulders of his brother, Prince El Hasan, then Crown Prince, the responsibility of overseeing Jordan's preparedness. King Hussein called for a national meeting to which dignitaries of the country were invited. They represented elected officials and Jordanians from all walks of life. His Majesty delivered a speech in that meeting. He provided convincing arguments, quoting verses from the Qura'an, that supported the decision to join in the peace process and go to Madrid with the other Arab brothers. "We can not stay behind," the King said, "when the future of the region is being shaped. We have to join our brothers and go together to the peace conference." The King formed a new Cabinet headed by Mr. Taher al Masri, a Jordanian of Palestinian origin and a long time minister, who had held different portfolios, since 1973. Masri's government took the formal decision to go to Madrid.

After much convincing, the Minister of Water and Irrigation, Samir F. Kawar, contracted the services of the former President of the Jordan Valley Authority, Munther J. Haddadin, to prepare a position paper on water that the Jordanian negotiators with Israel would use as background. The Royal Court and the Government Ministries started to work diligently to prepare for the peace negotiations. By early October 1991, it was announced that the Peace Conference on the Middle East would convene by the end of that month in Madrid, and would be attended by the core parties, Lebanon, Syria, a joint Jordanian-Palestinian delegation, and Israel; and by parties representing regional groupings in the Middle East and North Africa. These were: Egypt for the Arab Cooperation Council; Saudi Arabia for the Gulf Cooperation Council; and Morocco for the Maghareb Union. Representatives of the European Union and of the United Nations would also attend. The sponsors of the Peace Process, the United States and the Soviet Union, would chair the international conference. The conference convened in Madrid as scheduled.

3. The Bilateral Negotiations Start

The Palestinians participated in the Process as part of a joint delegation under the Jordanian state umbrella. The Palestinian delegates, Israel insisted, were to be chosen from those Palestinians legally residing in the occupied territories outside Jerusalem. Palestinians in the Diaspora were not to be represented in the delegation. However, and although the PLO was not recognized, nor was it a formal party to the peace process, it was a known "secret" that it actually chose the Palestinian delegates or approved of them. Israel kept a watchful eye on the choice of those delegates and retained what amounted to a veto power over their participation. Jordan compensated for the absence of representation of Jerusalem Palestinians. She included in her delegation two Jordanians of Palestinian origin from Jerusalem's prominent families[291]. The Joint Palestinian-Jordanian delegation was formed, headed by the Jordanian Foreign Minister, and participated in the Madrid Conference on 31 October 1991. The Joint Delegation was composed of fourteen members, seven each for the Palestinians and Jordanians. The other delegations were headed by the Foreign Ministers of their respective country with the exception of Israel whose delegation was headed by the Prime Minister, Mr. Yitzhak Shamir. The Madrid Conference adjourned after the opening and yielded a two-track approach to the peace process: a Bilateral track of negotiations in which Israel and each of the other core parties would negotiate solutions for their issues of conflict separately but simultaneously, and a Multilateral track in which the core and regional parties along with countries of interest in Middle East peace would seek co-operative solutions to certain problems in the region, solutions that would reinforce peace.

The first meeting of the Bilateral track (first round) was held in Madrid on 1 November the day after the International Conference was opened. Speeches were made in the Jordanian-Palestinian--Israeli track, and it was agreed that the delegations to their bilateral negotiations would be composed of five members for each of the three parties: the Israelis, Jordanians and Palestinians. No agreement, however, was reached on a venue for the Bilaterals. The Israelis wanted a venue in the Middle East, and the

---

[291] Mr. Anwar el Khatib, former Governor of Jerusalem, and Professor Walid el Khalidi.

Arab parties wanted to be away from the region, to exclude the possibility of normalizing contacts with the Israelis before the negotiations yielded fruits. The parties agreed to leave the choice of the venue to the sponsors. The United States and the Soviet Union consulted with each other and decided that it was most important to maintain the momentum of direct negotiations, and therefore decided that the bilateral negotiations should start in Washington, D.C. on 4 December 1991, and proposed topics for talks on all tracks of bilateral negotiations (see Letter of Invitation, Appendix 6).

The Jordan Delegation was composed of 14 delegates[292] and nine supporting staff. The members of the Joint Delegation left Amman on Monday, 2 December 1991, on board a special Tristar flight which the King had ordered for the purpose. In Washington, D.C., the Jordanians stayed at the Willard Intercontinental Hotel and the Palestinians, eager to emphasize independence, stayed at the Grand Hotel. The Israeli delegation did not show up at the State Department on the specified time, but arrived in Washington about a week later, on Sunday, 8 December 1991. The Palestinian and Jordanian delegations took their seats, at the specified time and date, in the rooms assigned for their respective negotiations with Israel. They waited for 34 minutes, each knowing that the Israelis were not coming. The press was not allowed in the negotiation wing on the ground floor of the U.S. State Department to take photographs of the vacant Israeli seats.[293]

---

[292] In addition to Dr. Abdul Salam Majali, the members were Dr. Walid Khalidi, a prominent political scientist and scholar oringally from Jerusalem, who was teaching at Harvard; General Abdul Hafez Marea Ka'abneh, Assistant Chief of Staff of the Jordan Armed Forces; Dr. Fayez Tarawneh, a former Cabinet Minister; Dr. Jawad Anani, a former Cabinet Minister; Dr. Abdallah Toukan, Science Advisor to His Majesty King Hussein; Dr. Munther Haddadin (the author), former President of the Jordan Valley Authority; Ambassador Talal Al Hassan, Jordan's Ambassador to Belgium, the Netherlands and the EU; Ambassador Nayef Al Qadi, Jordan's Ambassador to Qatar, Awn Al Khasawneh, Adviser to Crown Prince Al Hassan and a member of the International Law Commission of the U.N.; Dr. Musa Breizat, a political scientist at the Ministry of Foreign Affairs, Dr. Mohammad Bani Hani, Secretary General of the Ministry of Municipal and Rural Affairs and the Environment; Dr. Marwan Muasher, Director of the Jordan Information Bureau in Washington, D.C.; Anwar Al Khatib, former Governor of Jerusalem, who did not travel with the delegates but participated in its meetings in Amman; and Dr. Ahmad Katanani, of the University of Jordan.

[293] The State Department Spokesperson said that allowing the press to take photos of the rooms with the Israeli seats vacant would be "hitting below the belt".

The second round of bilateral talks (the first was in Madrid), which started on Tuesday,[294] 10 December and lasted until Wednesday, 18 December, was spent hammering out matters of form that implied important substance.[295] The three heads of delegation argued over issues of representation of each delegation and the format of negotiations. Their meetings were held on a sofa in the corridor of the ground floor of the State Department's west wing, and no agreement was reached. Israel was trying to expand the role of the Jordanians in the negotiations and have them participate in negotiations of the Palestinian affairs, something that the Palestinians resented and the Jordanians rejected altogether. They maintained that the Joint Delegation was only the mechanism by which Jordan provided a State umbrella for the Palestinians to participate in an international conference, nothing more. The Israelis understood the Joint delegation as one party, with which Israel should negotiate peace.

As indicated in the Letter of Invitation to come to Washington, Jordan was prepared to have a Palestinian, even two, attend the negotiation sessions on the Israeli-Jordanian track and, likewise, the Palestinians were prepared to have one or two Jordanians attend their track of negotiations. The Israelis, however, wanted the negotiations to be formally conducted between Israel and the Joint Delegation in a plenary meeting, and to discuss and debate matters of substance in two tracks: one takes up Israeli-Jordanian issues, and the other looks into Israeli-Palestinian issues. Both the Jordanian and Palestinian delegations refused the Israeli approach. The second round ended without much progress, not even getting into the negotiation rooms.

4. A Water Meeting with the Americans

It was noted that the letter of invitation to the bilateral negotiations did not suggest water as a topic for negotiations between Jordan and Israel. Rather, it mentioned the "joint management of wastewater facilities." The Jordanian delegate who was assigned the water file, Dr. Munther Haddadin, was eager to have a first-hand discussion with Ambassador Richard Armitage who, since 1988, had been acting as an intermediary between Jordan and Israel for the Wehda dam construction. The meeting took place in Mr. Armitage's office at the State Department on Friday, 20 December 1991.

---

[294] The Israelis wanted to start on Monday, 9 December, but the Arab delegations declined because that day was the fourth anniversary of the eruption of the Palestinian *Intifada* in 1987.

[295] The Israelis wanted to negotiate with one Joint Jordanian-Palestinian delegation, and have the different topics negotiated by committees. The Jordanians and Palestinians were prepared to negotiate in accordance with the terms spelled out in the Letter of Invitation, i.e, negotiate in two tracks, one Jordanian and another Palestinian.

Armitage talked about his efforts to have the Wehda dam on the Yarmouk River financed and built, and to narrow the differences between Jordan and Israel over water shares. He was well versed on the issues and on the on-going Syrian activities of dam building that were violating even the 1987 Jordanian-Syrian Agreement. He suggested that something had to be done to stop the Syrian expansion in building dams in the Yarmouk catchment and using more water. He thought the World Bank could be the mechanism to arrest that expansion. The Jordanian delegate noted that Syria was defaulting on their World Bank loans and the Bank's relations with Syria were not at their best. The Jordanian delegate also sensed that Mr. Armitage was eager to continue his mission and, if successful, would have made the breakthrough that his predecessors, Eric Johnston (1953-54), Philip Habib (1978-81) and Richard Murphy (1984), had failed to achieve.

In that regard, the developments in the Armitage mission were not heading toward Jordan's advantage, especially in water sharing. There were concessions to Israel that should not have been made.[296] The Jordanian official in charge had earlier made more concessions to Syria[297] as the 1953 Agreement was replaced by a new one in 1987. In a way, the start of bilateral negotiations with Israel was timely for Jordan, as far as her water issues with Israel were concerned. It was possible for her to freeze the efforts of the American intermediary and negotiate directly with Israel.

The Jordanian delegate expressed to Armitage the view that a new era of historic significance, an era of peace, was dawning on the Middle East, and that it might be a worthwhile idea to re-examine the previous plans of development of the Yarmouk in that context, and review the wisdom of building dams in Syria, whose declared purpose was to win a race with the enemy (Israel) in maximizing the use of Yarmouk water. There should be no more enemies when the peace process succeeds, and the doors should be wide open to co-operation among former adversaries.

---

[296] Yarmouk floods that could not be diverted by the diversion weir at Adasiyya were conceded to Israel. These floods would come from the intercatchment between the Wehda dam and Adasiyya, and from any spills from Wehda dam. Israel's share in the River was set by the Unified Plan at 25 mcm per year, and the rest of the flow at Adasiyya belongs to Jordan.

[297] The Jordanians conceded to Syria in 1987 the right to retain water impounded by 26 dams, which Syria had built in the Yarmouk catchment, with a total storage capacity of 156 mcm. These waters were destined, under the 1953 Agreement, to the Wehda (Maqarin) reservoir. Syria's share would not exceed 90 mcm under that agreement or under the Unified Plan. The new Agreement of 1987 conceded to Syria another quantity which is the annula yield of the above dams, with an estimated average of 120 mcm/yr. Neither Agreement addressed the abstraction from the groundwater aquifers feeding the Yarmouk springs, especially the lower springs below elevation 250 m.a.s.l., which are part of Jordan's share.

Armitage and his guest touched on the history of water sharing and on the principles observed in that. The Jordanian delegate noted that the "Harmon Doctrine", which advocates absolute sovereignty over water resources, was made obsolete even in the lands in which it originated (the United States). Mr. Armitage noted that the Arab states did not endorse the Unified Plan, and thereby its applicability could be questionable. That was a repetition of the opinion of Mr. Philip Habib, in his last visit to Jordan in 1981.

"This was on the political side," interjected Dr. Haddadin, "but on the technical side, you know that the Arab Technical Committee had accepted the Unified Plan."

"Besides," he added, "Jordan has committed herself to the Plan in response to an American request to that effect in 1958. The memo of the Jordanian Foreign Minister No. 58/14/6719, dated February 25, 1958[298] was explicit in that matter. Similarly, there was an Israeli commitment to the same effect on their part before they qualified for financing of the Tiberias-Beit Shean pipeline."

Mr. Armitage did not comment on those interventions, but asked his guest what he thought would be Jordan's reaction if he obtained the Israeli consent to build the Wehda dam.

"It is the Diversion Weir that Israel should be concerned about, not the dam itself," noted Dr. Haddadin, "but frankly," he continued, "I do not know what Jordan's reaction would be."

Members of Armitage team were present in that meeting. The meeting touched on the Multilateral conference that would convene in Moscow soon, and on the role water would play in that conference. The Multilateral talks would address the issues of water resources, environment, regional economic development, refugees and regional security and arms control. The participants in the meeting all agreed that for the multilateral conference to be meaningful, progress on the bilateral track should be made.[299]

---

[298] This note acknowledged and specifically accepted the understanding conveyed to the Foreign Minister on 24 February 1958 by the American Charge d'Affaires that American assistance in financing the first year costs of the proposed Yarmouk Diversion Project and the East Ghor Canal "will be extended provided there is an explicit undertaking from the Hashemite Kingdom of Jordan that it will not draw from the Yarmouk River more water than the share allotted it under the Unified Development Plan."

[299] Memo dated 27 December 1991 from the author to Dr. Majali reporting on the meeting. Records of the Peace Talks, Ministry of Foreign Affairs, Amman, Jordan.

5. Developments in the Bilateral Track

The following round of negotiations (third round) was scheduled for the 7th of January 1992. The Arab delegates delayed their arrival in Washington in protest against Israel's expulsion of twelve Palestinians from the occupied territories. The negotiations started on 13 January and ended on the 16th, a period of four days. The joint Jordanian-Palestinian delegation and the Israeli delegation succeeded in overcoming the difficulties of the second round, and were able to enter the negotiation rooms. Two tracks would handle the negotiations, one for Jordanian-Israeli matters, and another for Palestinian-Israeli matters. The number of delegates in the Jordanian track would be nine Jordanians and two Palestinians, and the Palestinian track would include two Jordanians as observers in addition to nine Palestinians. A general meeting was also decided on, in which the representation was (12 +1) Israelis,[300] 11 Jordanians, and 8 Palestinians. The general meeting would discuss matters of common interest, but would not be a substitute for either of the two negotiation tracks.

In the Jordanian track, Jordan presented her "Vision of Peace" in a speech made by the head of delegation, Dr. Abdul Salam al Majali, in the negotiation room. Israel reciprocated the following day by surprising the Jordanians with a "Peace Treaty". The Jordanians refused to receive the peace treaty, and represented that it was like putting the cart in front of the horse. Israel then presented its contents as their own "Vision of Peace". The round ended with each delegation presenting its proposal for an agenda to be discussed and jointly adopted for their negotiations. The gap between the two proposals was very wide. Appendix 7 presents the two texts of the proposals for common agenda by Jordan and Israel.

Dr. Majali and Ambassador Elyakim Rubinstein, head of the Israeli Delegation, agreed that their respective groups on water, energy and the environment could discuss informally those issues. The three topics were summed together in one group, based on a Jordanian proposal. The Jordanians thought that Israel had an interest in bilateral cooperation in the fields of environment and energy, while Jordan would be prepared to cooperate only after the water conflict with Israel was resolved. The integration of the three topics would handle the three issues in a fair way, and serve the interests of both parties.

The initial informal meetings were held between one Jordanian delegate, Dr. Mohammad Bani Hani, and one Israeli delegate, Dr. Elyahu Rosenthal. This meeting was followed a while later by another, where three Jordanian delegates participated: Haddadin, Toukan, and Bani Hani, all of

[300] The twelve Israelis would sit at the meeting table, and the thirteenth would sit in a chair in the back row behind them.

whom met with the same Israeli delegate, in the office reserved for the use of the co-sponsor, the representative of the Soviet Union.[301]

The conversation was mostly to get to know each other. They had a chat about the inter-relationships between the three topics they were going to discuss. They alluded to the topic of the environment in general, and to the need to control the level of the Dead Sea by linking it to the open seas. The Jordanians preferred the Dead Sea linkage with the Red Sea, and the Israelis had the Mediterranean-Dead Sea canal in mind. The idea of linkage was not new, and the need for it had become more prevalent since the diversion of the Jordan River and the installation of the potash plants by Israel and Jordan.

The fourth round of negotiations started on 24 February and followed from where the third round ended. It lasted until 4 March 1992. It focused on the composition of the drafts of the common agenda, setting out separately the agreed items and the items on which there was difference of opinion or disagreement. An attempt to form a special group from the two delegations to explore the understanding of U.N. Resolution 242 leaked to the press and was misinterpreted by many, especially the Palestinians. The attempt was aborted by instruction from the Jordan government and no such group was formed. There was no question in the minds of the Arab parties as to what 242 actually meant. It was withdrawal from the territories occupied in the 1967 war, nothing less.

It was jointly agreed between the heads of the two delegations that a meeting of experts on water, energy and the environment be held to explore the possibility of agreeing on issues to be negotiated. The Jordanian side had composed the item pertaining to water that could be inserted in the common agenda, and proposed it for discussion of the expert group the following day. In the evening the text of the water item was cleared with Dr. al Khasawneh and Dr. Majali. It read:

"3. Water
a) Securing the rightful shares of the riparians in the Yarmouk and Jordan Rivers.
b) Searching for ways to alleviate water shortage.
c) Controlling the level of the Dead Sea."

The text formed the basis of discussions in the informal meetings of experts, who came close to an agreement on a common text that should be adopted by the two delegations in a track session. It did not take much time to agree informally on a text; it read:

---

[301] Soon after, the Soviet Union disintegrated into the Russian Federation and the Commonwealth of the Newly Independent States.

"3. Water

a. Securing the rightful water shares of the two sides in the Yarmouk River and the Jordan River.

b. Searching for ways to alleviate water shortage."

This agreed text differed from the Jordanian text in that it addressed the rightful shares of the two sides instead of the riparians, which included more than the two sides. The control of the Dead Sea level was shifted to another item in the common agenda, as shall be seen later. It was further informally agreed that the parties would negotiate, in addition to the water sharing of the joint watercourses, the environmental protection of water resources. The informally agreed text was kept away from publicity, and it was a good start.

The fifth round started on 20 April 1992. The Jordanian delegation focused on preparing a new draft for a common agenda, building on the debates that had taken place in the bilateral track thus far. A new draft was produced on 26 April,[302] submitted to a meeting of the Jordan delegation at the Willard Hotel and adopted, with a few changes made as a result of their discussions. The draft was more detailed than any other previous proposal, and constituted the base for Jordan's arguments in the succeeding two rounds. A copy of the new agenda proposal was handed over to the Palestinian delegation that same day.

In addition to the debate over a common agenda, water was a major theme for the fifth round. The progress made thus far justified emphasis on the water issue and an opening statement on water was made by the Jordanian delegation in the track meeting. The statement is shown in Appendix 8, and spells out Jordan's goal in water negotiations with Israel.

In the break after the presentation, the head of the Jordanian informal group, Dr. Munther Haddadin, talked with the Israeli Director General of the Ministry of Energy, Dr. Uri Wurzburger, about the control of the Dead Sea level, and advocated its linkage with the open seas at Aqaba. Wurzburger was open to the suggestion, and was prepared to have it formally studied in a joint effort. On the part of the Jordanians, they were not prepared to launch a cooperative study on the subject at that early stage of negotiations. The preparedness of the Israelis to study this Jordanian project was indicative of possible endorsement.

On 29 April, Dr. Rosenthal of the Israeli delegation asked Dr. Haddadin in the corridor, what he considered as Jordan's rightful share of water. Haddadin's answer was "that a compromise solution was reached in 1955 through the American envoy, Ambassador Eric Johnston. Jordan

---

[302] The new draft was produced by Dr. Munther J. Haddadin, based on the debates in the negotiation room and on the position Jordan took regarding the negotiation topics.

committed herself to her share under that compromise, and Israel did likewise as a condition to have her water projects in the basin financed. We do not intend to turn our backs to commitments we made unless Israel turned back on her commitments first, in which case we will have to start from the very beginning."

The Jordanian water negotiations strategy aimed at securing the rights of the Kingdom as it existed in 1955, i.e., the East Bank and the West Bank. Haddadin thought that strategy served two purposes:

1) It would avoid splitting the water rights of the Kingdom between West Bank and East Bank,;and,
2) It would put the Palestinian water rights in the Jordan River basin in Jordan's hands, a much more secure place than Israel's, where they actually were at the time. Additionally, the Palestinians could not negotiate their water rights with Israel until the Final Status negotiations, five years down the road, as stipulated in the Letter of Invitation to Madrid.

Besides, he thought, displaced Palestinians who live in Jordan needed water. Should any of them be allowed to go back to the West Bank, he or she would be allowed to transfer his drinking water share to the West Bank at the rate of 50 cubic meters per person.[303]

In a meeting on Thursday, 30 April 1992, of the group on water, energy and the environment, six Israeli members came into the room, and only three Jordanians were attending. The Israelis had their legal advisor, Dr. Robby Sabel, with them. He asked for any Jordanian proposal to resolve the issue of the rightful water shares. The intent, obviously, was to hear in the group meeting what Rosenthal had heard in the corridor.

"I suggest that we do not try to re-invent the wheel," responded the head of the Jordanian team, "to make a long story short," he added, " we are prepared to pick up where we had agreed, in 1955, to the compromise formula proposed under the Unified Plan. Failing that, I am afraid we will have to disregard all our past commitments and start from the very beginning."

"You mean the Johnston Plan of 1955?" Asked Dr. Sabel.

"Yes I do."

"Well, our time now is limited. The plane is going to depart sooner than we expected, we will come back to this particular issue in the next round."

---

[303] This strategy, however, was later modified by instructions from the King on 1 June 1994, after the Palestinians, under PLO leadership, conducted their negotiations with Israel separately pursuant to the Oslo Accord of 13 September 1993. Jordan thereafter confined her negotiations to the rights of the East Bank only.

The round ended at that, with some progress achieved on the wording of some items of the common agenda drafts. It sounded like the informal talks would get into substantive discussions soon.

6. The Multilateral Conference: The Vienna Chat

The Multilateral Conference convened in Moscow on Tuesday, 28 January 1992. Syria and Lebanon did not participate in that conference, because they considered it an attempt at normalizing contacts with Israel before peace was reached. The Palestinians demanded parity with the other delegations (seven delegates) and the inclusion in their delegation of the Palestinians in the Diaspora. The Foreign Minister of Jordan demanded the attention of the United States Secretary of State, Mr James Baker, as he stood in the reception line welcoming the delegates to the Conference. The efforts of the Jordanian minister lasted for about 15 minutes, but were not successful to convince the sponsors to accommodate the Palestinian demands. The opening of the Conference was delayed because of that, and the Palestinians did not go into the Conference hall.

The Jordan delegation was[304] seated next to the Israeli delegation. Mr. David Levy, the Foreign Minister of Israel, exchanged a few words with a Jordanian delegate through an interpreter. He expressed his admiration of the Jordanian efforts to develop the Jordan Valley and referred to some technology transfer from Israel to the Jordan Valley.

"A little perhaps," said the Jordanian delegate, "but the Israelis have exaggerated it a great deal. Drip irrigation, for example, got to us from Australia via the Gulf States and Lebanese entrepreneurs, but the Israelis claim it was they who invented and introduced the system."

"There is some truth to that," said the Minister.

"There may have been some truth to it later when we started importing hardware of drip irrigation from Cyprus. There is a strong Israeli element in their manufacturing in Cyprus. But also later, we started to make the hardware in Jordan ourselves."

"Please do not be offended if I make reference in my speech to the fruits of potential co-operation, and mention that as an example," said Mr. Levy.

"You see, Mr. Minister, I was afraid you might claim credit for some of our achievements."

Speeches were made by representatives of the Sponsors, the European Union, Canada and Japan, the core and regional parties. Minister Levy's

---

[304] The Jordanian Delegation was headed by the Foreign Minister, Dr. Kamel S. Abu Jaber and comprised Dr. Fayez A. Tarawneh, General Abdu Hafez Marie, Dr. Munther J. Haddadin, Dr. Abdallah Toukan, Dr. Mohammed A. Adwan, and Ambassador Saleh Kabariti.

speech in the opening session alluded to agricultural technology and the benefits of co-operation in that and other fields. He cited the story of the Jordan Valley as an example, and the Jordanian delegate shook his head in amazement.

The Conference broke up the following day into six groups, one each on water resources, regional security and arms control, regional economic development, environment, refugees, and a steering group. An agreement was reached on the agenda for the works of the Water Resources working group, and a proposal was made for the venues of the next meeting. These were Turkey and Vienna. Jordan opted for Vienna, because they did not want to be in Turkey while efforts were being made to have Syria join the Multilateral Conference. At the end of the Conference, the Jordanian Minister of Foreign Affairs, Dr. Kamel Abu Jaber, appeared in a press conference with the Jordanian heads of the working groups. A correspondent asked when they expected normal cooperation with Israel to materialize.

"The sequence is," responded Haddadin, "Restoration, Mitigation, and Co-operation." He explained that this meant restoration of rights, mitigation of adverse impacts resulting from the creation of Israel, and then cooperation would start. These became buzzwords for the Jordanian delegation.

The second round of multilateral talks on water resources was convened in Vienna on Tuesday, 12 May 1992. The Israeli Water Commissioner, Dr. Dan Zaslovsky, headed the Israeli delegation that also included Dr. Elyahu Rosenthal. The Palestinians participated in the joint delegation and had a coordination meeting with the Jordanians the day before. Their presentation was, as expected, critical of the Israelis. The Jordanian presentation was very much supportive of the Palestinians and openly critical of Israel. The Arab delegates, including Kuwait, loved it. At the beginning of the coffee break, Dr. Rosenthal approached Dr. Haddadin and asked if Dr. Zaslovsky, the Israeli Water Commissioner, could have a word with him in private. He accepted and went to talk to Zaslovsky in an empty corner of the hall[305].

"The Palestinian accusations in their presentation are not true. They claim we are stealing their water. I am the Water Commissioner and I am telling you we are supplying them with water from Israel," said Dr. Zaslovsky. At this point, Dr. Rosenthal, who was aware of the Jordanian proposal to resolve the dispute, joined them.

"Your response to their presentation," Dr. Haddadin answered, referring to the Palestinians, "was anything but accommodating. Remember that they are under Israeli occupation, so do not expect them to hug and kiss you."

---

[305] "Memorandum to the File," by Munther J. Haddadin, dated 19 May 1992.

"You can come and see for yourself; I invite you to come and you will be my guest. No one would know about your visit and I will keep it a secret," said Dr. Zaslovsky.

"Unlike others who fear violence in retaliation for contact with the enemy," said Haddadin, "I have no fears of that. However, throughout my life, I have never done a thing of which I am not convinced, and when I did what I believed in I did it overtly and in the open; I will not come to you unless you provide me with a good reason to make the trip in public".

"You mean your demands in the bilateral negotiations on water rights?" Asked Zaslovsky.

"Precisely, there is just no way to interact through mutual visits before our rights are recognized and settled. To repeat, 100 mcm from the Lake, and your share in the Yarmouk confined to 25 mcm annually."

"For the sake of peace," said Zaslovsky, "I am willing to go along. I have two politicians that I must convince. I hope things will work out, and that we will have an agreement on the next round of Multilateral talks in Washington. These were the shares specified in the Johnston Plan."

"For your information," Haddadin resumed, "Jordan undertook to abide by her share under the Johnston Plan, and committed herself to that in fulfillment of a condition for financing the East Ghor Canal Project in 1958. Israel, too, committed herself to abide by her share in the Plan, as a condition to receive American aid to build the Beit Shean pipeline and the National Water Carrier. The consent of both countries to the Plan boils down, in effect, to an agreement reached between the two parties through the United States. If you are not willing to honor that agreement, it would be a serious handicap."

"Do you mean that the Armitage efforts and talks are not valid anymore?" Asked Zaslovsky.

"Listen my friend, the good offices of the United States were the only avenue we both could pursue during the past era. I had worked with Philip Habib and with Dick Murphy before Armitage succeeded them. The efforts of the United States are applicable in a 'state of war'. Now we are negotiating face to face and this is different. The Armitage efforts will be picked up should you choose to have the peace talks collapse. We both should keep the Armitage efforts on 'hold' as an alternative to bilateral negotiations."

"We hope we will respond positively to your positions because there is a lot we can do together for mutual benefit," said Zaslovsky.

"I consider your preparedness to honor the old commitments a confidence building measure. It would encourage further progress in the bilateral negotiations."

"We are also willing to go along with you on the Red-Dead project. Our idea is to build a desalination plant in Wadi Arava for the benefit of both sides. We are sure we can bring the cost down to $0.65 per cubic meter. We are about to reach a breakthrough in desalination technology. A team from the United States will come to us next month before we can make the announcement."

"Look at the future potentials therein. Let us create the environment for cooperation. Let us build a lasting peace based on justice. Let us make water a vehicle for cooperation. Start by restoring our rights as a confidence building measure," commented Haddadin.

"Let us hope that we succeed," said Zaslovsky.

"By the way of confidence building," Haddadin said, "you know that Amman gets water from the King Abdallah Canal among other sources. Despite this year's good precipitation, Amman will be short of water in the summer. Water shortage is bad for people, and we need their support of our efforts. An immediate confidence building measure will be to make more water available to Amman from the Yarmouk."

"We have done that before, and we can do it again," said Zaslovsky.

"Maximize. Retain only the flow necessary to sustain the farmed areas in the Israeli Yarmouk Triangle that can be irrigated only from the Yarmouk."

"We will do our best to maximize," said Zaslovsky, and Rosenthal nodded in confirmation.

At this point, an Austrian usher interrupted the conversation and asked Haddadin to come to the press outside. Haddadin excused himself, but Rosenthal followed and stopped him for further talk.

"Under the Johnston Plan, the 100 mcm from the Lake were assigned to the West Bank," he said. "How much are you going to give to the Palestinians?"

Rosenthal was actually right, but what he came up with was not mentioned in the Plan per se, but was the basis for calculating Jordan's needs to irrigate the West Bank. What was more important to Haddadin was the strategy he had developed for the water negotiations, i.e., to negotiate for the rights of the West Bank and the East Bank, and return the Palestinian rights in due time after they negotiate their final status with Israel.

"If Jordanians gave their blood in support of Palestine," Haddadin said to himself, "would they deny water due to the Palestinians?" Haddadin believed he had to shut that door that Rosenthal just opened. "Besides," he continued, "you are wrong about West Bank allocations."

"But Vardi has the documents," insisted Rosenthal, referring to the Israeli water expert, Yacov Vardi.

"Vardi must be confusing issues," Haddadin answered, "he must be looking at the first draft of the Johnston proposals which was composed by Chas T. Main under which the West Bank would take its water from the Lake through a western conduit parallel to the Jordan river. This was later modified."

Haddadin was actually trying to confuse the issues himself, in order to implement his strategy and secure the Palestinian rights along with the Jordanian rights.

"How much would you give to the West Bank?" Rosenthal asked.

"This is purely a bilateral Jordanian-Palestinian matter, although you are somehow involved. I will give you the privilege to stand out there and watch."

"How are we involved to your thinking?"

"Palestinians who have taken residence in Jordan after 1955, including the Palestinians displaced during the war of 1967, have to have their water share, at least to drink. It will all depend on where those Palestinians will finally take up residence. You are involved because their return, I presume, will be the subject of negotiations with Israel."

The essence of this conversation was transmitted to Jordan's Minister of Foreign Affairs and to the Jordanian High Steering Committee on Peace Negotiations on 19 May 1992. Haddadin expressed hopes that Zaslovsky would deliver on his pledge by the next round of multilateral talks scheduled to convene in Washington, D.C. in September, if not before.

Again, it sounded like the water front was really making good progress.

7. Substance in the Bilaterals

The sixth round of Bilateral negotiations was held in Washington, D.C. between 24 August and 23 September 1992. The round was conducted in two parts, interrupted by Labor Day in America and by Jewish holidays. The June elections in Israel yielded a victory for the Labor coalition led by General Yitzhak Rabin, and the rightist government of Yitzhak Shamir of the Likud coalition was replaced. The mood was that of high expectations.

Progress was made in the Jordanian-Israeli track regarding agreement on some items of the common agenda. An updated Jordanian draft was composed on 23 August and presented to the Israelis. The Jordanian wording for the water item was finally accepted, as was the wording for the item on cooperation within a regional context, in which the control of the Dead Sea level was included. Several other points, in the items of security, refugees, and occupied Jordanian territories, remained unsettled. Side meetings on water, energy and the environment were held and substantive talks were conducted. In the first informal meeting of the expert group, Dr. Robby Sabel of the Israeli delegation had a written text from which he read.[306]

"The Johnston Project was vague," read Sabel, "what is your understanding of it and your position regarding it? You said that you have a

---

[306] Letter no. MS/6/18 dated 27/8/1992 from Head of the Jordanian Delegation to the Peace Negotiations addressed to the Prime Minister of Jordan. Files of the Peace Negotiations, Ministry of Foreign Affairs, Amman.

document that shows your commitment to it. Can we have a copy of that document?"

"Tell us," he continued, "what is the share of Syria in accordance with Johnston Plan, and are the Syrians abiding by that share?"

"You concluded an agreement with Syria in 1987. How much did you allow the Syrians to withdraw from the Yarmouk flow? How do you conclude an agreement over the Yarmouk without the approval of Israel who is the third partner?"

At this point, Haddadin remembered vividly his meeting with the Jordanian Prime Minister in July 1987, in which he opposed the replacement of the 1953 Jordanian-Syrian Agreement on the Yarmouk, and his expectation that such a replacement would be a great service to Israel. Sabel continued:

"How do you compare the share of Syria in the 1987 Agreement with her share under the Johnston Plan, and with the quantities they currently abstract from the river?"

"Did you take into any consideration the interests of the West Bank in the 1987 Agreement?"

"What was the quantity of flow that Jordan was supposed to deliver to the West Bank through the siphon across the Jordan River? What was the capacity of that siphon?"

"What is the percentage of salt water that the 100 mcm specified by the Johnston Plan be drawn from Lake Tiberias?"

"Since Jordan did not deliver any water to the West Bank, where is that water now?"

The Israeli accent was stiff and dry; Haddadin responded: "I am surprised that the Israeli side is posing questions the answers to which they know very well. The presentation sounds like that of a plaintiff in a court of law [Sabel was the legal expert in the Israeli delegation]. We are prepared to present the defense of the defendant if this were a court of law in which case we need a jury and a judge."

"They are close," he continued, "they must be in this same building [referring to the United States officials]. Let us bring them in so that they hear the charges and the defense, and we both should accept the verdict."

Sabel understood the sarcastic response, declined the proposal and politely apologized for the accent of his presentation.

"Your repeated reference to Syria puzzles me," Haddadin resumed, and, remembering that the bilateral negotiations were an Israeli demand, he added, "do you want the Syrians with us here or don't you? They are negotiating with Israel in the other wing of this same building. We are prepared to work either way."

"If there are parties who would testify to our consideration of the rights of others on the Yarmouk, Israel should be the first such party," Haddadin continued, "Israel should be the last to complain about ignoring her in any venture Jordan had undertaken on the Yarmouk. The detailed designs of the Wehda dam were forwarded to you via the American intermediary from the outset. We also came to know that our consultant, Harza Engineering Company, hosted a delegate from your Water Commission to explain the designs to him and answer his questions. That, however, was done without our consent, and we did not fire our consultant for being that liberal in the disposition of our material. We understand the factors and the pressures that prompted them to abide by the requests of outsiders."

"But that was..." interrupted Dr. Sabel.

"Now, please," interrupted Haddadin firmly, "I gave you all the time you needed without any interruption. I expect the same from you and your colleagues."

There was extreme silence that was obviously to Haddadin's advantage, and he deliberately became quiet for a few more seconds, and then resumed:

"As for the 1987 Agreement with Syria, its cornerstone is the Wehda dam, and such agreement should be of value when the dam is built. We will be prepared to look into any grievance you have when we come to build that dam. Therefore, it is a waste of time to enter into a debate over an agreement that has no effect unless it is implemented. Besides, we are not responsible for what Syria does in the river basin, and we have informed them of our objections on several occasions before 1987.

"With respect to the Palestinian rights under the 1987 Agreement, Syria has never recognized the West Bank as part of the Hashemite Kingdom. Therefore, Syria's talk with Jordan does not apply to the West Bank. As a matter of fact, Jordan and Syria both consider the PLO as the sole legitimate representative of the Palestinian people."

The head of the Jordanian group, Dr. Haddadin, continued, answering the several questions posed by Sabel:

"Jordan did not deliver any water to the West Bank because the share of the West Bank was to come from Lake Tiberias which has been under Israeli control. The Palestinian share would be delivered to the KAC from Lake Tiberias and Jordan, in turn, would deliver it to the West Bank through a siphon under the Jordan River. So you see that the share of the West Bank under the Plan is still in Israeli hands."

"We never have disregarded the Palestinian rights, but we have refused, and still do, that Israel speak on behalf of the West Bank. As for the quantities we intend to give to the Palestinians, it is for us and the Palestinians to decide, although Israel has a role there pertaining to the number of Palestinians who would exercise the right of return.

"Israel's protest over ignoring her in the 1987 Agreement would have been sustained were the contracting parties not in a state of war with her."

"Syria's share in the Yarmouk under the Johnston Plan is 90 mcm per year, which was the same share implied in the 1953 Agreement we had with them.

"The percentage of salt water to be included in the 100 mcm from Tiberias is not more than 15 mcm."

"We are prepared to exchange documents with you. We are ready to give you a copy of our commitment to abide by our share in the Unified [Johnston] Plan provided you give us a copy of your commitment to your share under that Plan. You issued such a document as a condition precedent to receiving financial aid from the United States to build the Tiberias-Beit Shean pipeline, and the National Water Carrier."

The Israeli side appreciated the Jordanian explanation, and insisted that Israel's share in the Yarmouk of 25 mcm was only the summer allocation! A share in the winter months has to be negotiated and agreed upon. That was a repeat of their old position, which was based on their Memorandum of Understanding with Johnston dated 5 July 1955, in which they insisted their annual share was 40 mcm. The Jordanian side rejected thei claim and insisted that the annual, not the summer, Israeli share in the Yarmouk was 25 mcm.

Detailed discussions ensued about the flow of the Yarmouk and its recession over time. The Israeli and Jordanian estimates of the average annual Yarmouk flow remaining at Adasiyya were only 0.3% apart, 269 mcm against 270 mcm respectively. Quality of the water and the changes that had occurred were touched upon. Reasons for the quality changes, although not admitted in the discussions, were the substantial increase in Syrian uses upstream where the water is fresh, as compared to the springs downstream had a higher salinity.

The Israeli side suggested that the Jordanian side forget about the past and look to the future potentials for cooperation. This is implied that Jordan should forget about the Unified Plan, over which Israel had disagreements with the United States anyway. They proposed the impoundment of Yarmouk floods in reservoirs to be built on the Jordan River. The Jordanians turned down that suggestion, on the ground that storage places on the Jordan have very limited capacities, that there were better locations for storage, that the water quality would be degraded if stored on the Jordan river course, and that energy would be needed to pump it for use.

The Israelis did not give an answer to the Jordanian proposal on water sharing, and suggested that the talks start from where the efforts of Mr. Armitage had ended. This was the third time that the Israelis indicated their preference to the "progress" achieved through the efforts of Mr. Armitage. Throughout his intermediary efforts, the Jordanians agreed that Israel could use the spills of the Yarmouk floods over the Adasiyya diversion weir to be built across the Yarmouk. These floods, under the Johnston Plan, were meant to be stored, for Jordan's benefit, in Lake Tiberias or in any other economic location. The Israelis had other benefits from Armitage's work, and the Jordanians were amateurish in giving their consent. The Jordanians who handled the technical aspects of the Armitage mission were not very well equipped to handle the task. They were, incidentally, the same individuals who had agreed to replace the 1953 Agreement with Syria over the Yarmouk, and to concede to Syria the right to the water impounded by the dams that they had built in contradiction with that Agreement. The loss to Jordan because of their incompetence could have been double edged: a loss of water to Syria, and an imminent loss of water to Israel. For Jordan the commencement of the bilateral negotiations was timely as far as the Israel and the Yarmouk were concerned. It gave Jordan a renewed opportunity to put the Armitage efforts on hold and start direct negotiations with Israel instead. This might have left sour feelings on the part of at least some of the Armitage team members. They were hoping to work out the agreement themselves and leave on record a greater success than Eric Johnston had accumulated in his famous mission.

The Israeli team was eager to start talks on topics of the environment and energy. The Jordanian team, with their priority given to water, were willing to listen. The items of Israeli interest were environmental, particularly the protection of the Gulf of Aqaba against such potential environmental hazards as oil spills from ships, the joint management of the wastewater treatment plants in Aqaba and Elat, the birds feeding on grain from the silos in Aqaba and bringing their droppings on hotels and the beaches of Elat, the mosquito problem in the Aqaba-Elat area, and the housefly problems in the Jordan Valley. The Jordanians said that action on such topics require cooperation, the grounds for which must be laid down by first making progress in the bilateral negotiations, especially over water sharing.

During the first half of the negotiation round, the Jordanian team leader pressed for adjustment of the current percentage sharing in the Yarmouk water at Adasiyya. Jordan was in need of more water in the KAC, some of which would be pumped to Amman for municipal uses. He brought up the subject in the group's meeting on 25 August 1992, and explained the tight water situation in Amman. He reminded Rosenthal of his Vienna chat with the Israeli Water Commissioner, who had pledged to maximize Jordan's share in the Yarmouk that summer.

"But we delivered on our pledge," said Rosenthal, "and we have increased the share to Jordan since late last May as we pledged. There are violations by Jordanian farmers who started pumping water from the Yarmouk downstream from the diversion, in effect taking water from the Israeli share. We then stopped our measure that increased your share," said Rosenthal, promised to consult with his superiors so that he could provide an answer.

The Head of the Israeli delegation, Ambassador Elyakim Rubinstein, found in watersharing an opportunity to put the Jordanian transient need as a topic on the negotiations agenda to take advantage of it. He wrote a letter to Dr. Majali about the subject, but Majali rejected the letter and explained that the matter of the current water sharing was not on the track agenda, but was rather a temporary measure that could be worked out between experts. The Jordanian team leader took up the issue with Rosenthal in three successive meetings and succeeded in producing two documents, shown below. Non-Paper # 1 was the first document signed by an Israeli official testifying that the "ongoing diversion regime" allows only 30% of the summer flow to bypass for Israeli use. Non-Paper # 2 paved the way to the construction of a "temporary" diversion weir, something Jordan had been attempting for so long. The contents of the first non-paper were implemented promptly, but the contents of the second were not taken up by the Jordanian officials in charge as promptly as was hoped. A series of water measurements was to be made before action could be taken to build the weir.

"NON-PAPER[307] # 1

September 3, 1992

To: Dr. M. Haddadin, Jordanian Delegation
From: Dr. E. Rosenthal, Israeli Delegation

1. Israel is willing to have the Yarmouk flow diverted to Jordan to a maximum limit of 3.5 million cu.m during the period ending November 15, 1992.
2. This arrangement is temporary. It will not disturb the ongoing diversion regime of the Yarmouk waters under which 30% of its flow in summer is allowed to bypass the Adasiyya tunnel for Israel's use. It shall also not prejudice the ultimate water shares that either party will be entitled to upon full development of the river.
3. The means of implementation of the increase in the flow diverted to Jordan shall be as will be agreed upon by experts in the field.

Signed
E. Rosenthal"

---

[307] The term "non-paper" was used to avoid any commitments that may ensue because of the contents, and to avoid formal exchange of memorandums between the two adversaries before a peace agreement was reached.

"NON-PAPER # 2

September 3, 1992

To: Dr. M. Haddadin, Jordanian Delegation
From: Dr. E. Rosenthal, Israeli Delegation

1. The following is meant to reflect the understanding that we have under which the experts shall jointly examine and initiate measures needed to have a temporary weir constructed across the river at Adasiyya to improve the efficiency of controlling the diversion of its summer flow and its regulation.
2. If this structure will prove not to be compatible with the requirements of diversion under the current field conditions or the conditions of successive development of the river, it shall be dismantled and replaced with another suitable structure.

Signed
E. Rosenthal"

8. The Washington Encounter

The third round of multilateral talks on water resources was due to convene in Washington, D.C. on 15 September 1992 when the Bilateral negotiations were also in session. The Israeli team in the Bilaterals assured the Jordanians that the Water Commissioner, Dr. Dan Zaslovsky, would head the Israeli delegation to the third round and would bring with him to Washington the associated maps and data for resolution of the issue of permanent water sharing. The news was encouraging because they reflected what Zaslovsky had pledged to Haddadin in the Vienna Chat four months earlier. If the two sides succeeded in resolving the water issue, they would achieve a breakthrough.

Haddadin reported this to His Majesty King Hussein, as the Jordanian delegation visited him in his house in Maryland on 4 September 1992 after he was discharged from the Mayo Clinic where he had stayed for surgery to remove the left ureter.

"Sire," Haddadin said to the King as members of the Jordanian delegation were bidding him farewell, "it sounds like we are likely to make good progress on water in a few days. It may be possible to settle our water dispute with Israel on basis of the formula under the Johnston Plan."

"This is good news," said the King and he looked at Majali, "good news indeed."

"I had talked the thing over with the Water Commissioner of Israel during the second round of multilaterals in Vienna, and it looks like he will be prepared to strike a deal when he comes here on the 13th of this month."

"Good, I think you should have that reported to Sidi Hasan in Amman. Since there is now a recess in the bilaterals, it would be a good idea if you reported that in person."

"Yes, Sire," Haddadin abided.

Awn al Khasawneh was also instructed to go back to Amman and report the status of the other talks. The two delegates boarded a British Airways flight to London, on 5 September, and then on to Amman via Berlin with Royal Jordanian Airlines. They briefed the Crown Prince, who brought to the meeting the senator and former Prime Minister, Taher Masri, and other officials. The mood became optimistic, and they all were anticipating progress in the negotiations, as the Labor coalition had taken office in Israel. The two delegates returned to Washington, D.C. on Saturday, 12 September, ready to resume the Bilateral negotiations. Simultaneously, Haddadin had to head the Jordan delegation to the Water Resources Working Group of the multilaterals.

In preparation for that round of multilateral talks, Haddadin had written four position papers in Jordan's name on the various topics on the agenda of the multilaterals, and submitted them to the Washington Round. It turned out that Jordan was the only participant who had prepared detailed position papers, and fully responded to the decision taken in Vienna by the Water Group to have such papers prepared. The Jordan delegation to the multilaterals[308] arrived in Washington, D.C. and checked in the Washington Hotel next door to the Willard, for lack of space at the Willard where Jordan's bilateral delegation stayed.

On Sunday 13 September, Dr. Zaslovsky called on his Jordanian counterpart at the Willard. Haddadin was embarrassed to be seen meeting with the head of the Israeli delegation outside the negotiations halls. He went down to the Lobby and politely declined Zaslovsky's proposal to hold a preliminary meeting on the multilaterals, and to talk about the bilateral negotiations on water. Zaslovsky put his counterpart in an awkward position. Jordanians never met with the Israelis outside the negotiation venue, nor were they about to start. Such moves would indicate a willingness to normalize relations between enemies before peace is achieved.

At the State Department, the Jordanian head of the multilateral group, Haddadin, had an appointment with the gavel holder of the Water Resources Working Group, Mr. Alan Keiswetter. The gavel holder normally had prior

---

[308] The delegation was composed of Dr. Munther J. Haddadin, as head, and members, all of the Ministry of Water and Irrigation, were: Dr. Kamel Radaydeh, Mr. Zafer Alem, Mr. Ziad Elias, and Dr. Hazim el Nasser.

talks with the parties to go over the program of the multilateral sessions, and to make sure there would be no surprises.

"I understand," said Mr. Keiswetter, "that there is movement toward progress in the bilateral negotiations. If this is the case, I think it calls for parallel movement in the Multilaterals in the direction of approving some joint projects."

"I believe it when I see it," Haddadin answered: "If you have news of a forthcoming movement on the bilaterals I would appreciate it if you would share it with me."

"No, not really. I was just repeating a rumor I heard. It could be anything."

Haddadin sensed that Mr. Keiswetter was eager to score some progress in his Working Group of the Multilaterals. Such progress would be to his credit, undoubtedly. The Jordanians were eager to help, but not at the expense of any of the principles that guided Jordan's conduct in the Multilaterals. He reminded Kieswetter of Jordan's slogan, raised in Moscow in January: "Restoration, Mitigation, and Cooperation".

The Bilaterals were resumed the next day, Monday. The expert group on water listened to what Mr. Zaslovsky had to say, and he did not have any outstanding new positions. He talked about water sharing in a context of joint cooperation in the field of water resources, and engaged with one of the Jordanian team members in a debate of fruitless results, and rather harmful to the chances for progress. Zaslovsky carried no maps with him nor did he propose a solution to the water sharing issue, as he had pledged in Vienna.

"You are asking me for 170 mcm per year of water," said Zaslovsky, "this is way too much." He was referring to the share earmarked for the Kingdom from Lake Tiberias (100 mcm) and to the amounts in excess of 25 mcm the Israelis were using from the Yarmouk water (70 mcm).

"I am not asking you for charity, Dr., I am demanding that you stop using water that is rightfully Jordan's," Haddadin responded.

"Besides," resumed Zaslovsky, "who told you that our share in the Yarmouk is only 25 mcm per year. Our share is almost double that quantity."

"If this is your position, it definitely is a non-starter for negotiations," said Haddadin, and the meeting ended with disagreement and disappointment.

Zaslovsky's position had changed. It was unlike his talk in Vienna on 12 May before the Likud lost to Labor in the June elections. Something was the matter. Outside, the Jordanians heard from the Israeli delegates that Dr. Zaslovsky would be replaced by a new Water Commissioner, and the name of Shaul Arlassaroff, a competent Israeli engineer then working with the World Bank, was circulating as the likely replacement. Arlassaroff had served as Deputy Water Commissioner in Israel, and resigned his post after the Likud coalition took office in 1977. If Zaslovsky had become a lame duck, he would be unable to force his views on the political leadership.

The multilateral meeting convened on the following day, on the same floor of the State Department where the bilateral negotiations between Israel and the Palestinians and Jordanians were simultaneously taking place. The seating arrangement in the Multilaterals allowed for the Israeli delegation two seats at the table for two heads, one for the incumbent and the other for the new head that would succeed him. When the Jordanians demanded similar treatment the Israelis adjusted their seating arrangements to have only one head sit at the table, the incumbent.

The opening speech of the gavel holder hinted at "the positive atmosphere" in the Bilateral negotiations, and so did the speeches of the delegates of the European Union, Japan, and Canada. The Jordanians felt that the speakers had coordinated their speeches in praise of progress made in the Bilaterals to put pressure on the Arab side (the Jordanian-Palestinian Joint Delegation) to accept joint projects with Israel within the Multilateral conference in that third round. Then came the speeches of the core parties. The Israeli speech was optimistic and gave the impression that things were going on in the bilaterals without difficulties, and that progress was anticipated. The head of the Jordanian delegation had a prepared speech that contained reservation against any claims of a breakthrough.

"We believe it is only fair for each party to take of the peace responsibilities his own share. We further believe that any attempt to ignore the causes of conflict in our region is an attempt to retard our work in this group. Consideration is to be given to the prerequisites for regional cooperation which are, as we have stated in Moscow and in Vienna, restoration of rights and mitigation of damages."

"Unfortunately, despite the positive tone we have just heard, we were until yesterday not closer to these pre-requisites than we have been since the Madrid Conference."[309]

"Unless," he added on, "we consider our having coffee together with the Israelis and exchanging a few jokes as progress!" The Jordanian reservation concerning progress in the Bilaterals did not please many.

The Palestinian speech flagged the monopoly Israel had been exercising on water data and water management in the occupied territories. In the discussions that ensued, the Israeli head of delegation, Dr. Zaslovsky, kept referring to the occupied West Bank as "Judea and Samaria". Arabs are sensitive to Israeli usage of such biblical names, because such use indicates Israel's intention of changing the Arab character of Palestine and replacing it with a Jewish character. In particular, the issue of the occupied Palestinian

[309] "Opening Statement," by Munther J. Haddadin, Head, Jordan Delegation to the Water Resources Working Group, Multilateral Peace Talks, Washington, D.C., 15-17/9/1992. Records of the Peace Process, Jordan's Ministry of Foreign Affairs, Amman, Jordan.

territories had to be kept alive. The term "West Bank" has always been used to describe the Palestinian part of Palestine that was rescued in the 1948 war by the Jordanian and Iraqi armed forces. The Jordanians were not about to accept the description of the West Bank as Judea and Samaria.

The head of the Jordanian delegation urged the head of the Palestinian delegation to respond, and not let this Israeli terminology take root in the multilaterals.

Al Khudary, head of the Palestinian delegation, asked for the floor, and made an intervention refuting the term that Zaslovsky had used in reference to the West Bank. His colleague, Dr. Marwan Haddad did likewise, and finally their colleague, Dr. Abdul Rahman Tamimi, made a third intervention that indicated displeasure with the term Judea and Samaria in reference to the West Bank. The discussions continued and the Palestinians told the audience that Israeli officials deny them access to water data and that they were at an obvious disadvantage. That was why they could not prepare the working papers that the Vienna meeting requested. Data availability was a primary item on the standing agenda of the meetings.

"All that the people of Judea and Samaria have to do to get data," said Zaslovsky defiantly, "is to just ask for them!"

The Jordanians thought Zaslovsky went too far by stressing the use of Judea and Samaria despite Palestinian protests. They were aware of the limitations that the Palestinian delegation had. The Palestinians were under occupation and were subject to bad treatment by the Israeli authorities upon return if they did not behave well. The head of the Jordanian delegation, Dr. Haddadin, asked for the floor.

"Gentlemen," he addressed the attendants, "we have heard the distinguished representative of Israel use archaic terminology in modern time. What we all know as the West Bank suddenly became Judea and Samaria. The recognized owners of the land have protested at no avail. I seek your understanding and patience, because if he repeats the use of that terminology, I will have to refer to the entire lands between the Mediterranean and the Jordan River as Palestine."

That intervention was like a bomb. The Israelis and the Jordanians exchanged the floor and their points of view, mainly political, were aired. The debate lasted about twenty minutes, after which a member of the Israeli delegation from their diplomatic corps, Ambassador Amos Ganor, took the floor.

"We are sorry," said Ambassador Ganor, "we did not have in mind the political meaning that Dr. Haddadin interpreted in our use of the term. We just used the term that is normally used by our people."

Haddadin took the floor and responded, "I accept the apology. But I also assure you that I have used the term that is normally used by our people: Palestine."

The debate ended with the gavel holder announcing that sensitive political issues should be avoided, and announced the break for coffee.

The efforts of the first day materialized into agreement on the installation of two desalination pilot plants. The Japanese expressed readiness to finance one in Jordan and the European Union expressed their intention to sponsor one in Deir el Balah, in the Gaza Strip.

As head of the multilateral team on Water Resources, Haddadin reported to the Foreign Minister. Simultaneously, he was a member of the bilateral delegation under Dr. Majali, and Majali had no part in the multilateral talks. It was late in the evening when Dr. Majali sent for Haddadin.

"What happened today?" Asked Majali.

"It was multilaterals, not bilaterals," Haddadin answered jokingly.

"No, seriously, I want to know."

"But why?" Haddadin responded, equally seriously, "I just faxed my report to the Foreign Minister."

"Because what happened in the multilateral session this morning affected our work in the bilaterals."

"Seriously? But how?" And then Haddadin remembered that an Israeli delegate had left the hall when he was delivering his opening speech.

"Yes," said Majali, "Rubinstein and I were having our morning-one-on-one meeting. He spoke highly of you and your abilities and said, among other things, that he wished he had two like yourself in his own delegation. I thought that was a nice compliment."

"He must be after something. Israelis do not distribute compliments for free," Haddadin commented.

"Suddenly," resumed Majali, "there was a knock at the door, and someone called him. He excused himself for a few minutes and came back, visibly angered and his face turned pale."

"What is this Dr. Haddadin of yours!" he protested, "he is saying in the Multilateral session that you and I are wasting our time sipping coffee and exchanging jokes!"

"Rubinstein then suspended the meeting," resumed Majali, "and went to consult with his delegation. He made a few contacts before he emerged into the corridor where he met Dr. Tarawneh and told him that he would file a formal complaint with the sponsor against Haddadin."

"I do not work for the sponsor," Haddadin said, "and if the sponsor accepts such complaints it would be to our advantage. I would file a daily complaint about the Israelis hurting our feelings because of their conduct in the occupied territories and in Southern Lebanon. The sponsor then has to have a special office for complaints!"

Haddadin briefed Majali on what happened, and gave him a copy of his speech with the comment about the coffee and the jokes added. This encounter triggered a clash with Majali the next morning, as the bilateral delegation had its morning session before going to work. Majali wanted to discuss what happened at the multilaterals in the meeting of the bilateral delegation, and Haddadin refused, because the Multilaterals and any of the issues therein were under the jurisdiction of the Foreign Minister.

"Remember Pasha, we are negotiating with the enemy. I do not expect any of the Israelis to wish either of us any success, because our success will not be to their credit. Besides, they do not deliver on their pledges, and they tend to change their mind."

The Bilateral round ended on 23 September with no breakthrough of the type that was expected. The Jordanians were actually disappointed, and braced themselves for an extended era of negotiations.

9. A Visit by the United States Ambassador

The Ambassador of the United States to Jordan, Mr. Richard Harrison, paid a visit to Haddadin at his home on 6 October 1992. His aim was to see how prepared Haddadin was to attend an international water conference, organized in Jerusalem by the Truman Institute in Israel, to discuss the water matters between Israel and the Palestinians. Haddadin had already advised the Minister of Water and Irrigation, Mr. Samir Kawar, to object to an international conference in occupied Jerusalem, and both were of the same opinion. The organizers therefore were considering a different venue for the conference.

The Ambassador was ready to have the conference shift venues to Switzerland, if Jordan would attend. But Haddadin was of the opinion that it wasn't right to involve Jordan and Israel in a seminar over Palestinian issues, and recommended that the conference be a local seminar involving Palestinians and Israelis.

"Could you probe the position of the government if the conference was held in Austria?" Asked the Ambassador.

"We are confused here, I admit. Who is going to extend the invitations to attend the conference? Is it the United States? Is it the PLO? Or is it the organizers [the Truman Institute]?"[310]

The Ambassador was taken by surprise at the question. "I want the conference to be held with Jordanian participation, and the PLO has nothing to do with it. The Jordanian participation, I think, is a must because the solutions will be within the Jordanian context."

"Please say that again," Haddadin requested.

---

[310] Obviously, Jordan would not respond to an invitation from an Israeli party before peace was reached and diplomatic relations established.

"I do not know if I made a mistake," said the Ambassador, and he did not repeat what he said.

Ambassador Harrison shifted to ask about the last bilateral round of negotiations and Haddadin expressed his dismay at the Israeli failure to deliver on their pledges.

"Do you trust their experts?" Asked the Ambassador.

"No, I do not."

"What makes you so negative about them?"

"They have not responded to the Jordanian proposal that calls on both parties to accept the compromise solution that Ambassador Eric Johnston worked out in 1955. They even insinuate that they do not accept it."

The Ambassador stressed that the United States did not support the implementation of the Johnston Plan because the concerned riparian parties did not accept it. The Ambassador's remark was an explicit expression on behalf of the United States, distancing herself away from the Unified Plan that she so carefully had engineered.

"Jordan," Haddadin responded, "had bent under American pressure and committed herself in 1958 not to withdraw from the Yarmouk more than the quantities allotted to her under the Johnston Plan. That was an American condition precedent to extending aid to build the KAC project in Jordan. The condition was met, and American assistance was given in grants to build the irrigation infrastructure. The farmlands were developed using private Jordanian capital, and hundreds of millions of Jordan Dinars were invested over the years. I cannot imagine that the United States would so casually turn her back to the Plan that formed the basis of her involvement in financing water projects in the Jordan basin, both in Jordan and in Israel. The United States advocated the Unified [Johnston] Plan as the best available plan for equitable utilization of the waters of the Jordan River basin."

"But I am sure the United States does not support the Johnston Plan anymore," insisted the Ambassador.

"How would you differentiate between the United States as a partner in the development of water resources of the Jordan basin, and the United States a peace sponsor?" Haddadin asked.

The Ambassador did not answer, but one of his assistants, Mr. Stewart, volunteered an answer by saying that the United States was an honest broker.

"I will keep up my efforts to get an explanation of how the United States reneged on her announced positions regarding the Plan she promoted and we adhered to. On that basis, a huge amount of Jordanian capital has been invested. I will start a campaign in the United States and seek finance for it from NGO's. I just cannot imagine Jordan settling for less that her share according to that American plan."

"Failing that," Haddadin continued, "we will have to start from the very beginning, and see then how long it would take us to make peace. It looks like we have to start by negotiating with the United States first."[311]

Haddadin was distressed to realize that he had to fight an uphill battle not only with Israel, but also with the United States. It was evident to him that the involvement of the United States in the water issues would be to the benefit of Israel. He decided to do his best to keep the United States as far away from the water negotiations as he possibly could. He also decided never to sound weak when talking to the American officials, nor to ask for their help to overcome the Israeli stubbornness.

He thought that if the Israelis and the Americans do not come to terms, he should be in no hurry to make peace. Actually, he'd rather wait until the situation of the Palestinians clears up, instead of Jordan making peace with Israel before the Palestinians did. For if this happened, the Palestinians would be left out in the cold.

The words of Ambassador Harrison rang in his ears: "the solutions would be within the Jordanian context!" and, "I do not know if I made a mistake saying that!"

"Could it be that we will be dragged to take part in the settlement of the Palestinian-Israeli dispute?" The Jordanians knew, from the Israeli attempts to keep Jordan linked with the Palestinian track, that the Israelis favor Jordanian involvement in the Palestinian solution, but the Jordanians rejected all such attempts. For such a notion to come from the United States was strange, and did not speak for the possibility of Palestinian independence, nor did it speak well for the honesty of the broker!

"What did we do to deserve all this?" Haddadin thought to himself, "We are losing our water to Syria; we are losing it further to Israel; the United States is now reneging on the Plan that made us invest so heavily in the development of the Jordan Valley depending on the Yarmouk river. I do not know how the Palestinians feel about my strategy to recover their water rights from Israel and return it to them when they are ready, they probably would resent any Jordanian involvement in their affairs. I know Syria would not love what I have in mind in terms of argument against the 1987 Agreement. Whom do I turn to for support?"

He briefed the Minister of Foreign Affairs about the situation and, being recruited from outside government ranks, Haddadin did not want to carry the entire responsibility for drawing up Jordan's water negotiation strategy, especially after he heard what the American Ambassador had to say.

---

[311] Memorandum dated 8 October 1992 by Munther J Haddadin on the meeting, addressed to the Minister of Foreign Affairs of Jordan, copied to the Head of the delegation for bilateral negotiations, and to the Minister of Water and Irrigation.

"Excellency," he said to the Minister of Foreign Affairs, "I am afraid you have to issue me instructions on what to defend against whom as far as water is concerned. I want to know what you expect me to do."

"Well, my friend," said the Minister, "If we are to instruct you on water negotiations, we have to first seek your advice on what to tell you. This means we will follow what you design for the country as far as water is concerned. No one can advise us better than you can. It is all yours."

This added to the burden on his shoulders, and he had to look in all directions to defend Jordan's water rights. He also hoped that the Jordan Government would support him all the way. The Minister's answer, at least, seemed to justify his hope.

CHAPTER IX

The Common Agenda

1. Need for a Breakthrough

The seventh round started in Washington on 21 October 1992 and ended on 19 November 1992, with an interruption between 30 October and 9 November because of the Presidential elections in the United States. The first anniversary of the Madrid Conference was quickly approaching without any breakthrough in sight. Dr. Majali longed for such a breakthrough to mark the first anniversary to give the peace process a boost and avoid stagnation. During this 7th round, it felt like a new phase could be in the offing. The Republican President George Bush lost the elections to the candidate of the Democrats, William J. Clinton. This loss meant that the champions of the peace process on the American side, President Bush and Secretary James Baker, would soon be out of office, and the commitment of the new administration to a just peace in the Middle East was not yet tested or known. In the Middle East, Israel escalated the military activities against Southern Lebanon, and violence erupted in the occupied Palestinian territories (West Bank and Gaza). These activities created hardships on the negotiation tracks with Syria, Lebanon and the Palestinians, and irritated the Jordanians accordingly. The Jordanian delegation focused on the draft Common Agenda, in the hope that its conclusion would provide the needed breakthrough.

On Tuesday, 27 October the Jordanian-Israeli track was to meet in the afternoon. Majali and Rubinstein agreed in their preparatory meeting to have a small group of the two delegations sit down to negotiate the Common Agenda draft, for as long as was necessary to come up with an agreed version. Rubinstein brought with him into the negotiations room Robby Sabel, the legal expert, and Ahaz Ben-Ari from the military judiciary branch. Majali took with him Haddadin, who had composed the first draft of the Common Agenda back in April, and Awn al Khasawneh, the legal expert. The proposals for a Common Agenda had been going back and forth between the two delegations since the fifth round in April.

Agreement on the first section was reached after a short debate. This section (A) of the draft set the goal of negotiations as arriving at a just, lasting and comprehensive peace between the Arab States, the Palestinians, and Israel in accordance with U.N. resolutions 242 and 338 in all their aspects. The Palestinians were mentioned as a party to the comprehensive peace. Section (B) had the title: "Components of Jordan– Israel Peace Negotiations", and had nine items under it. The first concerned searching for practical steps to arrive at a state of peace, as envisioned in Security Council resolutions 242 and 338. The second item was 'Security', which had five sub-items. The third was

'Water' and had three sub-items. The fourth was 'Refugees and Displaced Persons' with two sub-items. The fifth was 'Borders and Occupied Lands'. The sixth was 'fields of potential cooperation'. The seventh pertained to phasing of the implementation and the establishment of appropriate mechanisms for negotiations. The eighth was the discussion of matters related to both tracks (Jordanian and Palestinian) to be decided upon in common, and the ninth item of Section (B) addressed the conclusion of a 'peace treaty'. There was no third Section at the outset.

There was a long debate over a sub-item of the 'Security' section that addressed weapons of mass destruction, and it was left for a revisit before the end of the session. A problem erupted when the parties took on item B-5, 'Borders'. The draft contained language that defined the borders between Jordan and Israel with reference to the British Mandate borders between Transjordan and Palestine, (defined in 1922 and amended in the south near Aqaba in 1946). The language went on to say at the end of that sub-item, "Both sides shall respect these borders." It was mostly a Jordanian text.

"Ambassador," said Haddadin, "I have a problem with the word 'respect' in this text"

"But it is a Jordanian text!" Said Rubinstein.

"I realize that, but I have a problem with it."

"What is your problem," said Rubinstein, with a resentful tone.

"I know that you have borders with Egypt that you respect; I came to note yesterday's statement by Mr Lobrani, the Israeli head of delegation in the Lebanese track, in which he said that Israel respects her international borders with Lebanon. My problem is that your tanks are running across those borders every other day despite the respect you have for them!"

"What!" Exclaimed Rubinstein, "Do you compare Lebanon with Jordan? Lebanon is run by Syria. Let me tell you what the Lebanese could not do. I was there during the drafting and negotiation of the Israeli–Lebanese Accord of 17 May 1983. The Lebanese agreed to it, but they were never able to have it ratified by their President because of Syrian pressure. Lebanon cannot decide for itself. Lebanon is not a country!"

"Lebanon," Haddadin interrupted, "is as good a country in the region as any other. We do not allow, nor do we accept to hear statements like this said about Lebanon in our presence. Ambassador, you should take back your statement!" And he leaned over to Majali, and whispered, "Pasha, he had Lebanon for breakfast, and he will have Jordan for lunch if we let his statement go without contesting it." And Majali nodded.

"Rubinstein," said Majali, "this is going way too far. We will not allow the scratching of the sovereignty of Lebanon or the dignity of her people. You should retract what you said."

"I am saying the truth, and the truth can not be masked by pretending that Lebanon today is sovereign. Pasha, Jordan is different. Jordan has leadership admired by the whole world. Please do not take offense in what I said."

"What you said is beyond the permissible levels in any form of diplomacy. You should 'withdraw' what you said," insisted Majali.

"Pasha, for heavens sake," said Rubinstein, who often cracked jokes, "do not bring us back to the stories of withdrawal. We have had enough debate over withdrawal under Resolution 242. Now, please, let us get on with our business."

Majali, "If you do not retract what you said, I am afraid I will not continue participating in this session."

After a brief silence, Majali stood up, "I am going to the men's room," he said and walked out of the room. Majali did not return to the negotiation room, but headed for the Jordanian delegation room where the rest of the delegates were waiting. He showed mixed signs of relief and worry.

"What is the matter?" asked Dr. Anani.

"A minor problem," replied Majali.

"Aren't you going back in?" asked Al Qadi.

"No, no," said Majali, and continued with his Jordanian southern Karaki accent, "I left them with Munther *mistabeehum fi hjar* (throwing stones at them) with visible success. I will let him carry on!"

And so it was. Al Khasawneh and Haddadin played complementary roles in concluding an agreement on the Common Agenda. Haddadin acted the tough senior negotiator, while al Khasawneh played the role of a diplomat who was after results without undue hardships.

The last sub-item (number 9) in Section (B) spoke of concluding a peace treaty after agreement was reached on the 'above items' (the eight items listed in Section B). Agreement on Section (A), the goal, was thus not included. In a sneaky move, Haddadin marked that sub-item as C instead of the number 9. In other words, he made it a third section of the Common Agenda after Sections A and B. Reference in it to agreement on 'the above items' would now include Sections A - the achievement of a just, lasting and comprehensive peace, and all eight items under Section B that the draft actually intended. That meant that a peace treaty would be concluded only after agreement on both Section A and Section B was reached. The Israeli delegates did not take note of this change, i.e., the changing of the number 9 into the letter (C), nor did they bring up any points related to it. Hence, as the Agenda ultimately indicated, a peace treaty between Jordan and Israel could be concluded only after the achievement of a just, lasting and comprehensive peace as specified in its Section (A), in addition to reading agreement on the 8 items of Section (B).

This marathon session lasted five hours. The agreed Common Agenda was redrafted in clean handwriting, and the Jordanians called in the Palestinian delegate, Abdul Rahman al Hamad, to consult with him before they initialed the text. The Palestinian delegate said he had comments that involved the Jordanians and the Palestinians only, and not the Israelis, and would say them in private.

The doors of the room were opened, a sign of completion of the session. Majali entered with the rest of the delegation. He read the text, consulted with both al Khasawneh and Haddadin, and was clearly very pleased.

The handwritten text was initialed (see Appendix 9) and copies were kept by each delegation. It was after nine o'clock in the evening by then, and the Press was waiting outside, by the C-Street Entrance, for their usual briefing by the heads of delegations. Majali headed towards the C-Street exit, followed by the members of the Jordanian delegation. Haddadin stayed behind to talk to al Hamad in private, and took him into the Palestinian delegation room across the hall from the negotiations room.

"What is it that bothered you in the Common Agenda?" Haddadin asked.

"Oh, nothing, it is just item (B-4)- Refugees and Displaced Persons," said al Hamad. "Why should Jordan negotiate their case?"

Haddadin understood the dilemma, and assured al Hamad that the sub-item, as it read, called for resolution of the *bilateral aspects* (i.e., Jordanian-Israeli aspects) of the problem of refugees and displaced persons in accordance with International Law.

"Don't worry, we will not step on anybody's toes," he noted to al Hamad. "We will negotiate only the bilateral aspects of that problem as the text reads. Remember, you have had this text since last April, six months ago, why haven't you raised a question about it all this time? Is this now the proper time to raise questions- six months later?"

Al Hamad was not convinced. "Why should Jordan negotiate for the refugees and displaced persons?" He asked.

"The bilateral aspects refer to the refugees in Jordan, and as you well know they carry Jordanian citizenship. If Jordan does not speak for them, who would?"

"The PLO," responded al Hamad.

"I agree with you but only with one condition. You and I will go door to door knocking in all the refugee camps in Jordan. Those who choose the PLO to speak on their behalf are entitled to their choice. You will have to buy him and his family the airplane tickets, and you shall take them to Tunis, where the PLO resides," Haddadin said.

Al Hamad was not happy with the explanation, nor was Haddadin happy with the attitude. He hoped that the view al Hamad expressed was personal, just like his own view was, and that it was not the formal position of the PLO, and that the Jordanian leadership would have the same view that which Haddadin had expressed.

2. Publication of the Draft Common Agenda

With the conclusion of the Common Agenda, Majali had his great moment. Finally, a breakthrough was achieved under his leadership to mark the first anniversary of the Madrid Conference. His briefing to the press by the C-Street entrance betrayed his satisfaction, despite his attempts to conceal the achievement. The governments in Jordan and Israel had to be informed first, and their approval of the text of the Common Agenda had to be secured.

Back at the presidential suite that Majali occupied at the Willard, the activity was like that of a beehive--people on the phone, others drafting reports, and others recollecting what had been attempted to produce a Common Agenda since the beginning of the negotiations. Anani was drafting the report to Government for Majali to sign, and a question was put to the Prime Minister in Amman. The question was:

"In our press conference to break the news of the agreement on the Agenda tomorrow. What is your instruction? Shall we play it with a low, medium or high pitch?" There was no question in Majali's mind, or in the mind of any of the delegates, as to the need to break the news before 31 October, the Madrid Conference anniversary.

In the meantime, Majali got busy with Marwan Muasher, the delegation spokesperson, on making arrangements for a joint press conference on the following day where Majali, Rubinstein and Djeridjian, the U.S. Deputy Assistant Secretary of State, would appear and break the good news to the world. Marwan was asked to make arrangements for, and work out the details of, that press conference. After midnight, when the delegates were gone from Majali's suite except for Marea, the Assistant Chief of Staff, and Haddadin, the latter looked at Majali and wondered.

"Don't you think, Pasha, that our public in Jordan should hear the news from our own media, and not through the international press? I say so in particular because we have Jordanian correspondents accompanying the delegation, I think we should give the edge to those correspondents who are with us here. We owe it to them and to our own people," he said. Majali looked at him with a big smile, and agreed.

On the way to his room at about 1:00 a.m., Haddadin came across Caroline Faraj, a clever Jordanian journalist, in the hallway. She was the correspondent of the leading Jordanian daily, *Al Rai*. Having obtained the consent of the head of delegation, he handed her a copy of the Common Agenda, and asked her not to have it published until further instructions. He also passed by the correspondent of *Petra*, the Jordan News Agency, left him a similar copy under his door, and retired for the day.

The next track meeting was scheduled at 4:00 p.m. the following day, Wednesday, 28 October. The two delegations were to ratify the initialed Common Agenda. At about 10:00 a.m. that morning, Marwan Muasher came into Majali's suite with disturbing news from the region. Israel had been shelling from ground and sea, and bombarding from the air, several towns in South Lebanon and the mood in the region was that of anger. For how dare the Israelis commit such atrocities while, at the same time, negotiate peace with the Arabs, including Lebanon. Muasher was of the opinion that a press conference of the type Majali was proposing would not be in Jordan's best interests in light of these developments, especially as there had been no progress achieved on any of the other Arab negotiation tracks. There was debate, in which other delegation members participated, but the result was inconclusive. It sounded like Majali was eager to break the news, at least through his regular briefing of the press after the track session.

The mood in the delegation was not an easy one. Haddadin then called the Jordanian press correspondents to whom he had given a copy of the Common Agenda and instructed them to freeze everything until they receive a call from Muasher or himself, in due time that afternoon, as to the publication of the news or not. He wanted to be sure of the decision to announce the Common Agenda.

There was no answer back from the Prime Minister in Amman. Majali contacted the Prime Minister in the afternoon before the track session started, and he was told that the agreement on a Common Agenda had to be kept discreet, and the press conference was not approved. The Jordanian press correspondents were instructed accordingly and the Common Agenda was taken back from them. In the track meeting, the two parties reviewed the typed Common Agenda and made sure it coincided with the handwritten one. The meeting approved it and mutually agreed not to leak a word about it to the press. The Jordanians were fearful of the Israeli media, to which all the news of the negotiations was consistently leaked.

Early Thursday morning, Majali was awakened by a call from Dr. Kamel Abu Jaber, the Minister of Foreign Affairs. Abu Jaber was upset because the Common Agenda, over which no word was to be leaked, was published in full in the Jordan Times, and was translated and published by *Al Rai* newspaper into Arabic. Majali was surprised that such a thing would happen contrary to the understanding shared by the delegates the day before. It turned out that Faraj, *Al Rai* correspondent, had wired the Common Agenda right after she laid hands on it after midnight. When she contacted the Chief Editor of her paper the next day to tell him of the new instructions, he defied the instructions and had the Common Agenda published.

The publication of the Common Agenda was a big surprise to all. Most surprised was Fayez Tarawneh, who was in Paris attending the meeting of the REDWG (Regional Economic Development Working Group), of the

Multilateral Peace Talks. As he came out of the meeting for a recess, the press was waiting for him.

"Dr Tarawneh, how easy was it for you to come to agreement with the Israelis on the Common Agenda?" Tarawneh was able to hear among the confusing shower of questions.

"What Common Agenda? What agreement?" Tarawneh responded.

"Sir, it has been published, Jordan and Israel agreed to a Common Agenda for negotiations," said one correspondent.

"Really? Seriously?" asked the bewildered Tarawneh, who was supposed to be the deputy head of delegation; another embarrassing situation indeed.

Rubinstein was very unhappy. While he publicly denied any substantial progress with the Jordanians as agreed, *Al Rai* newspaper carried the news to the contrary, and the Israeli press was up in arms because they were denied an equal opportunity.

Haddadin explained to Majali what happened, and took the blame for the situation. Majali was called to Amman to participate in meetings with leading PLO figures and defend the text of the Common Agenda.

3. The Reaction in Jordan

The *Jordan Times* reporter Ayman Al Safadi, in the 8 November 1992 issue, reported the reaction of the Moslem Brotherhood bloc in the Jordanian Parliament. Safadi wrote, "In a strongly-worded statement against what it called 'The Jordanian-Israeli agreement', the parliamentary office of the bloc said the informal draft agreement reached by Jordanian and Israeli negotiators in Washington on October 27 bodes ill for our national unity and inter-Arab relations."

The bloc's statement refused the draft agenda, calling it purposefully an "agreement"; on the grounds that it surrendered Arab rights in Palestine, squandered the rights of Palestinian refugees and helped realize the Zionist dream. These statements by the Moslem Brotherhood parliamentary bloc were made while four Palestinian leaders were reviewing the text with the Jordanian government. On 8 November, Jordanian Prime Minister Zaid Ibn Shaker received the Palestinian delegation, which consisted of Yasser Abed Rabbo, Yasser Amr, Suleiman Najjab and Faisal Al-Hussaini. The meeting was the last of a series of earlier meetings that took the draft agenda to task, going over every detail.

Majali participated in the meetings in Amman to discuss the concerns of the PLO and Chairman Arafat over the Common Agenda. The PLO officials came to Amman and voiced their objections to the speed at which Jordan was progressing in negotiations with Israel, and to three items in the published Common Agenda. The first pertained to Item 3 on Water, where they claimed that the Palestinian rights were overlooked.[312] Majali assured them that the matter was purely bilateral between Jordan and Israel, and that as such, the Palestinian rights were not on the negotiation table. The second was on Item 4 - Refugees and Displaced Persons, where the PLO thought that they should be the party to negotiate their problems. Majali presented that the Agenda item, addressed the bilateral aspects of their problems, and stated that the matter is Jordan's concern, especially as these refugees all have acquired Jordanian citizenship. Jordan will negotiate their rights, and if the PLO later would bring more rights to the other Palestinian refugees, who are outside Jordan (in the West Bank, Lebanon, Syria, and Egypt), such gains would apply to those in Jordan. The third and last point was on Item 5- Borders. The Common Agenda specified the demarcation of the borders between the two countries with reference to the Mandate borders, "*without prejudice to the status of the territories presently under Israeli military administration.*" The PLO presented that such a text jeopardizes the status of Jerusalem, because it was not under Israeli *military* government control. Majali counter-argued that the international legitimacy of Israeli existence rested on U.N. Resolution 181 of 1947, the Partition Resolution. Any Israeli presence outside the territories allocated for the Jewish State in that resolution was the result of *military* expansion and *military control*. The PLO officials were not satisfied, and demanded that the text be amended to read "without prejudice to the territories that came under Israeli military government control in 1967."

Finally, on the issue of the speed at which Jordan was progressing, Majali said that the Common Agenda was the beginning of negotiation, not the end. The Palestinians should not be surprised at this, because they had two delegates in the Jordanian track who were handed all the documents that were exchanged with the Israelis, and Majali shared with them and the other Arab delegations all the developments on the Jordanian track.

The dialogue with the PLO in Amman settled on the need to change the text of the borders as outlined above. The attempts to have it changed would take time, which catered to the PLO desire to have Jordan slow down. Majali tried hard to obtain the government's final approval of the Common Agenda, but did not succeed. Another development factored in the Jordan government's reluctance. It was a domestic factor, and it pertained to the arrest of two members of Parliament, when it was in recess, and their trial on serious security charges.

When the new Parties Law was passed in August 1992, thus allowing the party system to be set in motion after an over thirty-five years freeze, the

---

[312] The first attempt ed draft by the Jordanians was to address the "rightful water shares of the riparians on the Yarmouk and Jordan Rivers", but the negotiations over the Common Agenda led to wording referring only to the two sides (Jordan and Israel), and not to all riparians.

public mood relaxed and hopes for democracy increased dramatically. However, that mood was in contrast with some Government actions. Security forces arrested two members of Parliament on charges of alleged involvement in underground armed organizations that were banned under Jordanian law. According to the Ministry of the Interior, Member of Parliament Yacoub Qarrash, originally from the West Bank, was arrested for "his involvement in leading the Islamist group calling itself the Shabab El Nafir [Youth of the Clarion Call], found to possess weapons and explosives."

After a few days, another deputy, Leith Shbeilat, was arrested and charged with allowing his car to be used for transporting explosives. Shbeilat was an outspoken Member of Parliament and a critic of alleged government corruption and was on the parliamentary opposition. Members of Parliament are not immune to arrest when the Parliament is in recess. They would otherwise enjoy immunity, and it would take a majority vote in Parliament to lift the Parliamentary immunity off any of its members.

The trial of members of Shabab Al Nafir, or the Youth of the Clarion Call, had captured Jordanian public interest to a level similar to the American public interest in the trial of O.J. Simpson. When the draft agenda news broke out in November, the trial was in its final stages and the verdict was to be taken by the three-man State Security tribunal, which usually handled such cases. The eruption of the news regarding the draft agenda put the government between a rock and a hard place. It became easier for the Jordanian government, at the time, to suspend approval of the Agenda in light of the opposition to it which was championed by the Islamists and found many supporters in the Public. The public was reluctant to accept progress made by the Jordan Delegation while all other tracks yielded no progress. Majali's efforts to convince the government to have it approved met deaf ears. It was then that he contemplated resignation, because he thought that the hard work that he and the delegation had put in was squandered on the political conveniences of those who did not appreciate it.

When the second leg of the seventh round resumed on 9 November Majali did not return in time to assume leadership of the Jordanian delegation.

4. Israeli Reaction

Tarawneh, in the absence of Majali in Amman, was asked to lead the delegation, after his return from the second round of the Regional Economic Development Working Group of the multilaterals that met in Paris. Tarawneh, assisted by senior members of the Jordan Delegation, did not find it difficult to thrust the Jordanian position before the disbelieving eyes of Rubinstein. The Jordanian delegation needed to change the language of the "without prejudice" clause in Item 5 of the Common Agenda pertaining to borders and territorial matters, as agreed with the Palestinians in Amman. Rubinstein kept insisting that his government was unhappy with Jordan's reluctance to approve the Agenda.

When Majali finally arrived on 15 November, he explained all the circumstances that had led to the delay in the Jordanian governments approval of the Agenda. In fact, Rubinstein was deeply unhappy because he considered both parties to have worked out a magnificent negotiation document. Furthermore, he believed that the peace treaty concept included in the draft agenda was a precious gain that was similar, but not equal to, Israel's success in arriving at peace accords, if not a treaty, with Egypt thirteen years earlier, when Rubinstein served as a young and promising negotiator. However, it escaped Rubinstein that the treaty provision, contained in Section C of the agenda, was conditioned upon achieving the other two of its sections, A and B, the first of which was achieving comprehensive peace i.e., involving all tracks—Syrian, Lebanese, Palestinian).

"You know, Pasha, our government has approved the Common Agenda," said Rubinstein, "I want to admit now that the Jordanian text as it now exists under item B-5, Borders, is stronger than the requested amendment. I will support the amendment because it will pull me out of a deep hole dug in the existing text, the issue of legitimacy and military occupation," said Rubinstein.

"Then there will be no problem," said Majali, "let us go for the amendment."

"In the meantime, I do not think it pays to engage in a language battle, nor would it change the current status. It will take a decision of our government to accept the amendment, and I think that will be forthcoming."

The two sides agreed, instead, to break into non-official expert groups to continue discussions on a range of bilateral issues, such as security, water, environment, energy, economic cooperation and the re-opening of Jordanian banks in the West Bank. Such meetings started on Monday, the 16th, with a position paper on water read by Dr. Majali (Appendix 10).

In his paper in response, Ambassador Rubinstein mentioned that both Jordan and Israel are located on the outskirts of the desert, and that there were no known water resources on which each can call upon for future increases in demand. The two sides should focus on creative means to make ends meet, and not to engage in lengthy discussions about the old ways and principles of water sharing. "It is the forward look that will benefit all," he contended. Obviously, the Israelis did not favor a consideration of the past, in which Jordan's water rights and the rights of the other riparians were rooted.

5. Israeli Proposal for Water Agenda

In the late afternoon after the track session ended, the expert group on water met. Dr. Rosenthal handed the Jordanian team leader a typed document that he said carried their proposal for topics to negotiate (Appendix 11). The document was not read in the meeting, but conversation started over the statements of positions read by the two heads of delegation in the meeting of the track session. The positions of the two sides were obviously divergent, and it was the job of the expert group to make them converge through negotiations. The Jordanian team leader, Dr. Haddadin, told the Israelis that he would come back to them with a sub-agenda for negotiations over water, energy and the environment, after the Common Agenda was put in its final text with the required amendment included. The Jordanian position was to have the Common Agenda agreed to, before they engaged in negotiations on any of its components.

"Your side referred to unilateral actions on water several times," said Rosenthal, "I want to emphasize the fact that it was not only Israel that undertook unilateral actions. We know of other parties that have taken things into their own hands and taken water.[313] What we have to do today is to face the reality and find solutions." Obviously, Rosenthal was referring to the Syrian unilateral actions on the Yarmouk.

"I assure you, Dr. Rosenthal," responded the Jordanian team leader, "that we shall never plead 'not guilty' to things we have done ourselves. We, however, are responsible only for what we do, and expect you to take responsibility for what you have done. Actions by others are not our responsibility."

The Jordanians studied the Israeli proposal that evening, and again the following morning Tuesday, 17 November. The written Israeli proposal had three elements: one on cooperation in the exploitation of water resources along their common boundary, another on transferring water across their boundaries, and the third was to refrain from construction and operation of water projects that would change existing water arrangements, such as the Karama dam in Jordan.

While the second item was unclear, the third was clear but invalid. It was an attempt to interfere in Jordanian water policy. The team leader, Dr. Haddadin, remembered the visit in 1984 of Mr. Richard Viets, the United States' Ambassador, who came to his office protesting JVA's attempts to study the feasibility of the Karama dam on the grounds that it was against the provisions of the Johnston Plan, which it was not. Now the Israelis are doing the same thing, but the Jordanians would not entertain their goal. When the group meeting convened on the afternoon of Tuesday, 17 November, they thanked the Israeli team for their initiative in proposing a sub-agenda for talks on water.

"Thank you for your initiative," said the Jordanian team leader, "but when the time comes, you will discover that our proposal for a sub-agenda is more comprehensive and more clear. The right time for our submission will be after we are through with the Common Agenda. In the meantime, it does not hurt to exchange views about matters that touch on our negotiation themes, as we enter a new negotiation era."

"Thank you, Dr. Haddadin," responded Dr. Rosenthal, "and may I protest the use of the expression "an era" to describe our negotiation period

[313] That was clear reference to the Syrian activities in the Yarmouk catchment.

because we believe that we will soon come to an understanding. It should take little time to come to positive conclusions, not a whole era."

"Good, our side is more than prepared to make the negotiation era only a short time," said the team leader, "but note that people always address the water issue in terms of shortage in water resources, as though the resources are to blame. I have not heard them blame the population levels. I think if we both agree to address these factors and review what has happened in the past that affected the water situation; review the conflicting plans and how they went into a phase of reconciliation in the past and into implementation at varying degrees, we will come to fruitful conclusions. We will realize how Jordan has been in the unfortunate position better described by the American saying, 'at the bottom of the totem pole'. Things have to be rectified and their negative impact mitigated."

"Let us exchange views on the document we gave you," said Rosenthal.

"Well, half of it is a repetition of the relevant text in the Common Agenda. The other half has three points. I have a question about the second, and a response to the third."

"Please go ahead."

"What exactly do you mean by transboundary transfer of water?"

"We mean to convey water across the borders between our two countries."

"Like where?"

"We are pressed for water in the Arava, and we know, and you say it, that you are pressed for water in the north. We could help each other in that respect."

"Thank you, this is made clear. Now I turn to Item 3 in which your proposal crossed red lines. We will never entertain such interventions in our internal affairs. Do you realize that your suggestion to stop work on the Karama dam is an attempt to infringe on the sovereignty of Jordan? This, simply, cannot be entertained at all."

"I am sorry, but that was never our intent. I give the floor to my colleague, Professor Uri Shamir."

"What we intended to set out was that it was necessary to agree on the shares of both parties in the Yarmouk river before structures are built to harness its waters, not the other way around," said Dr. Shamir, a professor at the Technion, Haifa.

"Don't you see that your proposal carries conditions before negotiations start. Did not we come here to negotiate without prior conditions? Wasn't that the Israeli demand before the bilaterals started? Do you see where you contradict yourselves? Besides, you always said that you wanted to help Jordan with water, but you turn around and demand that we stop the Karama dam, which will bring us more water for use. Don't you see the contradiction between what you say and what you demand from us?"

"Secondly," continued the Jordanian team leader, "You are mistaken. The Karama dam does not wholly depend on the Yarmouk River. Its own catchment contributes a modest flow; the catchments of other side wadis contribute different quantities, and the rest will come from the Yarmouk floods. The dam is on Jordanian territory, and it is nobody's business how we manage our own affairs. Keep your camel's nose out of our tent!"

The Jordanian side was clearly on the offensive, and the Israeli side was politely on the defensive. The Jordanians took advantage of the error the Israelis had committed, and went all the way criticizing them until the end of the session at about 8 p.m. The Israeli side assured the Jordanians that their proposal was never meant to interfere in Jordan's affairs, and asked that they be supplied with data about the hydrology of the Karama dam.

On Thursday, 19 November, a strange thing puzzled Haddadin, the Jordanian team leader. He was in the corridor of the negotiation wing when Dr. Robby Sabel, the Israeli legal expert, and Aharon Ben Ari, the legal expert in the Ministry of Defense of Israel approached him. These were the legal experts who accompanied Rubinstein in the final session for negotiation of the text of the Common Agenda. They told Haddadin what they had heard, that he had problems with his government!

This was shocking to Haddadin, to hear from the Israelis that he had problems with his own government. He had no idea about that. Dr. Majali had come back four days earlier and he did not tell Haddadin anything except that the government was not very happy, and that he himself thought of resigning.

6. Replacement in the Delegation

Back in Amman, after the seventh round ended, Haddadin received in his private office on Tuesday, 24 November, a visitor from the office of American Senator Patrick Moynahan, accompanied by a political officer of the United States' Embassy in Amman.

"You are the head of the delegation to the Water Resources Working Group of the Multilateral Conference," asked the visitor.

"Yes I am, and also a member of the bilateral delegation in charge of the expert group on Water, Energy and Environment."

The guest looked at his companion inquisitively, and his lady companion gestured back like she did not understand. They had a half an hour conversation about the multilateral talks only. The next day, Haddadin's former boss, Omar Abdallah Dokhgan, told him that he would be replaced, and would no longer be a member of the Bilateral Delegation. On the American Thanksgiving Day, Thursday, 26 November, the Foreign Minister, Dr. Kamel Abu Jaber, called Haddadin to his office.

"I asked to see you, Dr. Munther," said Abu Jaber, "to ask you who do you recommend to put in your place in the delegation. I am sorry to have to do that, but it is some s.o.b's that are to blame for this."

"Please do not feel bad about it. I am but a soldier in service of my country in these difficult times. You tell me get in, I abide, and you issue an order to get out, I too abide without asking questions."

Abu Jaber, "I am really sorry, because you are one of the most efficient and effective members of the delegation. By the way, you will stay as head of the Jordan delegation to the Water Resources Working Group of the multilateral talks."

"As I said, you issue the orders, and I am the soldier who never backs off."

"Who do you propose to succeed you in the bilaterals?"

"I nominate Omar Abdallah Dokhgan. He is another Jordanian who knows the details of our conflict with Israel inside out."

As Haddadin was leaving the Foreign Ministry, he realized that the Israelis and the Americans knew of that move before he did. That was more hurting than the move itself was. He wrote a detailed report to Majali on the status of the progress of his share of the negotiations with Israel, and outlined the Jordan position on water and the basis on which that position was taken. Majali was pleased with the report and Haddadin bid his Chief farewell. It turned out, however, that his absence would not last for long.

The delegation was formed by decision of the Council of Ministers for the following (eighth) round of negotiations. Dokhgan and Mr. Mohammad Saleh Keilani, a former water minister, replaced Haddadin on the delegation. The eighth round did not last for long and ended without entering into substantial negotiations. The negotiations became stalled because the Israeli government deported over 300 Palestinian activists from their homes in Gaza and the West Bank to Marj al Zuhour in South Lebanon. The delegations were called home as a result of that action, and because of the Christmas holidays.

The Common Agenda, however, was the subject of subsequent deliberations to remove from it any ambiguity related to Jerusalem. The disclaimer clause, i.e., the no prejudice provision in Item 5 addressing the borders, would be applied to Jerusalem as part of the occupied territories. This issue was settled in the ninth round in May 1993. The revised text of the saving clause read like this:

"....without prejudice to the territories that came under Israeli government control in 1967."[314]

[314] This text replaced the one arrived at on 27 October 1992 that read, "without prejudice to the territories that are presently under Israeli military administration."

The official initialing of the Common Agenda awaited progress on the Palestinian track, which progress surprised the world from Oslo in early September 1993. The PLO-Israel Accord was arrived at through secret negotiations outside the format of the peace process. The Oslo Accord was signed between the PLO and Israel on the White House lawn on 13 September 1993, and the Common Agenda for the Jordanian-Israeli track of negotiations was initialed in the U.S. State Department on 14 September 1993 (Appendix 9).

CHAPTER X

The Process Continues

1. Rejoining the Delegation

Jordan wasted no time in getting prepared for the coming rounds of negotiations. During the interruption in the peace process caused by the Marj al Zuhour deportees, the Jordan Delegation managed to catch its breath and get things organized. The Government established headquarters for the peace delegations at the Ministry of Foreign Affairs and offices were assigned to the delegates on the third floor.

Having appraised the performance of the delegation in the eighth round, the King ordered that Haddadin be reinstated in the bilateral delegation.[315] The Minister of Foreign Affairs sent after him.

"Tea or coffee?" asked a smiling Dr. Abu Jaber as Haddadin took a seat.

"Tea, please! What can I do for you?"

"Well, it has been decided that you rejoin the delegation. What do you say?"

"I say now what I said to you last time I was here. I am but a soldier, and will execute the orders of my superiors."

"Thank you," said the Minister.

"But for this soldier to fight with high morals, may I know why I was wanted out six weeks ago and I am now wanted back in?"

"I told you why you were out. It was the works of some guys unworthy of respect. You are now back in by orders of His Majesty. Does that satisfy you?"

"It does me a great honor, but does not answer my question. So, I would be grateful if you could arrange a meeting for the purpose with the Prime Minister. I want to see him with you."

"Okay, I will ask for an appointment."

2. Validation of the Water Strategy

Haddadin met with the delegation head, Dr. Majali, and with the Minister of Foreign Affairs. He intended to formalize the Water Negotiations Strategy and have it validated officially. Several water officials and other former officials were invited to a series of meetings at the Ministry of Foreign

---

[315] Details of the session that the King chaired were related to the author by one of the attendants.

Affairs. Four such meetings discussed the strategy at length, starting on 6 February 1993 with the discussion of a paper by Omar Dokhgan, followed by a paper on Sunday 14 February by Haddadin. The meetings culminated on Wednesday, 4 April 1993 with a consensus. The strategy spelled out Jordan's position on water matters toward Israel, the Palestinians, the United States, and Syria. This strategy was very close to the basis on which Haddadin built the negotiation policy, tactics and various contacts. The consensus was documented as follows:

"Summary of the Conclusions of the Water Meetings[316]

Four meetings were held at the Foreign Ministry to discuss the bases of the Jordan position in negotiations over water. There was agreement on these bases. They are as follows:

1. Adoption of the Johnston Project for water sharing.
2. Amendment of the 1987 Agreement with Syria to utilize the Yarmouk waters.
3. Adoption of the Mukheiba site in lieu of Maqarin site for the construction of a dam with suitable storage capacity.
4. Doing away with whatever results came out of the Armitage intermediary efforts in connection with the Wehda dam.
5. Discussing the rights of the West Bank and the Palestinians in the Jordan Basin.

Wednesday 7/4/1993."

The consensus provided a solid base for the negotiators. It validated the adoption of the Unified (Johnston) Plan for water sharing; amended Jordan's stand on the storage site on the Yarmouk, ignoring whatever results came out of the Armitage efforts; and authorized discussing the rights of the West Bank in the Jordan basin. The amendment of the 1987 Agreement with Syria was another point of clarification for the negotiators. This issue would have a dual advantage: The first advantage would be in amending the effectiveness date of the Agreement with Syria and the water sharing wherever possible. It cannot be made effective before the Wehda dam was built; because any expansion of Syria's uses as stipulated in the Agreement prior to the construction of the dam, will cause appreciable harm to the water supplies to the Jordan Valley. The second advantage lay with the Jordanian

[316] Details of the Strategy are kept in the records of the Ministry of Foreign Affairs. A copy was retained by the author and can be released only by written permission of the Foreign Ministry.

arguments in the negotiations with Israel. It would provide evidence that Jordan was having second thoughts regarding the 1987 Agreement with Syria that contradicts the Unified Plan in the first place. On the Armitage efforts, they pertained to the Wehda dam and never touched on the Mukheiba dam, nor did any of his predecessors. His talks would thus be redundant in view of the shifting away from the Wehda site to the Mukheiba site.

Other issues in the discussions included water quality, water storage policy, groundwater, droughts, and co-operation. The discussions and the consensus helped chart the way Jordan would approach a water agreement with Israel.

3. A Meeting with the Prime Minister

A week later, a meeting was set with the Prime Minister. Dr. Majali was present, without any previous arrangement. Attending also were the Foreign Minister and the Director of the Special Office at the Foreign Ministry, Mr. Abdul Ilah al Khatib.

"I want to tell you, Dr. Munther," said the Prime Minister, "that the leaking of the Common Agenda to the press caused us a lot of problems. We could have done without it," and he elaborated on the problems the government had had with that incident, especially with the PLO who sent a delegation to discuss it and demand changes to the border text. Haddadin asked if he could tell his side of the story.

"Yes, I am all ears," said the Prime Minister.

"I am surprised, Excellency," he said, "to hear what you had to say for two reasons. First, that you think for a moment that I am a rookie at a time when you yourself know how I handled the Yarmouk throughout my tenure as President of the JVA, and how I managed its delicate affairs without the closest person to me knowing about it, and, second, because of the reaction your government had towards what you termed as a 'leak' to the press that I was obviously blamed for. It appears as though you tell us to go and fight alone with God's support; when we do, the credit is yours and if we make any mistake, knives come down on us from every direction. This is not a comfortable feeling that a fighter hopes for, and you, as a professional soldier, know that."

"I now, with your permission, turn to tell my story," Haddadin continued and turned to Dr. Majali.

"Here is the man for whom I hold a great deal of respect. I ask him to correct me if I make one mistake in this presentation of mine."

Haddadin told the details of what happened the night of Tuesday, 27 October 1992, and how he told Dr. Majali that Jordanians should hear the news of the Common Agenda, for which he labored most, from Jordanian papers and not from the Western press agencies.

"I instructed the correspondents not to publish the news in accordance with your instructions, which we received on the afternoon of Wednesday the 28th. The man in charge of *al Rai*, refused to honor our instructions, and he went for the scoop."

"I am no rookie, Mr. Prime Minister, and you know how much confidential information I harbored without anyone knowing about it, outside the ranks of the military and the bosses. I did not leak the news. I merely gave a hard copy of it to two correspondents with the consent of my boss, and things got out of control because, also, your instructions came a bit late."

"At any rate," said the Prime Minister, "you are now a senior member of the delegation, and we look forward to your contributions. Bygones are bygones."

Everyone was pleased that the atmosphere was cleared. Dr. Majali was very gracious. He did not interrupt, nor did he correct one word of what was said. He left for an Arab coordination meeting in Beirut before Haddadin could see him and, from there, Dr. Majali headed for Washington, D.C. to head the delegation in the ninth round of bilateral negotiations. He stopped in London to see His Majesty King Hussein before the ninth round started.

4. Resumption of the Peace Process

The Security Council issued a resolution requesting that Marj al Zohour deportees be repatriated. The United States supported the resolution, and the deportees' problem was on its way to be resolved. The sponsors and the core parties decided that the process could resume on the fronts of the Multilaterals and the Bilaterals.

The fourth round of the Water Resources Working Group of the Multilaterals was held in Geneva, Switzerland. The Jordanian delegation[317] arrived at the Intercontinental Hotel on Sunday, 25 April 1993, and left a message for the Palestinian delegation which arrived later, to arrange for a coordination meeting with them first, before meeting with other Arab delegations. On the following morning the two delegations met accidentally, in the restaurant at breakfast, and had a casual discussion. The Palestinian delegation had with them to direct their efforts Mr. Ahmad Qurei' (Abu Ala') of the PLO Executive Committee. He would not attend the meetings, but would direct the Palestinians from his suite.

The Jordanians organized a good coordination meeting with the Arab delegations. Egypt did not attend, because they had a meeting with their Ambassador that afternoon. Besides, Egypt was never used to playing secondary roles in Arab meetings. Jordan had set the tone for that meeting and

---

[317] Heading the delegation was Munther J Haddadin, and members were Zafer Alem, Ziad Elias, Ali Ghezawi, and Hazim Nasser. All members were from the Ministry of Water and Irrigation.

was leading the discussions. The meeting went very well, and all the delegations pledged support to the Joint Jordanian-Palestinian delegation.

In the multilateral Group meetings the following morning, the Arab delegations succeeded in having the head of the Palestinian delegation be recognized separately as such. The gavel holder would say, "I recognize the head of the Palestinian delegation of the Joint delegation." Al Khudary, the head of the Palestinian delegation, would take the floor and thank the gavel holder before he ended his recognition statement! That is, before he said the words "joint delegation." Somehow, the Israelis were not as uptight about recognizing a separate Palestinian entity as they had been heretofore!

Business in the Multilaterals went on as usual. The Palestinians did not want to deny the participants the excitement of watching clashes between the Joint delegation and the Israeli delegation. That had become part of the show since Madrid. They and the Jordanians clashed with Israelis in Moscow, Vienna, and Washington, and the Palestinians did that in Geneva. They did so by insisting on discussing "water rights" in the Multilaterals. The topic is of a bilateral political nature, and belongs to the bilateral track of negotiations. However, the Palestinians insisted on dealing with it in the Multilaterals because their bilateral track was confined to the issue of the Interim Self Government Arrangement, ISGA, and all the other issues would be on the agenda of the Final Status negotiations that were to follow five years later. There was a standoff and the gavel holder announced a recess until the matter was resolved outside the conference hall.

The gavel holder used all his skills and called on assistance from the Jordanian delegation to get over that barrier. They finally proposed a text, to be included in the gavel holder statement, by which the Israelis undertook to discuss the Palestinian water rights in the bilateral track of negotiations. The Israelis cabled the text to Tel Aviv for clearance, and the Palestinians asked the head of the Jordanian delegation to help them clear the text by their director, Abu Ala'.

Al Khudary and Haddadin went upstairs to the suite where Abu Ala' was staying. Bodyguards met them at the entrance and escorted them to Abu Ala'. al Khudary briefed him on the developments downstairs and handed him the text that was worked out.

"The Israelis," said Al Khudary, "asked for time to have the text cleared from Tel Aviv."

Abu Ala' looked at Haddadin. "We appreciate your support," he said, "but the Israelis are not trustworthy. They stab in the back, and they do not honor their pledges. I do not have trust in them, and I will be careful. We count on your support."

"But the text is clear. It puts the water rights on the agenda of the bilateral negotiations between the Palestinians and the Israelis. I do not see anything wrong with that," Haddadin said.

"Oh! They just say that. We cannot believe them, we have to emphasize our rights in every forum, and this multilateral forum is a good one for the purpose."

"You know that your current negotiations with the Israelis in the bilaterals are confined to ISGA. Adding water to the agenda is quite a gain," Haddadin noted.

"We like to keep up the pressure on the Israelis until they recognize our rights," said Abu Ala'.

"But, sir, with all due respect, we would sound unreasonable if we turn down such an opportunity for agreement on a very sensitive issue. The gavel holder and most of the other delegations will think that we are not serious, and that we deliberately put obstacles in the way of our joint work with them. Next time, senior delegates like the ones we have now on the various delegations will not be eager to come at this high level of representation, they would rather send junior representatives. Actually, some of them have told me that."

"No, no, this is too important for them to turn away from," said Abu Ala'.

"But this is exactly why we should make it attractive for them to keep coming. Remember, this international forum was the Arab demand, and not the Israeli's. If we help in marginalizing it, we will not be serving our purpose."

"Still, we are of the opinion that we keep our pressure, and not be fooled by the Israelis."

"Sir, with all due respect, I pledge my support to the maximum. However, may I note that you do not have much to lose in terms of reputation. The PLO has been termed as a terrorist organization by the western powers, what more can they call you? But I have a state behind me whose reputation I have to protect and enhance. The best you can expect from me in support of your refusal of this text that I wrote in cooperation with your delegates is 'silence'. I will not be able to say a word."

"We hope you can do better than that!"

"If I open my mouth, I will have to speak out my conviction. I believe the text is very good for advancing the cause of Palestinian water rights," said Haddadin.

The meeting ended at a standoff, but on good terms. It had not yet been revealed to the world that the PLO had started secret talks with the Israelis in Oslo.[318] Actually, Abu Ala' had come to Geneva from Oslo where the Working Group on Refugees had its session, and he knew about the secret

---

[318] Secret talks started in Oslo in 1992, with Abu Ala's participation, under the auspices of the Norwegian Foreign Minister. See Uri Savir's, "The Process: 1,100 Days That Changed the Middle East," Random House, New York, 1998. pp. 341-2.

talks that were being conducted with the Israelis. The news about those meetings broke out four months later.

The Israelis accepted the text after consultation with Tel Aviv, but the Palestinians refused to go along, and the concluding statement of the gavel holder did not carry any exciting developments.

5. Water, Energy and Environment

The Bilateral negotiations had started in Washington on the same day that the Multilaterals did in Geneva. The Jordanian delegation took residence at the Ritz Carlton Hotel, by Tysons Corner in Virginia because the Willard was all booked on occasion of a big convention that week.[319] The informal talks of the expert group on water, energy and the environment started on Monday, 3 May 1993. The talks covered general topics, in which the experts tried to explore what could be included in a sub-agenda or terms of reference for the group. Practically all the experts of the two sides participated, and it was helpful to survey the viewpoints in the ranks of each team. The Jordanians got busy on Tuesday, 4 May 1993 to prepare a draft for a common sub-agenda for negotiations of the two sides in the group. Joining Haddadin in drafting the proposal were Dr. Hani Mulki (energy), Dr. Dureid Mahasneh (environment), and a legal adviser. They gave their first draft the number "Non-Paper 001."

"Why the 001? And not a simple 1?" Asked Mulki.

"It is to carry the message that I am prepared to make 999 proposals to come to a common sub-agenda!" Haddadin answered.

"And just how long do you think that would take?"

"Making the proposals? Or negotiating the items with the Israelis?" Inquired the team leader.

"Both," said Mulki.

"I do not care even it took from now to eternity. We should be in no hurry!"

The water issues to be negotiated with the Israelis, as far as the Jordanians were concerned, targeted the water sharing of common watercourses, water quality and protection, especially the water quality of the Lower Jordan, and the construction of the diversion weir on the Yarmouk at Adasiyya and of other storage facilities. Cooperation in the alleviation of the impacts of water shortage was on their agenda, as well as other cooperation issues in the fields of environment and energy, and in the development of the Jordan Rift Valley. The Jordanians wanted to stay within the bounds of bilateral issues and keep Israel away from involvement in any of the issues that pertained to Syria, Lebanon, and the Palestinians.

---

[319] It was the Gay Rights Convention, which began on Monday, 26 April 1993.

The members of the Jordanian team[320] did the reading of the draft in a meeting on Wednesday, 5 May. The head of delegation, Dr. Majali, cleared the text after he amended item IV (d) (Appendix 12).

The draft proposal 001 had a preamble and five sections, and was more elaborate than the Israeli proposal forwarded in November of 1992. The sections addressed the issues that had to be negotiated:[321]

A. Surface water basins
B. Shared groundwater aquifers
C. Alleviation of water shortage under efficient management
D. Potentials for regional cooperation within a regional context
E. Jordan Rift Valley development.

The above would form the subject of negotiations through February of 1994.

In the delegation's in-house meeting, Dr. Majali emphasized Jordan's strategy in negotiations. "There shall be no entertaining of a bilateral peace treaty with Israel outside of a comprehensive peace. We shall therefore urge the Israelis, despite all the current and past difficulties, to make progress in their negotiations with the Palestinians in particular, and in the other two tracks as well."

In his meetings with Ambassador Rubinstein, Majali gave priority to such items of the Common Agenda as water, refugees, and the occupied Jordanian territories, and warned against linking the withdrawal from these territories to the conclusion of a peace treaty. Rubinstein, in turn, gave priority to security issues, and the items that would address matters of co-operation and normalization. To solve this impasse, the two heads agreed to group the works into three provisional working groups, without giving priority to the works of either. The groups were:

---

[320] The team leader of the Jordanian side was the author, and members were Ambassador Shaker Arabiyyat, Jordan's Ambassador to Bahrein; Dr. Mohammad Bani Hani, Secretary General of the Ministry of Municipal and Rural Affairs and the Environment; Dr. Dureid Mahasneh, Director General of the Ports Corporation; Dr. Hani Mulki, President of the Royal Scientific Society; and an environmental legal adviser. The Israeli delegation was headed by Dr. Rosenthal, and was composed of Eli Avidan, Deputy Director, Center for Political Research, Ministry of Foreign Affairs; Ambassador Robby Sabel, Legal Adviser, Ministry of Foreign Affairs; Yoav Sagi, Chairman, Society for the Protection of Nature; Arie Zohar, Deputy Government Secretary; and Ran Croll, Director General, Ministry of Energy and Infrastructure.

[321] The proposal contains many items with a language that points at the direction of resolution. Appendix 12 should be read carefully, in order to grasp the issues and the indication of their resolutions.

Group A: Water, Energy and Environment.
Group B: Security and Borders.
Group C: Refugees and Displaced Persons, Banks and Economics, Tourism, Health and Drug Controls.

The priorities soon surfaced in the talks, and each delegation tried to place its priorities first. As could be expected, the Jordanians emphasized the importance of dealing with Jordan's rightful shares of water, and the Israelis wanted to address the issues of environment and the interconnection between the two countries of electricity grids. Such topics require the bilateral cooperation that Israel favored.

In the meeting of Group A on Sunday afternoon, 6 May, after a short exchange of views on the "scope of work" of the group, or the sub-agenda, the Israeli side asked to break up the Group into smaller groups. They proposed one to talk energy, another to talk environment, and the third to talk water. This way, they reckoned, progress can be achieved on the items of their priority (environment and energy), and would be stalled on the items of Jordanian priority (water). It was a tactic to de-link the three topics that had been integrated in one group since the early rounds of negotiations. The Jordanians turned down the Israeli proposal, and pledged to present their own proposal of a "scope of work for the Group", on Monday".

On Monday, the Israeli side[322] opened the meeting and requested the Jordanian proposal for a "scope of work", and they demanded that the Group break up into smaller teams to discuss the three topics separately. The Jordanians refused their proposal and argued back; the atmosphere became tense.

"But then, where is the proposal for our work that you promised us?" Protested Noah Kinarti angrily.

"I will produce it in due time, why the rush?" Haddadin answered, fueling Kinarti's anger.

"If you do not want to discuss the topics, and do not have a proposal for a scope of work, what are we here for then?" Kinarti wondered.

"Remember, we went through seven rounds of negotiations before we finally agreed on a common agenda. Why the hurry?" Commented Haddadin who intended to drive the Israeli team leader into more anger.

"If you do not change your position, I will have to walk out," said an upset Kinarti in protest, and he stood up and started to walk out. His team members followed him.

The Jordanians guessed that the Israelis would complain to their head of delegation, who would, in turn, complain to the Sponsors. The Jordanian

---

[322] The head of the Israeli side became Mr. Noah Kinarti, Advisor to the Minister of Defense. He joined the Israeli delegation in the seventh round, after the Labor Coalition formed the Israeli Government.

team leader walked over to the door and met Kinarti just before he left the room.

"While you are at it," said Haddadin, "why don't you take this with you and read it," and gave him a copy of proposal 001 for a scope of work. Kinarti took the document and walked out angrily, followed by all the members of the Israeli team. The door was kept open.

"No one is allowed to leave his seat, guys," Haddadin told his colleagues, "not even to go to the rest room. This is the first time that I can remember where the Israelis walk out of a meeting with an Arab party. It used to be the Arabs who consistently walked out and boycotted Israeli speeches and meetings. This is a very good sign. The Israelis must have caught our virus!"" But it was not quite a joke.

Officials of the Sponsors, going back and forth in the corridor, could see that the Jordanian delegation was occupying the right hand side of the room, properly seated and looking smart, and opposite them were empty seats. The Israelis were not there, and why? It was in protest; just like what the Arab delegations used to do in international conferences to protest Israeli speeches.

In the room across the hall, Kinarti was talking to Rubinstein, and other delegates were listening in. Rubinstein would raise his eyebrows and say something to Kinarti. Pretty soon, the American contact with the Israeli delegation appeared by request of the Israelis. He had a conversation with them, apparently with back and forth arguments; them telling him something, and him asking questions. Then Kinarti opened the document that Haddadin had given him. The American official started to read it. He apparently told the Israelis that the document met the specifications of a draft proposal for a scope of work for the Group. Soon, Rubinstein and the American official left the room of the Israeli delegation. Minutes later, Majali sent after Haddadin.

"What happened?" Asked Majali.

"I just do not like the arrogance in the tone of the Israeli delegates, their team leader included."

"But what was it?"

"Honestly, I was looking for an excuse to embarrass them, to drive them nuts like they did to us on so many occasions," Haddadin answered.

"But why? Does that serve our purpose?" Asked Majali.

"When do we expect the Israelis to serve our purpose? We have to help ourselves."

"I did not hear them say any nasty word," said Dr. Mahasneh who was new in the Jordanian delegation. "Dr. Munther was not kind to them."

"I leave the kindness to others. I have a mission to accomplish and this is the way I will conduct my business."

"Come on!" said Mahasneh, "I bet you will yell at them even if they said 'good morning' to you."

"Dureid, you will gain experience in dealing with the Israelis as we go along. If you decide to be kind to them it will be your own decision. Just exempt me from that kind of protocol and its gestures."

"I have not yet heard what the matter was," said Majali impatiently.

"Well, for one thing, they are demanding that we break up into three groups. The three topics are interconnected, and I am not about to compromise the integrity of the group to please them," Haddadin said.

"One good thing worked for us. Kinarti complained that you did not honor your promise to present them with a proposal for a scope of work. It turned out that he had it in his hand, and that eroded whatever credibility their complaint had with the Sponsor."

"Sir, I do not work for the Sponsors. As a matter of fact I consider that they are here to work for us, all of us. It is in their interest, and ours, to get these negotiations to bear fruit."

"I appreciate that, but would also like to keep that thin thread intact. If they pull we should give in a bit, and if we pull, they are apt to give in a bit. The sponsor will help maintain that thread. Let us have the group split in teams for less that two minutes, and then you regroup." Advised Majali.

"Just for your sake, not theirs. You take care of that thread. I never developed liking for threads since I was a child, when my mother used to have me put the thread through the needle hole. They were tiny and my eyes got tired doing that job more than three times a day!" Haddadin said jokingly.

Majali then led the Jordanian team and Rubinstein led his. They announced that they could break into smaller teams for just a very short while. The Group broke up into three teams that occupied three corners of the same room. Kinarti and Haddadin stood by the coffee pots; talked, and had coffee. When they finished coffee, Haddadin clapped three times to draw attention:

"Okay guys, the break is over!"

The following three days, the Group's discussions focused on the Jordanian proposal. Dr. Robby Sabel, the legal expert of the Israeli group, objected to the text of the preamble, and to items I.A and I.B that addressed the rightful water shares for each party in the Jordan and the Yarmouk.

"These issues come as a culmination of negotiations, not at the outset," he said.

"For Christ's sake," said Haddadin, "if we drop these items from our scope of work, what is there left to negotiate over? Do you expect the environmental issues, or the electricity interconnection to be so contentious that we would spend a long time negotiating over them? I say that these matters are issues for cooperation, and they, not the water issues, come as a culmination of negotiations."

The talks continued, in an attempt to converge on a common sub-agenda. Before the round was over on Thursday, the Jordanian proposal was the document on the table, and the items that were contested were I.A and I.B, on water sharing, and item IV.D of the environment section relating to nuclear radiation. The rest of the items were not rejected by the Israelis, but were the subject of non-substantive amendments. Sabel objected to the nuclear radiation item because, he said, "it is not within our jurisdiction."

"Israel," he added, "is co-signatory to the Nuclear Accidents Convention, according to which Israel is obliged to notify adjacent countries in case of any nuclear accident. This obligation is self explanatory, and is enough to replace item IV.D."

6. Majali the Prime Minister

On Tuesday, 11 May 1993, Haddadin walked into the restaurant for breakfast, on the ground floor of the Ritz Carlton, and joined Ambassador Nayef al Qadi, the Jordanian member of the delegation, who was sitting eating breakfast.

"It is now certain that Dr. Majali will succeed Sharif [later Prince] Zeid Ben Shaker as Prime Minister soon," al Qadi said.

"How did you know? How can you be so sure?"

"I have my sources," he said, "and Majali is taking Jawad Anani with him as Minister."

"You are pulling my leg!" Haddadin exclaimed.

"No, I am serious."

"Come to think of it, there is merit to the rumor. Majali had stopped in London and conferred with His Majesty King Hussein before he came here. Last Sunday, he and I were sitting alone in the suite, and he was telling me of the need to amend the Elections Law so that we shift to the one-man, one-vote system," Haddadin told al Qadi.

"You see," said al Qadi, "why should he think of amending the Elections Law unless he will be involved in such a thing shortly?"

"Okay, Nayef," Haddadin said, "Let us find out for sure. This is what we will do at the meeting of the delegation after breakfast." And they divided the labor between the two of them.

"Pasha," said al Qadi, addressing Dr. Majali before he opened the morning meeting as agreed with Haddadin, "before we start, I want to tell you that we heard a strong rumor that you will be our next Prime Minister soon. The rumor originates not on the street, but in the ranks of the Israeli delegation!"" Majali smiled in satisfaction.

"What is this *ya Abu Khalid*"" Haddadin said addressing Nayef al Qadi, "do you believe everything the Israelis say? What do they know about our politics in the first place?"

"No, no," interrupted Dr. Majali, "they know, they do know!!"

This reaction from Majali was sufficient evidence that the rumor had roots. Majali was sworn in as Prime Minister on 29 May 1993. He had a team composed heavily of technocrats. Dr. Jawad Anani was appointed Minister of State for Prime Ministry Affairs and was given the file of the peace process. Anani's appointment was a relief to the delegation, because one of their colleagues, Anani, was put in charge of their affairs. The ranking military officer in the delegation, General Marie, was appointed Army Chief of the General Staff, and was replaced in the delegation by General Tahseen Shurdum. Mr. Marwan Doudin, a former Minister, replaced Dr. Anani.

The tenth round of negotiations was due to start on Monday, 14 June.[323] Dr. Fayez Tarawneh, Jordan's Ambassador to Washington since September 1992, and a member of the delegation since the beginning of the Bilaterals, was appointed head of the Jordan delegation succeeding Majali.

The delegation met under the new leadership on Sunday, 13 June. The Raporteurs of the three Groups were named. Haddadin for Group A, "water, energy and the environment;" Dr. Abdallah Toukan for Group B, "security and borders;" and Marwan Doudin for Group C, "refugees and displaced persons, health, tourism..." On the Israeli side, Rosenthal was named Rapporteur of Group A, Ambassador Eitan Bentsur for refugees, and Dan Rothschild for security and borders, aided by Amos Gilad and Nachman Tal.

7. Substantive Water Talks

In the delegation meeting on 14 June, Dr. Musa Breizat asked about the conclusion of the "water strategy" paper that was discussed in four meetings with other experts and the Ministers of Water and of Foreign Affairs.

"The latest development was that the strategy was approved, and that the former Minister of Foreign Affairs wanted us to talk to the former Prime Minister," Haddadin reported.

"But I did not attend any of the meetings," protested Dr. Bani Hani.

"I was not the one who called for the meetings, nor was I the one who selected the participants."

The negotiations round started on Monday, and new delegates were introduced by each side. The groups then started their meetings, and Group A had discussions over water from the point they ended last round. There were also discussions over electricity grid connections, and on generation of energy from indigenous sources. The environment had a good share of the talks, focusing on the usual topics of the Gulf of Aqaba, the problems of pigeons, mosquitoes, houseflies, and the clean up of the Jordan River. On water, the Israelis gave a counter proposal that had a preamble of its own, and listed the

[323] Mr. Marwan Dudin succeeded Dr. Jawad Anani on the delegation. Dr. Tayseer Abdul Jaber joined the delegation. Other experts joined, including Dr. Ibrahim Badran (energy), Safwan Toukan (economics and banking), and Dr. Abdallah Toukan, who left the delegation for a short while and then returned.

points of agreement with the Jordanian proposal. The Jordanians had submitted a proposal under number 002, and had rearranged the items under the environment section, but there were no substantive changes from the proposal numbered 001. The preamble to the Israeli proposal revealed their preferences as follows:

"Whereas the parties wish to achieve a settlement of all outstanding water issues between them and,

Whereas the parties wish to increase the availability of water resources for the mutual benefit of both parties, and,

Whereas the parties have agreed to cooperate in the protection and improvement of the common environment, and,

Whereas the parties have agreed to cooperate for the mutual benefit in the future development of energy resources, and,

Whereas the parties have agreed to cooperate in the protection and development of the Jordan Rift valley,

The following common agenda has been agreed:..."

It was clear that the Israeli side was more interested to jump into cooperation before the water dispute was settled, something that the Jordanian side did not accept.

During this round, His Majesty King Hussein made an official visit to the United States. Majali, the Prime Minister, accompanied him. Tarawneh advised us that the negotiation's tone should be positive, at least as long as His Majesty was in Washington.

Actually, the King had been trying hard to mend fences with the United States over the Kuwait invasion by Iraq and Jordan's stand concerning it. The Kingdom was practically under siege since then. The port of Aqaba, Jordan's only seaport, was under surveillance by the United States Navy. Ships suspected of carrying embargoed goods to Iraq were diverted for inspection in the next ports of Jedda or to the Gulf ports. The markets of the surrounding countries were closed to Jordanian goods and services. The foreign airlines stopped their service to Amman and Royal Jordanian was the only carrier servicing the country. The King resumed contacts with the Bush Administration, after he was discharged from the Mayo Clinic in September 1992, and kept the dialogue going with the American Administration. The economic situation in the country was suffering from lack of growth and the country embarked on an economic recovery program involving economic structural adjustments. In short, the blessings of the United States were needed to alleviate all these hardships, and Jordan's performance in the Peace Process weighed a great deal in that regard.

"I have a message for you, Munther," said Tarawneh to Haddadin, in private. "Please be as tactful and diplomatic in the negotiations as you possibly can. You can go your usual way after the King leaves."

"It is widely known by now," he added, "that the rough going has been in your department. Lots of complaints were filed about you. Please let us avoid collisions and sharp assaults!"

"Will do," Haddadin answered, "but that will not bring the end of the story. We are just beginning."

The meetings continued. By the end of the round, the Israelis came back with a proposal to make Jordanians move a bit away from their stiff stand. Ambassador Sabel offered a package that included:

- No preamble,
- Merge the items of both proposals (Jordanian and Israeli) for an agenda,
- Combine items I.A and I.B of the Jordanian proposal in one item reading; "securing the rightful shares of the two sides in the waters of the Yarmouk and Jordan rivers in all their aspects."
- Changing item IV.a.2 to read: "Air quality including man made hazards."

The Israeli package came closer to the Jordanian proposal, while maintaining the Israeli proposed agenda items. Still, no agreement was reached on that package.

"I am sure you can do better than that," Haddadin said to Ambassador Sabel, "we expect a better counter-proposal next round."

On the multilateral side, there was a consultation meeting in Washington during this round of bilateral negotiations. The consultations normally take place by invitation of the gavel holder, and the core and regional parties participate in addition to the representatives of the other gavel holders of the multilateral working groups. The gavel holder, Mr. Keiswetter, informed Haddadin in private that the Palestinians have changed their mind over the text that was worked out in Geneva, and that they now approve it.

The bilateral round ended on the first of July. Jordan emphasized the need to make progress on all tracks of negotiations, and that no progress could be made on their own track without the others, especially the Palestinian. More attention was paid to the Palestinian track in Washington, D.C., but no one was aware that secret meetings were taking place in Oslo between high officials of the PLO and the Israelis.[324]

8. The Oslo Accords

The following round was scheduled for Tuesday, 31 August 1993. Haddadin took a Royal Jordanian flight from Amman to New York on

---

[324] Since December 1992, Secret talks were being conducted between two Israeli academics, Yair Hirschfeld and Ron Pundak, and three senior PLO officials: Ahmad Qurei (Abu Ala), Hassan Asfour, and Maher al-Kurd, all close associates with Yasser Arafat. The talks became official with Uri Savir heading the Israeli side on 20 May 1993. See Uri Savir, 1998.

Saturday 28 August. He ran into the Palestinian delegate, Sa'eb Ereikat, at the airport. On the plane, they had a good chance to exchange views. They went over the status of negotiations on both tracks, and agreed that the coordination between them, despite expressed desires and instructions by the Palestinian/Jordanian leaderships, was not up to the desired level, and that something should be done to improve it. Each undertook to be the liaison member in his own delegation to promote coordination, and scheduled a working dinner on Monday evening at the Willard, to be followed by another working dinner on the following day, at the Grand Hotel where the Palestinian delegation took residence..

Ereikat disembarked in Amsterdam, where the Royal Jordanian flights stop to refuel and let passengers off and take others on board. He was due to join Hanan Ashrawi and Faysal Husseini, who were scheduled to arrive in Amsterdam from Tunis, and catch a KLM flight to Washington, D.C. Haddadin arrived in New York and spent the night there. He took the shuttle from La Guardia on the following morning, and picked up the *New York Times* that carried an exciting front-page headline.

The headline was too good to believe. There was a breakthrough in the search for Middle East peace. The PLO and Israel have arrived at a Declaration of Principles after secret talks in Oslo. Israel had been talking with the "terrorists" and will exchange formal recognition soon. The stories in the *New York Times* killed the flight time for Haddadin, from La Guardia to National Airport in D.C., and gave him a feeling of relief. The most complicated knot in the Middle East peace process had, at last, been untied. The reports carried feelings of astonishment and surprise from different circles in the Middle East, including King Hussein. In Washington, the Jordanian delegation was excited, but very confused.

Ereikat did not show up at the Willard on Monday as had been agreed. He did not even call to cancel or postpone the dinner date he had set with Haddadin. He must have been equally confused.

"This changes the whole format of the delegation composition," said Tarawneh. "I do not think that we will be a 'Joint delegation' any more. The Palestinians are now as independent as any other delegation. This resolves their dilemma of being tied to a Joint Delegation and wanting to be independent at the same time."

The Palestinian delegates were stunned. There they were, laboring in negotiations with Israel under the guidance of the PLO, only to be excluded from a historic deal that was struck in Oslo, outside the frame of reference of the Peace Process. The head of the Palestinian delegation, Dr. Hydar Abdul Shafi, felt very awkward, not knowing what was going on, nor was he sure about what to do in the impending round of negotiations. The Jordanian delegates were equally confused. The Acting head of the Israeli delegation,

Ambassador Eitan Bentsur, shared with Tarawneh what he knew of the exciting historic developments.

The bilateral meetings with the Israelis were to start on Tuesday. Sa'eb Ereikat showed up a bit late to represent the Palestinians in the Jordanian track. As he approached the wing where the talks were conducted, on the ground floor of the State Department, he practically shouted from a distance to tell Haddadin, "I swear by the Almighty that I did not know a thing about Oslo. I was briefed about it by Ashrawi after I had left you in Amsterdam. She, in turn, knew about it in Tunis, where she and Faysal Husseini stopped over on their way here. We are not extremely happy about the way we were bypassed."

"But you did not show up yesterday as agreed to have dinner and co-ordinate," Haddadin said.

"So much for co-ordination. I was as surprised as you guys were, and so were the other members of our delegation."

There was no meeting of the track; the delegations did not know what to talk about in the track meeting. But that did not prevent pursuing the work of the Groups. Five meetings were held in that round for group (A) on Water, Energy and the Environment, and the Israelis moved closer to Jordan's position as stipulated in the proposal 002. On the issue of nuclear radiation, they agreed to put it on the agenda of the multilateral Working Group on the Environment. They tried to de-link the Jordan River north and south of Lake Tiberias with the Lake itself.

"But Jordan and Israel face each other along the Lower Jordan River," said Sabel, "and you have no frontage on the Upper Jordan north of Lake Tiberias."

"But the river is an integrated unitary whole. What is done upstream affects the downstream end of it, and thereby the need to agree on works or changes that the parties upstream do to the river. Such works could change the flow and the quality of water, and that concerns us very much," Haddadin responded.

The Israelis sounded like they would agree to a right that Jordan has on the river Jordan, and that it could be negotiated. This was a positive shift from their previous positions. The Group came very close to agreement on the wording of items I.A and I.B as proposed by the Jordanians. The Israelis wanted to change the words "rightful shares of the two parties" to "of all the parties." That was another shift from their position taken in the third round of negotiations, over a year ago, when they did not want to include the other riparian parties.

"What is up their sleeves?" Haddadin asked his colleagues, "we initially proposed the phrase that they are proposing now, and they refused it back in the third round. We then took their phrase to address the rightful shares of the two sides instead of the rightful shares of all the parties. The negotiations are bilateral after all, why should we deal with the rights of the other parties in the basin?"

"Obviously, we are trading positions. Their understanding of 'the parties' is Jordan, Israel and the Palestinians. Our understanding includes Syria and Lebanon," Haddadin suggested.

"We should be careful," said Toukan, "it looks like they want us to handle the Palestinian rights in the Jordan basin,"

"Bless your heart," Haddadin replied," this is a political trap. We have to build our own defenses."

"But how?" Asked Toukan.

"We will see. We can accept the wording they are proposing, and that would be of service to the other parties. However, we will negotiate our own rights without infringing on the rights of any other party, particularly the Palestinians."

A new gavel holder, Mr. John Herbst, had been appointed for the Water Resources Working Group of the Multilaterals. He asked to meet with the head of the Jordanian delegation to that Group to get acquainted. Haddadin went to the appointment at 3 p.m. on Thursday, 9 September, and Herbst came out of his office and greeted him in Arabic.

"*Ahlan wa sahlan,*" said Herbst.

"*Ahlan feek.*" Haddadin responded with obvious appreciation in his tone of voice, and they walked into Herbst's office.

"Interesting developments," said Herbst, obviously referring to the Oslo Accord between the Palestinians and the Israelis.

"I would say historic."

"You are going to sign with the Palestinians on Monday!" Said Herbst.

Haddadin did not appreciate the tone with which Herbst said his statement, nor did he appreciate him breaking such news to him in his office.

"Well, this must be American hopes; but brace yourself, we will not sign along with the Palestinians on Monday," Haddadin said.

"I bet you will," said Herbst while the two were still standing.

"May I sit down?" Haddadin reminded his host.

"Yes, please, I am sorry."

The meeting lasted for half an hour, mostly focusing on the Oslo Accord; Herbst trying to emphasize a role for Jordan, and Haddadin assuring him that Jordan could not afford to assume any active role, nor would she be a party to the Oslo deal. It was not a very good start for the gavel holder with Haddadin. It later showed in the works of the Multilateral Group meeting that was held three months later in Beijing, on 26-28 October 1993.

The following day, Friday, Tarawneh spoke to His Majesty King Hussein by phone, and received instructions to initial the Common Agenda on Tuesday, the day after the Palestinians and the Israelis sign the Oslo Accord.

Jordan would be represented at their ceremony by Ambassador Tarawneh and Muasher, the Jordanian officials stationed in Washington, D.C. The rest of the delegation was to leave Washington by Saturday.

The fallout from the Oslo Accord was hazardous to the peace process, but it gave hope that the core of the Arab-Israeli conflict was finally touched on. Israel and the PLO exchanged formal recognition, and Mr. Arafat, heretofore viewed as the head of a terrorist organization, became Mr. Chairman! The movement towards reconciliation started, and the PLO began to prepare itself to move to the parts of the occupied territories from which the Israelis would pull out under the Accord. Gaza and Jericho would be the first of the Palestinian territories to be liberated, and a Palestinian National Authority would be set up there. Soon afterward, Palestinian elections would be held to elect a legislative council, and the infrastructure of a Palestinian political entity was being set up.

Still, the fallout impacted the bilateral negotiations. Syria was upset about the developments and so was Lebanon. The bilateral negotiations were suspended until the shockwave of the Accord faded away. However, the suspension of the bilateral negotiations was overshadowed by events that accelerated in the wake of the Oslo Accord. A donor's conference was organized by the United States, to which donors who would contribute to the development of the Palestinians were invited. Jordan, the host of the highest percentage of Palestinian refugees, and the Arab country that kept the bridges open with the Palestinian territories in support of their people and economy, participated.

After the Oslo Accord was signed and the Common Agenda was signed between Israel and Jordan, Jordan's strategy on water negotiations was reviewed in light of these developments and a more detailed presentation of it was made to Jordan officials.

9. The Trilateral Economic Committee

On the occasion of the Donors' Conference, President Clinton made arrangements for a meeting at the White House between Prince El Hasan Ben Talal, then Crown Prince of Jordan, and Mr. Shimon Peres, then Foreign Minister of Israel. The meeting ended with an announcement that a Trilateral Economic Committee of the three countries would be set up, to look into economic projects in which the three countries could cooperate for the sake of peace making, peace building and cooperation in the region. The Committee provided yet another parallel forum to that of the bilateral negotiations.

The United States took on the challenge of jump-starting the suspended bilateral negotiations. Secretary of State, Warren Christopher, conducted rounds of consultations with delegates representing Syria, Lebanon and Jordan. The consultations did not succeed in having the bilateral tracks resume. Actually, the heads of the Syrian and Lebanese delegations got transferred to other positions. However, the Trilateral Economic Committee provided the opportunity to resume talks between the Jordanians and the Israelis, in February 1994, albeit in a trilateral forum. The talks and contacts between the Palestinians and the Israelis continued separately. Also the Multilateral talks, in which Syria and Lebanon never participated, were resumed. The next round for the group on Water Resources was to be held in Beijing, 26-28 October, and it was the first time the new gavel holder of the Group, Mr. John Herbst, participated and held the gavel.

Mr. Herbst surprised the Jordanian delegation prior to the beginning of the sessions. He told them that Mr. Edward Djerijian, Deputy Assistant Secretary of State, was determined to move the Multilateral talks to the region, and that the Sultanate of Oman would extend an invitation to host the next meeting of the Water Group.

The Jordanian delegation expressed the view that such a move was too early. The right timing would be when substantial progress was made in the bilateral tracks or, better yet, when peace is achieved. The Jordanians had no instructions from their government on how to respond to such a move, and therefore did not feel willing to cooperate in the matter. They first contacted the Saudi Arabian delegation. The Saudi delegation was strongly against the move, and undertook to talk to the Egyptian delegation. The Jordanians mobilized the other Arab delegations and some of their sympathizers, and talked to the other primary delegations. It was not enough to mobilize support for opposition to the move to the region, but it was necessary to find an alternative venue to host the next round. Canada, Haddadin thought, could be the choice.

The head of the Jordanian delegation approached his Canadian counterpart at lunch on Tuesday, 26 October. He detected that the Canadian had not yet been briefed on the move to the region.

"When are we going to have a meeting of the Group in Canada?" Asked the Jordanian as the two sat together for lunch.

"Not this year," said the Canadian, "but may be next year after the budget is approved."

"I think Canada has a lot to offer in terms of lessons in water resources management. You have the highest per capita share of fresh water in the world, and yet you have clear examples of water quality degradation because of urban pollution. We can learn a lot."

"I welcome the idea to convene a meeting of the Group in Canada."

"How about the next round? Around April of next year," suggested the Jordanian.

"April, yes, but not before, because the fiscal year starts on 1 April and the new budget would be operational. We do not have appropriations for this before April."

"April is fine with us, I guess."

Haddadin briefed the Saudis and the Egyptians. They were happy to know of the preparedness of the Canadians to host the next meeting.

"In the meeting this afternoon," the Jordanian advised the Saudi, "you propose that the next round should be after the first of April of next year. My turn would be after you."

In the coordination meeting that afternoon, attended by the gavel holders of the other Working Groups and the regional parties in addition to the core parties, the new gavel holder opened the meeting. The Omani delegate, not a member of the consultative group, was invited to attend. It was clear that he was invited to extend the invitation to the next round. The gavel holder asked for ideas! The Saudi was quick to propose that the next round should not be held before April of 1994, and there was agreement. Before the Omani delegate had time to extend his invitation, the Egyptian delegate said, "we understand that the Canadian delegate has something in mind to tell the group!"

"Yes," said the Canadian delegate, "Canada will be very happy to host the next round of Multilateral talks, and I hope the group accepts."

"This is very kind of you, sir," said the Jordanian delegate. "A meeting in Canada will have more than one benefit," and he started counting more benefits to be gained form Canada's experience in water management!

The gavel holder was cornered. He could not decline the invitation, but he could not accept it either.

"Is there any other party who is interested to extend an invitation?" asked Mr. Herbst as he looked at the Omani delegate.

The Jordanian looked at the Omani and gave him a "no" signal with his eyebrows. The poor delegate found himself torn between the gavel holder and the Arab delegates.

"I had in mind to extend the invitation. Actually that is why I am here for the first time. But after the Canadian delegate has invited you, I guess my invitation would be out of place!"

"Don't lose hope," said the Jordanian delegate, addressing the Omani. "Let us hope that peace would come soon after the Canada round, and we will then be very glad to all come to Muscat."

The Jordanians invited the other Arab delegations to a gathering that evening, reviewed the events, and did the necessary coordination. A few hours later, at about 03:30 after midnight early Wednesday, the head of the Jordan delegation was awaken by a call from Amman.

"This is Jawad speaking from Amman, how are you?"

"Fine, a bit sleepy!"

"What are you doing to the gavel holder?"

"Like what? Has he complained?"

"Yes, bitterly. He complained to Washington that you are making his life miserable. Washington turned the matter over to the American Ambassador here. He talked to us about the Jordan delegation in Beijing."

"Well, my O' my, whose conference is it anyway?"

"Listen, Munther, I am talking to you after we have contacted His Majesty in Paris."

"Please go ahead," said the head of delegation to the Minister of State for Prime Ministry Affairs.

"Can you make his life less miserable?"

"Give me one reason why I should."

"Aqaba!" said the Minister, "Aqaba," obviously referring to the American naval blockade of the Jordanian port of Aqaba. Jordan had been trying hard to lift the blockade which was imposed after the Iraqi invasion of Kuwait in 1990, or to change the regime of inspection of the imported goods.

"Okay, I understand, this is a good enough reason, I will see what I can do."

It was not easy for the head of the Jordanian delegation to approach the gavel holder and suddenly change course, but the reason was well worth it. He decided to think about it over breakfast, and maybe use the interventions of some members of the Jordanian delegation. While he was at breakfast briefing the members of the Jordanian delegation, Mr. Herbst walked into the restaurant, obviously looking for him. When he spotted the Jordanians, he made his way toward their table.

"Good morning," said Mr. Herbst.

"Good morning," replied all the Jordanians.

"Can I talk to you after breakfast?" Herbst asked their leader.

"But of course. You hold the gavel, I am willing to listen to you any time."

After breakfast, the Jordanian head of delegation took with him Mr. Ali Ghezawi, a Jordanian delegate, and they had a chat with Mr. Herbst. It was open and frank, and it provided a chance to speak freely.

"The best I could do," said the Jordanian head of delegation, "is to leave the venue of the next round open for the Steering Committee of the Multilaterals to decide." Herbst accepted, and the matter was resolved.

At that round of talks, the Jordanians presented a joint paper with the German delegation to study the supply and demand for water in the countries of the core parties (Jordan, Israel, the Palestinians, Syria and Lebanon). Germany undertook to finance the study. The Israelis contested the goal of the study and the Jordanians had to work hard to convince them. The Jordanians also talked with them bilaterally, on the side, about the Red Sea-Dead Sea canal, and the need to write a "Scope of Work" for consulting services to be acquired for that huge project. The two topics, the supply-demand study and the Red Sea-Dead Sea canal would be the subject of a trilateral meeting with the Germans that was held later in Frankfurt in December 1993.

The Steering Committee of the Multilaterals held its meeting in Tokyo, Japan, on 15-16 December 1993. The gavel holders (the United States and the Russian Federation) consulted with the Jordanians over the venues and the Jordanians went along. The Committee, among several decisions it took, decided that the next venue of the Water Resources Working Group would be Muscat, Oman. Haddadin, who was a member of the Jordan delegation to the Steering Group, headed for Frankfurt on Saturday, 18 December, to have talks with the Germans and the Israelis over the supply-demand study that Germany would finance and supervise, and over the Red Sea-Dead Sea canal. Mr. Ali Ghezawi, then adviser to the Minister of Water and Irrigation, joined him from Amman. Upon his arrival in Frankfurt, Haddadin set out to write a proposal for a "Terms of Reference" for the supply-demand study, and another for the Dead Sea-Red Sea canal.

The meetings started on Sunday morning. The German delegates distributed a proposal for the "Terms of Reference" for the supply-demand study.

"We have problems with this proposal," said Avraham Katz-Oz and Mr. Ilan Baruch, who represented the Israeli side.

"We have a few problems with it too," said Haddadin, "but we have another proposal that we think would do the job," and distributed the proposal he had written the night before. It was very well received, and formed the basis for the study that Germany financed and supervised.

10. The Jordan Rift Valley

The Jordan Rift Valley, a part of the Great Rift that extends between Turkey and Kenya and passes through Syria, Lebanon, Israel, Jordan, Palestine and the Red Sea, forms the interface between Jordan and Palestine and Israel. It was thought that the development of that region would be a starter for peace projects. Jordan's experience in the integrated development of the Jordan Valley dates back to the early 1960's, and the cumulative experience qualified Jordanians to suggest what could be done in the Rift Valley in peace time. A major undertaking would be the control of the level of the Dead Sea, whose recession has been causing groundwater to seep into the Dead Sea from the adjacent aquifers.[325] The control can be made by importing water to the Dead Sea from the open seas. Israel had been studying to link it with the Mediterranean,[326] and Jordan did some studies to link it with the Red

[325] By 1999, the seepage from the eastern mountain aquifers into the Dead Sea was causing local landslides on the shore of the Dead Sea.

[326] Mostly in line with the Hertzl idea, and the analytic work of Bourcart in 1899. It was later taken up and advocated by Shlomo Gur of Israel.

Sea at Aqaba.[327] The Israeli negotiators in the Bilaterals and Multilaterals agreed to have a joint study done of the Jordanian option, and to select the better of the two projects.

In the Frankfurt meetings, the Jordanians circulated "Draft Terms of Reference" for the Dead Sea- Red Sea connection. The draft was discussed and was viewed with favor by the Israeli participants. The Jordanians alluded to the need to expand the "Terms of Reference" to reflect the concept of integrated development of the Jordan Rift Valley.

"We want to call it the Valley of Peace," said Katz Oz.

"We will not disagree over the name, just let us get the study started and the projects implemented, and the name would impose itself on us," Haddadin responded.

With that beginning, and with the concepts voiced by Prince El Hasan Ben Talal, Haddadin started drafting a more comprehensive paper about the concept of integrated development of the Jordan Rift Valley. including the Red-Dead canal. He presented the concept of the Jordan Rift Valley and also the water strategy to King Hussein, in London on 14 January 1994. The King was pleased, and gave his blessings. Upon his return to Amman, Haddadin concentrated on the preparation of the "Concept Paper" for the Jordan Rift Valley, and completed it in early February. A ministerial committee reviewed the Concept Paper, and had it discussed in an evening session that was chaired by Prince al Hassan. Prime Minister Majali, dressed in Arabic costume, joined that session and blessed the effort too. The paper was approved, and Haddadin was instructed to present it to the forthcoming meeting of the Trilateral Economic Committee to be held in Washington, D.C. on 22-23 February 1994.

The meetings in Washington, D.C., started on 14 February, under the umbrella of the Trilateral Economic Committee. Group (A) picked up where they had left off in the eleventh round in September, and labored to come to terms on the sub-agenda for negotiations on water, energy and environment. They were able to reach agreement on most of the text. On 23 February, the Trilateral Committee met, and the American side was heavily represented, headed by Ambassador Dennis Ross. The Jordanians were three persons (Tarawneh, Fayez Al Khasawneh, and Haddadin), and the Israelis were ten, headed by Rubinstein. In the course of conversation, Rubinstein criticized Jordan bitterly:

"Jordan," said Rubinstein," is dragging her feet. They are not prepared to sign a peace treaty with Israel. If they do not want to sign a peace treaty, what are we aiming at in our negotiations?"

---

[327] The credit for initiating the Jordanian studies in 1978 goes to Omar Abdallah Dokhgan, and the credit to renew the Israeli studies in the 1970's goes to Shlomo Gur.

Haddadin was not aware of other meetings that had been held earlier in the State Department the day before, but felt challenged by Rubinstein's words.

"Ambassador Rubinstein," he said in response, "I challenge you to come and sign a peace treaty this afternoon. It is all ready to go. If you chose tomorrow, you can have a day as a grace period. But please do not say that Jordan is not prepared to sign a peace treaty. If you do not show up this afternoon or tomorrow, according to your choice, then it is obvious who is dragging his feet and is not prepared to sign a peace treaty. I just do not guarantee that you will like all of its articles!" That was the end of Rubinstein raising the issue. It turned out later that a meeting had been held at the State Department in which the Jordanians turned down an American offer to sign a peace treaty first and then negotiate the details afterwards, something similar to the Oslo Accords.

"I am afraid that this would put us in a dark tunnel," said Awn al Khasawneh in that meeting at the State Department, "something I am sure we want to avoid."

"What!" said Dan Kurtzer of the American peace team, "I am working my a… off on the Hill to get Jordan 35 million dollars in grants, and you tell me you will not sign a peace treaty?"

"Sir," answered Awn, "if you think that we would compromise Jordan's national interests and her future for the sake of your $35 million, or any other sum of money, you know what to do with them!"

After the exchange on the peace treaty inside the meeting of the trilateral economic committee, the Jordanians were given the floor to present the "Concept Paper" on the Jordan Rift Valley. It was very well received, and almost everyone was taking notes. Very favorable comments came out of the Israeli delegation in response to that presentation. Ambassador Dennis Ross talked to Tarawneh during the coffee break, and told him that the paper, to him, was the first thing he saw "with meat on it" since the beginning of the peace process, unlike the other presentations that were mostly rhetoric and void of meaningful interpretations. Tarawneh was very happy.

"Dennis wants a copy of the paper, and I have to have a copy." Tarawneh said to Haddadin.

"Yes, of course," Haddadin answered, "I just want to wait until we hear the Israeli review of it, or their counter proposal."

"When can I have a copy of the paper?" Tarawneh reminded Haddadin on Thursday the 24th.

"On Monday," Haddadin said, "you know we are moving out of the Willard to the Watergate Hotel tomorrow, and when I unpack, I will look for a copy for you, and one for Ambassador Ross."

The meetings never resumed on Monday 28 as was planned, because on Friday the 25th of February, an Israeli settler committed a crime against the worshippers in al *Haram al Ibrahimi* in Hebron, during the holy month of Ramadan. About thirty Palestinians were murdered in cold blood, while they were conducting their Friday prayer. That put an end to the trilateral sessions in Washington and, instead of going to the State Department on Monday the 28th, the Jordanian delegation headed for home.

Ambassador Ross followed them to the region within a week, to exercise damage control. He met with the Jordanian officials and, in the process, obtained a copy of the Concept Paper by order of Dr. Majali, the Prime Minister, and he carried it to Israel. By mid April, the Jordanians received an Israeli proposal for the development of the JRV, but it was not as good as the concept paper that Jordan had presented. Jordan's paper actually formed the basis of a lot of the activity in the Jordan-Israel peace talks.

11. Fast Turns in Events

The talks remained stalled for quite a while. On Saturday evening, 28 May 1994, the Prime Minister told Haddadin and Dr. Michael Marto, deputy governor of the Central Bank, to get ready and travel to London to meet with King Hussein. Dr. Marto cared for the trade issues in the negotiations with the Israelis, and also talked over banking arrangements over the Governor of the Central Bank of Israel. The next day, there were more delegates at the airport than just Marto and Haddadin. The entire delegation was going to London to meet with the King, and from there they would proceed to Washington, D.C. for a round of Trilateral talks. Dr. Jawad Anani, Minister of State, followed them to London on Monday. It looked like more new ideas were developing, and a dynamic change in the plans was in motion.

"But all I have is 8,000 Sterling," exclaimed Marto when he saw the crowd growing without funds on them.

"You will do okay," Haddadin said, "you have many friends at the Bank of England. You can borrow money from that Bank to pay the hotel. Keep the cash because we might need it for meals and emergency!"

More Jordanian delegates converged on London from the United States. Ambassador Tarawneh and Dr. Abdallah Toukan arrived on Tuesday from Washington, D.C. and Cape Cod respectively. The delegation met with the King at the Claridges Hotel on Wednesday, 1 June 1994. Prince El Hasan was attending too. The meeting dealt with the potential developments in the peace talks with Israel, and His Majesty listened to presentations from some of the delegates.

"Dr. Munther," said His Majesty, "would you brief us again on the strategy of negotiations over water. What are we after, and how do we go about it."

Haddadin immediately sensed there was something inconveniencing the King about the water talks. He had presented the strategy on water negotiations through position papers to His Majesty on two previous occasions, and a presentation as recent as 14 January 1994. It was clear that the King wanted him to present it in front of the delegation.

"Sire," said Haddadin, "we are after the rightful share of the Kingdom as it was defined back in 1955 by the Arab Technical Committee, as a result of their negotiations with Ambassador Eric Johnston Plan. The share so defined included the West Bank as part of the Kingdom. We are including the rights of the West Bank in our pursuit for two reasons:

a) We stand a much better chance to rescue the water share of the West Bank currently totally utilized by Israel from Israel, and,
b) We have hosted hundreds of thousands of Palestinians from the West Bank since the rights were defined in 1955. They own part of these rights."

His Majesty looked serious and said, "*arjouk ya akhi* (Please) take good consideration of the decision of disengagement with the West Bank. Palestine has its own men to defend Palestinian rights. We limit ourselves to the Jordanian rights as Jordan stands today."

Although it was highly unusual to argue with the King, his instructions came as a shock to Haddadin, and as a sudden change in strategy.

"But, Your Majesty," Haddadin answered, "those West Bankers living amongst us have to drink. They can do so using their own water share in the river basin. Whoever of them exercises the right of return that may be granted to them in the future will take back with him his own water share."

The justification did not appeal very much to His Majesty, who interjected: "But there is land on the West Bank that did not cross over to the east, and that land has irrigation needs. Please disengage!"

"Yes, Sire, I will."

After the meeting ended, the King and the Crown Prince conferred with Dr. Tarawneh, the head of the bilateral delegation, before the King bade the delegation farewell. The participants then went back to resume the meeting.

"It looks like you are not convinced with His Majesty's argument," said the Crown Prince addressing Haddadin.

"Who am I, *Sidi*; I am but a delegate who implements the policy of the country, and we have heard it from the source. What else do I need? There was no ambiguity in His Majesty's instructions," Haddadin answered, "and I shall abide."

The Trilateral meetings started in Washington on 5 June. Group (A) finalized the sub-agenda on Water, Energy and Environment, after which the parties would be ready for substantive negotiations. On 6 June the Prime Minister of Jordan approved the wording of the last negotiated item, 1.A, and the sub-agenda became a common sub-agenda (Appendix 13).

There was talk about moving the negotiations to the region, and the talk had a serious tone, more than it ever had before. It was announced at the end of the Trilaterals on 7 June that agreements were reached on several topics of trilateral nature, and also on a common sub-agenda to negotiate water, energy and the environment, including the Jordan Rift Valley. Announcement was also made to move the negotiations to the region; an objective Israel had been after since the opening of the peace conference in Madrid. A sharp turn of events was taking place.

12. The Bilaterals in The Rift Valley

Pursuant to the announcement to move the Jordan-Israel negotiations to the region, preparations were made to hold them in an appropriate place. The first meetings would be held on an uncontested portion of the Armistice Line between Jordan and Israel. This meant a location in Wadi Araba, because the other segments of the Armistice Line lie on water in the middle of the Dead Sea and the midstream of the Jordan River. The place was chosen, and the task of making the place ready to host the negotiation sessions was given to the Jordan Armed Forces and to the Israel Defense Forces, each working on their side of the Armistice Line. The bilateral negotiations were set for 18 July 1994. Simultaneously, arrangements were being made for a historic meeting at the White House between King Hussein and Yitzhak Rabin, under the auspices of President Clinton. It looked like a big push was in the making for the bilateral peace talks, and for the Peace Process at large.

In line with these preparations, and upon the recommendations of the United States, the Trilateral Economic Committee was to hold a meeting at the Dead Sea Spa Hotel on the Jordanian shore of the Dead Sea. The United States Embassy in Amman forwarded a memorandum to Jordan's Ministry of Foreign Affairs on 12 July to that effect. The Trilateral meetings would start with a ministerial level meeting and continue therafter at the expert level. The ministerial level meeting meant that Mr. Shimon Peres, the Foreign Minister of Israel, would come to attend the meeting along with Dr. Majali, the Prime Minister and Foreign Minister of Jordan, and Secretary of State Warren Christopher.

Dr. Tarawneh headed the bilateral meetings in Wadi Araba, the first to be held in the region, on the morning of Monday, 18 July 1994, in an atmosphere of optimism. It was circulated that His Majesty King Hussein would soon meet with Prime Minister Rabin at the White House, under the auspices of President Clinton.

A big Army tent was pitched right on the Armistice Line. Support facilities were housed in tents on either side, and telephone lines were provided. In the central section of the big tent, a table in the middle was set for the delegation meetings. They convened for a short while and posed to the cameras of the press before they split up in the usual Groups.

Group (A) on water, energy and the environment occupied the central table. The Jordan delegation would occupy the chairs on the east side (on Jordanian soil), and the Israeli side would occupy the western side of the table. Despite the familiarity with the Israeli delegation, the Jordanians felt the tension of that occasion because the meetings were taking place for the first time in Jordan and Israel (the two countries still had not exchanged formal recognition). The tension caused the Jordanians to explode at any insignificant error made casually by any Israeli delegate.

The Group started from where they ended in Washington, D.C. The Jordanians stuck to the water items of the Common Sub-agenda, and the Israelis wanted to address the other items of cooperation in the fields of environment and energy. The topics of each were listed in the Sub-agenda. On the water side, the Jordanians wanted to pave the way to an agreement on the principles of water sharing, and to draw attention to the problems of water quality and water shortage. The Israelis wanted to deal with the issue of transfer of water across the common boundary. They wanted to get water from Jordanian territories in Wadi Araba, and they were prepared to trade that with water transfers from Israel to Jordan in the north. The negotiations culminated on the following day with two common documents, pertaining to water and the environment respectively, and they were initialed and announced on 19 July 1994 (Appendix 14). Obviously, there was some progress prompted and enhanced by the optimistic attitudes and raised the forthcoming Trilateral Committee meetings at the Dead Sea Spa Hotel, and by the forthcoming meeting at the White House.

As the Wadi Araba meetings ended on the late afternoon of 19 July, the delegation had to hurry back to Amman, rest for whatever was left of the night, and wake up to be driven by bus the next morning to the Dead Sea Spa Hotel, in the Rift Valley, for the Trilateral Economic Committee meeting. In the American initiative to hold that meeting, the Jordan Rift Valley development occupied a substantial space in the agenda. Another topic was to look into projects that were not capital intensive, but cooperative in nature, to have them encouraged and started. This kind of approach, although coming from the Peace Sponsor, responded to the Israeli interest. On the part of the Jordanians, they did not favor cooperation before the causes of conflict were addressed and resolutions found. When that is achieved, they would favor cooperation on projects of worthwhile economic and social returns.

The trilateral meetings started on 20 July at the Dead Sea Spa Hotel. Secretary of State Christopher headed the American delegation,[328] the Jordanian delegation was headed by the Prime Minister and Foreign Minister, Dr. Abdul Salam Majali,[329] The Israeli delegation was headed by Minister of Foreign Affairs, Mr. Shimon Peres, and included Ambassador Rubinstein and other delegates of the bilateral negotiations, and a score of journalists to cover this occasion on which an Israeli official was visiting Jordan.

The meeting reviewed the progress made over the past two days, and the delegations agreed to prepare a Draft Master Plan for the Development of the Jordan Rift Valley and on other matters.[330] The results of the Trilaterals added to the atmosphere of optimism, and attention became focused on the forthcoming meeting at the White House between His Majesty King Hussein and Prime Minister Rabin a few days later.

The historic meeting between the Jordanian Monarch and The Israeli Prime Minister took place at the White House, on 25 July 1994, and the "Washington Declaration" was announced. The King and the Prime Minster committed themselves to peace and pledged their personal efforts to overcome obstacles in the way of reaching peace. The Declaration gave a role to Jordan in caring for the Islamic holy places in Jerusalem, which triggered strong criticism from the Palestinians. They contended that Jordan had nothing to do with Palestinian territories and displayed this sensitivity at each turn of events. Jordan responded by explaining that it was imperative not to leave any vacuum regarding the holy places that were exempted from the disengagement decision of 1988. Any vacuum, Jordan contended, would leave Israel free to do whatever she wanted with the holy places in Jerusalem. The Washington Declaration gave a push to the momentum of peace-making between Jordan and Israel, and the negotiations resumed, not in the tents, but in a five star hotel, the Moria Plaza, on the Israeli side of the Dead Sea, by the Dead Sea Works.

13. Water at the Moria Plaza

In preparation for the sessions at the Moria Plaza Hotel, the Jordanians organized their agenda and coordinated positions in a meeting held at Aqaba airport, where key members of the delegation met with Crown Prince El Hasan on 8 August 1994. The talks with the Israelis started the next

---

[328] Included Under Secretary of State for Economic Affairs, Ms. Joan Spero, Deputy Assistant Under Secretary of State for Near Eastern Affairs, Ms. Tony Verstandig, and Regional Affairs Officer for Near Eastern Affairs, Mr. Jake Walles.

[329] Included Minister of State for Foreign Affairs, Mr. Talal El Hassan, Jordan's Ambassador to Washington, Dr. Fayez Tarawneh, Dr. Munther Haddadin, and Dr. Safwan Toukan, Under Secretary of Planning.

[330] See the joint statement (Appendix 15), issued on 20 July 1994.

day and were conducted under the auspices of the Trilateral Economic Committee, but also included topics listed on the agenda of bilateral negotiations. The Jordanian delegation flew to the Hotel by helicopters from Amman Airport in the morning, returned before sunset, and held meetings at the Ministry of Foreign Affairs to prepare the daily reports, compare notes and coordinate among the various negotiation groups.

On the second day of negotiations at Moria Plaza, the head of the Israeli water team, Mr. Noah Kinarti, approached his Jordanian counterpart, Dr. Munther Haddadin, as the two were having doughnuts with the members of their teams, before the sessions were due to start.

"I would like to talk to you by myself alone. Could you talk to me alone, one-on-one? I do not want my colleagues or yours to know that we are talking," said Kinarti.

That was the only time Kinarti wanted to talk in confidence. Haddadin sensed there was something serious Kinarti wanted to talk about. When he responded affirmatively, Kinarti asked where they could meet without triggering suspicions.

"Follow me," Haddadin said as he put aside his coffee cup, stepped out of the room on the sixth floor and headed for the elevators. Kinarti followed, and they took the crowded elevator down to the lobby where a large crowd of journalists and correspondents of the international press were gathered. There was a hall in the lobby where free coffee was served to the press correspondents, and a coffee shop next to it where customers could be served for pay. The team leaders entered the crowded coffee shop and headed for the only vacant table. Haddadin looked around in the fully occupied coffee shop and recognized five Jordanian journalists occupying the table next to them. Among the journalists was Caroline Faraj, the correspondent of *al Rai* newspaper. Her eyes were shining as she had looked at Haddadin and Kinarti as though she discovered something her colleagues did not notice. She probably wondered what the team leaders were doing in a coffee shop at the time their session was supposed to start.

"Caroline, come here please," Haddadin asked. And she quickly abided.

"What can I do for you, for both of you?" She asked.

"Leave me your lap top computer, and please go get us free coffee from the adjacent press hall."

"With cream and sugar?"

"No, just black," Haddadin said, and Kinarti had a similar order.

Caroline went outside and came back with two cups of black coffee, and she was happy to be in on what she thought would be an important meeting.

They thanked her kindly, and Haddadin asked that she stay as far away from them as she possibly could!

The bright journalist left the coffee shop as Haddadin had instructed her, and sat in the lobby where she could see them through a large glass window, and hoped to read their lips.

In the very crowded lobby of the Moria Plaza hotel, which was jammed with journalists, the two leaders of the water teams started their secret conversation.

"What is it that you want to talk about?" Haddadin asked.

"What else, I want to talk water."

"Drink your coffee!" Haddadin suggested.

"How much of the Jordan River are you asking for?" Asked Kinarti.

Haddadin remembered the impression he had gotten from the Israelis when the common sub-agenda was concluded, that they would be inclined to recognize a right for Jordan in the Jordan River. He also wanted to make sure that they were talking bilaterally about Jordanian-Israeli matters only.

"Listen, my friend," Haddadin said, "there shall be no serious talk before we make it very clear that our talk applies purely to Jordan and Israel only. Whatever we agree to shall in no way prejudice the rights of the other riparian parties on the Jordan River basin, particularly the Palestinians."

"But why, who would then talk about their share?"

"I am not authorized, nor am I willing to talk on their behalf. They have to do that themselves, and they are very able to defend their rights."

Kinarti argued back, and Haddadin was adamant in his refusal to include any Palestinian matter in their talks.

"Strange," said Kinarti, "you insist on their independence, and support their side, but they are not taking your side."

"Listen, Kinarti, I am always prepared to give them all the help they may ask for, and I stand to be supportive of them the best I can if they ask for it. This is Jordan's position and I am a strong believer in that."

"They came to me yesterday, and they asked me to speak on their behalf," said Kinarti.

Haddadin was shocked at first, but then reckoned that Kinarti might be trying to drive a wedge between the Jordanians and the Palestinians.

"Are you serious?" Haddadin asked.

"Yes, I am."

"Who came to you and authorized you to speak on their behalf?"

"I can not reveal the name."

"Fair," said Haddadin, "but our talks would not start before we have that settled. Neither the rights of the Palestinians, nor the rights of any other riparian party will be the subjects of our talks at any time. Is that clear and acceptable?"

"Okay," said Kinarti finally.

Haddadin opened the laptop computer, made a table of two columns, and typed a summary of what he said, and Kinart's responses. He read the text to Kinarti.

"I want you to fully understand what the text means. I do not want to hear an argument down the road that your English is not very good, and that this text does not fully reflect what you said," asserted Haddadin.

After reading the text, Kinarti and Haddadin got to the hard substance of negotiations over water.

"How much are you asking for from the Jordan River?" Asked Kinarti again.

Haddadin thought the answer that he would give could be self-incriminating. The shares of the Kingdom from the Jordan River according to the Unified (Johnston) Plan that the Arab Technical Committee approved were quantities earmarked for the irrigation of the West Bank, and that was a Palestinian share. But since there had been agreement that the negotiations were bilateral, Haddadin decided to give the justification for his request for any quantities from the Jordan River.

"You know, Noah, when the water rights were defined back in the 1950's, no consideration was made of the water requirements for municipal uses in the basin. The municipal needs then were little and Jordan had other sources of domestic water supply. Today, however, those needs are much higher." Haddadin explained. "Amman, as you know, lies in the basin of the Jordan River, and it depends on the water from the King Abdallah Canal today for municipal supply and you know that. So, based on the municipal and other pressing needs, I will say my figure."

"I understand," said Kinarti, "but how much do you want from the Jordan?"

"I will not be very demanding. I can live with 200 mcm per year, and that would be allocated, mind you, for municipal uses in the cities in the basin."

"Your share ranges between zero and 200 mcm per year. Neither figure is reasonable. So, it must be in between," said Kinarti.

"I will be comfortable with the 200 mcm figure, not more. I do not want to sound like an impossible negotiator," Haddadin answered. Kinarti suggested that the quantity be decided on basis of municipal needs in Amman, but Haddadin advocated adding the needs of irrigating some perennial crops, as well. Kinarti demanded that the existing uses by Israel of water from the Yarmouk should be maintained, a suggestion that would leave some 75 mcm (65 mcm according to Kinarti) in Israeli hands. Haddadin insisted that Israel's share from the Yarmouk should not exceed 25 mcm per year as per the

Unified Plan.[331] Kinarti wanted the surface water in the north (Jordan and Yarmouk) and the groundwater in the south (Wadi Araba) be dealt with as a package, and Haddadin was open to suggestions, provided they would not compromise Jordan's sovereignty over her resources. The conversation went on and on without agreement on the figures or the principles. But the talks addressed the substantive issue of water sharing for the first time.

Simultaneous with the meeting in the coffee shop, a separate meeting on the Jordan Rift Valley was being held on the sixth floor, Mr. Zafer Alem deputized for Haddadin to chair the Jordan side. In that meeting, agreement was reached on the concepts advanced by Jordan in the Concept Paper, and on the need to employ consulting firms to study the proposed scheme. The two sides agreed to prepare "Terms of Reference" for consulting services, each side would propose his own version and present it in the next meeting to be held in three weeks, on 29 August, at the Dead Sea Spa Hotel on the Jordanian side of the Dead Sea.

The next day, 10 August, the talks continued but the positions did not change. Jordan emphasized the independence of Palestinian water matters from Jordanian-Israeli bilateral matters. In that regard, Kinarti wanted to amend the record of the minutes taken the day before, regarding the Palestinians asking him to speak on their behalf.

"They actually did not say that I should speak on their behalf," said Kinarti, "they asked me not to forget them when I talk to the Jordanians."

"You see, Noah" Haddadin said, "remember that I read the text to you and asked you to fully realize what it meant, and not to take your English as an excuse."

"But that was what they said!" Contended Kinarti.

"Okay, Noah; I will correct the record in today's minutes if you tell me who it was that came to you on that mission, and told you not to forget them in your talks with us."

"J.T," and Kinarti mentioned the name of a high official in the Palestinian ranks. Haddadin was very surprised.

The sessions revolved around much the same issues, and the positions did not change much. Appendix 16 shows summary minutes of the sessions on 9 and 19 August 1994. The developments were reported to the High Steering Committee every day, and there was optimism that some kind of an agreement could be reached.

The Jordanians tackled the borders issue with the Israelis in plenary sessions at the Moria Plaza, Israel made a good presentation of the border issue as the Israelis saw it. They said that the phrase defining the borders in the Mandate text, "up the center of Wadi Araba," meant the median line in

---

[331] The Unified Plan water division was endorsed by the Jordanians in official meetings in March and April, 1993.

that Wadi. To them, the median line bisected the distance between the mountain- tops overlooking Wadi Araba from east and west. The Jordanians explained that the center of Wadi Araba meant the "thalweg" which was the line connecting the lowest points in the Wadi bed. The Israeli interpretation shifted the borders to the east towards Jordan. Discussions then started, and focused on the change to the Mandate borders near Eilat and Aqaba, a change that was effected in 1946 by the Director of Lands and Surveys in Palestine and his counterpart in Transjordan. Both were British officials by virtue of the Mandate. The transactions started in December 1945, a few months before Jordan gained independence, and were completed in September 1946, a few months after the country was pronounced an independent Hashemite Kingdom in May 1946. Both dates preceded the establishment of the State of Israel on 14 May 1948. The State of Israel had not been created when those border adjustments were made. In the discussions that followed, the Israelis sounded like the 1946 border adjustments were mutually agreed to between Jordan and Israel.

"Where exactly do you put Israel in all of these developments?" Haddadin asked the Israeli side across the negotiation table. It looked like the question came as a surprise to the Israelis.

Three Israeli lawyers, each with a clear British accent, took turns in explaining the involvement of Israel, but none was in line with the argument presented by his other two colleagues.

"I will answer you by the word of international law," said the third lawyer as he flipped through some pages in a file he had in front of him, "it is the rule...."

"The Rule of Succession of States," interrupted Haddadin.

"Exactly, sir," said the polite lawyer, "the Rule of Succession of States. Israel succeeded and inherited the government of the British Mandate."

"And I accept that reasoning," Haddadin said, "and appreciate the fact that you are quoting from the words of international law. I would like it stated on record that Israel, and based on the provisions of international law, abides by the Rule of Succession of States."

In that statement lay an implicit intent to pin down the Israeli delegation to abide by international law and the resolutions of the United Nations Security Council and the General Assembly, of which they are part. In particular, the Jordanian delegate was aiming at the resolution of Item 4 on the Common Agenda that addressed Refugees and Displaced Persons. U.N. General Assembly Resolution 194 of 1948 handled the refugees and Jerusalem, and U.N. Security Council Resolution 237 of the year 1967 handled the case of displaced persons. Resolution 181 of the year 1947 addressed the partition of Palestine into a Jewish and an Arab State and set Jerusalem aside for internationalization. The Palestinians, under the British

Mandate, were the responsibility of the British Government before Israel's creation caused their eviction and misery. The Rule of Succession of States would make Israel inherit that same responsibility that the preceding British Government had for the Palestinians, and therefore commit Israel to the repatriation of Palestinian refugees. Additionally, the provisions of international law would oblige Israel to conform to the above resolutions of the General Assembly and the Security Council.

Ambassador Rubinstein quickly grasped the potential trap his colleagues were falling into. He took the floor and asked to change the subject. That drew an angry reaction and a clash between Haddadin and Rubinstein, and the meeting was adjourned. Feelings ran high on both sides, but they were restored to normal two days later.

14. Terms of Reference, JRV

The Concept Paper of the JRV development gave the parties an opportunity to work on an uncontested topic. The Jordanians knew what was needed next and started to work on a "Terms of Reference for Consulting Services." Their text was ready on 27 August 1994, in time to present it to the following Trilateral Economic Committee meeting at the Dead Sea Spa Hotel on 29 August.

"Nice looking," said Prince El Hasan, when Tarawneh presented the document to him, "but of all the documents Haddadin has prepared throughout the peace process, this is the only one on which he attached his name."

"I asked him why, and he told me it was because the document had not been cleared by Government," said the Chairman of the delegation, Dr. Tarawneh, "and we are going to discuss it with the Israelis in two days. He did not want to commit the Government to his own views, and that is why he put his name as the author, which he is."

The Trilaterals were to meet in Jordan and resume in Washington D.C. a few days later. The Jordanian team presented the Terms of Reference document, and the Israeli team presented their proposal for the same topic. Dr. Hani Mulki and Dr. Fayez al Khasawneh, President of the Aqaba Regional Authority, cooperated to fill in for Haddadin, who was asked to travel to Germany on a mission connected to the Jordan Rift Valley development concept. The Israeli Military Industries had joined efforts with Noell, a leading German construction firm, and with a Japanese banking group to prepare a mega-project to link the Dead Sea with the Mediterranean at Haifa. This project was advocated by the Israeli engineer, Shlomo Gur, and had its roots in the early works of Zionism. The water intake would be on the Mediterranean at Haifa, pass under the Carmel mountain via a tunnel, and would be piped to the Jordan River at Beit Shean in the West Jordan Valley.

Reverse osmosis technique would be used to desalinate water.[332] The water would be stored in a reservoir created by a dam on the Jordan River, and remaining brine, the cooperating parties proposed, would be discharged in the Jordan River to flow down to the Dead Sea.

A project like this competed with the Red Sea-Dead Sea connection included in the Jordan Rift Valley development concept, which the Jordanians prefered. Each project would be barely feasible, if it could carry the entire amount of water proposed. If two projects were thought of to carry jointly that same amount, none of them would be feasible. There is room for only one project.

"How could we change the character of the Jordan River?" Haddadin asked; "it is an important element of Christian heritage and is holy because of the baptism of Christ. We should look for another alternative to dispose of the brine."

The cooperating parties suggested transporting the brine down to the Dead Sea via pipelines on the Jordanian side of the River. The proposal entailed additional expenses, as well as added environmental hazards. The discussions lasted for two days, in which the Jordanian participant pointed out the environmental hazards associated with the plan, ranging from potential contamination of groundwater inside Israel should the pipes from Haifa leak, to potential contamination of the Jordan River should the brine pipes break under an earthquake, an event for which the Jordan Rift Valley has been famous in the past.

The project calls for the storage of desalinated water behind a dam to be built on the Jordan River at Beit Shean. However, the storage capacity of any site on the Jordan is limited and would not hold the outflow from the desalination plant unless the plains on both sides of the river were allowed to be inundated, in which case the loss to evaporation would be substantial (about 2500 mm per year). Also, because the desalinated water would be stored in the open, floodwaters would blend with it, and it would therefore have to be treated for municipal uses, another added expense. The meeting concluded with the understanding that more studies, especially on the environmental impacts, should be done before the project could be proven feasible. This gave time to accelerate the steps to study the Red-Dead Canal in more detail. That study was included in the Terms of Reference for the Jordan Rift Valley consulting services that was formally discussed with the Israeli delegation at the Dead Sea Spa hotel on the same day.

---

[332] In the early version of the Project, conceived in 1899, electric power would be generated and used to pump water to irrigate the slopes of the western Jordan valley escarpment. See Bourcart's work, Chapter II.

The second leg of the Trilateral Committee meeting started on 31 August 1994, at the State Department in Washington, D.C.[333] The meeting was to continue the progress made two days earlier at the Dead Sea Spa Hotel, and to prepare for a ministerial meeting of the Trilaterals. Rubinstein gave a briefing on the progress at the Dead Sea Spa, where talks were conducted on tourism, civil aviation, border crossings, parks, transportation, and the Jordan Rift Valley. He commended the Jordanian document on the JRV and said that the difficult points in the bilateral talks were water and borders. No progress was made on water "because of the absence of Dr. Haddadin," he said. Tarawneh responded with an equally optimistic tone, and talked about the forthcoming ministerial meeting for which preparation of an agenda was necessary in the hope that substantive announcements would be made.

"Excuse me," said Yosi Vardi from the Israeli delegation as he was given the floor, "I am not a diplomat and I normally tell it as it is."

"Dr. Haddadin," Vardi said as he looked at him with the Terms of Reference document in his hand, "you made us ashamed of ourselves when we compare this document with these modest pages that we had prepared," and he waved the document in his hand, "We feel ashamed; your document is a masterpiece and we all endorsed it."

That was such an unexpected compliment. It was not expected that an Israeli would concede that the work of a Jordanian would be superior to the parallel work of Israelis, but Vardi said the unexpected.

15. Involvement of the World Bank

Ambassador Ross seized the opportunity in that Trilaterals meeting to make progress. He asked for ideas on how to proceed. The Jordanians suggested that the next step was to go out for bid based on the Terms of Reference already endorsed by the two sides, and acquire the services of a competent consulting firm. There was one obstacle pertaining to the legal entity of the contracting party that would assume the role of the "Employer" or the owner of the works. To go around that difficulty, the Jordanian team suggested that the World Bank be approached to play that role on behalf of both Jordan and Israel.

---

[333] Attending from the Israeli side were Uri Savir, Rubinstein, Etimar Rabinovich, who used to head the Israeli delegation in the bilateral negotiations with Syria, Yosi Vardi, Rafi Benvenisti, and others. From the American side, Ambassador Dennis Ross headed the team that included Tony Verstandig, Dan Kurtzer, Aaron Miller, and others. The Jordanian team was headed by Ambassador Tarawneh and included Dr. Fayez Al Khasawneh and Dr. Munther Haddadin. In the afternoon they were joined by Awn Al Khasawneh and Hani Mulki.

Ambassador Ross liked the idea. "But if the Bank joins," he warned, "they will take the driver's seat. Are you prepared for a drive steered by the Bank?"

"We do not have worries about steering the drive. We are used to the Bank and have a good working relationship with them. We trust their technical competence, and this is more reason to bring them in," Haddadin responded.

Ambassador Ross asked his assistants to make a call to the Vice President of the World Bank in charge of the Near East and invite him to the afternoon session. In the afternoon session, the World Bank Vice President, Mr. Caio Kochwesser, and the Director of the Near East Department, Mr. Ram Chopra, attended and expressed preparedness of the Bank to join the effort. They made arrangements for a Bank's Mission to go to the field and prepare a Project Paper that would form the basis of the Bank's involvement and the involvement of other donors. Some Bank people were already in the region at that time on a different mission.

"You know John Hayward," said Mr. Chopra, "he is out there with three others. We will contact him to get them ready to conduct that field mission, and will be joined by other specialists."

"Excellent, we will be pleased to invite him to the Trilateral meeting due to be held on 12-13 September at Beit Gibrael, if our Israeli friends agree," Haddadin proposed.

"By all means," said Rubinstein, and the process for studying the Jordan Rift Valley was set in motion.

16. Meetings in Beit Gibrael

A bilateral meeting was set for the review and clearance of the document on the JRV Terms of Reference. A small group of three members from each side met on Thursday, 8 September, and went over the document paragraph by paragraph in preparation for a plenary meeting of the Trilateral Economic Committee[334] on Monday, 12 September.

The Trilateral meetings on Monday and Tuesday, 12-13 September, in Beit Gibrael were successful. Bilateral meetings were conducted too. Lunch was served on a boat cruising in Lake Tiberias and the main dish was fish from the Lake.

"When we told Prime Minister Rabin that we will have lunch on a cruise in the Lake," remarked the host, Ambassador Rubinstein, "he thought that we selected the wrong place." Rubinstein quoted Rabin as saying that the

---

[334] On the Jordanian side were Haddadin, Mulki and Dr. Ibrahim Badran; and on the Israeli side were Yosi Vardi, Rafael Benvenesti, and Noah Kinarti.

cruise in the Lake would only make Haddadin more determined to demand more water from the Sea of Galilee!

"So, Dr. Haddadin," added Rubinstein as he pointed to the surface of the Lake, "this is not water." And the company all laughed.

In parallel with the extended meeting of the Trilaterals that discussed the JRV in Beit Gibrael, a bilateral meeting took place. This meeting was three on three. The Jordan side was led by Ambassador Tarawneh and included Awn al Khasawneh and Haddadin, and the Israeli side was led by Ambassador Eliakim Rubinstein and comprised Moshe Kochanovsky (borders) and Noah Kinarti (water). The bilateral talks displayed seriousness and linked, for the first time, the settlement of border issues with the settlement of water issues. "The solution for water in the north and south will be related to the issue of borders," said Kinarti.

The Israeli proposal to settle the borders issue south of the Dead Sea was to have a "generalized corridor" bound by two broken lines running north- south. The one to the east would be the current line that was fenced by Israel and infringed on Jordanian territories, and the line to the west would be the 1949 Armistice line. In between, a series of straight lines could be drawn and agreed to, and these would form the borders. Israel's intent was to keep on their side of the borders all the lands they had developed in agriculture, and the wells that supply them with groundwater from Jordanian territories, and to give back to Jordan about 50% of the occupied Jordanian lands. In Baqura, the occupied land in the north at the confluence of the Yarmouk and Jordan Rivers, Israel proposed joint tourism projects, but indirectly hinted that sovereignty over the occupied land would be returned to Jordan.[335]

"The more the border is to the east the more will be the water for you in the north," offered Kinarti. "If the package is worked out and agreed to," he added, "Israel will be able to help you next summer to meet municipal water shortages in Amman. When we say water in the summer, there is an equation. The value of summer water is a lot higher than the value of the same water quantity in winter."

It was clear that Kinarti wanted to trade water for land, and to have a compromise worked out on the borders in Wadi Araba and on water, simultaneously. He also wanted to assign values to water in accordance with the demand for it: higher in the summer than it is in winter[336]. Two issues were important to the Israelis, the borders between Aqaba and Eilat which

---

[335] The most likely reason for Israel to return the land to Jordan's sovereignty is the fact that more lucrative tourism attractions could be built under Jordanian sovereignty, away from the stringent ban on casinos and gambling.

[336] The notion of treating summer water differently was first introduced in the June 1963 report by Wayne D. Criddle, prepared for the State Department. See reference to the report in Chapter IV.

were delineated and demarcated in 1946, and the "de facto" borderline in Ghor Feifa south of the Dead Sea, where the Israeli saltpans of the Dead Sea Works infringed on Jordanian land east of the Mandate borders. "These are non-negotiable," said Moshe Kochanovsky. In effect, Israel wanted to swap land for water without going back to the principles that define the water shares of each party.

"Looking ahead is more fruitful than digging up the past," said Rubinstein, "We shall develop more water resources and Jordan will get most of what is to be developed."

The following day, the meetings resumed. The Jordanians felt like there was movement towards coming to terms with the Israelis. The tone of the bilateral talks was clear, and the agreement on the "Terms of Reference" document was reached, mostly along the lines of the Jordanian paper that was submitted in late August at the Dead Sea Spa Hotel.

17. The September Turn

Events took a sudden turn a week later. After the Beit Shean meetings, several Jordanian negotiators, including Tarawneh, Awn al Khasawneh and Haddadin, were included in the company of Prince El Hasan who was scheduled to address the General Assembly of the United Nations and to attend a high level meeting of the Trilateral Economic Committee at the White House with President Clinton, and Foreign Minister of Israel, Shimon Peres. Prince El Hasan was to head for America from Europe, where he was having a short vacation. Tarawneh had left for Washington, D.C. where his post was, and Haddadin, along with staff from the Office of the Crown Prince, left for New York on the morning of Friday, 23 September on board Royal Jordanian flight 261 via Amsterdam. The Station Manager of Royal Jordanian in Amsterdam surprised the staff of the Office of the Crown Prince with instructions that they should return to Amman, because the trip of the Crown Prince to the United States had been postponed. This charge was a big surprise to all.

"Any instructions for me, Ray?" Haddadin asked.

"No sir."

Haddadin thought for a moment and decided to return to Amman with the group. By 6 p.m. he was home again, after 12 hours spent mostly in the air. On the following day, Saturday, he did not succeed in getting an explanation of what was going on. To his surprise, he was told that the Director of the Office of the Crown Prince was on his way to New York that morning! Haddadin revived a commitment he had canceled to go to Cairo on a consulting mission for the FAO. While in Cairo, Haddadin was further surprised with a headline that the Cairo daily, *Al Gumhuriyya*, carried on Tuesday, 27 September. The sensational headline across its front page read:

"A Peace Treaty Between Jordan and Israel." The article spoke of a Treaty of Peace being negotiated between Jordan and Israel.

"Could this be true?" Haddadin asked the Jordan Ambassador in Cairo, Mr. Nayef el Qadi.

"You should know of such a development before I do," noted the Ambassador rightfully.

Haddadin did not have any idea, and decided to return to Amman that afternoon after his FAO mission was completed.

After he watched the evening news at home, he received a phone call.

"What are you doing?" asked the Director of the Communications Office of His Majesty the King.

"Eating delicious French fries, come and join me."

"Thank you, but would you be surprised if I told you that you are needed?"

"No, I would not be surprised."

"You did not ask when?" Said the Director.

"When?" Haddadin asked.

"Right now!"

"I am ready," Haddadin said.

"But you did not ask where."

"Okay, where?"

"Aqaba," came the answer.

"I just was flying over it, I have just arrived from Cairo, but I am ready all the same."

"Please make it to Amman Civilian Airport in Marka as soon as you can. There is a C-130 waiting for you and two others."

Haddadin made it in half an hour and met General Marie, the Chief of General Staff, and General Shurdum, his Assistant for Intelligence. They boarded the C-130 to Aqaba. They were driven to meet with the Crown Prince, at his home in Aqaba. With him were the Prime Minister, Dr. Majali, the Legal Expert and Negotiator, Awn al Khasawneh, the Prince's Economic Advisor, Dr. Ahmad Mango, and two international legal experts[337] on whose services Jordan called throughout the Peace Process. The newcomers were briefed on the recent developments. Israel had handed over to Jordan on Monday, 26 of September, a draft Peace Treaty. Awn al Khasawneh and the international legal experts worked non-stop until they composed a counter proposal -- a Jordanian draft based on the items of the Common Agenda and the Jordanian negotiating position. The Jordanian draft of the Peace Treaty was communicated to the Israeli team shortly before the newcomers arrived, and the Israelis were expected to come to Aqaba the following morning for

[337] Professor James Crawford (Australian) of Cambridge University, and Professor Bernard Graefrath (German).

negotiations. The Jordanian team stayed up late talking about these developments, and spent what was left of the night in the chalets of the Royal Palace, next door to the residence of the Crown Prince.

The following morning, Wednesday the 28th, a small Israeli team, headed by Ambassador Rubinstein, and included General Ehud Barak, then Chief of the General Staff, came to Aqaba.[338] The talks started on the front porch on Security, after which General Barak left.

"We are sorry we did not bring our water people; we were not aware that Haddadin would be here," said Rubinstein.

"He will fit practically anywhere," said the Prime Minister.

Haddadin drafted an Article on Water to be included in the Treaty, had it typed and gave it to Rubinstein, who promised to have it reviewed and negotiated in the following day. On Thursday, Kinarti came along with the Israeli team. It took little time to agree on the text of the water article, which contained general terms and avoided the areas of contention. It referred to an Annex that would detail the agreement on water issues. Other issues, however, were of contention between the two parties. Most notable was the issue of "Refugees and Displaced Persons," the subject of item 4 on the Common Agenda.

"These issues will be dealt with in the Final Status negotiations with the Palestinians," insisted Rubinstein.

"But we will not be at the negotiation table," contended al Khasawneh.

The arguments continued. al Khasawneh was adamant and un-nerved his opponent. Rubinstein, who had quit smoking for quite a number of years, looked at al Khasawneh: "May I have one of your cigarettes?" He asked.

"Please!" And Awn gave him a cigarette and lit it for him. Rubinstein could not stop al Khasawneh from including the issue of refugees and displaced persons in the draft treaty.

Rubinstein then referred to item 6 on the Common Agenda, particularly to the part of its text stipulating "co-operation within a regional context," and took that as a Jordanian tactic to shy away from direct bilateral cooperation with Israel. More effort was made to agree on the texts of some other articles of the draft treaty that dealt with co-operation. Differences persisted concerning the article on borders. Israel was aiming at keeping the border areas she developed in agriculture and also the lands where groundwater wells were drilled to abstract water for Israeli uses. They wanted the borderline to be east of those lands, but this was not acceptable to the Jordanian team.

---

[338] Accompanying him were General Ehud Barak, the Chief of the General Staff, Efraim Halevi, of the Mossad, Moshe Kochanovsky, and Chaim Yisraeli, both of the Ministry of Defense.

Little was achieved before the King arrived in Aqaba for a meeting with Prime Minister Rabin that evening. In the spirit of the 25 July Washington Declaration, issued when the King and Rabin met at the White House, the two leaders met to facilitate agreements on issues standing in the way of achieving a Peace Treaty. The meeting lasted for a few hours interrupted by dinner, and some progress was made. Rabin alluded to the status of peace-making with the Palestinians, the Syrians and the Lebanese, and said that "there is no way Israel can take back the Palestinian refugees," and added that the issue of "displaced persons will be taken up by the quadrilateral committee formed in the wake of the Oslo Accord."

"Jordan has every right to seek an equitable solution to the problem of refugees and displaced persons," said King Hussein, "and expects to get her rightful share of water."

On Water, Rabin said that Jordan would be getting more than 100 mcm per year as a result of the Treaty. To avoid misunderstandings later, Haddadin handed the Crown prince a note.

"Rabin may be talking about the Yarmouk floods when he spoke of 100 mcm, but those floods are ours anyway," Haddadin wrote in the note. The Crown prince then stressed that "the figure that the Prime Minister mentioned should not include water that is Jordan's anyway."

After dinner, Rabin and the King had a short one-on-one meeting. In the mean time, Barak proposed to the Crown prince that the Israeli forces could pull back in Wadi Araba, halfway to the Armistice Line, provided that some joint cooperation would start soon afterward. When asked for an opinion, Haddadin said he would accept a pull-back, provided that a final withdrawal is achieved shortly after that. At that point, the King and Rabin were walking down the stairs from the second floor when the Crown Prince mentioned the Barak initiative in Wadi Araba. "Things will be better in the next round of negotiations when Prince El Hasan returns from his trip to the United States," said the King.

A statement was released to the Press after the meeting. After Rabin left, the Jordanians sat with the King and reviewed the progress. Shortly before midnight, the King let his brother take the Royal plane, the *Hashemite Kingdom of Jordan*, to return with the team to Amman. On the way, the Crown prince told al Khasawneh and Haddadin that they should be ready to join him and his company to the United States in a few hours!

On Friday, 1 October 1994, the Crown prince and his company boarded the King's plane to go to New York. After take-off from Amman Airport, the aircraft took a path that was unlike the regular path the planes took to go to Europe. Haddadin looked out the window, and the plane was heading west, over the Dead Sea. The passengers were excited to see Jerusalem from the air. This was the second time this very jet had flown over

Israel. The first was on 28 July 1994, when the King returned from his trip to the United States after he met with Prime Minister Rabin.

The plane landed in Shanon, Ireland for refueling, resumed the flight to New York and arrived in the evening. The company was driven from JFK to St. Regis Hotel on 55th Street and Madison. They started to work the following morning, in a meeting chaired by the Crown Prince. The tone of the visit was set, and Haddadin was asked to prepare a paper on water for discussion with Mr. Peres, Israel's Foreign Minister. Before the meeting ended, Awn al Khasawneh came in and asked for a word with Haddadin.

"You have to prepare the Water Annex to the Peace Treaty within 24 hours," said Awn.

"Who says so?" Haddadin asked in surprise.

"The Crown Prince," answered Awn.

"But I am not carrying any documents related to the subject, no references," Haddadin said.

"You have composed all the documents, you do not need references."

I guess you are right."

After the meeting, Haddadin prepared a paper for the Crown Prince to serve as a basis for "water talks" with Peres, and proceeded to prepare the Water Annex to the Peace Treaty. On Sunday, 3 October the Jordanian company met in the morning with the delegation accompanying Mr. Peres and talked over the forthcoming meeting between Clinton, the Crown Prince and Peres, and about the announcement to be made after the meeting. Uri Savir, of the Israeli delegation, mentioned the need to make a study for water supply and demand, and Haddadin referred to the study to start soon with sponsorship from Germany. He alluded to the effort he had had to make to convince the Israeli delegation to the Water Resources group of the Multilaterals in Beijing to accept the German-Jordanian initiative, which aimed at conducting a study for water supply and demand.

The Crown prince and Peres joined the meeting. Peres spoke of the need to start work on the Jordan Rift Valley development and on the Mediterranean-Dead Sea connection. Haddadin was asked to make a short presentation and he briefed the meeting on the results of his deliberations in Wuerzburg, Germany, in August, over the Med-Dead Canal, for which more work was needed to assess the environmental cost. He requested that the Red-Dead connection be given equal opportunity to be studied and compared with the Med-Dead Canal. In that meeting, the Jordanian Minister of Transport, Mr. Samir Kawar, handed out copies of Jordan's position paper on Transportation, which Haddadin had prepared for the Ministry in mid-September. The paper was well received. After the meeting, Haddadin went back to work on the draft Water Annex, and finished it in the early afternoon. He reviewed it in a meeting with Awn al Khasawneh and the two international legal experts. Very little amendments were proposed, and Haddadin completed the draft Annex II-Water Related matters on Monday, 3 October, and had it finalized with Professor Graefrath, the international legal adviser, on Tuesday, 4 October 1994. He then gave a copy of it to the Crown Prince.

That evening, Haddadin attended a reception made for the delegations by a group of potential investors, and he approached the Crown prince to be excused to go visit his daughter in Seattle.

"When would you be back in Amman?" Asked the Crown prince.

"I have to stop at the World Bank on my way back to sort out some differences over the contents of the Bank's 'Project Identification Paper' on the Jordan Rift Valley. It looks like our ranks in Jordan are not unified over the Red-Dead Canal, and that is causing us some problems. I will be in Amman next Wednesday."

"Not tomorrow, Wednesday?" joked the Crown Prince.

"No Sir, next Wednesday. I have to convince the Bank people."

"You find your own way to convince them," said the Crown Prince, "the latest you should be in Amman is Sunday the 9th."

"Yes Sir, will do."

"Say hello to Sumaya for me!" Said the Crown Prince who named Haddadin's eldest daughter when she was born in 1972.

Haddadin excused himself and hurried to the airport to go to Seattle, where he stayed until Friday the 7th, headed back to New Jersey to attend his nephew's wedding on the 8th, and left JFK for Amman on that same evening. He arrived in Amman on the evening of Sunday the 9th, and headed for a pay phone to call his wife. He was expecting messages to go to some destination for the negotiations, but there were no messages.

He cleared immigration and customs and took a taxi to go home. By the time he got there, several messages had been left for him. The most important was to head for Aqaba and be there that same evening. Luckily, there was a plane ready to fly to Aqaba in one hour. That gave him time to pick up another suitcase and he made it to the Aqua Marina II hotel in Aqaba that same evening. Tarawneh was there to head the delegation to Eilat the next morning, and the final days of difficult negotiations were about to start.

CHAPTER XI

The Peace Treaty

1. The Bilaterals Merge with the Trilaterals

Dr. Tarawneh headed a large delegation to the Trilaterals to resume talks in Eilat, Israel, on trade, agriculture, health, tourism, culture, education, Jordan Rift Valley and other topics of interest to the Trilateral Economic Committee. The Jordanian delegations crossed to Eilat via the new crossing point inaugurated under the auspices of the Crown Prince and Minister Shimon Peres two months earlier. They headed for the Moria Plaza Hotel in Eilat on Monday, 10 October 1994, where the negotiation sessions were scheduled to take place. There was excitement on both sides, and there was heavy security in and around the hotel. The talks went on smoothly, but many of the Jordanians were facing Israelis for the first time. Veteran negotiators had to go over the work of the rookies and participate in their separate group meetings. There was intensive review of the works of the groups in the evening, after the delegations returned to Aqaba. Consultations were maintained with the World Bank concerned officer to finalize the Project Identification Paper of the Jordan Rift Valley

The Jordanians crossed to Eilat the following day, Tuesday, for more talks over items on the Trilaterals' agenda. At lunch break, Kinarti approached Haddadin in the lobby of the Moria Plaza.

"Why don't we, you and I, sit down and draft the Water Annex that is supposed to be part of the Treaty?" suggested Kinarti.

"Oh, I am glad you brought that up. In my spare time over the past few days," said Haddadin, "I managed to draft the Annex in an attempt to save time."

Haddadin opened his briefcase and gave Kinarti two copies of Annex II-Water Related Matters, the draft he had prepared and cleared with his colleagues in New York. Kinarti was surprised and said that he did not expect the work to be done that fast.

"I will be ready to talk it over with you whenever you are ready," Haddadin assured Kinarti, and they both proceeded to lunch. The Trilateral talks proceeded without difficulties because they addressed Item 6 on the Common Agenda, topics for bilateral co-operation.

There were intensive talks on civil aviation: the flight corridors and the altitudes. No agreement was reached. Other intensive talks were conducted on trade and, again, very little progress was achieved.

More of the same talks were conducted the following day, Wednesday. During the break, Noah informed Haddadin that the water issues "will be resolved by the two leaders." He meant Prime Minister Rabin to whom Kinarti was very close, and His Majesty King Hussein. After he had lunch with some of his Jordanian colleagues, Haddadin moved over to the table where Tarawneh and Rubinstein were having lunch, along with other members of their delegations.

"I read your Annex, Dr. Haddadin," said Rubinstein, "It gave me a heart attack and I fainted."

"Oh no, this is the last thing we would want to be -- a cause for any sudden illness. We need you today more than we did before. We will be ready to mobilize our best heart doctors to come and examine you, and take care of you," Haddadin said.

"Heart doctors will not be enough, I will need a psychiatrist," said Rubinstein.

"That too, but I am afraid these would take some time to check the history, the potential causes and to diagnose."

"No, it will not take long. I know the causes and they center around you!!"

All laughed at Rubinstein's joke, as they often did, and asked him for more jokes.

"I sure would like to be the second Israeli Ambassador to Jordan," he said.

"But why not the first?" Asked Tarawneh.

"Because the first Israeli Ambassador will be shot!!"

And all laughed again, before Rubinstein was called to the phone, waved from a distance to the crowd and left in a hurry. The Jordanians left the restaurant and noted that some Israeli delegates looked serious as they hurried out. In the lobby, the press correspondents, who apparently had seen some Israeli delegates leave the hotel, circled Tarawneh and Haddadin.

"Why are the Israelis leaving?" Asked one correspondent.

"Is it true that you, Dr. Haddadin, made them angry and they left in protest?" Another asked.

Many others asked similar questions and they demanded answers. The Jordanians were equally surprised, because they did not know what was going on. They assured the press that they had nothing to do with the Israelis leaving, if they actually were leaving. It turned out that some of them, including Rubinstein, left the hotel, but others stayed behind to complete the work of the day.

"Where is Awn?" Tarawneh asked Haddadin.

"I do not know. I remember seeing him as we started lunch."

Another member of the Jordanian delegation was missing. It was Nidal Saqarat, the engineer with the Geographic Center, who was in charge of the maps used to delineate borders.

The Jordanian delegation returned early to Aqaba. Tarawneh and Haddadin watched the 8 o'clock evening news and were surprised to see Mr. Yitzhak Rabin, Mr. Shimon Peres, General Ehud Barak, Rubinstein and others visiting Hashemiyyah Palace west of Amman, where they met with a Jordanian delegation headed by King Hussein.

"So that is why Rubinstein left in a hurry," Haddadin noted to Tarawneh.

"And I bet that is the reason that Awn and Saqarat had departed in a hurry," he commented, "But do you like it that we know about these developments from the public media?" he wondered.

"At least it is Jordanian media," Haddadin said, inferring that it could have been worse if they knew about it from Israeli sources, instead.

2. The Visit to Hashimyyah

At the Hashimyyah Palace west of Amman, the atmosphere was relaxed as His Majesty King Hussein welcomed the Israeli guests. They all were seated on comfortable sofas in the reception hall at the ground level. His Majesty had a few words with Rabin in the area adjacent to where the guests and the Jordanians were seated, and then both joined the other delegates.

At Hashimyyah, Rabin assured his host that he did not intend to keep one inch of Jordanian territory, nor a drop of Jordan's water. The two leaders agreed that minor adjustments could be made to the borders in Wadi Araba whereby Jordan and Israel would exchange territory with exactly the same area. Under the exchange, Israel would retain Jordanian lands that her farmers had been exploiting on the border zone, and Jordan would be compensated with exactly the same land area of Israeli border, uncultivated lands. On water, Israel offered a water package with three components of fifty million cubic meters each. The first fifty million component would come from what Israel had been using, the second fifty would be from dams that the two countries will jointly build, and the third component of fifty million had no specific source.

Other talk went on over the need to arrive at a comprehensive peace, and the need to stop wars and bloodshed.

The Israeli group departed, and the Crown Prince flew to Aqaba. Awn al Khasawneh and Nidal Saqarat returned, and Marwan Qassim, Adviser to His Majesty, joined the Jordanian team also. Thursday morning was spent in Eilat with the Israeli counterparts. In the afternoon, the Jordanian delegates went back to Amman, and Tarawneh, al Khasawneh, Shurdum and Haddadin stayed in Aqaba. The Crown Prince sent after them, "How are you folks!" the Crown Prince greeted them as he parked his jeep.

"Al Hamdu Lillah, Sidi, (praise be to God, Sir) we are happy when you are happy," Tarawneh said.

The Crown Prince disembarked, and walked with the group to his house. He gave them a briefing of what happened at Hashimyyah. The Crown Prince then asked Haddadin to step outside with him, into the back yard. They strolled until they halted under a tall palm tree, at the southwest corner of the porch.

"They are offering us a water package of three components, fifty million cubic meters each," said the Crown Prince, "It is thought that this package may be suitable," and he explained the above package.

The words of Noah Kinarti the day before rang in Haddadin's ears. "The water issues will be resolved by the two leaders," Kinarti had said.

"Sidi, I am happy with the last fifty million component despite the fact that we do not know its source; but with regard to the Yarmouk and the two other components of 50 mcm each, I think I can do better through rigorous negotiations." Haddadin responded.

"Actually Peres made the offer," said the Crown Prince, "when one Israeli asked Rabin where this third component would come from, he gestured with both hands indicating he did not know."

"I hope we will soon find out."

"They are coming soon, tonight. So be prepared," said the Crown prince.

"Yes Sir."

The Prince explained the deal on the minor adjustment of the borders.

"This is good news," Haddadin said, "at least they now recognize the line of lowest points in Wadi Araba is its center as is meant by the text of the Mandate definition of borders, not the median line as the Israelis always insisted. They tried to throw that line to the geographic mean, which would have taken quite a bit of Jordanian territory, with fossil water aquifers under it."

"Rabin said he was not interested in one inch of our land nor in one drop of our water," said the Crown prince.

"We will soon have that tested, Sidi."

3. A Review of Potential Positions

It was evident that the moment of decision was approaching. All the nice things and agreements thus far had never touched on the hard issues of the water dispute. Matters like allocation, sharing, storage, groundwater, water quality, and others were never seriously negotiated except in broad general terms in the common sub-agenda. The two sides were soon to tackle these problems and come up with acceptable results. Haddadin stayed in the backyard after the Crown Prince went back into the house, and started imagining what the forthcoming confrontation was going to be like, and how he should call upon all the patience he could muster. It was true that Jordan's position was stated in the draft Annex II, which he had handed over to Kinarti on Tuesday, but he remembered what Rubinstein had to comment about it.

Haddadin tried to sum up the Israeli position, as it stood when the efforts of the United States ended just before the June war of 1967, and how those were somehow amended in 1979-1980 during the mission of Phlip Habib. The last efforts were those of Ambassador Armitage, which were interrupted by the current Peace Process and were to be disregarded as the Jordanian official meetings on water decreed (March-April of 1993).

One important position that Israel always took was that their negotiators evaded any international rule as far as water matters were concerned. The Israelis did not consider as part of International Law the United Nations resolutions when all other countries did. The bilateral negotiations adopted in the Peace Process upon Israel's insistance were based on the premise that no prior conditions would be set. Additionally no mention of international law or legitimacy was emphasized. Such notions served the interests of the more powerful in the absence of a balance of power.

Haddadin recalled the approved Jordan strategy in these negotiations, which abided by International Law and adopted the Unified (Johnston) Plan sharing formula. The instructions of His Majesty the King on 1 June 1994 at the Claridges Hotel in London rang in his ears.

"Please, *ya Akhi,* abide by the decision of disengagement," were the King's words to him in London. That instruction was translated into separating the shares of the East Bank from the shares of the West Bank, as per the Unified Plan figures and the works of the Arab Technical Committee (see Chapter III). The East Bank had its requirements (479 mcm) satisfied from the waters of the east side wadis (175 mcm), groundwater (8 mcm), and the Yarmouk (296 mcm). Fourteen mcm were set aside from the Yarmouk to account for evaporation from the Maqarin Dam, so that the total sum due to Jordan from the Yarmouk in accordance with that Plan was the sum of 296 mcm and the 14 mcm, or 310 mcm total. No shares were allotted to Jordan's East Bank from the Jordan River (Tiberias), but the Kingdom's (as it existed in 1955) share in Tiberias was justified to irrigate the lands in the West Bank. Thus, and based on the Unified Plan that Jordan had adopted for a strategy in water sharing, and on the 1988 disengagement with the West Bank, Jordan does not have rights on the Upper Jordan (from Lake Tiberias). Jordan was utilizing the water from the side wadis and the groundwater without any foreign competition. Only Jordanian actions affected the flow of these sources from the side wadis and the groundwater, and any recession in their flow was Jordan's responsibility. The Yarmouk alone was left in a contest with Israel (and with Syria). The Lower Jordan was not a subject in the Unified Plan because, with the waters of the tributaries diverted for use before their waters got to the Lower Jordan, it would be transformed into a drain, unusable for beneficial purposes at the time.

It was not the time, nor was it the forum, to address Syrian actions on the Yarmouk. Rather, the time and the forum were for the conflicts with Israel, not Syria, to be resolved through bilateral negotiations. In other words, the Syrian infringement on Jordan's share, as stipulated in the Unified Plan, was not a subject to be settled in the bilateral negotiations with Israel.

The Israeli position in the wake of the Unified Plan insisted that their share from the waters of the Yarmouk was 40 mcm per year, 25 mcm of which would be summer share. That had been reiterated by Ambassador Robby Sabel, the Israeli legal expert, in the negotiations in Washington, D.C. as the group of experts discussed the Unified Plan. The rest of the flow of the Yarmouk, after deducting 90 mcm for Syria and only 25 for Israel, the Plan ruled, belonged to Jordan. The Syrian uses, the Plan estimated, would send some 39 mcm back to the Yarmouk as return flow, so that the Syrian depletion was only 51 mcm. However, both Israel and Syria violated their respective shares. Israel installed pumps off the Rutenberg pool and was pumping whatever she could of the Yarmouk before its confluence with the Jordan, and that was estimated at 75 mcm (going as far up as 95 mcm in good years). The average increase that Israel took from the Yarmouk over and above her share was about 50 mcm per year (believed to be the subject of the first component of the Israeli offer).

Syria expanded its use of the Yarmouk springs in her territories and used them to the fullest. Syria also built a series of earth dams in the Yarmouk catchment to impound the winter flow of the Yarmouk tributaries. By the time of the negotiations with Israel, the Yarmouk flow average at Adasiyya was 270 mcm per year compared to an annual average natural flow of 467 mcm and an agricultural return flow from Syrian uses of 39 mcm, a total availability of 506 mcm according to the Unified Plan. Thus the Syrian depletion at the time of negotiations was 197 mcm per year compared to a calculated depletion of 51 mcm under the Unified Plan. Syria had by then gone over her planned depleting uses by 146 mcm per year on the average. Of the Syrian increased depletions, and based on the Unified Plan calculations, 81 mcm belong to the West Bank, and the rest, or 65 mcm belong to Jordan.

The residual flow at Adasiyya, after deducting 25 mcm for Israeli uses from the total average flow of 270 mcm, belongs to Jordan; the average Jordanian share was thus reduced, by Syrian and not by Israeli withdrawals, to 245 mcm. When the Jordanian share of the Syrian depletions (65 mcm) is added, the Jordanian share becomes 310 mcm, the figure stipulated in the Unified Plan. However, of her remaining share, Jordan was using an average of 105 mcm through diversions in the KAC, Israel was infringing on 50 mcm on the average, and the remaining flow (90 mcm) gushed down to the Jordan river and the Dead Sea in winter. In short, Jordan's Yarmouk share (310 mcm) was distributed between Jordan (105 mcm), Syria (65 mcm), Israel (50 mcm) and floods gushing down to the Dead Sea at 90 mcm per year. Of the Israeli part, about 12 mcm can be used directly by Jordan without need for storage, because that was summer flow. The rest of the Israeli part of Jordanian water, or 38 mcm, along with the 90 mcm lost to the Dead Sea, a total average of 128 mcm need to be stored for later use, and a diversion weir need to be built across the Yarmouk to assure their diversion to Jordan. About 50 mcm could be stored in the Karama reservoir inside Jordan, leaving an average of 78 mcm to be stored elsewhere. The better location for that was the Mukheiba dam site, and that was what the official meetings in Jordan decided in April 1993 for their negotiations strategy. The dam should have double that capacity to yield the above amount, and the site has about 200 mcm storage capacity, as determined by Energoproject in 1963. Arrangements for the use of the Mukheiba site should be made between Jordan and Syria, with notification served to Israel. However, nothing could be done in the bilateral negotiations with Israel about the expanded Syrian use. That issue could be handled only through bilateral talks with Syria, something that Jordan failed in miserably in 1987, when the 1953 agreement between the two countries was replaced.[339]

The impending negotiations with Israel thus had many challenging objectives, which can be summarized as follows:

The first challenge was to bring Israel's share in the Yarmouk back to what it should be: 25 mcm per year. This meant reducing the Israeli use of Yarmouk water by an average of 50 mcm per year, or more. The Israeli Water Commissioner estimated what Israel had to give up in accordance with Jordanian demands at 75 mcm.

The second challenge was to render Jordan's share all usable. Jordan's share in the Yarmouk was winter and summer flow. The summer flow was no problem, because it could be used instantaneously as it came. The surplus flow comes in winter and would require storage facilities to make it available at times of demand, in other words storage reservoirs that could impound Yarmouk floods. The alternatives were the Maqarin site (Wehda dam), which was handicapped by a reduced Yarmouk flow due to the above Syrian abstractions; the Khalid dam at the confluence of Wadi Raqqad and the Yarmouk, and the third was the Mukheiba site. The Jordanian strategy, as

---

[339] Jordan conceded to Syria the right to use water stored in 26 dam reservoirs Syria had built in her territories in the Yarmouk catchment, and to use 10 mcm per year from the yield of the Wehda dam. Those dams impounded floods earmarked for the Maqarin dam under the 1953 agreement. The Syrians moved later to build more dams than the new agreement entitled them to.

outlined in April 1993, was to opt for the third site, Mukheiba. All three sites were on the borders between Syria and Jordan, and the last two sites are located downstream of Maqarin where the Syrian bank of the Yarmouk is part of the Israeli occupied Golan Heights, and were never brought up in talks with Syria as long as the Syrian Golan were occupied. However, in the Ionedis Plan (1939) for the development of the Jordan Valley, a site was identified at Adasiyya.

The task of the Jordanian negotiator was to find a storage reservoir for the Yarmouk floods that could be negotiated with Israel away from any occupied Syrian territories. The first choice was Lake Tiberias, at least temporarily, until Syrian occupied lands were returned to Syria under the peace process, at which time the Mukheiba site could be negotiated with Syria. On the Lake Tiberias option, Israel's position before the 1967 war was reluctance to accept storage of Arab waters in Lake Tiberias, thus the storage in the Lake was effectively off the table as far as Israel was concerned. The second choice for storage that could be negotiated with Israel was the Ionedis site at Adasiyya, and the third choice was a couple of small dams on the course of the Jordan River. Storage sites inside Jordan would not be negotiated with any party, but it was better to talk about storage sites off the river course. This meant storage on the side wadis, inside Jordan, to which Yarmouk water could be pumped from the KAC, like the raising of Wadi al Arab Dam.

The third major challenge was the construction of a diversion weir on the Yarmouk at Adasiyya. It was the weir included in the Baker-Harza Master Plan of 1955, and was later designed by Harza as part of the Maqarin/Wehda dam project. Jordan was never able to build it because of Israel's objections since the early 1960's. That weir was the primary focus of the diplomatic efforts of the American intermediaries, Habib, Murphy and Armitage. It became worth combining the diversion facility with the Ionedis dam site to have it perform a dual function; modest storage and effective diversion. The importance of the diversion weir parallels the importance of securing Jordan's rightful share in the Yarmouk because, without the weir, the diversion efficiency of Jordan's share would be impaired, and part of Jordan's water would keep flowing westwards towards the Jordan River and the Dead Sea, exposing it to pumping by Israel, as she had been doing for years.

The fourth challenge was to negotiate with Israel the improvement of water quality in the Jordan River. Israel had long diverted the saline springs to discharge into the Lower Jordan instead of their natural discharge into Lake Tiberias. This system was envisaged at the time of the Johnston's negotiations and after them. Because the fresh water resources of the basin would all be diverted for utilization before any of those resources got to the Lower Jordan, the latter was bound to become a drain for the agricultural return flow from the East and West Jordan Valley. However, with the modern advances in irrigation technology since the Unified Plan, and applications of genetic engineering to plant production, it became possible to use slightly brackish waters for plant production. Moreover, since dams were contemplated on the Lower Jordan, the river course had to be cleaned up and be managed to allow beneficial uses. Israel also discharged municipal wastewater into the Lower Jordan. By so doing, Israel had exacerbated the quality degradation of the Lower Jordan. The general issue of the protection of water resources from environmental degradation was embedded in this challenge.

Finally, there was the issue of groundwater in Wadi Araba, and that had to be carefully negotiated. It depended on how the Jordanian occupied lands, farmed by Israel, would be treated. The water pumped from Jordanian territories was needed for those farmed lands, and supply of water to them would have to continue regardless of who would eventually own and operate them.

There were other topics of joint interest; especially those pertaining to cooperation for the alleviation of water shortage, and the mechanisms for joint cooperation and follow up.

Most of the above challenges and topics were addressed in the draft Annex that the Jordanians had prepared and shared with their Israeli counterparts on Tuesday, 10 October 1994. No figures of water sharing were inserted in the draft; rather they were left blank because they were meant to be the subject of negotiations.

All these thoughts passed through the mind of the Jordanian negotiator as he strolled in the backyard, facing the waters of the Aqaba Gulf, and he was able to pull his ideas together, mobilize his capacities and call on all the courage that he could muster. He could feel the tension inside him as he felt the moment of decision was fast approaching.

He aimed at maximizing the gains for Jordan and to maintain, as much as that would allow, credibility with the Israelis. The Unified Plan was the minimum position he would accept as decreed by the Jordanian official water meetings of March/April 1993. He recalled his proposal to the Israelis for a Jordanian share from the Jordan River on account of municipal water needs. This position had been explained to Kinarti in August, and the Jordanians had then asked for 200 mcm for that purpose.

4. Serious Negotiations Commence

4.1 In Aqaba

In the early evening of Thursday, 12 October, the Israeli delegation of five men arrived by bus to the home of the Crown Prince. Negotiations started soon after their arrival and the water negotiators, Kinarti and Haddadin, occupied the eastern table on the back porch, facing the water of the Gulf of Aqaba.

"I suggest that we go over the draft Annex that I gave you day before yesterday," Haddadin suggested.

"Yes, but I noticed that you did not pay attention to our need for groundwater in Ha Arava," said Kinarti.

"Easy, we can accommodate whatever we agree on and have it inserted in the text. Let us just start, and inspect the Articles of the Annex in order."

There was a clear advantage to Jordan to have the draft Annex as the takeoff point for the negotiations. Kinarti inquired about the figures to be inserted in the share for Israel from the Yarmouk, and the answer was 25 mcm split up into 12 mcm during the five dry months (15 May - 15 October), and 13 mcm for the rest of the year.

Kinarti did not accept the Jordanian answer, and insisted that the 25 mcm was Israel's share in the dry period only and that an additional quantity was due to Israel during the rest of the year. This stance was similar to the position Israel had taken all along, since the completion of the Johnston mission. Kinarti even asked for more; he demanded that another 20 mcm (and not 15 as Israel always had claimed) be added to the 25 mcm to yield Israel's share, making the annual total 45 mcm instead of the 40 mcm that Israel had been claiming. The negotiators debated for quite some time.

"Listen, Noah," said Haddadin, "we can play this music forever with no conclusion, but we should save time. Your share was based on the contribution of the Yarmouk to the irrigation needs of the Yarmouk Triangle in Israel. The Triangle has frontage on the Jordan River and on Lake Tiberias, in addition to the Yarmouk. The Yarmouk contributes 25 mcm to irrigate 16,000 dunums in the Triangle. That is an annual rate of 1562 cu.m per dunum and you know this is high already. The balance of its irrigation needs for an equal area would have to come from the other two water fronts, Tiberias and the Jordan."

"True, but Israel never conceded to the 25 mcm. Show me one evidence of that."

"I have two pieces of evidence, Noah. The first is the declaration by the Prime Minister of Israel, Levy Eshkol, in the Knesset in 1964 that Israel accepted the Johnston Plan, and the second evidence is the Plan itself. Check it and you will see that Israel's share is 25 mcm per year. I have a third, verbal evidence that is just out of the oven, fresh and smells good. Prime Minister Rabin said yesterday to King Hussein that Israel was not interested in one drop of Jordan's water. You put two and two together, and there you are".

"But the Israelis never accepted that share, and you say that asking for our share of 45 million is infringing on Jordan's water?"

"Yes, exactly, this would be infringing on Jordan's water, not by my own claim, but by what the US emissary ruled in September 1955. So let us not waste time."

Haddadin knew that the Israelis never conceded to the 25 mcm in their talks with Johnston over the period 1953-1955, and demanded 40 mcm as their annual share from the Yarmouk. The minutes of the meetings of those talks revealed that position. The two negotiators argued for a long time. Haddadin knew that the Israelis needed to use the Yarmouk water because of a certain dependency they had created on it since the early 1930's. There were areas in their Yarmouk Triangle that could be watered only from the Yarmouk. In addition, Israeli had expanded the use of the Yarmouk waters to include areas in Beit Shean and storage in Lake Tiberias.

"Okay, Noah, I am willing to help you with this. I can let you have the additional 20 mcm from the winter flow of the Yarmouk that you are asking for, provided you would give us in return 20 mcm from Lake Tiberias in the summer."

The Yarmouk winter flow that Israel would be allowed to use bypasses the Jordanian intake anyway, and is of no use to Jordan in winter before storage facilities and a diversion weir are built. Conceding 20 mcm of it to Israeli use in return for Israel supplying Jordan with 20 mcm in the summer, when Jordan badly needs them, was a big gain for Jordan. In effect, it is as though Jordan had found a storage facility for that much winter quantity, without any loss to evaporation, seepage or other.

The Israeli share outside these mutual concessions would be only 25 mcm, and the remainder of the flow belongs to Jordan. This was what the Arab Technical Committee finally had agreed to in its negotiations with Eric Johnston in 1955. Since then, the Arabs had suffered major setbacks in their confrontation with Israel. To enforce the 1955 water allocation would be a major achievement for Jordan. After much debate, Jordan was able to state in her water agreement with Israel what the Arabs had accepted back when they were united in their stands and apparently strong in the region.

When Haddadin's proposal fell on receptive ears, he wanted to go even further.

"Noah, I could use more summer water if you are willing to take more water in winter. Your installed pumping capacity from the Yarmouk before the confluence with the Jordan exceeds 50 mcm, maybe a bit more. How about trading the entire pumping capacity from Yarmouk winter flow for an equivalent amount of summer water from Tiberias. Let us make the concession figure 50 mcm instead of 20 mcm," he suggested.

"No, no," said Kinarti, "what would I do with 50 mcm winter flow. Besides, I really need all the water stored in Tiberias."

Haddadin tried to convince Kinarti of the wisdom of water trading, and its positive impact on bilateral cooperation, but Kinarti would not budge. He was receptive to the 20 mcm, and no more.

Kinarti wanted to switch to Wadi Araba and the groundwater there. He knew that Israel would pull back from the Jordanian territories she had occupied since 1949 and thereafter. There are wells Israel had drilled in those occupied territories on which Israeli farms of the moshav[340] partially depended. The moshav there needed more water to upgrade the cropping patterns and to expand their farms.

Kinarti demanded, in an undiplomatic way, that the wells Israel had dug be kept by Israel as her property after she had withdrawn. Haddadin responded, even more undiplomatic, that no parts of the Hashemite Kingdom could be conceded to any other country, and that the Jordanian Constitution so dictated. Kinarti wanted the wells as the property of Israel to guarantee the water flow forever, and Haddadin wanted a way to link that water usage with irrigation of the farms, no matter who owned them, and for a water exchange as well. On the guarantees of water flow from Jordan, Haddadin cited Jordan's reliance on waters from Tiberias as agreed above without guarantees. So, why should Israel ask for guarantees when Jordan did not? The mutual dependence, he suggested, was the best form of guarantee. In effect the two arrangements of cross border transfers would be the mutual guarantees of continued flow.

"Let me assure you that we cannot concede to any outside party any jurisdiction inside Jordanian territories, none whatsoever," said Haddadin at the end of a long argument.

"If you do not agree, I will not negotiate with you," said Kinarti.

"Listen, Noah, if you think I am in a hurry to get to terms with you, you are mistaken. I have all the time in the world."

Kinarti stood up in protest, and walked toward the other table where Rubinstein sat. He said something to Rubinstein, who hurried and talked to the Crown Prince. The Crown Prince summoned Haddadin, and stood with him under the same tall palm tree by the corner of the porch. The Crown Prince comprehended the water conflict and had trust in Haddadin's ability to defend Jordan's rights. He exchanged a few words with him and confirmed that the results of the negotiation have to be acceptable to Jordanians.

"Let us avoid the sharpness in exchanges that may lead to breakdown of negotiations, like what we have now at hand," said the Crown Prince.

"Sir, Problems do not come by on their own," Haddadin said. "It takes two parties to have a problem at hand. I guess it is the language we use in negotiations, English, which is not the mother tongue of either side, and the tensions we harbor, make us at times sound edgy when we don't mean to be."

---

[340] Moshav is the Hebrew name for the Israeli collective farming settlement.

"Then see what you can do to resume negotiations."

"*Elli shabakna ykhallisna* (the one that created the problem shall solve it). I am confident, *Sidi*, that if you are happy with the results that I come up with, it will be acceptable to the majority of Jordanians, and His Majesty will be very pleased with them too."

"Will you be able to convince the Government?"

"Yes, *Sidi*, I will be."

"Including Dr Abdul Salam Majali?"

"Yes, *Sidi*, including the Prime Minister, Dr. Abdul Salam Majali."

"Bless your efforts, now show me how you can have the negotiations resume."

"Don't worry, *Sidi*, Kinarti and I got used to each other, and I will get him back to talk."

By that time the two Jordanians had strolled long enough to reach the beach, and they turned back to the porch where the others were sitting and negotiating. The security and border group occupied the annex to the house. The stroll with the Crown Prince helped a great deal in impressing on the Israelis that he had given Haddadin his support. It also helped in getting Kinarti to pay attention to Haddadin's overture to him and get him back to the negotiation table. The rest of the evening did not produce much progress. Haddadin approached Kinarti and reminded him that they both were wasting valuable time, and suggested that they go back to work. Kinarti concurred and they resumed negotiations. Haddadin brought up the subject of the three components of the water package offered to Jordan at Hashimyyah, and focused on the third component, the 50 mcm whose source was not specified. He asked where that water could possibly come from. Kinarti tried to enlist brackish water in the East Jordan Valley (Jordan) among the resources where more water could be made available to Jordan to fulfill that third component of 50 mcm pledged by the Israelis, but he did not succeed. Haddadin suggested Lake Tiberias but he, too, did not succeed in convincing Kinarti.

The following day, Friday 14 October, the talks went a bit smoother, but not much of progress. The issues were known, the positions of each party became clear, and each was trying to make the best out of the preparedness of their two leaders to remove obstacles. Awn al Khasawneh insisted that the issue of refugees be addressed in the treaty, and then he pulled another issue from his sleeve: Arbitration as the means for settlement of disputes.

The Israelis, after they lost in the Taba arbitration with Egypt, were adamantly against arbitration. They insisted that disputes be resolved amicably between the parties through negotiation. Al Khasawneh insisted in return that the disputes be resolved through arbitration. That exhausted the negotiators, consumed time, but neither party would budge. Decent progress was scored on the border delineation following the principle that Jordanian borderlands developed by Israelis in irrigated agriculture would be traded with an equal area of Israeli land across the border, with the borderline adjusted accordingly.

In a short break after lunch, a Bedouin-dressed soldier brought coffee to the team that was breaking at the western table on the porch.

"Say, Fayez," said Rubinstein, "our Foreign Minister, Mr. Peres, brought back with him from New York a good piece of work by Jordanians. It was a position paper on transport. We never thought that Jordanians could produce such quality work." Rubinstein was referring to the document that Minister Kawar had handed over to Peres and to Tarawneh in New York, about two weeks earlier.

"Thank you for the compliment," said Tarawneh.

The Bedouin-dressed soldier gave Rubinstein a cup of Arabic coffee while he was asking Tarawneh:

"Who wrote it in Jordan?"

"Dr. Haddadin did," answered Tarawneh.

At this moment, Rubinstein could not swallow the sip of coffee he had just put in his mouth! He almost choked and made no further comments.

Kinarti and Haddadin clashed again over the Wadi Araba groundwater and the ownership of the wells that were located in Jordanian territory. Kinarti sought the backing of Rubinstein and others but that got him nowhere.

The negotiations went on and on with modest success in the basic issues of water, the text of the Annex was cleared where agreement was made and the points of difference were postponed until later. On the other Articles of the draft Treaty, the issue of settlement of disputes remained unresolved. The reference to the problem of refugees was inserted in the treaty, and consideration was taken of the future negotiation on the final status of Palestinians, in which the issue of refugees would be taken up by Israel and the Palestinians and, simultaneously, by Jordan and Israel for the refugees in Jordan. For the displaced persons, reference was made to the quadripartite committee stipulated in the Oslo Accord. As such, it was agreed between the two parties, Jordan and Israel, that the issue of displaced persons be resolved through the framework of a quadrilateral committee (Jordan, Israel, PLO and Egypt), as stipulated in the Oslo Accord.

The teams took a break on Friday evening and Saturday 15 September, and returned to work on Sunday the 16th at the house of the Crown Prince in Aqaba. The group on borders and security continued its delineation of the borders without many problems.

"Careful," al Khasawneh said to Haddadin, "history will not pardon any of us if we make a mistake or compromise on the basic rights."

"Don't you worry, Awn; in my case, and judging from what has been achieved so far, I am sure history will judge my work positively. I wish you good luck on your work."

The Israeli water negotiating party became two people. Major Daniel Reisner, a lawyer in the Israel Defense Forces, accompanied Kinarti. The new addition was a real gentleman and was knowledgeable. The Jordanian side continued to be represented by Haddadin alone. The water negotiation on Sunday shifted to the table on the western porch facing Eilat. The talks covered a lot of ground on the issues of pollution, clean-up, and protection of water resources, but agreement was not forthcoming on the Wadi Araba groundwater, nor on the third 50 mcm component. At one point of dispute, Kinarti exercised his favorite reaction, that of going to Rubinstein to report to him how stubborn Haddadin was. The Crown Prince was sitting with Rubinstein, Tarawneh and the negotiation team on the southern porch facing the water. Kinarti came back walking like a peacock, and signaled with his right hand over his shoulder, meaning that Haddadin was wanted back there where he made his complaint.

"He wants to talk to you," said Kinarti as he was walking back to the table.

"Listen Kinarti, I am here to negotiate with you," Haddadin shouted angrily.

There was silence for a few minutes, before Dr. Ahmad Mango, adviser to the Crown Prince, appeared from around the corner of the porch. Before Mango had a chance to say a word, Haddadin shouted in clear English:

"Listen, Ahmad, if you have anything to say, you come here and take my place and carry the responsibility. Otherwise I shall accept no s--- from no one!"

Mango was surprised at Haddadin's outburst, but noticed that his presence had broken a deep silence. He also noticed the tense atmosphere, and signaled to Kinarti, who excused himself and walked to talk to Mango. The two stood under the tall palm tree at the corner of the two porches.

"This Haddadin is driving me crazy," said Kinarti, "he must be crazy too. You see how he does not listen to the Crown Prince."

"You have not seen much," said Mango who saw an opportunity to disarm Kinarti, "I have seen Haddadin say no to His Majesty face to face."

"I can not negotiate with him," resumed a surprised Kinarti, "why don't you bring Dr. ---- ----?"

"It is too late, we can not do anything about this. I am afraid you have to find your own way with him," said Mango.

"Can you help bring us back to talking terms?"

"I will try."

Mango came and sat next to Haddadin. Daniel Reisner who, sitting across the table during all this, witnessed his first show, looked so puzzled and did not know what to make out of it. Mango tipped Haddadin in Arabic and was able to bring the negotiations back on track. From that morning hour on, negotiations went smoothly and a lot was accomplished by noontime when the teams assembled for lunch. Work resumed immediately after lunch with an intention to achieve more. The negotiators were told that His Majesty and Prime Minister Rabin would arrive at around 5 p.m., and that they should be ready to give briefings. By 5 p.m., Kinarti and Haddadin had achieved a lot. However, the plan to meet with the leaders abruptly changed, and the teams took a C-130 military transport plane of the Royal Air Force from Aqaba to Amman Airport, from where they boarded two helicopters to fly over to Hashimyyah Palace in West Amman.

4.2 At Hashimyya West of Amman

As the teams arrived, the three Jordanian principal negotiators[341] and the Head of the delegation, Dr. Fayez Tarawneh, went into conference with Crown Prince El Hassan. He reviewed with them the extent of the progress they had made thus far. Before that was finished, His Majesty King Hussein arrived and joined the meeting.

"How is it going?" asked the King.

"Pretty well," said the Crown Prince, and asked each of the negotiators to brief His Majesty. Awn al Khasawneh made a briefing and identified the major problem he was facing, that of the mechanism of settlement of disputes. Shurdum gave a concise presentation and singled out the problem they were facing in the border adjustments. It was a piece of land developed by Israel, about 300 hectars in area and located about four kilometers east of the Mandate line in Wadi Araba, inside Jordanian territory. To adjust the borders there was not easy without giving the borderline an odd and awkward shape. Haddadin gave a short briefing on the progress on the water issues.

"Things are going well, Sire," He concluded.

"Are you happy?" asked His Majesty.

"Yes Sire, I am. We have finished about 70% of our work, and I hope to finish the remaining 30% tonight."

"Good, bless your efforts, all of you young men!" Said the King.

The Israeli leaders arrived from Israel shortly after the teams arrived from Aqaba. They conferred with their teams as well. The official meeting between the two sides[342] started with a warm welcome from the King, who soon gave the floor to Prime Minister Rabin.

---

[341] General Tahseen Shurdum, Awn al Khasawneh, and Munther Haddadin

[342] His Majesty King Hussein headed the Jordan side. It comprised HRH Crown Prince Al Hassan; Prime Minster Abdul Salam Majali; Chief of the Royal Court Sherif Zeid Ben Shaker; Minister of Information Jawad Anani; Adviser to His Majesty, Marwan Kassim; Chief of Staff General Abdul Hafez Mari'e Al Ka'abneh; Head of Delegation Ambassador Fayez Tarawneh; Dr. Munther J. Haddadin, General Tahseen Shurdum; Ambassador Awn Al Khasawneh; and Dr. Michel Marto.

After a nice response, Rabin went over the obstacles facing the negotiators over the Articles of the Treaty. He alluded to al Khasawneh's insistence on arbitration for the settlement of disputes and criticized him. Throughout the peace process, Haddadin happened to be the target of Israeli complaints all along. This time, he was glad to hear the Israelis complain for once against someone else, al Khasawneh, but he knew his turn would be coming up next.

His Majesty gave the floor to al Khasawneh. Awn defended his case very politely and in a highly professional manner. His focus was on arbitration as the method of resolving any conflicts that might arise out of the implementation of the treaty, or through disagreements on the meanings and intents of the text of any article. Rabin did not comment further on Awn's response as he took the floor again.

"And, Your Majesty, we are not sure if Dr. Haddadin has been helpful either. He has been putting obstacles in the way of negotiations so much so that we are not sure he wants peace."

His Majesty looked at Haddadin and nodded in a sign asking him to come to him. Haddadin walked over to listen to the King.

"What is this that I am hearing," said His Majesty, "I thought you were happy."

"I am, *Jalalet Sayyedna*. It is him who is obviously not happy," Haddadin said referring to Prime Minister Rabin. The King could not hide his smile.

"What is it all about?" asked the King.

"Sire, they pledged to Your Majesty fifty million cubic meters per year in the third component of their water package; and they have been trying to pull out from that pledge. They should deliver if we are to come to an agreement."

The King was very kind in his comments as he responded to Rabin. He said that the negotiators should be given more time to come to acceptable terms. In that session, the Israelis were not as organized and disciplined as were the Jordanians in the presence of Royalty. They interrupted one another, talked to each other in Hebrew, and took the floor without discipline. In contrast, the Jordanians were disciplined, polite and very organized. Finally, His Majesty proposed to go over to the hall across the corridor, where the maps of Wadi Araba were displayed. It was necessary to find a high-level solution to the remaining problem in the border adjustment.

Prime Minister Yitzak Rabin headed the Israeli side, which comprised Minister of Foreign Affairs Shimon Peres; Chief of Staff General Ehud Barak; Head of delegation Ambassador Eliakim Rubinstein; Deputy Chief of the Mossad; and the negotiators on water, borders, and security.

The two delegations crossed the hallway over to the opposite hall. The map showing the 300 hectares land developed by the Israelis inside Jordan was laid out on a large coffee table. Rabin stood to the right of the King, and to his left stood Crown Prince El Hasan, Haddadin, Majali, Marie, Shurdum and the others, not in any order of protocol. Some stood behind in a second row.

"If we adjust the border to give that piece to the Israelis," whispered Majali as he leaned toward Haddadin, "it would look like the long p.... of a donkey!" Majali was obviously not happy with the idea of modifying the border at that location (Al Ghamr/ Zofar). Rabin did his best to convince the King to have the borderline adjusted, but the King did not favor the idea.

"Why don't you, Your Majesty, lease it to us for some time?" asked Rabin.

"The notion of lease is not on the table," said the King, "but for how long do you want to stay in it?"

"Say, 25 years, renewable by mutual consent."

"This sounds reasonable."

The King looked to his left, "any additions?" he asked without addressing the question at any one person in particular.

There was pause. A condition flashed in Haddadin's mind. It was a condition contained in *Al Uhda Al Omariyyah* (the Pledge of Omar) issued to the Patriarch of Jerusalem by *Amir Al Mu'mineen* (the Prince of the Believers), Omar Ibn Al Khattab, in the year 637 C.E as he rode on camel back from Madeena in Hijaz to be handed the keys of the gates of Jerusalem.

"No Jews shall be our neighbors in Eliyya (Jerusalem)," said the condition in the Pledge of Omar. As such, Jews were allowed into the city during the day, and were not to stay in it overnight.

Haddadin broke the pause. "I have one, if I may Your Majesty."

"Go ahead," permitted the King.

"That the Israelis cross into our territory and out of it during day light hours only. They shall not stay in it overnight, and shall therefore not build houses on that plot of land." His worry over Israelis building houses there was the future difficulty of uprooting them.

The King looked at Rabin, and the Prime Minister accepted the condition, and the obstacle was overcome. The deal was not that of a lease, it was a Jordanian permission for the Israeli farmers to keep utilizing that plot of land for 25 years, renewable by mutual consent. The duration was the normal economic life assumed for agricultural projects, like the one the Israelis had developed in that area. The King treated the guests to a nice but late dinner, after which the negotiation teams went back to resume work.

The water group occupied a meeting room in the south wing of the Palace. The other teams occupied rooms in the northern wing. The rest rooms were on that side too. For a water negotiator to go to the men's room, he had to walk across the lobby to which a nice terrace was attached. The terrace faced Jerusalem. Every time Haddadin made that trip to respond to the call of nature he saw the Israeli Mossad official standing on that terrace with his cellular phone glued to his ear communicating, presumably, with Israel.

The teams worked through the night to complete their tasks. The water negotiator feared that his two Israeli counterparts, younger and stronger, would outdo him. He looked at his watch and it was 3:45 a.m., with issues still remaining to be resolved.

"Who would get exhausted first?" He wondered, as he typed the agreed paragraphs on his laptop computer.

At about 4 a.m., al Khasawneh passed by the water negotiation room. He looked tired and sleepy.

"I am through with the Articles of the Treaty, and I need to get a couple of hours of sleep. The borders group is not doing fine. I was instructed to ask you to take care of Annex I (a) to the treaty that handles the borders. They are across the hall, you know where they are."

"Yes, I know where they are meeting. I will attend to that piece of work as soon as I am through with Annex II, the Water Annex," said Haddadin.

"Please. The borders' Annex is important. Give it all that it takes."

"Don't you worry; have a good sleep."

By 6:15 a.m., the Water Annex was done. With the issue of the farmed 300 hectares in Wadi Araba resolved as above, the use of the wells that supply them with groundwater was allowed to continue. Israel agreed to supply Jordan with 10 mcm of desalinated water in the north, and Jordan agreed to have Israel increase the abstraction of groundwater in Wadi Araba by 10 mcm per year pursuant to a determination, through a hydro-geological study, that the increased abstraction could be permitted. Until the desalination plant in the north is installed and made operational, the 10 mcm due to Jordan would be supplied from Lake Tiberias. The ownership of wells remained Jordanian. On the issue of the third 50 mcm it was agreed that the two countries would cooperate to find the source of their supply within one year. It is important to know that this 50 mcm was additional water that Jordan became entitled to through these negotiations. The quantity was never part of Jordan's share in the Unified Plan, but was over and above the allocation under that Plan. The agreement was bilateral and nothing in it would jeopardize the rights of any third party, particularly the Palestinians.[343] Because the additional water was new and not part of the established rights defined under the Unified Plan, had no specified source, and it was not possible to agree on a source. Haddadin tried to name Lake Tiberias as a

---

[343] This point was agreed to and documented in the negotiations on 9 and 10 August, 1994 at the Moria Plaza Hotel on the Israeli side of the Dead Sea.

source for it, but the Israelis refused that. No agreement was reached on paying the costs of certain projects, and it was agreed to forward that difference of opinion to the leaders, as soon as they were ready that morning. The two leaders finally resolved it to the satisfaction of the negotiators. Annex II- Water Related Matters, in its final form, is shown in Appendix 17.

As the water negotiations ended that Monday morning, Kinarti went to shave and so did Daniel Reisener. Haddadin went to the cafeteria downstairs to have a bite. Just as he started to eat, Awn al Khasawneh showed up, still looking tired.

"Haven't you gone to the borders negotiations?" Awn asked Haddadin before he wished him good morning!

"I will, right now. You do not look like you had any sleep."

"Not a minute. I could not," answered Awn.

Haddadin left his breakfast and hurried to the room where the borders were being negotiated. On one side of the table, facing the entrance door, sat Nidal Saqarat. Opposite him were three Israeli negotiators headed by Moshe Kochanovsky. All were quiet and waiting.

"Where are the rest of the Jordanians?" Haddadin asked Saqarat.

"They went to sleep," answered Saqarat.

"And how come you did not go to sleep like they did?"

"I am afraid someone could tamper with those maps. I stood guard to safeguard them."

"Bless you," said Haddadin, "I wish we have a dozen people of your dedication. If it makes you feel any better, I have not had a minute of sleep either. This is a historic night for all of us."

Haddadin reviewed the draft Annex I (a), and started to clean up its English language in agreement with the Israeli side. It needed a lot of editing.

"Who wrote this?" Haddadin asked Saqarat who was educated in France.

"Don't ask me please. My English vocabulary is hardly enough to get me home if I were lost in town!"

There was a very important issue regarding the borders. The West Bank has a front on the Jordan River facing Jordan, and extends between Wadi Yabis in the north to the Middle of the Dead Sea in the south. The rest of Jordan's western borders abut Israel proper (1949 Armistice Line). The part of the Jordan separating her from the West Bank cannot, in Jordan's view, be recognized as part of her border with Israel. Otherwise, Jordan would give legitimacy to the Israeli occupation of the West Bank or, at least, to Israel's permanent presence on a strip of the western bank of the Jordan River. The issue was debated with the Israelis over the past months, and Jordan succeeded in having a disclaimer inserted on the maps to indicate that the status of the Occupied Territories would not be prejudiced. The disclaimer started with the words "This international line…etc ..." in reference to the centerline of the Jordan River separating Jordan from the West Bank.

"Who wrote this disclaimer?" Haddadin asked Saqarat again.

"Awn al Khasawneh and the legal advisors," said Saqarat.

Haddadin called for a break, left the room and headed to the reception room next to the lobby to look for Awn al Khasawneh. He found him in a meeting with the Crown Prince, and signaled to him for a word. Al Khasawneh came.

"I am afraid I can not live with the text of this disclaimer," Haddadin said.

"I know what you mean; but this was the best we could do."

"Awn, when you refer to the line separating us from the West Bank as 'international' it means there is a State on each of its two sides. We know Jordan is a State on the east side of the line, but, to me, there is no sovereign State on its west side. It will be construed by all who read this disclaimer that it means Israel. In effect we will be recognizing sovereignty of Israel over the West Bank at a time when we seek the disclaimer to drive away that notion. This will put us in trouble with so many Arab parties, on top of whom you know who."

"I know, I know, and fully agree. We tried hard and we could not. Try your best and I wish you good luck."

Haddadin crossed to the southern wing to use the computer of His Majesty's secretary. He tried six different texts for the paragraph of a disclaimer, and finally decided on one that he felt could be negotiated. He printed that text in three originals and came back to al Khasawneh.

"What do you think of this text?" Haddadin asked.

"Oh, I do not think there is a chance of it passing. We tried."

"I am seeking your clearance of the text as the legal expert in the delegation."

Al Khasawneh read it carefully, cleared it and wished Haddadin good luck. Haddadin returned to the negotiation room and had the meeting called to order. Two more negotiators joined the Israelis. The Jordanians embarked on a few details of the Annex before they came to the important point.

"Now we come to the disclaimer. Here is our proposal for a text," Haddadin said as he handed over to the Israeli side one hard copy of the text he had prepared a half hour earlier. The two sides were clearly tired and the text was somehow approved.[344] That particular issue was the only one remaining before the treaty would take its final form.

---

[344] The Israeli side thought the text was the one they had negotiated with the Jordanians earlier; most likely they were under the impression that Haddadin's job was to edit the language of the Annex on borders.

"Thank you all," Haddadin said, "I think we can call this a success, and call it a day or a night!" He invited the group to coffee, socialized with them, and then hurried to al Khasawneh.

"Here it is, Awn, I had good luck after all."

Awn could not believe it. He was very happy and he explained the development to His Royal Highness.

"It appears that the draft treaty will be initialed today, Awn," Haddadin said.

"Yes, I have been so advised," said al Khasawneh.

"We have completed Section B of the Common Agenda, but what about its Section C, which stipulates that a Treaty would be concluded only after successful completion of Sections B and A. Section A, as you know, has the comprehensive peace provision. The parties should achieve the goal of comprehensive peace before the Treaty is signed."

"Remember the Arab coordination meeting in Damascus in 1992, and the one in Beirut in 1993. It was agreed there that each party could proceed with his negotiations with Israel at a different pace. Syria's Foreign Minister was quoted as saying: "each party shall take out his own thorns with his own hand!" Said Awn, "the second paragraph in the preamble of the Treaty, however, confirms the goal of comprehensive peace."

"I am afraid we have to brace ourselves for increased Syrian criticism, and for criticism from PLO supporters," Haddadin said.

"We did not criticize Oslo, and we will be happy for a Peace Treaty between Syria and Israel."

However, the two negotiators had a feeling that Jordan was under some kind of pressure to conclude a peace treaty with Israel. They felt it, but did not talk about it.

5. The Peace Treaty Concluded

The Hashimyya Palace began to receive more Jordanian officials. All the advisors to His Majesty arrived, and the senior staff of the Royal Court. The Prime Minister arrived at about mid-day, and spoke to Tarawneh and to Haddadin as they both stood in the Lobby close to the entrance.

"How did things go?"

"Very well," said Tarawneh.

"Is the Treaty ready?"

"Soon it will be, sir."

"Do you have any comments on it, Munther?"

"No, Pasha, it is acceptable. It is finally in good shape."

"Who is supposed to initial it?" asked Majali.

"You as Foreign Minister and Shimon Peres," said Tarawneh

"*Ala Barakat Illah*" [with God's blessing], said a very brave Majali.

The dining hall downstairs was set up for the occasion. Shortly after lunch, the ceremony of initialing the peace treaty would take place. The press correspondents filled that hall. A limited number of chairs were put to seat the Jordanian and Israeli officials on the left side of the desk on which the Treaty was to be signed. Haddadin went downstairs with the others to attend the ceremony. He looked for his name on one of the chairs set for the Jordanians, but did not find it on any chair. As he was retreating, the Chief of the Royal Protocol, Mr. Ayman al Majali, brought him a chair and squeezed it in the back row.

His Majesty entered the hall with Rabin, and the Prime Minister with Peres. The dignitaries took their seats. Haddadin looked around for the negotiators, looked for Awn al Khasawneh, for the guys who labored all these years to make this conclusion possible, and particularly for the very few who did the negotiations over the past few weeks, but he could not see them. At some distance, he spotted Awn al Khasawneh in the midst of the crowd of press correspondents, not giving press interviews, but standing in the crowd like he was one of them.

After the ceremony ended at about 4 p.m., Dr. Michael Marto and Haddadin headed for the door to go home and get some sleep. The Crown Prince stood at the door to say goodbye to the Israeli guests and he looked happy. Haddadin approached him.

"Congratulations, *Sidi,*" he said.

"*Ye'teek el Afiah* [May God give you the vigor]."

"I have an apology to make; I think we have been a burden on your home for over a week, and you have been very gracious with us despite our clashes with the guests."

"You know," said the Crown Prince, "I used to hear about how tough you were in negotiations with the Israelis; but hearing is not like watching you in action. More impressive yet was the fact that you got what you wanted."

"*Al Hamdu Lilah* [praise be to God] it is over."

"Not yet, you are to appear at the Royal Court at six o'clock this evening to brief the advisors to His Majesty, ministers and parliamentarians. I guess there is no time to sleep!" said the Crown Prince.

"Don't worry, *Sidi*, I will be there."

Haddadin went home and lay down for one hour, and then hurried to Basman Palace at 5:30 p.m., briefed the officials in the main meeting hall, and answered their questions.

That was on Monday, 17 October 1994. The Treaty was initialed and the news was carried on all the wires. It was a day to celebrate for many, but was not good news for some others. Some, like Syria and Egypt, picked on errors of press reporters carried by the wires, especially those talking about "leasing" the Jordanian lands to the Israelis. There was not a lease arrangement in the treaty, and leasing of Jordanian lands was not on the table. The Israelis were allowed to use lands in Jordan, some of which (Baqura) had been owned by Jewish entities since the 1920's, for a specified period of time. On Tuesday, 18 October, Shurdum and Haddadin appeared at the meeting of the Council of Ministers to explain the provisions of the Treaty before the Ministers discussed it in their meeting for approval. Haddadin was not surprised at how little the ministers knew about Jordan's water rights.

"Do you mean to tell me that we do not have water rights on the Jordan when our country is named after it?" Protested Dr. Hashim Dabbas, Minister of Communications and Post.

"The water of the river is actually the water of its tributaries. I say, "*Al Nahru Bi Rawafideh,*" [the river is in its tributaries]. We get our rights from its tributaries, cheaper and cleaner," Haddadin answered.

Other ministers asked about the storage arrangements, and Haddadin read from the Water Annex. He was surprised at the comment of the concerned minister.

"I think the Johnston allocations were better for us!" Said the Minister of Water and Irrigation, Dr. Saleh Rsheidat.

"No, sir, they are not better than what we got when we respect the decision of disengagement. The share of 100 mcm from Tiberias earmarked for the Kingdom in the Johnston Plan belong to the West Bank, and the reduction in our flow from the Yarmouk is due to Syrian abstraction. Israel's share in it is now confined to 25 mcm per year, split into 12 in the summer and 13 in winter, better for Jordan than the Johnston Plan by which the Israelis claim their share to be 40 mcm."

"Besides," Haddadin continued, "we are now entitled to 50 mcm per year of water of drinkable standards. This is above what the Johnston Plan earmarked for Jordan."

After discussion, the Council of Ministers approved the Treaty including all its annexes.

On Sunday, 23 October 1994, an Israeli team crossed over to Amman to have both sides proofread the Treaty before it would be signed on Wednesday 26 October. The work was done at the Jordan Geographic Center in Jubeiha of West Amman. Tarawneh headed the Jordanian side.[345] The work continued the following day, 24 October. A few disagreements over the format of maps and names of places were resolved, and the Treaty was put in the final form ready for signature.

---

[345] The Jordanian team was composed of Tarawneh, Haddadin, al Khasawneh and Dr. Abdallah Toukan.

Another Israeli team[346] crossed over to Jordan on Tuesday 25 October to resume talk over the Jordan Rift Valley development. The meeting was held at the Hashimyyah Palace, and the two sides agreed on the next steps to be taken. After lunch, their Jordanian host, Dr. Haddadin, decided to show them Amman, as it was their first visit to it ever. He took them on a drive to the Amman Citadel. From there, they overlooked the Roman amphitheater, as they stood next to a group of tourists who had a tour guide with a detectable Palestinian accent.

The Amman Marriott Hotel was vacated of its guests that day to make room for the American delegation that accompanied President Clinton to Jordan to attend the signature of the Treaty the following day. Preparations were also made for President Clinton to address a joint session of the Upper House and the Lower House of Parliament, after the signing ceremony, in the evening. The tourists at the Citadel sounded unhappy, because they were among the many expelled from the Marriott five star hotel to make room for the presidential delegation.

"You are mad that you were kicked out of your hotel for one evening!" wondered their Palestinian tour guide, as the Israeli visitors listened in.

"Of course we are," said several tourists in protest, with a clear American accents.

"I have been kicked out of my home in Palestine all my life. What should I say? Think about that!" said the tour guide, and the Israeli guests looked at each other and at their host. It was a different kind of reasoning that they were hearing, not the kind they were used to hearing in Israel, about the Palestinian refugees.

As they looked across to the amphitheater downtown, they could hear the chanting of a modest demonstration in the Hashemite Plaza next to the amphitheater. It was basically an anti-peace demonstration of about 1000 people chanting slogans against Israel and the peace treaty. The Jordanian host drove his guests down to the amphitheater, stopped for a while to watch the protesting demonstration, and then drove off, all with the surprise of the Israelis who thought he would not have the guts to come that close to the demonstrators.

The next day was the historic day. Arrangements were made for a major ceremony in Wadi Araba north of Aqaba for the signing of the Peace Treaty between Jordan and Israel. The location was convenient in several ways. On the one hand, it was not readily acceptable to sign the treaty in either Amman or Tel Aviv; nor was it viewed appropriate to have it signed on the territories of either country. A border point was the more suitable location

---

[346] The Israelis were Oded Eran, Yossi Vardi and Rafi Benvenesti. The Jordnian side was represented by Munther Haddadin.

and, obviously, it would be difficult to have it on the centerline of the Jordan River, or the Dead Sea. On the other hand, the desert-like terrain of Wadi Araba would trigger the interests of the leaders to support the integrated development of the Jordan Rift Valley. The location, chosen to the north of the Aqaba Airport, was the border point where the adjusted borderline of 1946 met with the original Mandate borderline of 1922, and it was used as a site for the first meetings of bilateral negotiation in the region on 17-18 July 1994, some three months earlier.

Aqaba Airport received more planes that day than it ever had. Three planes left Amman Airport for Aqaba carrying Jordanian dignitaries and their spouses to attend the event. United States President William J. Clinton and the foreign ministers of Russia, Germany and the United States attended. The organization was superb, and the signing ceremony was very well conducted. President Clinton made a speech in appreciation and support of the Treaty, and he spoke of transforming the desert-like Wadi Araba into a Valley of Peace!

6. The Middle East-North Africa Economic Summit

On Friday, 28 October a Jordanian delegation met an Israeli delegation in Bonn to make a joint presentation to the Chancellor's office and the Foreign Ministry key staff. The Federal Republic was the head of the Troika representing the European Union. The hosts were pleasantly surprised when they saw the Jordanians and the Israelis sit at the same side of the meeting table. The presentation was well received and the hosts asked questions about what could be done to reinforce the peace that had been achieved.

Three days later, the Israelis, the Jordanians, the Palestinians and a score of delegations from the Middle East and North Africa, Europe, Japan, United States, Canada, and the rest of the world gathered in Casablanca, Morocco, in the first Economic Summit for the Middle East and North Africa. His Royal Highness Crown Prince El Hassan Ben Talal headed the Jordan delegation.

The Jordan delegation did very well on all working groups. They made good presentations on economic cooperation, the Jordan Rift Valley, water and the environment, and transport and trade. The plenary session was so crowded that there were not enough seats for all the delegates, nor were the good hotels of Casablanca sufficient. Several important side meetings were held between the delegations of the core parties (Jordan, Israel, the Palestinians) and the delegations of advanced countries. The delegation of the E.U. was eager to hear about the water agreement, and the Jordanians gave a good presentation. The E.U. were prepared to support the study on the water peace projects, and required a document on the "Terms of Reference" on which the consulting services would be solicited.

Other side meetings were conducted with the World Bank. One such meeting was with Caio Kochwesser, Vice President of the World Bank. It dealt with the feasibility study for the Jordan Rift Valley that the Bank was sponsoring on behalf of Jordan and Israel.

"We, at the Bank, are concerned that you would be biased for your own ideas, primarily the Red-Dead Canal. What if it will not be feasible?" Asked Mr. Kochwesser.

"No sir," Haddadin said, "we will not advocate a project that would be a white elephant. The fact that we stand behind the idea of the Red-Dead canal does not mean that we want to build it come hell or high water. We will abide by the feasibility analyses."

'Good," said Kochwesser, "this will prompt all parties to look at the scheme objectively. Our team is working hard to finish its work on the JRV before the end of the year."

"But please be assured that we are neither biased nor do we consider ourselves the fathers of the idea. The Red-Dead linkage was introduced last century by some French engineer. Omar Abdallah Dokhgan pursued it diligently and I was his helper. The initial study showed hope it could be the best alternative for the control of the Dead Sea level," Haddadin resumed.

The Summit was the first such economic show for the region. It was an opportunity for introductions of people, businesses and projects. It was considered a big success and gave further momentum to peace.

7. At the Multilateral Working Group on Water Resources

The multilateral working group on water resources was to convene in Athens on 8 November 1994, shortly after the Casablanca Summit. The Jordanian head of delegation took a flight to Athens from Casablanca via Rome on Wednesday, 2 November, to attend a seminar organized by the University of California at Los Angeles ahead of the multilaterals. The seminar was held in the same hotel complex as the Multilaterals. Its topics focused on peace in the Middle East.

"Hello there, Haddadin!" yelled an Israeli as the Jordanian head of delegation entered the seminar room.

"Oh! Shaul! Good to see you," he responded.

"I want to congratulate you and ourselves. You have done a great job. Congratulations."

"Thank you, Shaul, and how have you been?"

"You remember that long night of 16-17 October, when you were doing the last stretch of negotiations?"

"Yes I do, what about it?"

"Well, they contacted me and asked about you."

"Who are they?" Haddadin asked.

"I will not say who they were, but they contacted me."

Haddadin remembered the cellular phone that was glued to the ear of the Mossad officer standing on the terrace of Hashimyya Palace that night. Shaul continued:

"Are you a friend of Dr. Munther Haddadin?" the caller asked.

"I know the man," answered Shaul, "but I can not claim I am his friend."

"We have been negotiating with him and he is like a solid wall that we can not penetrate. Do you know any weakness in him?" asked the caller.

"Like what?" Asked Shaul.

"Like women, money, stature, etc.."

"No sir, I have not heard a thing like that about him. Usually, at the World Bank where I met him, the word goes around about any such qualities in the character of high officials in a developing country that the Bank deals with. Nothing like that circulated about Haddadin."

"Lead us to some inroads to him," said the caller.

"You are negotiating with him over water, aren't you?"

"Yes, we are," said the caller.

"I will tell you what I heard about him at the World Bank. They said he was the only Jordanian who had guts to tell the Syrians off over the Yarmouk in their own capital, Damascus. My advice to you is: 'any Jordanian who has the guts to tell the Syrians off over water should not be expected to be kind to Israel.'"

"I have an idea who called you," Haddadin said after he finished his story, "but shame on you, Shaul, you denied me a chance of the company of one of their beautiful agents!" Haddadin said jokingly.

In that seminar, an American of the Jewish faith made a presentation that was skewed in favor of Israel, and threw suspicions at the intentions of the Arabs. It was more like the presentation made by Israelis before the peace treaty was concluded. The speaker did not update his attitudes nor, perhaps, was he willing to.

"I can see that our job in making peace with Israel is not complete," Haddadin addressed the audience, "It looks like we ought to extend our efforts to make peace with Jews all over the world!"

The Palestinian delegation arrived for the multilateral peace talks. They, too, were unhappy with the Jordanian-Israeli treaty. They felt that any peace with Israel concluded by any Arab country before they came to peace with Israel was eroding their assets. They had told the E.U. delegation that visited Gaza that the Jordanians had recovered their water rights from Israel at the expense of the Palestinians. Such allegations were printed in their newspapers published inside the Territories, as well as in the ones that sympathized with them in London.

"But how?" Haddadin asked the head of the E.U. delegation to the multilaterals in Athens.

"We don't know, but we have been told that by the highest official in the Palestinian Authority," and he named the official.

"Don't tell me. This can't be. I just can't believe it," Haddadin said.

"I am telling you the truth of what we were told."

Haddadin sat with the Palestinian delegation and had a chat with them. No one in the delegation knew what were the bases of the Palestinian water rights in the Jordan River basin. They just threw charges that were politically motivated, to make noise and to smear the Jordanian achievement. That was to be expected, because the pace of peace on the Jordanian track had became much faster than the pace on the other tracks

The Jordanian government forwarded the Peace Treaty to the Parliament for ratification and enactment as law. Abdul Karim Kabariti, a skilled Jordanian politician, was the chairman of the Parliamentarian Committee on Foreign Affairs, when was in charge of appraising the terms of the treaty. They wanted explanation of the water annex, and Kabariti was sending questions by fax to Casablanca, where the Minister of Water and Irrigation, Samir Kawar, was attending the Economic Summit. More questions came to Athens, and were all answered. The vote in the Lower House of Parliament was 54 votes for and 25 votes against, with one member absent, and the Treaty was approved. The Upper House of Parliament also approved the Treaty, and a Royal Decree had the Treaty issued into law. The Treaty was ratified by the two countries in a ceremony at Beit Gibrael on the southern shores of Lake Tiberias on 11 November 1994. It became effective as of that date.

8. Public Campaign

The Jordanian negotiators were hosted by several TV shows, and gave interviews to the foreign press, explained the Treaty, and responded to questions from participants. They also made speeches all over the country, especially on campuses and in social clubs. The Treaty was well received among an enthusiastic Jordanian public, and the approval rates were reported very high.

The press carried analyses of the Treaty, and columnists contributed to its appraisal. There was acceptance by the elite that Jordan had extracted from Israeli controls more than was expected under the circumstances.

"So what did the Treaty bring us in Jordan?" asked a caller to one of the TV shows.

"If you can bear with me, I will explain it very briefly," said Anani who was hosted on the show with al Khasawneh and Haddadin.

"First: Jordan has recovered every inch of territory that was controlled or occupied by Israel. Israeli farmers are permitted to continue their activities in two farms under Jordanian sovereignty and Jordanian law.

"Second, Israel's borders with Jordan have been defined for the first time, and that will put an end to the aspirations by Israeli extremists to expand eastward at Jordan's expense.

"Third, Jordan has recovered its rightful share of water at a time we are badly in need of more water.

"Fourth, the Treaty confirms the rights of Palestinian refugees and displaced persons to repatriation or compensation in accordance with International law.

"Fifth, it consolidates collective security in the region, and the stabilization of peaceful relations. Such a state will attract foreign investment, reduce unemployment, and facilitate the establishment of a Regional Development Bank.

"Sixth, Jordan qualified for some foreign debt forgiveness, especially our debts to the United States. The debt rescheduling with other lenders will be a lot easier.

"Seventh, there will be cooperation in the integrated development of the Jordan Valley. The majority of the benefits from such an undertaking will accrue to Jordan.

"Eighth, we have ended a state of siege under which we have been since the Gulf War of 1990. We can reach out now with ease to other countries in the region and in the world at large."

There was another caller:

"Is Israel committed to the implementation of the U.N. resolutions concerning the refugees and displaced persons?"

Anani asked al Khasawneh to take that question.

"The Treaty stipulates, as indeed did the Common Agenda, that the bilateral aspects, i.e., the Jordan-Israel component of the case of refugees and displaced persons will be resolved in accordance with international law. There is a Security Council resolution, number 237 of 1967, requiring that displaced persons be allowed to return to their homes and properties, and there is a score of General Assembly resolutions confirming the rights of the 1948 refugees. These resolutions are part of international law."

A third caller asked:

""Why should we allow the Israelis to stay in Baqura and the other piece of land in Wadi Araba for 25 years?"

"The other piece of land is a 300 hectares area developed by the Israelis and is called Al Ghamr," Haddadin said, "but let me address the Baqura area, if I may."

"Baqura and another stretch of land next to it on the east side of the Jordan River and on the west side of it was owned by the Jewish Palestine Electric Company since 1926. Some private Jewish entities also owned some land parcels in Baqura since the twenties and the thirties. We, in Jordan, respect private property, and we have assigned a guard to the enemy property since 1948 so that no one would trespass against it. We expect the Israelis to reciprocate. If we do not respect Jewish property inside Jordan, Israel would be happy to reciprocate and not respect the property of Jordanians inside Israel, and there is a lot more Jordanian property there than there is Israeli property in Jordan."

"The important thing is that the two areas are returned to Jordan's sovereignty. Jordanian law applies to both no matter who owns the land. The Israeli farmers were allowed to stay for 25 years, which is the average lifetime of agricultural projects. After that period, the matter would be negotiated between the two sides. Actually, the Treaty provides for transforming the Baqura area into a tourism area by mutual agreement. Now, Jordanians can access all their territories which they were not able to before the Treaty."

Another caller asked:

"What did we get in terms of water compared to our water rights?"

"Our rights, pursuant to the disengagement decision," answered Haddadin, "are those of the East Bank as defined by the Technical Committee of the Arab League. We have been instructed to respect the Disengagement Decision and to leave the rights of the West Bank for the PLO to defend and recover. Accordingly, our rights are in all the side wadis that discharge into the Jordan River from the east, and in the Yarmouk River. No one shares with us the flows of the side wadis, but the Yarmouk River is shared with Syria and with Israel. It is important to remember that the negotiations and the Treaty were concluded with Israel, and that we do not in this context address any Syrian-Jordanian affairs over the Yarmouk. The Treaty addresses only Jordanian-Israeli affairs. The Arab Technical Committee finally gave its consent in 1955 to an Israeli share in the Yarmouk at 25 million cubic meters per year, and said that the balance of the Yarmouk flow at Adasiyya, after deducting 90 mcm for Syrian uses, is for Jordan. These have been the rightful shares of the parties in the Yarmouk. What we have achieved in the Treaty is:

"First, we confined Israel's share to 25 mcm per year in lieu of some 75 mcm they have been using. The Israeli share is now exactly equal to what the Arabs said in 1955 should be Israel's share.

"Second, we allowed Israel 20 mcm of winter water in the Yarmouk that we cannot use, and obtained 20 mcm from Lake Tiberias in the summer, at the time when we need the water most. This is analogous, but not quite, to storage of Yarmouk floodwater to use it in the summer without any loss to evaporation.

"Third, we have also traded 10 mcm per year of groundwater of marginal quality in Wadi Araba with an equal amount from Lake Tiberias, which is of superior quality.

"Fourth, we have become entitled to 50 mcm of water of drinkable quality from sources in Israel to be defined through cooperation between the two countries. This quantity, to say the truth, is over and above what we consider our established right. Its validation does not affect the rights of any other riparian party in the basin.

"Fifth, we succeeded in sharing the remaining flow of the Jordan River, excluding the flow of the Yarmouk of course, in equal amounts with Israel along our borders with it. This, too, was not the subject of previous allocations and rights because, under the Unified Plan, the river would have been transformed into a drain for brackish water.

"Sixth, we succeeded to get Israeli acquiescence to build a diversion/ storage dam on the Yarmouk River at the Adasiyya tunnel location, something we have been trying to do for ages [since 1960] without success.

"Seventh, we got Israel to agree that we build a dam or two on the Jordan River to store any excess flood for the benefits of Jordan, and also to store such floods off the course of the River. We will be able to do that via the King Abdallah Canal and store Yarmouk water in reservoirs built on the side wadis.

"Eighth, the two parties undertake, each in her own territories, to clean up the Jordan River and protect the water resources against pollution and against tampering with the installations on their respective territories that belong to the other party.

"Sir, it is worthwhile to note that point one above comprises the water rights of Jordan, i.e., the side Wadis and the Yarmouk residual flow after the Syrians and the Israelis take their respective shares. All the other points above have been wrestled with Israel through hard bilateral negotiations. None of our points would prejudice the rights of any other party, particularly the Palestinians."

A call from a peace supporter came in and praised the achievements.

"May I interject," asked al Khasawneh, "a primary achievement is the security undertakings in the Treaty. The two sides undertake not to use force or the threat of using it against one another, and, the two sides undertake to work together to make the Middle East an area free from the weapons of mass destruction."

"One last thing, sir," Haddadin added, "please note that, after a series of Arab defeats since 1955, Jordan was able to achieve in this Treaty more than the Arabs said would be our water rights before those defeats."

"There are many more corollary achievements and potential achievements for Jordan. I think we have entered a new era that just dawned.

Thanks to the efforts of His Majesty King Hussein, and to the efforts of His Royal Highness Crown prince El Hasan," concluded Anani.

The hopes for progress on the Palestinian track became high as the Treaty gave the peace process a shot in the arm. The same hopes were also attached to the Syrian track and to the Lebanese track so that the goal of comprehensive peace will be achieved. It did not take long before Syria and Israel resumed bilateral negotiations at Wye Plantation in Maryland, and information leaked about Israel's preparedness to 'withdraw' to the lines of 4 June 1967. However, elections in 1996 in Israel brought the Likud coalition back to power, and hardships developed on the Syrian track of negotiations. These hardships stalled progress on the Syrian track, and more emphasis was put on the Palestinian track.

Pursuant to the effectiveness of the Treaty, Jordan and Israel started working out detailed agreements on the various fields of their dealings. Eighteen agreements were concluded in such fields as transport, agriculture, trade, aviation, health, education, information, tourism and others. The borders between the two countries were demarcated and border markers installed. In the field of water, the Water Annex had enough details to start implementation. A Joint Water Committee was established, and two subcommittees under it were set up. The venue for the meetings of the Joint Water Committee alternated between Israel and Jordan. A pipeline was installed to convey water from Lake Tiberias to the King Abdallah Canal in Jordan, and the King inaugurated the project in 1995. The water sharing in the Yarmouk River was adjusted to conform to the Treaty articles. The water thus gained enabled Jordan to fully operate the water system that supplies Amman with municipal water from the King Abdallah Canal in the Jordan Valley. This was a timely gain for Jordan, without which the hardships in supplying water to Amman would have been immense.

The Multilateral working groups received a shot in the arm as a result of the Cairo Agreement and the Jordan-Israel Peace Treaty. The Water Resources and the Environment Working Groups held a joint meeting in Amman in June 1995 that was addressed by Crown Prince El Hassan. The other Working Groups scored progress in their works, and the atmosphere was conducive to peace related activities.

His Majesty King Hussein had the Jordanian negotiators decorated with medals commensurate with their respective ranks in government, or the equivalent of such ranks if they were in the private sector. The heads of the delegations, Tarawneh and Rubinstein, were honored in Beit Gibrael by the two leaders. They received financial awards and Tarawneh donated his to the Ministry of Social Development of Jordan to be spent on the orphans and the needy. On the return trip from Beit Gibrael, Tarawneh missed his helicopter ride and had to come home in an Army jeep!

"So you went there without even asking for your colleagues to attend the ceremony with you," Haddadin noted to Tarawneh after he congratulated him the following day. "Eleven members of Parliament were taken, senators too, but not a single member of the delegation!"

"It was never my arrangement, nor was I consulted," assured Tarawneh.

"Were there any Israeli negotiators in that ceremony?"

"Yes, practically all of them that we know, many asked about their counterparts. Noah cannot wait to see you!"

"I bet! At least their government remembers them. Blessed be Awn al Khasawneh, who once told me that our peace-making could be a thankless job. I hope we will not end up as defendants in a court room for what we have achieved for the country."

"Come on, you must be kidding."

"I am, but someone out there may not be!"

The negotiators disappeared from the scene, and others took over. Majali and Anani left government in January 1995. Haddadin left the Process in March 1995. Awn left it by the time the Treaty was signed. People changed, but the Treaty survived.

CHAPTER XII

The Challenge of Implementation

1. The Mechanism of Implementation

The provisions of the Treaty were to be implemented through the conclusion of several agreements in the sectors listed in the Treaty articles. Article 6 on water and Annex II detailing its intents were exceptions. A Joint Committee was stipulated under Article VII of Annex II to be established "for the purposes of the implementation of this Annex." The Committee may form, as it deems necessary, a number of specialized sub-committees and assign them technical tasks. The mechanism of implementation was therefore stated in the Treaty and did not need further agreements.

In a follow-up on the meetings in Casablanca with the European Union during the Economic Summit, Jordan prepared the needed "Terms of Reference for Consulting Services," based on which the engineering consulting services would be solicited. The document was reviewed with a visiting Israeli delegation headed by the diplomat, Oded Eran, in late November 1994, and a joint document was agreed.

The Jordanians, headed by Crown Prince El Hasan, and the Israelis, headed by Foreign Minister Shimon Peres, presented that document to the E.U. when they were on a joint visit to Brussels on 28 November 1994. The studies would cover the storage dams, the conveyance systems, and wastewater treatment in the Jordan basin and its safe disposal. The E.U., as a result, agreed to contribute 11 million ECU to finance these consulting services.

The first meeting for the Joint Water Committee was scheduled to take place at the Moria Plaza Hotel in the city of Tiberias in December 1994. It was a pleasant occasion in which Jordanians were glad to get to meet their counterparts.

2. Resolution of Points of Contention

"I noticed, on the way here," said Haddadin to Kinarti, across the meeting table at the Moria Plaza, "that you are not pumping from the Yarmouk and it is winter already." He actually did not visit the pumping site on the way to Tiberias, nor did he know whether or not they were pumping, but threw the comment in to test the Israeli reaction.

"You did not pay us for the electricity; pay us the cost and we will pump," said Kinarti.

"Why should we pay? It is water that Jordan conceded to you to pump. We do not know, nor do we care, where you take that water or what you do with it. It is yours in return for 20 mcm of summer water from Lake Tiberias."

There was silence, and the Israelis looked at each other.

"Why are you perplexed?" asked the Jordanian team leader, "do you have a copy of Annex II?" And he pulled out his own copy, on which he and Kinarti had their initials.

"Yes, we do," answered Kinarti.

"Okay, show me where does it say we have to pay the cost. We only pay for the transfer of our water from Tiberias to Jordan as per Article I.2.a. Wherever we have to pay it is so mentioned," he explained.

Kinarti and company examined the Article and other paragraphs, and they talked to each other in Hebrew, then they asked for a recess at that early stage of the meeting. They retreated to the lobby of the hotel and had their own consultations. Professor Shamir stayed with the visitors. The recess lasted for more than an hour.

"Okay, you win," said Kinarti, as they took their seats to resume the meeting. The two sides discussed the provisions of the Annex and the Jordanians pointed out that a list of the operating wells in Wadi Araba had to be prepared and agreed by 31 December. The Israelis promised to do their best to have that list and maps prepared by the specified date. But that was the first slip of the dates. Jordanians later came to realize that dates in the eyes of the Israelis were not sacred. Prime Minister Rabin made such an announcement when the schedule of events with the Palestinians started to slip because of Israeli delays.

The second matter to which the Committee had to turn its attention was the quantity Israel was using from the Lower Jordan. Jordan was entitled to equal amounts, provided that her uses would not jeopardize the existing Israeli uses (Article I.2.c of the Annex). The Joint Committee was entrusted with the task of surveying the existing Israeli uses, but the Israeli side dragged their feet and submitted what the Jordanians considered inflated figures of Israeli uses. The matter dragged on through 1995 and an accurate joint survey was not actually agreed on.

The third matter on the agenda was the implementation of the studies that the E.U. pledged to finance. Several detailed matters had to be attended to and those took a good part of the meetings. The talks covered such needs as site investigations on the Israeli side, movement of equipment, entry of personnel, escorting by liaison officers, custom duties, etc. The Israeli treatment of the matter, added to the time the E.U. procedures consumed and the Jordanian slow performance, forced delays on the implementation of the needed studies.

The fourth, and most important matter on the agenda, was the construction of the diversion/storage dam on the Yarmouk at Adasiyya (Article II.1). The Israelis offered to do the designs of the facility, but the Jordanians turned their offer down. The facility was as important to Jordan as the acquisition of water rights, and the Jordanian action as recommended to the Government was to proceed with its construction without delays, using Treasury funds. Any involvement of outside donors could complicate the matter, given the status of the lands on the northern bank across the river. These lands were part of Mandate Palestine and were part of a demilitarized zone under the 1949 Armistice Agreement between Syria and Israel. The lands came under Israeli military occupation in June of 1967, and were still occupied. The construction of the diversion facility will not cause any harm to, nor change the nature of these lands, and the diversion weir was part of the Maqarin/Wehda dam, the drawings and specifications for which were approved by Syria. The Jordanian intent in the Joint Water Committee was to secure Israeli approval of the design of the facility, and to arrange for its construction, without any predjudice to the status of the occupied land on the northern bank.

The Government of Jordan took immediate steps to build the conveyance facilities to transfer water from Tiberias (the Jordan river from a point immediately upstream of the Degania gates) to the KAC. That was done in record time and the facility was finished in July 1995.[347] King Hussein inaugurated the project on 5 July 1995, and water started flowing from Lake Tiberias in Israel to the King Abdullah Canal in Jordan. It took fast steps to have the diversion/storage facility on the Yarmouk designed. Two types were proposed.

a. A diversion structure at the intake location where the sand bags were put every summer to control the diversion of water and the proportions thereof (see Chapter VII), and,

b. A small retention dam downstream, capable of storing about 8 mcm of water. It would retain Yarmouk floods for a limited time and have the water diverted to the KAC at the maximum permissible flow rate. Water then could be pumped up for storage in dams on the side wadis.

The Jordanians had preference for the second option, because it could provide Jordan with about 25 mcm more water than the pure diversion dam would. Their talks with the Israelis focused on that option but the Israelis brought up many factors to stall the progress. Their attitude toward the retention dam was not very favorable because it gave Jordan more water. This meant less water overflowing to their pumps off the Rutenberg pool,

---

[347] The Engineer was Dar Al Handasah, and the contractor was Issa and Naffa', a Jordanian construction firm.

especially in winter. On the dam site, the Jordanians insisted that the dam be located such that it would abut the banks of the river where the lands are Jordanian on the southern abutment, and Israeli, not occupied territories, on the northern abutment. There was little physical space to maneuver there, because of the topography. Almost at the location of the 4 June line, west of which Jordan recognized Israeli territory, the terrain flattens out and there is no site for a dam.

In March 1995 the peace teams were reorganized. It was time for the Ministry of Water and Irrigation to assume responsibility for the talks with the Israelis in the bilateral relations and in the multilaterals. Haddadin, coming from the private sector, resigned his services in all posts pertaining to the Peace Process, especially that the government did not pay for his services and only provided for his travel expenses. He stayed out of the whole process without being briefed on the developments therein for a whole year.

In just as much as there were changes in the Jordanian individuals who handled the Water Annex, there were also changes in the ranks of the Israelis. The take-over of the Likud in 1996 sent Kinarti and Gideon Tsur, the Water Commissioner, away from the scene, and brought Meir Ben Meir as Water Commissioner, reporting to a Minister of a new Ministry of National Infrastructure, General Ariel Sharon.

The site investigations for the retention dam revealed some tombs on the north side of the river, and ruins of what was believed to be one of the oldest human settlements, according to Israeli archaeologists. The Israelis insisted that the tombs be kept in place and demanded that the dam be shifted a bit upstream to clear the tombs. The upstream movement brought the right abutment of the retention dam to a point east of the June 4 line, i.e., it touched lands that were part of the demilitarized zone before the Israeli occupation in June 1967. This shift could potentially trigger criticism from Syria over and above her criticism of the Peace Treaty itself.

The other joint studies were not productive either. The study of the dams on the Jordan was awarded by the E.U. to a German consultant, and he had a hard time from the Israelis, who dragged their feet and gave him little cooperation. The study of the conveyance system was awarded to a Dutch consultant, but the finalization of their work depended on the finalization of the sources, or the dams, which were stalled. The wastewater treatment studies went on well, and several locations for wastewater treatment plants were identified.

On the groundwater front, things moved slowly, although Israel promptly delivered on its commitment to transfer to Jordan the off-summer flow of 10 mcm from Lake Tiberias The increment in groundwater abstraction allowed for Israel from Wadi Araba did not materialize, because the Jordanian side insisted on a study to determine whether or not such an increment was hydro-geologically feasible. The Israelis contended that they had conducted a study of their own, and concluded that the increment was feasible. The Jordanians insisted on a joint study performed by an independent consultant. That study was never tendered nor were its terms of reference prepared.

In contrast with the slow progress in the joint study of the water projects, the E.U. moved fast to secure its financial support. In early June 1995, Chancellor Helmut Kohl of Germany made a visit to Israel and Jordan. His Majesty the King and Prime Minister Rabin held a meeting with him at Baqura, the land that reverted to Jordan under the Peace Treaty, and the Chancellor announced that the E.U. and Germany were willing to help finance the construction of the water projects under the Treaty. The donors were forthcoming, but the efforts to have the projects ready for implementation were slow. Both sides were responsible for these delays. One simple example could be cited. It took the Jordanians three months of bureaucratic hassle to have the terms of reference for the consultants officially approved, despite the fact that these terms were prepared by their negotiation team. The E.U. complained of that bureaucracy which stood in the way of efficient utilization of their allocated funds.

3. Stalling factors

The public was enthused on 26 October 1994, when Jordan and Israel signed the Peace Treaty in the presence of President Clinton and the Foreign Ministers of the Russian Federation and Germany, which represented the E.U. The hopes were high that the long awaited peace would bring with it an end to confrontation and trigger development and prosperity. Polls showed that some 90 percent of voting Jordanians were for the Peace Treaty. International, regional and domestic awards were bestowed on leaders and peace-makers respectively.

That enthusiasm for peace started to fade away, in less than a year after the Peace Treaty was signed. The factor that had the maximum impact was the lack of progress on the Palestinian track of negotiations. After the Gaza-Jericho First Agreement, signed in Cairo in May of 1994, the Palestinians and the Israelis had to go through hardship in further negotiations that left their imprint on the Jordanians. The Syrian and Lebanese tracks were no less discouraging. The Peace Process was dealt a severe blow when an Israeli assassin opened fire at close range and killed Prime Minister Rabin on 4 November 1995. This was a violent terrorist act by the opponents of peace in Israel. The opponents to the same peace on the Palestinian side were no less violent, and car bombs became part of the scene. In Jordan, the opposition to peace started to grow, and calls against normalizing relations with Israel became louder.

Rabin had an impressive funeral attended by many world leaders, including President Clinton and former presidents of the United States. King

Hussein and President Mubarak of Egypt participated and the King delivered an elegy, an unrehearsed speech, in which he expressed his deep sorrow for losing a partner and a friend. After Rabin, Shimon Peres became Prime Minister. He soon called for general elections in the hope that Israeli and world sympathies for the death of the Prime Minister would bring a decisive victory for the Labor coalition.

Feelings in Jordan ran high in April 1995, when the Israeli Army shelled a United Nations barrack in Qana of South Lebanon, where hundreds of Lebanese were taking refuge. Over 250 Lebanese were killed in that raid, and the Arabs, including Jordanians, were very angry. The peace opponents had a golden opportunity to criticize peace with Israel. The developments on the Palestinian track added insult to injury, and the peace proponents had little ammunition to use in return. It became clear that the extremists on both sides were reinforcing each other, and the moderates were losing ground.

The Qana attack cost the Prime Minister, Shimon Peres, his Premiership in the early elections he called for. The Israeli Arabs who had been traditional supporters of Labor did not cast their usual votes. Benjamin Netanyahu won the election as Prime Minister by less than 30 thousand votes, and the Likud Coalition secured more seats in the Kenesset. The power in Israel shifted to the right. The peace process shifted accordingly, from under a dripping roof to the rainstorm outside!

The progress toward peace was stalled. It even went backward after the Likud took office. President Clinton, King Hussein and Mubarak of Egypt exerted intensive efforts to save the process, but the Likud, led by Benjamin Natanyahu, went the other way. In violation of the Treaty and the spirit of peace, the Israeli Mossad engineered and executed an assassination attempt to knock out, in broad daylight and on the streets of Amman, a Jordanian of Palestinian origin.[348] Peace advocates in Jordan lost more ground to the opposition, and the region's mood became a reminder of the no peace-no war era that preceded Madrid. On the water front, no progress was made on the identification of the source of the additional water of 50 mcm (Article I.3), nor was there any progress toward cleaning up the Lower Jordan. The meetings of the Joint Water Commission became intermittent and less productive. Some progress, however, was made in the studies of the JRV

---

[348] In September 1997, the Israeli Mossad waged an assassination attempt against Khalid Masha'al, the head of the Political Bureau of Hamas, the Palestinian Islamic Resistance Movement. He was injected with a lethal drug by Israeli agents, who were chased by a Masha'al bodyguard, and arrested by the Jordanian police. The attempt caused rage in Amman and King Hussein made it clear to Netanyahu that if Masha'al was not rescued, the Peace Treaty would be in jeopardy. Masha'al was rescued with Israeli help, and the Hamas spiritual chief, Sheikh Ahmad Yassin, was set free from his jail in Gaza in return for the arrested Israeli agents.

integrated development. The Jordanians and Israelis worked together and appointed a consultant to do the studies. The consultants completed the work on the pre-feasibility study, and a session of the Trilaterals was held at the Dead Sea Spa hotel to review the findings. The private sector was invited to attend, and the meeting was chaired by Crown Prince El Hasan in February 1997. Dan Meridor, Israel's Minister of Finance, represented the Israeli side. It became clear in that meeting that the Israelis were used to looking westward for investment and trade, and it would take them time to get used to looking eastward.

4. The Ministerial Confrontation

On 19 March 1997, Dr. Abdul Salam Majali was appointed Prime Minister for the second time. The Jordanian water negotiator, Dr. Munther Haddadin, was appointed Minister of Water and Irrigation. Three other key ministers were chosen from among the peace team of Jordan. Anani became Deputy Prime Minister and Minister of State; Tarawneh became Minister of Foreign Affairs; and Hani Mulki became Minister of Trade and Industry. With Majali heading the Cabinet in which four key ministers were from among the peace-makers, the government appeared as one in support of the peace that was retreating on the streets.

The Minister of Water had his priority set on the regional front to pick up progress in the implementation of the Water Annex provisions. He instructed the Jordanian side on the Joint Water Committee to promote progress there and gave them direction on what to do. By early April, a Jordanian soldier lost his balance in Baqura and fired his machine gun at Israeli school students, killing eight girls and injuring others. The atmosphere became tense and King Hussein diffused the tension through a special trip to Israel to pay condolences to the families of the victims.

Israel organized a program to commemorate the 30th day of the death of the victims in Baqura, in accordance with the Jewish tradition. A request came from Israel for high level participation by Jordanian officials. Crown Prince El Hasan considered the possibility of participating himself, and when the Minister of Water was asked for an opinion, he recommended that the participation by the Prince would be appropriate, if a sign of progress in peace-building was made after the commemoration. Such a sign would be to announce an agreement on the implementation of some difficult provisions in the Peace Treaty. Ambassador Dennis Ross was expected in the region, and he could participate in the commemoration, if a deal was worked out with the Israelis.

The Minister made contacts to meet with his Israeli counterpart, Minister of National Infrastructure, General Ariel Sharon. His intention was to review progress in the implementation of the Water Annex, press for the delivery of 50 mcm of additional water, the source for which still had to be defined, and to press for the construction of the diversion/storage dam on the Yarmouk. Minister Haddadin thought that if progress was scored on at least one of these two topics, it would be worth the high level participation in the commemoration ceremony on Tuesday 6 May 1997.

The visit by the Minister to Israel was set for Sunday, 4 May 1997. He set out heading a delegation of three officials from Amman and were joined by the Jordan Ambassador to Tel Aviv. Israeli protocol met them at Ben Gurion Airport. They were driven to the Ministry of National Infrastructure, where Minister Sharon had his office. His reception was cordial and they went into the conference hall after they had a short visit with him in his office. The conference table was covered with plates of pastries and sweets. On the other side of the table sat the Director General of the Ministry of Infrastructure, the Water Commissioner, and others. Dori Gold, the Director of the Office of Prime Minister Netanyahu was there too and he also attended with Sharon the initial courtesy meeting in the office when the Jordanians first arrived. Minister Sharon welcomed the Jordanian guests in the meeting, and was very courteous. The Jordanian Minister reciprocated. General Sharon spoke of the general atmosphere and about the peace process, especially with Syria.

"The Golan to us is a strategic area." Said General Sharon, "We can never go back to the status that existed before 1967. This would jeopardize the lives of Jews," he said, "and I will never condone any step that carries with it any threat to the life of any Jew."

"Excellency, why are you saying this to us?" interrupted Haddadin, "are you under any illusion that I may take a step that would jeopardize the life of any Jordanian?"

"No, this is not on my mind, but I meant to say that the life of any Jew is so important in our approach to peace."

"Please also bear in mind the equivalent importance I place on the life of each and every Jordanian. This brings us to the subject, water, which is the commodity indispensable to the life of man, Arab, Jew or any other nationality," responded the Jordanian Minister.

The exchange of views appealed to Minister Sharon to the extent that he called his wife on the telephone and asked her in Hebrew to "Come down and meet this wonderful Jordanian Minister." Mrs. Sharon appeared afterwards in the meeting hall and the Jordanians had the honor of meeting her.

The two sides discussed the standing issues in the Water Annex, pending implementation. These were:

a. Two dams on the Jordan River, which the Israelis considered as being not feasible. From the Jordanian viewpoint, these dams would not be needed when storage facilities are provided to fully impound

Yarmouk floods, because, after the Yarmouk is fully regulated, the floods in the Jordan would be reduced to spills from the Yarmouk storage facilities and spills from Lake Tiberias, which were infrequent.

b. The supply to Jordan of the "additional water" of 50 mcm per year, and the source of this quantity had to be identified. During previous contacts it was understood that the source would be a desalination plant that would process brackish water on the Israeli side, and the processed fresh water would be delivered to Jordan. There was no progress in that matter and the Jordanians were in need of more water. The Jordanian Minister was shooting for Lake Tiberias as a source of supply, until the desalination facility became operational.

c. The construction of a diversion/storage facility downstream of the Adasiyya intake. The Jordanians were shooting for the storage option, and to have it built as soon as possible.

d. Increase of groundwater abstraction for Israel from Wadi Araba. Israel had been supplying Jordan from Lake Tiberias with 10 mcm per annum during the seven off summer months. The Jordanians were waiting for the additional water in (b) above before they would give the green light to tackle the issue of groundwater in Wadi Araba.

The talks went on and on. The Jordanian Minister then took the initiative and made the following proposal:

"Mr. Minister," Haddadin said addressing General Sharon, "I have a proposal to make. I propose that Lake Tiberias be open for storage of Yarmouk floods on a temporary basis. This would do away with the immediate need to build the two dams on the Jordan, and would alleviate the impact of water shortage in Jordan."

Minister Sharon excused himself to talk in Hebrew to his staff. It took him a while, during which the Jordanians could see the staff arguing and the Water Commissioner making counter arguments.

"It sounds like you have thrown a bomb," whispered one of the Jordanian delegates in Arabic.

General Sharon then looked at the visiting Minister. "Mr. Minister, can I talk to you in private in my office?" Asked General Sharon.

"Sure."

The two Ministers retreated to Sharon's office, but Dr. Dori Gold followed Mr. Sharon and came into his office.

"Why don't I ever get the chance to negotiate one on one with the Israelis?" wondered Haddadin to himself, "they have always outnumbered me throughout the peace talks," and there was a time when they were 22 people and the Jordanians were one assisted by another whose task was to take down the minutes of the meeting in November 1994.

"What you are proposing," said General Sharon, "has never been accepted by any Israeli Government since the early 1950's. No Labor government, no Likud government, and not a coalition government ever accepted that proposal."

"But that is why I came to you!" Haddadin commented.

"I cannot, on my own, accept what all Israeli governments never accepted. I have to submit this matter to the Security Cabinet and get their decision."

"It is not my business how you conduct your own, but it is my business to tell you that I do not care where you take the issue provided that you endorse it yourself."

Haddadin's answer was not without a meaningful idea. The Security Cabinet in Israel had in it as members, in addition to Prime Minister Netanyahu, each of the Defense Minister Yitzak Mordachai, the Foreign Minister David Levy, the Finance Minister Dan Meridor, and Sharon himself. It was not expected that any of them would overbid Sharon in jealousy over Israel's security, nor in his zeal for Lake Tiberias to be fully Israeli. If Sharon adopted the idea, the others, most likely, would go along.

"Yes, but I need some kind of a presentation, a written paper to submit to the Security Cabinet."

"When is its next meeting?" Haddadin asked.

"On Wednesday, in three days."

"Your Water Commissioner is competent. I am sure he could prepare a paper for you."

"Yes, but I need to be assured that his ideas are acceptable to you, I do not want to clear that matter and discover that it did not quite reflect our mutual understanding."

"I am prepared to review with the Water Commissioner his paper, and make sure that it reflects my proposal."

"Good, I think you and Mr. Meir Ben Meir should sit and fine-tune the proposal. It is, as you know, very technical and I am no good at these technical matters like you are."

"But General," Haddadin said, "my proposal is not over. Lake Tiberias, I further propose, would be the source of the additional water of 50 mcm until the desalination facility is built and is operational. In other words, an arrangement similar to the supply of the 10 mcm that you are currently delivering to Jordan."

The Minister was taken by surprise, "We will think about that."

"May I say it frankly, Mr. Minister," Haddadin said, "our high level participation on Tuesday in Baqura in the commemoration ceremony will be more meaningful, if we could sign an agreement indicating progress on the water front."

The two Ministers returned to the conference room, but Dr. Gold excused himself and left for Jerusalem. The meeting was resumed and discussed the issue of additional water.

"His Excellency the Minister who, unlike me, is technically very competent and Ben Meir will work out something this afternoon," said Sharon, and the meeting ended. Ben Meir and Haddadin retreated to an office on the same floor and they talked about the role of Lake Tiberias.

"Listen, Meir," Haddadin said, "Israel's share from the Yarmouk is 12 mcm in the five dry months, and 13 mcm during the rest of the year as per Article I.1 of the Water Annex, and the rest of the flow is Jordan's."

"Yes, this is clear in the Annex".

"What I propose is to set aside some 100 mcm of storage capacity in the Lake for Jordan's use. We will give it up when we succeed in building a dam on the Yarmouk at Mukheiba after Israel and Syria arrive at a peace treaty. We will not need the Lake for storage of the Yarmouk floods after that, nor would we need to build dams on the Jordan to impound the Yarmouk floods that flow to it. Such dams would be needed for the spills from Lake Tiberias and from the dam at Mukheiba."

"True," said Meir, "but we do not have a spare capacity in Lake Tiberias to set aside for Jordan."

"You know, Meir, how to make it. I am not in a position to tell you how. You are my senior in age and involvement in the Jordan River basin issues. Surely you remember how to make that storage available," said the Jordanian Minister.

The two talked about the alternatives and they were attractive to Meir. The Jordanian Minister drew a sketch of how the conveyance could be made to the Lake and from it, made a copy of that sketch, gave it to Meir, and kept the original. Then they talked about the desalination facility and the raw water that would have to be desalinated.

"We will mobilize the necessary quantity of brackish water from the springs we have on the Israeli side extending from north of the Sea of Galilee down to Beit Shean," said Meir Ben Meir. "The flow in Beit Shean is being used to grow fish in fish ponds, but we will have that replaced with treated wastewater from the population centers on the Israeli side."

"We will not tolerate any treated waste water to be blended with the brackish water for desalination," Haddadin assured Meir.

"No need, there is sufficient flow of natural spring brackish water, and that would be the flow we would use."

"But until the desalination plant is up and ready, we need our additional water. We need that yesterday. So, the Lake is the only ready answer. I propose an arrangement like the one we have for the 10 mcm under Article I.2 of the Water Annex," Haddadin said.

"I am not sure, but a package of this kind could be worked out. It looks achievable, and I will write the paper for Arik [the nickname of Ariel Sharon] tonight. When will you be ready to review it and finalize it?"

"Any time you are. I do not want to come back to Israel. We can meet at the King Hussein Bridge tomorrow, and get the job done."

"Okay, I will call you in the morning to set the time, but please be prepared for a morning session."

"Will do."

The meeting ended, and the Jordanians flew back to Amman. Haddadin reported the trip and headed for his office. At about 8 p.m., he was called to be ready to go to the Jordan Valley, next to the King Hussein Bridge, to meet with Dori Gold and two other Israeli companions. On the Jordanian side, two other Ministers were present, Dr. Jawad Anani, the Deputy Prime Minister and Dr. Fayez Tarawneh, the Foreign Minister. The conversation with Dr. Gold focused on the water issue and the implementation of the provisions of the Annex. Haddadin was surprised that the issue be reopened after he had had that session with Sharon and his staff, in which Dori Gold also participated. He told Dr. Gold of the last exchange of views between him and Meir Ben Meir, and that they were due to meet the following morning, Monday, 5 May, on the Bridge.

"What is the likelihood of you and Meir agreeing tomorrow?" Asked Dr. Gold.

"A hundred and ten percent," Haddadin asserted in the presence of the other officials. Haddadin guessed that Gold was eager to know so that he could weigh the odds of the high level representation in the commemoration ceremony at Baqura on Tuesday. Other topics pertaining to trade with Israel and with the West Bank were also discussed with the visitors.

Haddadin waited for Meir's call the following morning to no avail. By 2 p.m. Haddadin called him on the cellular.

"Meir, what happened? How come you did not call?"

"I was in continuous meetings in Jerusalem. I am now in my car heading for home, and will call you from there. I want to fax you the text of my paper, but I can't take down your fax number because I am driving, but will take it down when I reach home."

Haddadin waited for two more hours but there was no call from Meir. At about 4 p.m., Haddadin received a call from the Director of the Communications Office of His Majesty the King.

"Get ready to fly in about one hour," said Brigadier Shukri.

"Okay, will do. Please inform my two other delegates, and I will inform the third delegate here at the JVA."

The Jordanians arrived at the Amman military airport, and were met by the Commander of the Royal Airforce.

"I will give you His Majesty's helicopter to fly you in, but it should return immediately after you disembark, because His Majesty is arriving this evening from Oman. I will then send you a fixed-wing aircraft to Ben Gurion Airport to bring you back. When do you want that aircraft to be there?" Asked the Commander.

"Please have it there at about 7 p.m. This is how long I expect the meetings to last."

The Jordanians boarded the Royal helicopter and headed west. When they approached the Jordan River, the pilot made a right turn to stay in Jordanian airspace. He then circled back and continued to make circles over Jordanian territory, without crossing over to the West Bank airspace.

"Officer, what is going on?" Asked one of the senior delegates on board.

"I am not receiving clearance to cross the River yet," answered the pilot.

The team wondered why. Then, when the helicopter crossed over the Jordan River, Israeli helicopters escorted it. They landed at the military airbase in Tel Aviv. Mr. Ben Meir was waiting for them in a van at the tarmac.

"Okay," said the Jordanian Minister to the two helicopter pilots, after thanking them, "you now head back to Amman as ordered."

"Yes sir," the senior officer answered.

Ben Meir received the team and apologized briefly for failing to contact Haddadin by telephone or fax, but he did not look good, nor was his reception as friendly as expected. He escorted the Jordanians to a barrack in the airbase, where they had a nice round table set with a tablecloth and soft drinks and biscuits. There were five chairs. Meir occupied one and the four Jordanians sat on the other four chairs. It did not look like others were supposed to join them. The meeting place was strange, given that the Ministry of Infrastructure should have been the logical venue, and it was not far away from the base.

"Where is your Minister?" Haddadin asked Ben Meir.

"He is not coming," answered Ben Meir with a dry unfriendly tone.

"Okay, where is your paper that we are supposed to review?"

"I have no paper for review," answered Ben Meir with the same tone of voice.

Haddadin started to lose patience with these successive surprises.

"But what about our agreement yesterday?"

"What agreement?"

"Storage in Tiberias, the additional 50 mcm and the other points."

"We had no agreement. I also want to know why you are here!!"

At that point Haddadin's patience ran out, and so did his politeness. He stormed the Israeli host with a shower of words that the man probably never had heard in his entire professional life, and did not spare Minister Sharon either. He then stood up and walked out. The other delegates followed him, and Meir headed for the telephone. Haddadin got out of the barrack to find himself on the street of the Israeli military air base.

"Please come back in," said one of the Jordanian delegates, familiar with the military arrangements, "You are not supposed to be out alone in an Air Force base. You have to have an Israeli escort."

Haddadin did not listen, but stood by the curb outside. Soon Meir appeared and came to him outside.

"Shall I secure your return to Amman?" He asked.

"Yes," Haddadin said with a reciprocal dry tone of voice.

He got in the van that suddenly appeared, and the other delegates joined. Haddadin thought the van and Meir with the group would have to be driven towards the King Hussein Bridge on the Jordan River, because by then, he thought, the Jordanian helicopter would be in the air halfway to Amman. Instead, Meir instructed the driver in Hebrew and the van turned left toward the tarmac.

"Gosh," Haddadin said to his colleagues in Arabic, "the helicopter is still on the ground!"

They got to the helicopter and boarded it without saying goodbye to Ben Meir.

"What happened?" Haddadin asked the senior officer-pilot, "you are supposed to be in the air close to Amman by now!"

"We were never given permission to take off," said the officer, "that was very strange."

"No, not very strange. I think they knew we would need to fly back much sooner than we thought we would!" Haddadin said.

Obviously, the trip was aborted and the Israelis expected it. Haddadin did not know how the arrangements for the trip and the meeting were made, but he appreciated neither the reception nor the venue of the meeting. He reported the outcome to the Crown Prince, and the Prince, in turn, reported it to His Majesty at the airport, upon his arrival that evening from Oman. Later that evening, the Prime Minister also arrived from a trip to Qatar. Haddadin went to his office to write the trip's report, and went home after midnight to get some sleep after that hectic day.

The following morning, while he overslept, was rich with broadcasts about an Israeli-Jordanian row. Haddadin had called in the early morning and informed the Deputy Prime Minister that he would not attend the scheduled meeting of the Council of Ministers on that Tuesday morning, and went back to sleep. The phone rang at 9:30 a.m.

"Did I wake you up?" Asked the Crown Prince on the phone.

"No, Sir, I am awake already."

"What are you doing at home? Come on, hurry to the Council of Ministers. They are waiting for you, to make them the same briefing you made here last evening. By the way, His Majesty was proud of your reaction, and he said that he would have been mad at you if you had behaved differently!"

Haddadin hurried to the Prime Ministry, unaware that the row was made public. As he entered, a group of journalists met him at the entrance. "*Himdillah As Salama* [thank God for your safety]" said some of them.

"I did not arrive from a trip abroad," Haddadin said, "it is the Prime Minister who was away, I was not with him."

Caroline Faraj, the correspondent of *Al Rai*, leaned over and whispered in his ear, "the story is all over the broadcasts. Your encounter in Israel has been on the radios and the wires!"

"Okay, folks, so I was in Israel on behalf of the Government. But I do not have anything to say now. You have to wait and see what the Government would announce." Haddadin said to the correspondents.

He then briefed the Prime Minister and advised that Jordan refrain from escalation. The proper reaction would be to avoid participating in the commemoration ceremony that day, [Tuesday] at the level desired by the Israelis. "Maybe the Governor of Irbid and some military officers should be enough," he suggested.

Ambassador Ross arrived that same day. The row was quickly handled and there was determination to control damage. Haddadin received a call from the Crown Prince Tuesday evening, informing him that he would be receiving a phone call and asking him to be very diplomatic. Soon the phone rang and Meir was on the other side.

"Would you accept an apology from an old veteran?" came the voice of Meir.

"Yes, of course. I hope things would be managed to avoid future misunderstandings and clashes."

On the evening of the following day, Wednesday, Haddadin was told to get ready to participate in a summit on Thursday evening in Aqaba. Dr. Jawad Anani and Haddadin flew to Aqaba for the occasion. Prime Minister Bibi Netanyahu arrived in Aqaba by helicopter, and an hour later came Sharon on a separate helicopter. Anani and Haddadin were asked to join the meeting and they walked in shortly after Sharon arrived.[349]

---

[349] His Majesty King Hussein led the Jordanian side that comprised HRH Crown Prince El Hasan, Prime Minister Abdul Salam Majali, Deputy Prime Minister Dr. Jawad Anani, and Minister of Water and Irrigation Dr. Munther Haddadin. The Israeli side was led by Prime Minister Benjamin Netanyahu and comprised Minister of National Infrastructure Ariel Sharon, Adviser to the Prime Minister Dr. Dori Gold, and Director of the Prime Minister's Office Dany Naveh.

"Here comes our villain," remarked the Crown Prince jokingly.

"Your Majesty," said Minister Sharon as they all sat down and he sat across from the King, "I have been described by your press, and the Arab press in general, with many descriptions. At times I am the bulldozer, at others I am the tough, heartless conqueror, a third I am even the murderer; but, Your Majesty, I found myself a very soft man compared to your Dr. Haddadin!"

The talks addressed the conflict the two sides had recently had and the issues to be settled. Haddadin stressed the issue of the 50 mcm additional water and that they should be supplied to Jordan from Israel instantly, because of the need for them that same summer. The issue of storage in Tiberias was also touched upon. Finally, it was agreed that Israel would supply Jordan from Lake Tiberias with a quantity between 25 and 30 mcm annually distributed evenly over the months of the year. The entire quantity of 50 mcm will be supplied from a desalination plant to be installed for the purpose.

"How long would that take?" Asked Prime Minister Majali.

"Three years," answered Minister Sharon. He mentioned that Israeli researchers were able to bring down the cost of desalination to affordable levels through new technologies. He also brought up the possibility of water transfers south of the Dead Sea, for use by the Israeli Dead Sea Works.

Minister Sharon reinforced the understanding in the meeting by writing a letter on 20 May 1997 to the Crown Prince, who responded affirming the understanding reached in the Summit. Sharon's letter of 20 May and the response on 21 May 1997 constituted an agreement between the two sides.

There was also agreement that Sharon and Haddadin would go on a tour to Europe and Japan, to explain the need for the desalination plant and raise the capital for its installation. No progress on that was ever made.

On 27 May 1997, Israel started delivery of additional water from Tiberias to the KAC, via the same pipe conduit that had been delivering the 20 mcm and the 10 mcm since 5 July 1995. The pipe conduit from Tiberias became a major water source, about 60 mcm per year are delivered to Jordan through it from Tiberias, half of which are delivered during the five months between 15 May and 15 October of each year, when water demands peak.

In early August 1997, another Summit was held at Aqaba. Haddadin joined the Jordan delegation and was among the ministers who received the Israeli delegation at Aqaba airport. Minister Sharon and Minister David Levy accompanied Mr. Netanyahu. Sharon got in the car with Haddadin at the airport to go to the Royal Palace.

"How is your good wife?" Haddadin asked the General for a starter.

"Oh! she is very fine. She is on the farm in the Negev."

"Nice place to live -- on a farm," Haddadin noted.

"You know, we wanted to invite you to visit the farm, but every time you came you get busy fighting with us!"

"Now that our differences are narrowed, we have a better chance to visit with each other," Haddadin said.

In that Summit, it was reaffirmed that Sharon and Haddadin would make a trip to Japan, in follow-up of a visit Netanyahu had just made to that country. Their task would be to talk to the Government of Japan about financing the desalination plant to supply Jordan with the additional 50 mcm of water, and with the current 10 mcm, so that supply from Lake Tiberias would stop and would be replaced with supply from the desalination plant at the rate of 60 mcm per year. The joint trip was never made.

5. The Diversion Weir and Lake Tiberias Storage

After the issue of the additional water was temporarily settled by delivery to Jordan, half of it from Lake Tiberias, the Jordanian Water Minister started pushing for the construction of the diversion/storage facility on the Yarmouk at Adasiyya immediately below the diversion point. The interaction with the Israeli side on this matter was done through the Joint Water Committee. As stated earlier, the centerline of the dam had to be shifted upstream to avoid demolishing some old graves and ruins on the northern bank. The shifting brought the dam about 100 meters into the DMZ of 1949, on the northern bank of the Yarmouk. There was no doubt that the diversion/storage dam was in the best interests of Jordan.

A leak of the understanding to shift the centerline of the dam was made from an Israeli source. The military analyst/strategist Ze'ev Schiff of *Ha'aretz* newspaper who was known to be hostile to Ariel Sharon, published an article on Friday, 15 August 1997. The article contained much exaggeration and claimed that Minister Sharon would soon build a dam on the Yarmouk that would infringe on the occupied Syrian territories at the foothills of the Golan. It took little effort to have the Jordanian papers echo that claim with a campaign to have it stopped by the Jordanians! Ze'ev Schiff hit more than one bird with only one stone. He embarrassed his target, Ariel Sharon, and he put domestic pressure on the Jordanians, the head of the negotiations team in particular, to stop that dam!! It was the first time Haddadin was stunned that the Jordanian media would not take his word and opted to echo the claims of Ze'ev Schiff. The headlines in many of the Jordanian papers were demanding the stoppage of the construction of the dam (which was to the benefit of Jordan). It was not to the detriment of Syria, but was not serving the interests of Israel. The Jordanian papers somehow bought the arguments of Ze'ev Schiff, and, unknowingly, sided against the interests of Jordan.

The opposition to the dam was actually a manifestation of the popular mood, which was disenchanted with the peace. The hoped for economic improvements did not materialize and criticism of the "peace dividends" was voiced by many. The peace opponents had lots of material to prove their point. Any positive outcome of the contacts with Israel was detrimental to their efforts, and they were careful to have these results short-circuited and criticized in the media. Haddadin had to bend to that storm and think of a solution that could disarm the opposition. Syria, he noted, approved the designs of the Maqarin dam, of which a diversion weir was part. As such, no one could claim that a diversion weir built at the diversion point would be against Syrian interests. *Al Dustour* daily newspaper[350] was the forerunner in that campaign, claiming that the storage dam would be a Jordanian recognition of the legitimacy of Israeli occupation of the Himmeh Strip and the Golan. The weeklies soon followed, especially the ones known to be allied with Syria, those papers that had lined up against peace from the beginning. Under such a mis-information campaign, Haddadin had to quiet things down and wait for the dust to settle. The efforts to build the retention dam came to a standstill, and Haddadin decided to wait for the right opportunity.

General Sharon never hid his desire to make an official visit to Jordan. Meeting King Hussein was a big plus to the credit of Israeli politicians, and Sharon was eager to make his first official visit. Sharon's desire was heeded and an official visit was scheduled to Jordan on Sunday, 22 February 1998.

Haddadin decided to take up the remaining water issues with Sharon before his visit would materialize. He made contacts and proposed an agenda for talks with Sharon. Arrangements were made for him to make a trip to West Jerusalem, on Wednesday 18 February 1998 to conduct talks with Minister Sharon on water issues and the Jordan Rift Valley. Haddadin led a delegation and headed for West Jerusalem on that morning. Sharon received him at the King David Hotel, where they held their meeting upon arrival.[351] The first item on the agenda that was faxed to Sharon beforehand by the Jordanian Minister was the construction of the retention dam.

---

[350] The Minister's efforts to educate their chief correspondent on the issue fell on deaf ears.

[351] The team that Haddadin led comprised Ambassador Omar Rifai, Jordan's Ambassador to Israel; Dr. Dureid Mahasneh, Secretary General of the Jordan Valley Authority; and Zafer Alem, Deputy Secretary General, JVA. The Israeli team was lead by Minister Sharon and comprised about sixteen delegates, among whom were Oded Eran, Israel's Ambassador to Jordan; Meir Ben Meir; Aharon Ben Ari; and others.

"I would like to have the dam built on Jordanian and Israeli lands like it was meant to be," said the Jordanian Minister. "The project is bilateral and we do not want to infringe on any non-Israeli or non-Jordanian territory."

"But the dam as it is designed now will be on Israeli territory on this side and Jordanian territory on the other," answered Minister Sharon.

"I mean territories not captured by Israel in the war of 1967, but rather 1948 territories," Haddadin said.

"Mr. Minister," said General Sharon, "all the stretch along the Yarmouk until the bridge east of Himmeh is Israeli territory. We did not capture all of it in 1948, but we tried in 1952 and we lost many soldiers, then we captured it in 1967. This sequence does not change the character of that territory, it is Israeli."

"I am here to talk bilateral issues. I am not authorized to speak on behalf of others. Should I start talking about the issue you just raised, which is none of my business, you will hear things you do not like, General," Haddadin said. General Sharon intended to take the matter further.

"You mean to tell me that the territory is Syrian? No it is not. Or do you mean to tell me it is the PLO that is entitled to it, I say no also."

"General, what I could politely say is that we do not recognize Israel's jurisdiction over any territory that she captured by war, this Himmeh strip included. We made that clear in our peace treaty with you as we had the borders demarcated. Please do not push me to indulge in someone else's business," responded Haddadin. "All that I seek is to have it built in the location that was meant for it, about 100 meters downstream. This way we avoid controversy and conflict between several parties," he added.

"But there are graves and archaeological reasons to keep the dam centerline where it is now, 100 meters upstream," said General Sharon.

"General, the Egyptians moved Abu Sunbul Monastery to build the High Dam at Aswan. What are a few old graves compared to Abu Sunbul?"

"You want it on Israeli and Jordanian lands, and, I say that its current location is in Israeli and Jordanian lands. This is very important to us."

"It is more important to us to stay away from non-Israeli territories."

There was silence that Haddadin broke:

"Okay, General, the floor is yours and please shift to the second item on the agenda."

The second item was on joint projects in Wadi Araba within the plan of integrated development of the Jordan Rift Valley. General Sharon was happy to talk about his favorite subjects and projects of cooperation. He smiled as he made his presentation and Haddadin just looked him straight in the eyes as he talked for about eight minutes. Sharon suddenly realized that his counterpart hardly blinked and had nothing to comment.

"You are truly a difficult man," said the General, "I want to go back to item number one."

"No, General, do not waste your time. I have heard what you had to say about it," Haddadin said.

"I insist," said Minister Sharon.

"If you insist on that, I insist on having only four eyes in this room," Haddadin answered. There were just too many delegates in the room and Haddadin was careful to prevent any leak that might be used politically.

The Jordanian delegates got the point, and they stood up and left the room. Sharon had a hard time, talking to his delegates in Hebrew to have them leave the room. They finally did.

"Can I call back Mr. Ben Meir? I am not as technically qualified as you are," asked General Sharon.

"Yes, please Mr. Ben Meir," Haddadin called as Meir was about to exit the door, "please come back, you can stay with us."

After the room was cleared of the other delegates, Sharon took the floor in an apologetic tone:

"I can not possibly, for political reasons, back down. The opposition will take it that I conceded that the strip of land is not Israeli, which is something against my beliefs and my public announcements."

"I understand, and I am not here to negotiate on behalf of the Syrians or the Palestinians. I have a proposal for a temporary solution that does not hurt either party, and keeps us away from criticism."

"What is your proposal?" Asked the General in an obvious eagerness to find a solution.

"I am willing to build a diversion weir in lieu of a retention dam temporarily. I will come back to build the dam after you reach a settlement with Syria and the Palestinians and have the fate of that piece of territory defined. However, I have to be compensated for the loss of the storage capacity through providing us with a storage facility in Lake Tiberias. These arrangements are temporary and they shall not change or modify the provisions of Annex II to the Treaty. This is my proposal."

General Sharon looked at Meir Ben Meir, and they talked in Hebrew.

"In principle, we agree to the construction of the diversion weir, but I do not know about the storage facility. This is in the hands of the Minister," said Meir Ben Meir. The two Israelis continued talk in Hebrew.

"I have to see for myself the location of the diversion weir," said Sharon, "as for the storage facility, I will have to think about it."

"You can think as long as you want, General," Haddadin said, "but let me assure you that an agreement on these two points is a condition precedent to your official visit next Sunday. No agreement , no visit!"

General Sharon had a surprised look on his face as he looked at Haddadin.

"We need to go to the site," he said.

Before he had time to propose the time, Haddadin interrupted:

"I am ready. Let us check the timing. Tomorrow, Thursday, I am busy with a visiting group of the World Bank Executive Directors. Friday, I do not work because it is the Sabbath in the Hashemite Kingdom. Saturday is your Sabbath and you do not work. This leaves Sunday morning, the day of your official visit."

Haddadin could actually have worked on Friday himself. He did that throughout his career more often than he took it as a holiday. Like Sharon, who observed his Jewish Sabbath, Haddadin, the Orthodox Christian Arab, insisted on observing the Moslem Friday, a national Sabbath for Jordan and the Arab world. But the Israelis has always insisted not to negotiate on Saturdays because of the Jewish Sabbath, and Haddadin wanted to make a similar emphasis on the Jordanian nationally Sabbath.

"Okay," said the General, "Sunday morning you could cross the bridge to our side. We will take a helicopter and fly to the Yarmouk. My official visit starts in the afternoon."

"I would rather meet you there, at the site, and will come to the Yarmouk from our own side."

The meeting was then resumed with the other delegates present and it lasted until late in the afternoon. There was wide press coverage, but the two sides kept their understanding away from the press. After the meeting, the Jordanians wanted to visit Old Jerusalem, which was captured by Israel in 1967. Their host offered to send escorts with them.

"No, please, I do not want to approach the holy city in the company of the occupying power. I know my way, and we want to go without escorts."

"You say we are occupiers, and I say we have reunited a once divided city," said Sharon.

"Peaceful unification with the consent of the inhabitants of both sides is welcome," Haddadin said. "What is not welcome is military occupation and forceful annexation."

After the Israelis left, Haddadin took the Jordanian delegates in his government car.

"Okay, folks, we start by paying homage to al Aqsa Mosque and the Dome of the Rock first; then we go to the Church of Holy Sepulchre," he instructed.

They parked the car by the Wailing Wall, and walked up to al Aqsa Mosque entrance gate. They were scrutinized by Palestinian guards because they were not announced to them, but the Palestinians welcomed the Jordanians heartily when they knew who they were. The delegation prayed in al Aqsa while Haddadin waited inside the mosque, close to where the late King Abdullah was assassinated on 20 July 1951. They then went to the Dome of the Rock, where the delegation members prayed again while Haddadin waited inside. From there they walked to the Church of the Holy

Sepulchre, where he prayed while the other delegation members waited inside!

General Sharon treated his Jordanian guests to a nice dinner at a restaurant in West Jerusalem, owned by the same woman who owns the Beit Gibrael complex on the shores of Lake Tiberias, where the Jordanians and the Israelis conducted Trilateral negotiations in September of 1994. Sharon told the Jordanians of his memories of the old days of confrontation with the Jordan Armed Forces in 1948.

"Your Army had quite a reputation," he said. "We were told that they did not take captives, but got rid of the enemy soldiers they got hold of in battle."

"That is an unfair reputation about the brave Arab Legion. They fought their battles in Palestine with limited resources and inadequate supplies, and did very well."

Sharon then spoke of his experience in 1948 when he led a platoon to retake the Latroun strategic site which was defended by the Jordanian Arab Legion.

"I led about 20 soldiers to that site and made a strong assault. The Jordanians fought back very bravely. We fought for every inch and lost people in combat. I was wounded and saw our losses in men. I do not know how I climbed over a wall and dropped myself over on the other side, away from the Jordanians. I guess it was the reputation your soldiers had that gave me the strength!"

"Oh, I bet it was not because they were unduly cruel," remarked Haddadin, "I bet if they ever got rid of enemy soldiers, which I doubt, it must be because of lack of food supplies!!"

"Whatever the reasons," Sharon continued, "I crept on my belly, wounded and bleeding, for quite a distance until I was aided by some Jews who took me to the first camp and I was treated. Our men were either killed or wounded in the battle for Latroun."

"This is one reason why we all should appreciate the peace we arrived at. It brought an end to war, and provided us with opportunities to make a better life," Haddadin commented.

"Now that we could communicate and talk to each other in peace, I want to ask you for a favor," Sharon said.

"Please, by all means".

"I want you to find me a Jordanian soldier who must be retired by now if he is still alive. I forgot his name, but I have his revolver that was given to him by the late King Abdallah for his bravery in the battle for Jerusalem in 1948."

"Whenever I think of him I feel ashamed that I betrayed his trust," continued Sharon, "I was the officer in charge of intelligence under Dayan.[352] Our headquarters were in Nazareth. Dayan came in one morning in 1953 with a gloomy face, and told me that the Jordanians had captured an Israeli soldier who went astray near Jenin in the West Bank. Efforts to have him released did not succeed."

"After a few days, when we lost hope of the Jordanians releasing our soldier, I set out to the Beit Shean Ghors with three other soldiers with me. We came close to the Sheikh Hussein Bridge and could see across the river a Jordanian army patrol. So I decided to do my part. I signaled to the Jordanian patrol and they signaled back in recognition of my overture, and they looked and sounded friendly. I invited them over for a cup of tea, and they talked to each other and decided to accept my invitation and to walk across the only girder left of that bridge."

"Have you seen three cows that were feeding around here?" Sharon asked the Jordanian soldiers as they sat for a cup of tea.

"No, we actually have not."

"Maybe that Jordanian police station over there got a word about our lost cows," Sharon suggested, as he pointed to a police station building across the river, "could one of you please go ask them if they did?"

"The helpful soldiers abided," Sharon continued his story, "two of them left us to go ask the police station. When they disappeared behind the bushes on the Jordanian side, I pulled my revolver and pointed it at the third soldier who was very trusting and sipping his second cup of tea."

"Drop your gun," Sharon ordered, and he took the poor soldier as prisoner back to Nazareth.

"We were able to work out a swap," Sharon told us, "it took us a couple of months, but we got our guy back. He spent his time at the Army Camp in Zarqa."

"When we released the Jordanian soldier," Sharon continued, "I kept his revolver. It is a military tradition that captives are stripped of their weapons and, when released, their weapons are not."

"I need to see that Jordanian soldier and give him back his revolver, and apologize to him" Sharon requested.

Haddadin promised to ask the headquarters of the Jordan Armed Forces to look for the man. When he asked the Assistant Chief of Staff for Intelligence to do the search, his efforts did not succeed in identifying or locating the man.

---

[352] Sharon was referring to General Moshe Dayan, who became Defense Minister during the June War.

As a result of their meeting that afternoon, Haddadin felt an achievement was on the way. On Saturday, 21 February he received a call from the Royal Palace.

"We have taken all the necessary steps for your meeting with Sharon tomorrow at 10 a.m. on the Yarmouk. We do not want to take a single security risk".

"Good, I will be there on time."

A half hour later the same caller made another call and told Haddadin that the meeting and Sharon's visit had been postponed, because of the likelihood of a strike by the United States against Iraq. Such strikes had become frequent since the Operation Desert Storm in January 1991, by which the allied forces drove the Iraqis out of Kuwait.

"We do not want to host an official visit for Sharon when missiles are landing on Iraq," said the caller.

"I see the point," Haddadin said, "and it is very valid."

Fortunately, Iraq was not hit and the visit was rescheduled for the following Sunday, 8 March 1998. When asked about the time for the meeting on the Yarmouk Haddadin said that it could be held at the same time of the day as agreed with Sharon the week before; Sunday at 10 a.m.

"Can we have it shifted to 12 noon?" Asked a Sharon aide on the phone, "because a 10 a.m. meeting will make the rest of the working hours unproductive."

"Okay, make the meeting at 12 noon with the same arrangements as before."

Haddadin briefed the Crown prince at 10 a.m. on what he was about to bring up with Sharon in the meeting, and what he expected. The Prince ordered a helicopter to take Haddadin to the site. When he arrived at two minutes before mid-day, he saw the Jordanian military officers on the Jordanian bank of the Yarmouk, shaded by a huge Eucalyptus tree, and on the opposite bank stood General Sharon under the sun, with his company, including Meir Ben Meir.

"We refused to allow them to cross over here before you arrived," said the chief Jordanian officer.

"But it is hot over there and the sunshine is strong at midday," Haddadin noted.

"We did not care. Their minister crosses only if our minister is here to receive him."

"Okay, no harm done."

The Jordanian Minister welcomed Sharon and his company of four. They sat at the "picnic table" where the Jordanian and Israeli delegates used to meet, under the auspices of the United Nations, to resolve water-sharing conflicts before the Peace Treaty. Haddadin was accompanied by the Deputy Secretary General of the Jordan Valley Authority. He carried with him the drawings of the diversion weir as part of the Wehda Dam project, to which Syria had given her consent in past years. This way, no one could criticize Jordan for building the weir in the same location where it was supposed to be built under the Wehda Dam project. The feasibility of the dam drops significantly if that weir were not built, so it made good sense to guarantee the construction of the weir before Jordan embarked on the implementation of the Wehda dam. With the weir built, the diversion efficiency would be highly improved and the feasibility of the Wehda dam would be assured. The two sides looked at the site, and inspected the maps. Sharon talked to Ben Meir in Hebrew and turned to Haddadin:

"We agree to have the dam built in accordance with these drawings and on this site," said Sharon.

"Thank you, Mr. Minister, this should be very helpful," Haddadin said as he thought of all the numerous attempts to have that weir built since 1960. He wanted the approval documented, but was more diplomatic than to ask outright for a written approval.

"Okay, Zafer," he turned to the Deputy Secretary General, "are you happy?"

Zafer was very pleased at the Israeli reaction. Haddadin then turned to Sharon.

"I have a problem with Mr. Alem here," Haddadin said referring to Zafer Alem, "he does not take verbal instructions as real and official. He always asks me for signatures."

"Okay, Zafer, here is my signature," and Haddadin signed on the drawing of the diversion weir after he wrote the word "approved", then turned to Sharon.

"Would you like to join me in signing this drawing?"

Sharon talked to Meir Ben Meir, and welcomed the idea. He signed the drawing below the word "approved", and then Haddadin asked Alem to sign it, and Meir Ben Meir to sign it too. The drawing had four signatures on it.

"Now we turn to the storage facility in the Lake," Haddadin said.

At that moment, Minister Sharon leaned over and asked that the topic not be discussed publicly. That stopped Haddadin's conversation and they talked about other things. The Jordanian Army officers were generous in offering the attendants coffee and tea, while they had their meeting under the Eucalyptus tree. By 12:30, the meeting was over and Haddadin walked with the General across the river over a little swing pedestrian bridge.

"We did not talk about storage in Lake Tibrias," he whispered in the ear of Meir Ben Meir away from the car that would take them back.

"How much storage do you want?" asked Ben Meir.

"Forty million," Haddadin suggested as a reasonable figure.

"Okay, you have it."

Haddadin approached the car where Sharon was standing waiting for Ben Meir.

"Did you agree?" Asked Sharon.

"Yes we did," they both answered, "and I will meet you at the bridge at 4:30 this afternoon," Haddadin added.

Haddadin realized how difficult it was for the Israelis to accept the principle of sharing the Tiberias storage facility with any other party, but was glad to get 40 mcm of storage facility. In their interpretation of the meaning of Article I of the Annex, the Israelis considered their pumping of Yarmouk winter flow in the amount of 20 mcm as a storage facility for Jordan because they have to release that quantity to Jordan in the summer. When Haddadin asked for 40 mcm, he had in mind 20 mcm over the 20 mcm that the Israelis consider were available for Jordanian convenience.

Minister Haddadin received Sharon at 4:30 p.m., at the King Hussein Bridge. They rode in a limousine straight to the Royal Palace. He had very few minutes to brief the Crown Prince before the meeting with Sharon and his company started.

The meeting was held at 6 p.m.[353] Sharon asked if Haddadin would brief the meeting on their agreement that morning of 8 March 1998. Haddadin preferred that they do the briefing instead.

"No, General; seniority counts," he answered.

Sharon asked Meir to do the briefing, and he did.

"We agreed to have the Jordanians build the diversion weir at the specified site, and to provide Jordan with 40 mcm additional storage facility in the Sea of Galilee," said Ben Meir.

"Excuse me, *Sidi*, but Mr. Ben Meir is too fast for me and I did not get the last sentence he said," Haddadin interrupted as he was taking notes down.

"I said that we agreed to provide Jordan with 40 mcm additional storage facility in Lake Kinneret," repeated Meir.

That was a pleasant surprise; the 40 mcm storage was then over and above the 20 that were available already. The total storage facility became 60 mcm, an amount equivalent to what the Arab Technical Committee once proposed for storage of the Yarmouk water in the Lake.

---

[353] The Jordan side was headed by Crown prince El Hasan and comprised Dr. Fayez Tarawneh, Chief of the Royal Court; Dr. Jawad Anani, Minister of Foreign Affairs; Dr. Munther Haddadin, Minister of Water and Irrigation; Mr. Omar Rifai, Jordan's Ambassador to Israel; and Dr. Ahmad Mango, Economic Adviser to the Crown Prince. The Israeli side was headed by Ariel Sharon, and comprised Meir Ben Meir; Mr. Katz the Director General of the Ministry of National Infrastructure; and Ambassador Oded Eran, Israel's Ambassador to Jordan.

The meeting touched on other subjects. When Haddadin asked that the draft minutes of the meeting be signed, Sharon consulted with his delegation and Ambassador Eran suggested that it be signed only after Prime Minister Netanyahu became aware of it and approved it.

Sharon met with His Majesty the King at the house of the Crown prince, and dinner followed. The King was very pleased with the water points of agreement.

Two days later, on 10 March 1998, the Crown Prince was scheduled to pay an official visit to Israel, and with him he took a large delegation composed of ministers, members of Parliament, and businessmen. In the program of the visit, Haddadin was to meet with Minister Sharon to finalize the agreement of 8 March and have it signed. They met in Sharon's suite at the Hilton Hotel at 5:30 p.m., and an official meeting was scheduled for 6:30 p.m., in which the Crown Prince would head the Jordan side and Prime Minister Netanyahu would head the Israeli side. As such, Sharon and Haddadin had only an hour to spend in water talks. Sharon had Meir Ben Meir with him, and Haddadin was on his own.

Sharon and Ben Meir wanted to have the Water Annex amended because they felt that the provision in Article I that gave Israel 25 mcm of Yarmouk water and allocated the residue to Jordan would jeopardize Israeli interests in Beit Shean. They also wanted Jordan to reconsider the Wehda dam in light of the storage facility now totaling 60 mcm in Lake Tiberias and, finally, they informed Haddadin that they would have to divert away from the Lake and from a point north of it a quantity equal to the storage facility given to Jordan. Those were major topics to be dealt with before they would sign the new agreement. Haddadin stuck to his guns and expressed his position on the Israeli points as follows:

a) There shall be no change to the provisions of Annex II-Water Related Matters;

b) The Wehda dam project is a Jordanian-Syrian matter, especially after Israel's share in the Yarmouk had been defined as per the Peace Treaty; and,

c) Jordan did not want to know the plans of Israel to secure the storage facility in Lake Tiberias for Jordan.

Their debate took a long time. They had not finished by the time the general meeting headed by the Crown Prince and Prime Minister Netanyahu was due to start, and they went to it 16 minutes late, after Haddadin drafted the agreement on the water points as shown below and had it signed by the parties.

"Dr. Haddadin never changed," said General Sharon as he took his seat in the meeting room next to Netanyahu, "he is as tough as ever."

The Crown Prince smiled and looked at Haddadin for a reaction.

"Mr. Prime Minister," Haddadin said addressing Mr. Netanyahu," General Sharon is unbeatable. No wonder he succeeded in making that crossing of the Suez Canal in 1973!" General Sharon was pleased at the comment. Haddadin walked over to the Crown prince to inform him of the outcome.

"*Sidi,*" he whispered in his ear, "do not care for what could be said about the water points. What is important is what is in my pocket," and he showed him the tip of the paper that comprised the signed agreement with Sharon and Ben Meir.

The meeting went well, and Haddadin made a copy of the agreement and gave it to Ben Meir after the meeting. The agreement read:
"This 10th day of March 1998, Israel and Jordan representatives agreed as follows:

1. In compliance with the provisions of Annex II to the Peace Treaty, and without any change to their respective shares in the Yarmouk River, a diversion structure will be built promptly, immediately downstream of the off-take into the King Abdallah Canal. The drawing of its location was signed on 8 March 1998.
2. A storage volume of 60 mcm[354] will be provided for the Jordan share in the Yarmouk flood. This volume is in the Lake of Galilee. Israel will return the full amount stored from Jordan's share in accordance with a protocol to be concluded for this purpose.
3. Nothing in this agreement shall effect any modification or change to the said Annex of the Peace treaty.

Signed	Signed	Signed
Ariel Sharon	Meir Ben Meir	Munther Haddadin"

The agreement helped drive away the fears of donors to the peace water projects. Germany and Japan allocated funds to double the capacity of transmission of domestic water to Amman from the Jordan Valley, and they demanded assurances that there would be sufficient water to pump in their expanded project. The above agreement provided the needed assurances.

---

[354] The text had a footnote that read:
"The 60 mcm is broken down as follows

20 mcm in use since the Peace Treaty was signed
8 mcm a replacement for the Adasiyya storage/ diversion dam
32 mcm a replacement for storage on the Jordan River.

Signed	Signed
Ariel Sharon, Meir Ben Meir	Munther Haddadin"

Haddadin forwarded the agreement to the Prime Minister via an official letter dated 15 March 1998, and gave a copy of it to the Minister of Planning later.

The Joint Water Committee worked hard to have the agreement implemented. Haddadin interfered on three separate occasions and asked Sharon for intervention to speed up the process of working out the details of the construction, operation, and the security measures needed. Finally, the Israeli approval came on Sunday, 9 August 1998, hours before Haddadin submitted his resignation to the Prime Minister, which was accepted by the King the following day. Haddadin's successor inaugurated the beginning of the weir construction with General Sharon in September 1998, and it was completed in November of 1999.

Finally, a long awaited structure was built in place (Figure XII.1)

6. The Fall of Majali's (Peace) Government

Actually, Haddadin's resignation came after a long public uproar stimulated by the press that was hostile to Dr. Majali's (peace) government. The uproar was first started by the press because of the government's declared intent to amend the Publications Law. A demonstration against the amendment took to the Prime Ministry and the police prevented the demonstrators from coming close to the building. Soon after that, another issue played into the hands of the press and politicians. It became known that the government based its economic restructuring measures on inaccurate figures of economic growth. The Parliament joined in the criticism of government, despite the fact that the growth figures belonged to the period before the government took office. Talk about unemployment, economic stagnation and poverty filled the papers, and the government was blamed. The uproar against the government intensified, on account of a draft Press and Publications law that the government forwarded to Parliament. Its discussion commenced after the Prime Minister left for vacation, and his deputy, the Minister of Information acted on his behalf. The Minister of Information was not very popular with Parliamentarians, and the Press Corps viewed him as responsible for drafting the new draft press law. The environment was not very friendly to government and the Cabinet members were not as cohesive in their defense of issues as they should have been. Parties and groups opposed to peace were only happy to see the government in trouble. Their reasons were diverse. The Islamic Front had never dropped their opposition to the government, whose Prime Minister introduced the one man-one vote Election Temporary Law, and refused to change it. The Islamic Front and the political parties in the opposition boycotted the Parliamentary elections in 1997 because of that law. Moreover, they all were against making peace with Israel. Politicians eager to promote their ascent to power were even happier, and played a subversive role from behind the scene.

The King had been suffering from occasional bouts of fever the causes of which were not properly diagnosed despite trips he had made to London for that purpose. He decided to make a trip to the Mayo Clinic for a check up and treatment, and departed on Tuesday, 14 July 1998. A week before the King departed, some consumers in the Jebel al Hussein Refugee Camp alerted the Water Authority to some taste and odor in the water serving their community. The source of that water was the KAC that carried Yarmouk water and water from Lake Tiberias, due to Jordan under the Peace Treaty. The contributions from Lake Tiberias were substantial and enabled the conveyance project to Amman to run at full capacity in the summer. Minister Haddadin was informed on the following day, Wednesday 8 July. He hurried to the labs, looked at the test results and made sure there was no contamination in the water that could be hazardous to human health. The Ministry of Health, whose Minister has the authority to stop water service, had been duly alerted. The water temperature was unusually high (about 35 degrees Celsius), and the water carried more algae than the treatment plant operators normally cope with. The traces of taste and odor were treated by adjusting the dose of activated carbon and the rates of inflow to the plant, and things were brought under control by Friday morning. However, because of the ration system of water service, consumers fill roof tanks with the pumped water, and the quantity thus stored would provide domestic water to the household for a week or more. The taste and odor therefore continued for another week to be felt by the consumers to whom that water was pumped. The Minister of Water and Irrigation appeared on television and explained the matter to the people and assured the public that there was no hazard to health. The Minister of Health decided to have tests done on the water at the laboratories of the Royal Scientific Society (RSS), as a double check on the results obtained by the laboratories of the Water Authority. Nothing was unusual until after the King departed. His brother, Crown Prince El Hasan was sworn in as Regent in the King's absence. It was customary for the King, as he bade farewell to the ministers, to chat with the concerned minister on any issue that could be of concern to him or to the public. He did not bring up the water subject with the Minister of Water, nor did the Minister raise it because it was not that critical of an issue.

The press got a tip on the fact that the Minister of Health had sent water samples to the RSS laboratories. They demanded from the Minister of Water to reveal the test results that had not been out yet. The results showed that the finished water had in it dead "free-living nematodes," which are not harmful to health even if they were alive. But that was beside the point to the hostile press. It was a golden chance to attack the peace water and the peace at large! Writers surfaced from the middle of nowhere, claiming water expertise and talked about nematodes, trihalomethanes (THM) and other "contaminants." They even had headlines of the "carcinogenic" effects of

THM's. The Minister of Health was shaken and asked for back-up from the WHO regional office in Cairo. He ended up advising people to boil the water before using it, and that was a devastating move. People were scared, and the Minister of Health demanded that the treatment plant be freed from nematodes, which had been accumulating in its filters for nine years since the plant was made operational. No sign of health hazard ever appeared in West Amman that could be attributed to water. At the insistance of the Minister of Health, the Water Authority ran the treatment plant at half capacity to allow for the cleaning of one track while the other was kept operational to secure some water to West Amman. The water availability therefore was reduced to one half, and people were getting water once every two weeks instead of once a week, as the rotation was before that issue came up. The result was acute water shortage and that sent the public complaints soaring.

The real technical issue was the occurrence of dead "free-living nematodes" in the finished water, and the subsequent order of the Minister of Health to clean the plant at a very inconvenient time, when the plant production at full capacity enabled rationing of once a week. The THM levels were within the specified limits as set by Jordanian specifications (150 ppm). There was absolutely no reason for panic or even concern. Instead, the papers carried horrendous claims by some parliament members, turned water experts overnight, that the water was polluted and blamed the matter on Israel and the Minister of Water and Irrigation. The foreign press correspondents joined in and the assault was orchestrated by politicians behind the scenes. A few hoped to replace Majali as Prime Minister, including some subversive elements inside his own Cabinet. Opposition groups claimed that Israel deliberately poisoned the water that she pumped to Jordan, and some Islamist writers testified that Israel actually was pumping to Jordan sewage water! The Minister of Information, who was Acting Prime Minister, did not bother to defend government, nor to instruct the State radio and television to do that. What was surprising was the convergence of the press media and the unanimity of the negative reporting on the Jordanian government, because of that water supply shortage and alleged contamination. The Peace Treaty gains were ridiculed, and Peace with Israel, the writers claimed, was nothing but a mirage in the desert. Some Parliament members stepped in and spoke like water experts claiming that there was water pollution and took irresponsible actions that hurt several officials of the Water Authority. Even the government-owned press joined in on the attacks on government!! This assault on the government by the "official press" could only be attributed to subversive elements within government itself.

All parties and corners of political inclinations joined in on the attacks. At no other time had the press been so united. The foreign press correspondents joined the crowd and the water problem in Jordan was carried by all the wires and the TV stations. The opponents to peace had their day.

They propagated through their licensed newspapers more mis-information than they had ever tried before. They claimed there was water "pollution" and blamed it on Israel who, they alleged, deliberately pumped sewage water to Jordan and caused the pollution. No wise man ventured to read the Treaty and tell the truth. What was surprising was the silence of people who knew the facts. They did not want to defend the case, for fear of being tagged that they advocated Israel against the interests of the Jordanian people. The assault on the Minister of Water, not the Minister of Health who is empowered by law to cut off the water supplies, lasted continuously for 34 days. Reason and logic had no chance. The Minister of Water and Irrigation, in the first place, was the Jordanian water negotiator with Israel, and this prompted the peace opponents to pursue every possible way to discredit him and the Peace Treaty. The King was apparently briefed at the Mayo Clinic and he sent a message of concern over the water situation and even mentioned in his message the reasons for the water problem. The Water Minister resigned, but that was not enough to ease the situation! The press attack gained momentum for ten more days, after which the government fell. The King asked that Dr. Tarawneh form a new Cabinet, and the politicians who played a role in the assault, hoping to benefit from it, were disappointed.

After those events, water from Lake Tiberias was looked upon by the public with suspicion. Peace with Israel lost most of its advocates. The issue of storage in Lake Tiberias was not followed up by the Jordanians, something that Israel became thankful for. Interestingly, after the government fell, the water problem suddenly disappeared! More interesting is the fact that the nematodes, that were the reason for assault on government, were found in more intensities in the finished water a year later, and some were even alive!

Events accelerated after Tarawneh became Prime Minister. It was announced that the King had a serious illness, cancer of the lymphoma. The King was treated with chemotherapy and stayed at the Mayo Clinic until January 1999, when he returned to Jordan, and opted to make a shift in the post of Crown Prince from his brother, El Hasan, to his eldest son, Prince Abdallah, a determination vested with the King by the Constitution. The King returned to the Mayo clinic for further treatment, but known treatments were not good enough to cure his terminal disease. He passed away on 7 February 1999, and was succeeded by his son, King Abdallah the Second.

However, the cooperation between Israel and Jordan, in general, was going downhill because of the Likud policy toward peace and the Palestinians. Water, on the other hand, continued to flow and proved to be an element of cooperation for the benefit of both sides. The water that the Treaty brought saved Jordan from devastating effects of drought that lasted from March 1998 to January 2000. Israel honored most of her commitments to

Jordan, at a time when she badly needed the water herself.[355] That cooperation proved the slogan Haddadin always held, in response to those who had been promoting the theory of water wars:

"Water, by its very nature, is used to extinguish fires, not to ignite them."

Figure (XII.1)- The Diversion Weir in Place

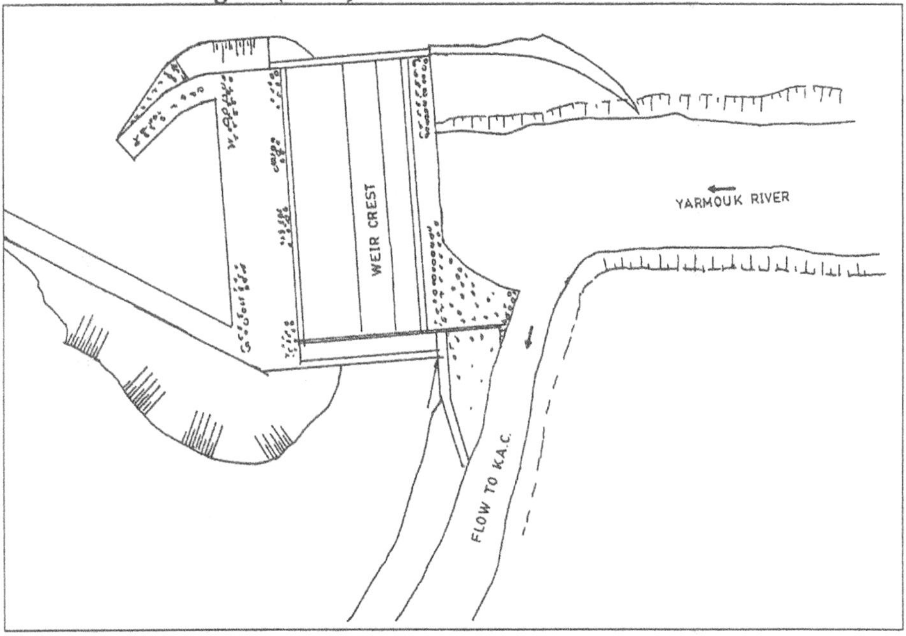

[355] The Israeli Water Commissioner announced, in April 1999, that he was cutting irrigation water to 40% of the usual quantities. He went on to say that he did not have water to give Jordan either. But the flow to Jordan continued until the level in Lake Tiberias got to the "Red Line."

Figure (XII.2)- Picture of the Diversion Weir in Place

Appendices

Appendix 1

Letter from Abraham Bourcart to Reverend John Wilkinson

Paris, 18th Febr. 1901
Basses Pyrenees (7 Place Royal)

Dear and Rev. Sir,

After having read your so interesting and well written book "Israel my Glory," I think you are able more than anybody to give me the council I want, and to help me, for you know the scriptures, the Jews, you believe in the restoration of Israel as a nation in their land, and in all the prophesies, some of which are still to be taken litterally [*sic*] as the scattering is litterally [*sic*] accomplished, and as our blessed Lord accomplishes them litterally [*sic*].

/ [*sic*] Last year, after a stay of 7 months in Palestine, I came to one of our Jewish colonies Rishon Lezion, near Jaffa, and found in the copybook where we were requested to write some note:
"I am highly pleased to see the prosperity of the colony, may the day be near, when England and Amerika together will help to transform the whole land in a garden like this. Sig: C. clergiman." I reproduce very approximately the words.

Now my studies in Palestine led me to the following discovery:

1. With a regular water supply we could produce a power sufficient for all kinds of industries, electric power and light all over the land, power for centrifugal pumps for irrigation (watering).
2. With a regular irrigation like in Egypt or India we could transform easily the whole land in a garden of Eden.

Now Ezechiel 47, which is generally taken in a spiritual sense, made me think of raising the level of the Dead Sea of about 300 feet, up to Jericho and to make it a Living Lake full of fishes, covered with boats, and with hotels and fresh gardens on its shore. This led me to look for more sweet water, which I indeed found in the river Litani or Leontes, flowing uselessly into the Mediterranean Sea. There is no village nor city on it.
So I went measuring the altitudes of the Litani and the Hasbani (1/3 of Jordan), to find the place where a tunnel could be pierced between both. I found the place North Djdeide near Barraj, the other end having to come out South of El Chijam near Hasbani in the hill. There we shall have a first waterfall.

The tunnel will not have a length exceeding 7 miles, the stone is soft limestone, and the costs of piercing will be about 250,000 pounds or less. Just before the rain in November the Litani had about 20 cub. Meter (m3), which supply we shall take through the tunnel during the whole dry season, having then the benefit of it through the whole valley of the Jordan.

So the waterfall of El Chijam gives us
Supply 20 m3 x fall 50 m (150 feet) .... 10,000 HP
With the Hasbani, before we reach the Leddan
And Banijas we shall have about 35 m3 x 200 m............70,000 HP
Then between the lakes Hula (Merom) and Tiberias in a
straight line of but 15 km a huge fall of
210 meters with a minimum supply with
the Litani water of 50 + 20 m3 = 70 m3 x 210m...... ....147,000 HP
South of the lake of Tiberias in El Abadije we
Have again by piercing a short tunnel
of about 30 met. A fall of 12 m x 770 m3.................... 8,400 HP
We have so the whole year round a minimum of.......... 235,400 HP
which in spring and summer may go up to 3 – 400,000 Horse Powers.
I could add various big water powers we have
in the Litani himself between Zahle and
Barraj (the tunnel); the Leddan + Banijas;
The Yarmouk koming down from +2000 feet to
Minus 660 feet = 2660 f. level difference; the
Dshalud of Besan; the Jordan South of
El Abadije (Tiberias); the Audsha strong
River near Jaffa; together more than .... 200,000 HP
Minimum Total = 435,400 HP

You see there is plenty to do in Palestine, there is no such a land on the whole face of the earth, God knew what he did in forming it as it is: high mountains and a Dead Sea 1,200 feet below the sea level.

The Zionist Congress was as much surprised as pleased when I showed them these treasures of their land, and they have now something more to stir the patriotism of them who think there is no hope for their barren land.

With 200,000 to 400,000 HP we have more than power enough to irrigate not only the low Jordan valley, but with the new centrifugal pumps who lift water up to 500 feet, the whole plain of Israel from Besan down to Haifa, a huge and good land, and others. Then we have powers enough to pump up out of the Jordan or better of Elisha's fountain or of the Fara river whole streams of good water up to Jerusalem, or Nablus or other places, from where the water would flow downhill again watering many gardens. The whole plain of Gaza-Jaffa would have pumps driven with small dinamos [*sic*]instead of mules, for their orange yards.

So my conclusion would be:

6. Would not England have all interest to purchase the Holy Land out off the hands of the Sultan, cheap, barren as it is?
7. to let the Jews go back to Palestine, to protect them there and sell them land to higher but reasonable prices;
8. to recuperate thus of the money lost in the Transvaal war;[356]
9. to merit of God a new blessing;
10. to hasten the easing of the torments of the Jews, and the coming of our Lord.

I may add that the whole Jordan valley is covered with a good 3-5 feet thick brown earth, which rationally watered gives 3 crops, and you may see in Jerico vines of 1 foot and more in diameter. Cotton, coffee, cocoa, oranges, palm trees a.s.o. would grow exceedingly well.

Fancy the revolution there, as soon as thousands of Jews will turn back in their land without having to fear any more the Turks!

Now, may I ask you to keep all this secret, because the zionists say the Sultan must not know anything of these projects, to let know the object to your Majesty the King of England, and the Lord who knows that the times are at hand, will make you succeed.

Should you not agree with my idea or desire, please let me know your objection, but I do hope you have none.

Meanwhile I ask you to kindly excuse my bad English, and to believe me

Yours very respectfully

Abr. M Bourcart

[356] The Transvaal War (1899-1902) was fought between Great Britain on one side and Transvaal and Orange Free State on the other. A war of the same name was also fought earlier between Great Britain and Transvaal, a province in NE the Republic of South Africa, where Pretoria the capital is located. Footnote by the author.

Map 1

Diversion of Litani

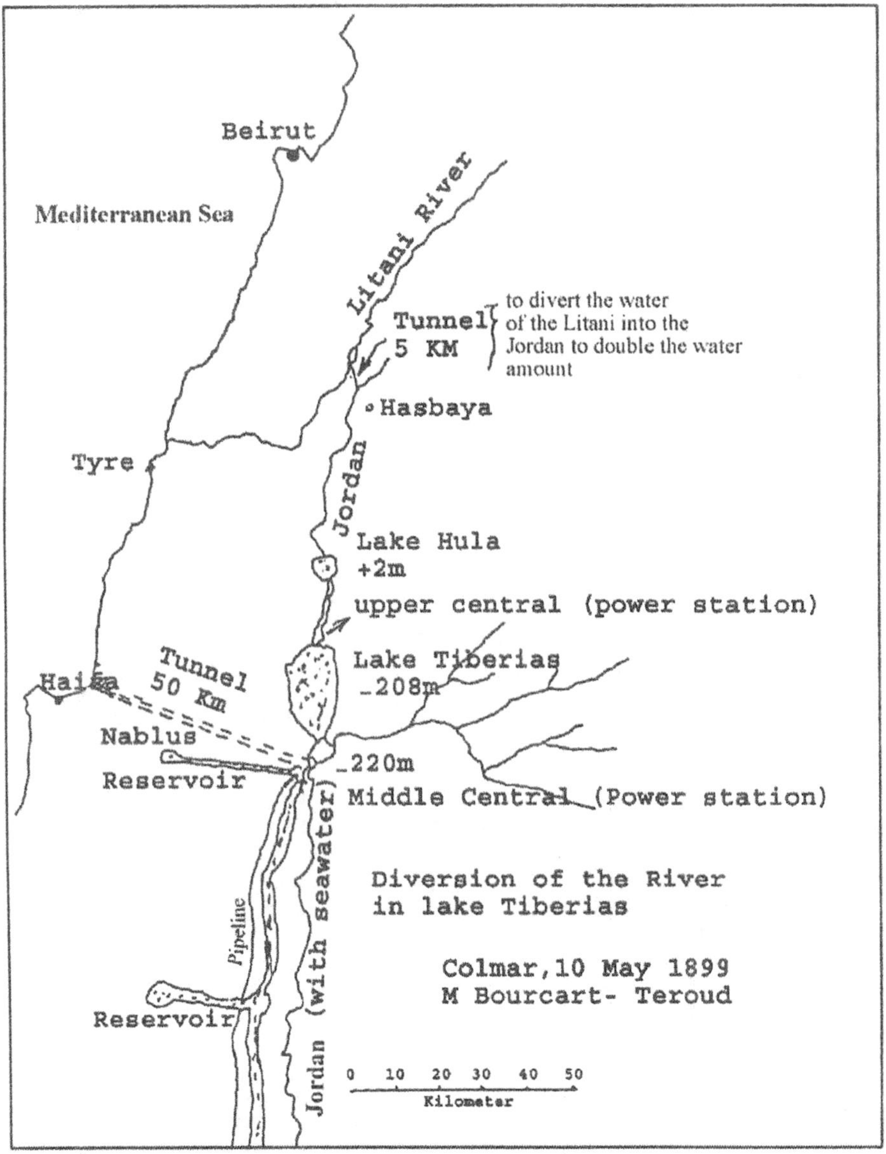

Map 2
Mediterranean - Dead Sea Linkage Via Jordan River From Haifa

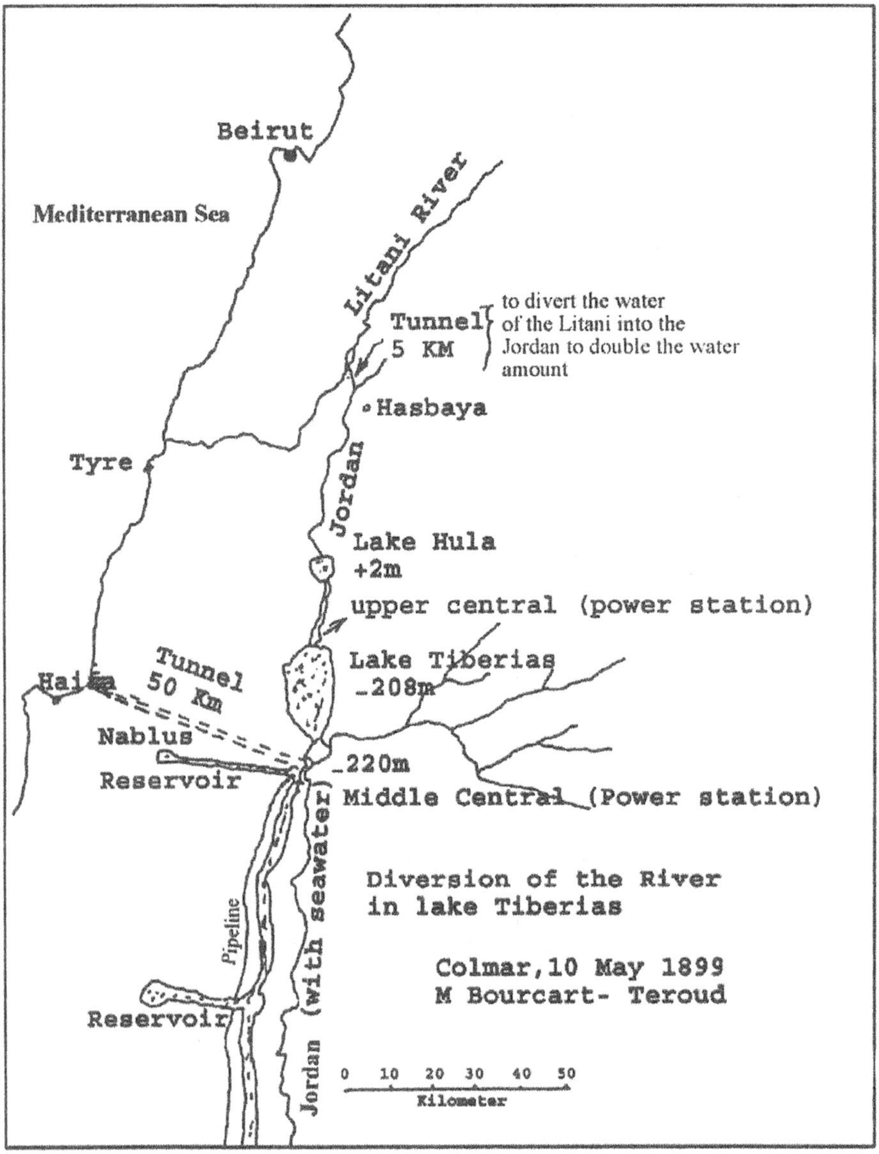

Map 3
Irrigation of Western Escarpment

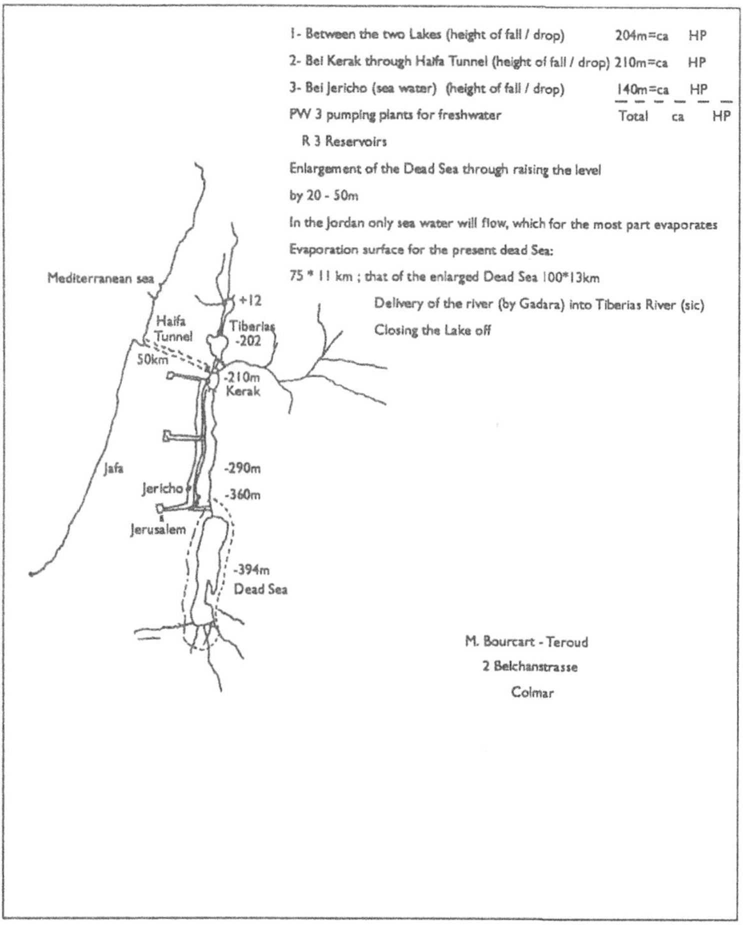

Map 4 Biblical References[357]

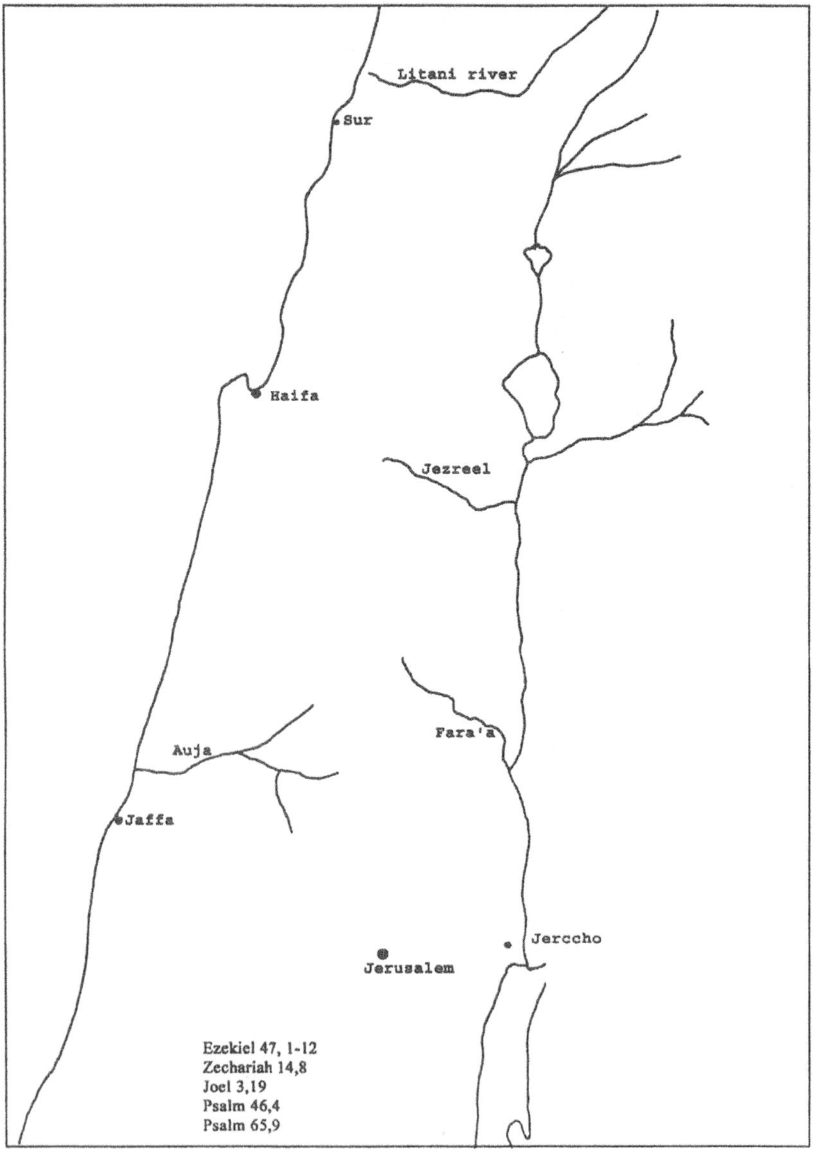

[357] Biblical references in Bourcart Letter that are listed on Map 4 are quoted from: Holy Bible Containing the Old and New Testaments Translated Out of the Original Tongues Being the Version Set Forth, A.D. 1611 Compared with the Most Ancient Authorities and Revised, A.D. 1881-1885. Newly Edited by the American Revision Committee, A.D 1901. Standard Edition. New York, NY: Thomas Nelson & Sons.

The Book of:
Eizekiel 47, 1-12 pp. 927-928

47 And he brought me back unto the door of the house; and,
behold, waters issued out from under the threshold of the
house eastward (for the forefront of the house was toward
the east); and the waters came down from under, from the
right side of the house, on the south of the alter. **2** Then he
brought me out by the way of the gate northward, and led
me round by the way without unto the outer gate, by the
way of *the gate* that looketh toward the east; and, behold,
there [1]ran out waters on the right side.
3 When the man went forth eastward with the line in his
hand, he measured a thousand cubits, and he caused me
to pass through the waters, waters that were to the
ankles. **4** Again he measured a thousand, and caused
me to pass through the waters, waters that were to the
knees. Again he measured a thousand, and caused me to
pass through *the waters*, waters that were to the loins. **5**
Afterward he measured a thousand; and *it was* a river
that I could not pass through; for the waters were risen,
waters to swim in, a river that could not be passed
through.
6 And he said unto me, Son of man, hast thou seen *this*?
Then he brought me, and caused me to return to the
bank of the river. **7** Now when I had returned, behold,
upon the bank of the river were very many trees on the
one side and on the other. **8** Then said he unto me,
These waters issue forth toward the eastern region, and
shall go down into the Arabah; and they shall go toward
the sea; into the sea *shall the waters go* which were
made to issue forth; and the waters shall be healed. **9**
And it shall come to pass, that every living creature
which swarmeth, in every place whither the [2]rivers

---

[1] Or, *trickled forth*.
[2] Heb. *Two rivers*.

come, shall live; and there shall be a very great
multitude of fish; for these waters are come thither, [3]and
the waters of the sea shall be healed, and everything
shall live whithersoever the river cometh.
10 And it shall come to pass, that fishers shall stand by it:
from En-gedi even unto En-eglaim shall be a place for
the spreading of nets; their fish shall be after their kinds,
as the fish of the great sea, exceeding many.
11 But the miry places thereof, and the marshes thereof,
shall not be healed; they shall be [4]given up to salt. **12**
And by the river upon the bank thereof, on this side and
on that side, shall grow every tree for food, whose leaf
shall not wither, neither shall the fruit thereof fail: it
shall bring forth [5]new fruit every month, because the
waters thereof issue out of the sanctuary; and the fruit
thereof shall be for food, and the leaf thereof for healing.

The Book of:
Zechariah (14,8) pp. 990.

14. Behold, a day of Jehovah cometh, when thy spoil shall be
divided in the midst of thee. **2** For I will gather….
3 Then shall Jehova…
8 And it shall come to pass in that day, that living waters shall
go out from Jerusalem; half of them toward the eastern
sea, and half of them toward the western sea: in summer
and in winter shall it be.
9. And…

---

[3] Or, *that all things may be healed and live*
[4] Or, *given for salt*
[5] Heb. *First-fruits*

The Book of
Joel (3,18) pp. 960

3 those days, and in that time, when I shall bring back the captivity of
Judah and Jerusalem, **2** I will gather all nations, and will bring
them down.... **3** and have cast..

18 And it shall come to pass in that day, that the mountains shall drop down sweet wine, and the hills shall flow with milk, and all the brooks of Judah shall flow with waters; and a fountain shall come forth from the house of Jehovah, and shall water [1]the valley of Shittim.

19. Egypt shall be a desolation, and Edom shall be a desolate wilderness, for the violence done to the children of Judah, because they have shed innocent blood in their land.

PSALMS
BOOK II (46,4) pp. 618

46. God is our refuge and strength,
A very present help in trouble.
2. Therefore will we not fear,.....
3. Though the waters thereof roar and be troubled,.......
4. There is a river, the streams whereof make glad the city of God,
The holy place of the tabernacles of the Most High,

Footnote, The Book of Joel:

[1]That is, *the valley of accacias*.

PSALMS

BOOK II (65,9)   pp. 629

65 [5]Praise waiteh for thee, O God in Zion;
And unto thee shall the vow be performed.

9 Thou visitest the earth, and waterest it,
Thou greatly enrichest it;
The river of God is full of water;
Thou providest them grain,
[9]when thou hast so prepared [10]the earth.

Footnotes, Psalms, Book II (65,9)

[5] Or, *there shall be silence before thee, and praise, o God &c.*

[9] Or, *for so preparest thou*

[10] Heb. *It.*

Memorandum of Understanding Between Ambassador Johnston and Israel

July 5, 1955

DRAFT MEMORANDUM OF UNDERSTANDING

1. From April to July of 1955, discussions were held in Washington between representatives of the Government of Israel and Ambassador Johnston relative to the development of the Jordan River Valley.
2. The purpose of this memorandum is to set forth the understanding ot the two parties as to the basic provisions of a plan, which if presented to Israel, might be accepted. This formulation is subject to approval by the Government of Israel.
3. Nothing in this formulation is to be construed as an agreement on the part of the Israeli Government or of Ambassador Johnston; it is not conclusive and is subject to further amplification and qualification.
4. It is understood that this understanding is subject to the conclusion of mutually satisfactory financial arrangements.

I. Storage

5. Water of the Jordan and Yarmuk Rivers will be stored and regulated through reservoir facilities to be constructed on the Yarmuk River and through operation of Lake Tiberias as a reservoir.
6. In order to determine more precisely the economic, engineering and operational feasibility of utilizing either Lake Tiberias or alternative sites outside Israel for the storage of Yarmuk floods, and the best use of potential storage for these flows, final decision as to the use either of the Lake or of alternative sites is deferred. It is understood that the present 700 mcm of live storage capacity in Lake Tiberias represents Israel's estimated requirement for storage in the Lake, and that no decision regarding storage of Yarmuk water in the Lake will reduce this capacity unless Israel determines that its need for storage is less than 700 mcm.
7. A final decision on this question will be made by the Engineering Board during the first six months of the year 1960. In

making its decision, the Board shall take into account requirements and supplementary water supply within the lower Jordan Valley, and the economic feasibility of utilizing Lake Tiberias or alternative sites for storage of the Yarmuk flows. The Board shall, within the context of the above considerations, consult Israel, Jordan and Syria and receive such comments and considerations as they may wish to present.

8. Should the Engineering Board's determination indicate the need for storage of Yarmuk water in Lake Tiberias, the Government of Israel will make available in the Lake such storage capacity as is determined by the Board, not exceeding 300 mcm, and will cooperate in implementation of the Board's decision

9. A declaration will be included in the agreements with the United States to the effect that Israel, Syria and Jordan will not use any provisions of the agreement as a basis for territorial claims, claims of frontier corrections or adjustments, or claims for participating in the operation of or control over any of the elements of the Jordan River system, which will be without prejudice to any rights or obligations which the states named have assumed under general armistice agreement, will also provide that the states named will refrain from invoking the provisions of either an armistice agreement or this agreement for the purpose of interfering with the operation of the other agreement.

II. Supervision

10. An Engineering Board shall be established for the purpose of performing the following functions:

[the text lists the scope of duties of the Engineering Boar in reviewing the engineering designs of the diversion structures and other project features on the Jordan River, the Yarmuk River, and Lake Tibrias should the latter be used as a reservoir for storage of the Yarmuk flows, with view to ensuring that such structures and other project features will not be inconsistent with the implementation of the agreement. The Board would make the necessary studies and decisions, and establish flow patterns for withdrawals of water.]

"If the parties desire and other arrangements are agreed for the appointment of a Watermaster," the text continued, "the Board may be abolished on the completion of the studies." The scope of work of the Watermaster was stated in the text.

III. Division of the Water

13. The following annual withdrawals and releases of water are agreed:

(a) From the Banyas: A withdrawal of up to 20mcm by Syria.
(b) From the Hasbani: A withdrawal of up to 35 mcm by Lebanon
(c) From the River Jordan: A delivery by Israel of up to 22 mcm to Syria for the area of the Boteiha farms.
(d) From Lake Tiberias: A delivery by Israel to the Kingdom of Jordan at a specified point south of Lake Tiberias of 100 mcm of which 70 mcm would be of average Lake Tiberias salinity and the remaining 30 mcm being derived at Israel's option from the diversion of saline springs around Lake Tiberias. The delivery of the agreed quantity will be made at predetermined constant flows according to seasonal schedules determined by the Engineering Board under paragraph 10 d, and will be affected through structures to be approved by the Board.
(e) From the Yarmuk: A withdrawal of up to 90 mcm by Syria, and a delivery by the Kingdom of Jordan of 40 mcm to Israel.

14. Except for the above withdrawals and deliveries, the waters of the Jordan River will be available for the unconditional use of Israel.

15. Should the Board require the storage of Yarmuk Water in Lake Tiberias in accordance with paragraphs 6-8, Israel will deliver such water to the Kingdom of Jordan at an agreed point, except insofar as the agreed storage capacity has been exceeded. This delivery will be made in periodically predetermined flows according to procedures to be agreed upon.

16. The principal Israel diversion structures will be located at or near Benot Yaacov bridge and the principal diversion of the Kingdom of Jordan will be located at or near Adasiya.

17. Israel agrees as an additional but temporary undertaking, to make available for use in the Kingdom of Jordan on an annual basis an additional quantity of water not exceeding 75 mcm annually during the period between conclusion of final understanding between Ambassador Johnston and states concerned and December 31, 1960. Determination of the flow volumes of such deliveries would be made by the Engineering Board from year to year after consultation with Israel, should the Board establish that the water resources of the Kingdom of Jordan are insufficient for irrigation and leaching purposes. These temporary deliveries will in no way prejudice agreed diversions as stipulated in the agreement.

It is understood that in any final agreement legal provisions satisfactory to Israel will be incorporated for the purpose of establishing the temporary character of the obligation in this paragraph and of safeguarding against any possible claims which might be made by Jordan for continuation of this obligation.

Appendix 3

The Revised Unified Plan for the Development of the Jordan Valley (Johnston Plan)

September 30, 1955
THE JORDAN VALLEY PLAN

I- The Role of the United States

The Government of the United States is interested in the Jordan Valley Plan for a number of reasons.

First, the United States is interested in the economic growth and independence of all of the nations of the free world. It is a basic aim of American foreign policy to assist less strong and wealthy nations to achieve higher standards of living in freedom and security. This policy has been pursued consistently throughout the world in recent years. The Jordan Valley project is not an isolated case; it is one of many similar projects in many countries to which the United States has offered and given its support.

Second, the United States has adopted, and seeks to implement, a policy of cordial and helpful friendship toward all the peoples of the Near East. Statements of the President and the Secretary of State to this effect have made it clear that the United States is especially concerned to help the nations of this vitally strategic and important area of the world. It believes this help will be most effective if it produces permanent sources of new wealth In the countries which accept it.

Third, the United States is interested in the Arab refugees from Palestine, whose plight must surely be relieved with as little delay as possible. Only recently, the Secretary of State expressed willingness to help resolve political problems of the utmost importance to the future of these people. In a wholly different context, the Jordan Valley Plan offers an opportunity to improve their social and economic status without impairing their rights either to repatriation or compensation.

Moreover, to the extent that some thousands of the refugees would be able to move into independent occupations on land and in ancillary pursuits in the Jordan Valley, the Plan would decrease the burden of international relief, now mainly borne by the United States.

The United States would feel that its contribution to the welfare of the refugees was being spent not merely to keep these unfortunate people alive but to help them achieve self-sufficiency through an undertaking of permanent economic importance to the nations concerned.

This is not to say that the interest of the United States focuses solely upon the Arab refugees or that the U.S. believes that newly created agricultural lands in the Jordan Valley should be available only to refugees. On the contrary, it recognizes that all of the people of the countries concerned are entitled to share increased opportunities for a better standard of life. The United States does feel however that a project supported by the international community should result in a substantial lightening of the relief load which the international community now bears.

Finally, the United States believes that the international resources vital to the growth and development of more than one nation should be peacefully and equitably shared. It does not believe that such resources belong to the strongest or to those who happen to be in a position to preempt them. It believes that its own experience in sharing rivers which flow between the United States and its neighbors may be usefully brought to bear on the problem of sharing the waters of the Jordan.

It was for all of these reasons that the President of the United States sent his personal Ambassador, Mr. Eric Johnston, to the Near East two years ago to offer the Governments of Lebanon, Syria, Jordan and Israel a plan for unified and comprehensive development of the Jordan Valley.

The role of Ambassador Johnston since then has been that of negotiator, seeking through the exercise of good offices, to work out a program for the Valley which could be accepted by all of the States having an interest in its waters. The United States does not regard itself as an advocate, pressing the Governments of these States to accept a plan which the United States believes to be to their benefit. On the contrary, the attitude of Ambassador Johnston is, and has always been, that the decision rests with the States themselves. He has merely tried in cooperation with the Governments themselves to work out a program to which all could subscribe without impairing their individual national interests.

Only in one sense, therefore, does the United States consider itself a "party" to the Jordan Valley Plan. This is in the sense that it has undertaken to support the project financially; and it is in this sense only that the United States has exercised any voice whatever in decisions made regarding the plan. Ambassador Johnston has been obliged to insist that the Plan meet certain conditions necessary to justify the expenditure of United States public funds.

One of these conditions is that the Plan must make full use of all of the available waters of the Valley without waste or extravagance. These waters are a resource of utmost value to the people of an arid area. Under no circumstances, would the United States feel itself justified in supporting a plan which did not conserve and utilize this resource to the utmost possible extent.

Another of the United States' conditions is that the Plan must be economically and technically sound. The construction of major installations to impound and distribute irrigation water throughout the Valley must be regarded as a permanent investment capable of producing a reasonable return for many decades to come. The cost must therefore be proportionate to the return by normal standards of economic value.

Third, the United States insists that the Plan must be equitable in its own judgment. It could not agree to support any project which might, because of basic inequity to one party or another, provoke disputes and possible termination of the understanding on which the Plan was based.

Finally, from the standpoint of the United Stated, the Plan must be accepted by all of the countries having an interest in or now using the waters of the River system. Without this acceptance by all concerned, implementation of the Plan would be impossible, the available waters could not be used to maximum benefit, the total area of irrigated land would be reduced with consequent imbalance in the economics of the undertaking. From the outset of his negotiations,

Ambassador Johnston has made it clear that United States support for the Plan was premised on the assumption that it would be accepted by all of the four states directly concerned.

This is not to say that agreement between these states is necessary. The political difficulties which make this impracticable are clearly recognized and Ambassador Johnston has made no suggestion that a compact between the Arab states on the one hand and Israel on the other be negotiated. So far as the United States is concerned, assurances of acceptance given to Ambassador Johnston by the Arab Governments, either jointly or severally, and by the Government of Israel, will be a sufficient basis on which to proceed with immediate implementation of the program.

In the view of the United States, moreover, the Jordan Valley Plan must be considered and decided upon outside the context of the long-standing political difficulties between Israel and the Arab countries. It offers an opportunity to move ahead at once with economic improvements of major importance despite existing political controversies.

Ambassador Johnston clearly reflected the U.S. position in this respect in his letter of August 30 to the Prime Minister of Jordan, in which he stated:

"I wish to point out and to reaffirm the distinct delineation between our economic negotiations and the political problems of which Mr. Dulles spoke. I have emphasized that may negotiations related to a program for the economic development of the area, and I insist that they remain in this context and not be linked with the settlement of any political problems."

II- Elements of the Present Plan

The Jordan Valley Plan now under consideration by the Governments of Lebanon, Syria and Jordan is the product of twenty-four months of detailed and painstaking negotiation. In contrast, the plan originally proposed by Mr. Johnston was no more than an outline of the idea. Through many discussions with officials and technical experts of the Arab countries this original plan has been refined and adjusted; compromises and concessions have been made; suggestions made by the Arab Technical Committee have been accepted on most if not all points. The result is a Plan for the Valley which, in the opinion of the United States, is equitable, workable, and economically justifiable.

The plan lends itself to consideration in several parts: Storage; Distribution; Division of the Water; and Supervision.

1.- Storage

A. The Maqarin Reservoir

The Plan envisions the construction of a dam 126 meter high on the Yarmuk River near Maqarin to impound 300 million cubic meters (mcm) of water for irrigation and make possible the generation of some 150 million kilowatt hours of electric energy a year.

Originally, a dam only 47 meters high, capable of impounding only 47 million cubic meters (mcm) of water, was proposed for the Yarmuk River at Wadi Khalid, on the assumption that as much as 550 mcm of Yarmuk water would be stored in Lake Tiberias. During the discussions with the Arab Committee, however, it became apparent that this arrangement, while technically adequate for irrigation purposes, would not satisfy the requirements of the Arab countries for electric power or for secure storage of their irrigation reserve.

Particularly in the case of Syria, though also in the case of Jordan, the argument in behalf of a larger dam to permit the maximum possible development of electric power was a convincing one. So was the argument of the Arab countries that as much as possible of the water needed for Arab crops should be under direct Arab control. Taking these two considerations into account, engineers determined that the 126 meter dam on the Yarmuk, as now proposed, would be justifiable both from the standpoint of economic soundness and practicability.

The Plan thus reflects an effort to meet the Arab position on the size of the dam required on the Yarmuk River, even though the larger dam calls for the expenditure of $21,000,000 more than the smaller one.

While deferring in this respect to the Arab experts, the United States has been obliged to reject an Arab proposal for a still larger dam, with a reservoir capacity of 460 mcm, on the Yarmuk River. The United States has also declined to support the construction of a dam with foundations sufficiently large to permit the Arab Governments to increase the structure to store 460 mcm if they should later decide to do so.

With respect to both of these suggestions, the U.S. attitude is the same. Competent engineering advice indicates that the erection of a structure to impound 460 mcm on the Yarmuk, while perhaps physically feasible, would neither eliminate the necessity of storing substantial quantities of water in Lake Tiberias nor add materially to the power potential of the development.

Flood waters of the Yarmuk averaging nearly 70 mcm a year would still spill form the larger reservoir and, unless captured below, would run to waste.

Thus, since Arab waters in substantial quantity would still have to be accommodated in Lake Tiberias, erection of the 460 mcm dam would affect the principle of storage in the Lake only in degree. The hydro-electric power potential would be increased only by some 25 million kwh annually, at a cost of $10 million.

The United States moreover is unwilling to involve itself in the expenditure of a substantial sum for enlarged foundations which might or might not be used. The investment of public funds on such a purely speculative basis would not be warranted.

Nevertheless, the Plan leaves it open for the Arab States to increase the height and capacity of the Maqarin Dam if they wish to do so at their own expense. The United States seriously questions the advisability or necessity of doing so; but in deference to the Arab position, it has been willing to agree that the privilege should be retained to the countries directly involved. In view of its own financial connection with the project, however, the United States has felt obliged to insist that the Arab States make known their firm decision on this matter within the net five years, before expensive installations involved in the use of Lake Tiberias are undertaken.

B. Lake Tiberias

Every plan advanced in recent years for the development of the Jordan Valley has recognized the utility of Lake Tiberias as a storage reservoir for the waters of the river system. Its natural advantages for this purpose were affirmed in the Arab Plan produced by the Arab Technical Committee.

The Plan proposes to utilize these advantages to a limited extent for the storage of Yarmuk flood-flows which cannot be economically captured elsewhere. In flood years, the Yarmuk flow will greatly exceed the storage capacity of any reservoir that has yet been proposed at Maqarin or elsewhere on the stream.

Unless this flood flow is caught and stored for use as needed during periods of low streamflow, millions of cubic meters of water will be lost. This water is absolutely essential for complete irrigation of Arab lands. The normal flows of the river are not sufficient; if flood waters are not impounded, the possibilities for Arab agriculture in the valley will be reduced.

Theoretically, it might be possible to construct a second dam on the Yarmuk below Maqarin to receive and hold these flood-waters. But the expense of building two reservoirs would push the project beyond the limits of economic practicability. The ratio between cost and benefit would be so unfavorable as to make the undertaking economically unjustifiable. This is especially true in view of the availability of adequate storage space in the natural reservoir of Lake Tiberias.

The Plan, therefore, contemplates the storage of Yarmuk flood-flows in Lake Tiberias. Averaged out over a period of years, these flood flows will amount to approximately 80 million cubic meters a year. Allowing for variable annual streamflows and needs for irrigation, it has been determined that storage space for about 300 million cubic meters will be required in Lake Tiberias. The Plan would assure the Arab states of space up this amount when needed.

The Yarmuk flood flows spilling from the Maqarin dam would be diverted at Adasiya into a conduit leading to Lake Tiberias. They would be released, as needed for irrigation in the lower valley, through another conduit leading from the southern end of the Lake to an appropriate point on the East Ghor canal.

The water deposited in the Lake to the credit of Jordan, so to speak, would be recorded by automatic gauges under the supervision of the Watermaster. The total amount entering the Lake during any period of time would be precisely known. Withdrawals from this "account" would also be gauged automatically under the supervision of the Watermaster, and would be made in seasonal patterns pre-determined to meet the needs of the irrigation system in the lower valley. The full amount of water entering the Lake form the Yarmuk River over a period of years would thus be released to Jordan under neutral supervision as needed. No charge for evaporation or other losses would be made against the quantity of Yarmuk water stored in the Lake.

The plan thus proposes to provide the total storage space required for Jordan's irrigation needs through the construction of a 300 mcm reservoir on the Yarmuk and through the utilization of approximately 300 mcm of storage capacity in Lake Tiberias.

C. Deferred Use of Tiberias

At the request of the Arab Technical experts, however, the Plan now envisions a delay of five years before Arab water is actually stored in Lake Tiberias. This is compatible with the normal "phasing" of the total development program and will involve no extra delay in the completion of the valley plan.

At the end of this five year period, the neutral Engineering Board (see below) would determine the necessity of storing Yarmuk flood waters in Lake Tiberias or whether more feasible and economical storage might be found elsewhere. In making this determination, the Board would base its decision solely on irrigation requirements of Arab lands in the lower Jordan valley. The decision would be final; and the Arab states and Israel would be committed in advance to accept.

By placing responsibility for this decision on the Board the question of the use of Tiberias is taken out of the hands of all of the interested parties and left to the judgment of impartial, neutral experts. This is necessary so far as the United States is concerned, in view of its commitment to support the development program financially. Obviously the Government of the United States cannot undertake to implement a later decision which might be made on the basis of political or purely national considerations without regard to economics and technical feasibility. In assuring its support, the U.S. must also assure itself that the Plan will be carried out in accordance wit the sound engineering and economic principles.

D. The Hasbani

In recognition of Lebanon's interest in increasing its irrigated lands, the Plan now provides for an immediate survey to obtain hydrologic and land-use data in the watershed of the Hasbani river in Lebanon. Funds for this survey in the amount of $250,000 would be made available by the United States. The information thus obtained would be used to determine the necessity of constructing a storage dam on the Hasbani to assure that water allocated for Lebanese lands could actually be made available. The decision as to the kind and the size of the dam to be erected on the Hasbani would await the conclusion of the survey.

2. Distribution

Once stored, the waters of the Valley must be conveyed, under careful regulation, to the lands they are to irrigate. The Plan therefore contemplates the installation of the following facilities to transport water to Arab areas:

a) A diversion dam near Adasiya to supply the East Ghor canal and, if necessary, to divert excess flood waters to Lake Tiberias for later delivery to Jordan;
b) A main canal network in Jordan, including:
  (1) The East Ghor Canal running from Adasiya southward to the vicinity of the Dead Sea.
  (2) A siphon or other structure for conveying water from the East Ghor to the West Ghor.
  (3) The West Ghor Canal in Jordan feeding from the siphon.
  (4) A feeder canal from Lake Tiberias to a junction with the East Ghor Canal.
  (5) A canal form Adasiya to Lake Tiberias, if necessary to capture and store Yarmuk flood flows in the Lake,

c) A distribution system to convey waters form the main Ghor canals to the farm lands;
d) Pumping plants to raise water to lands above the main Ghor canals;
e) Generating plants on the main canals to supply power for pumping. Inasmuch as these power installations will not produce excess power for sale and are necessary to pump water above the canal, they are considered integral parts of the Plan;
f) Main drainage facilities for removing excess water and salts from irrigated lands;
g) Regulating and control works on Lake Tiberias if the Lake is used to store Yarmuk flood flows;
h) A new diversion structure and canal from the Jordan River to Boteiha Farm in Syria, together with 50 K.W. of electric power to replace water power;

i) A diversion structure north of Lake Tiberias to permit Israeli withdrawals from upper Jordan;
j) A main canal from this upstream diversion to irrigable areas in the Galilee hills;
k) A short canal from Lake Tiberias down the west side of the river to serve irrigable lands in the Beisan area in Israel.

3. Division of Water

International law recognizes that each of the nations sharing an international river system has a right to a portion of the water. There is no single, generally accepted principle, however, on which the division of the water can be based.

In these negotiations, the basic principle was adopted of assuring to the Arab States enough water to meet the needs of all their lands that can feasibly be irrigated. There has been no change from this principle. The division of water set forth in the present Plan fully accomplishes the objectives which the negotiations have had from the very beginning.

In the original proposal offered for discussion two years ago, no specific provision was made for an allocation of water to Lebanon because of the lack of engineering data. The Arab Technical Committee provided information as to Lebanese needs and recommended and allocation of 35 mcm from the Hasbani River. This full amount has been accepted in the Plan.

The Arab Technical Committee recommendation has also been accepted with respect to the allocations for Syria. The first estimates discussed did not take into consideration the possibility of expanding the irrigable area in the vicinity of Boteiha Farms, nor was adequate information available at that time as to irrigation prospects form the Banyas River and on the Yarmuk Plateau. The Arab Technical Committee recommended allocations for Syria of 20 mcm. from the Banyas, 22 mcm frm the Jordan for the Boteiha Farm area; and 90 mcm form the Yarmuk. All of these allocations have been incorporated in the Plan.

The allocations to Lebanon and Syria are all within the upper basins, and water which is diverted but not actually consumed will return to the river system, either by flow underground or surface run-off. Thus, while the allocations may be liberal, any excess will not be wasted since it may be re-used down-stream.

The allocation to the Kingdom of Jordan has been based on a comprehensive land classification survey and a scientific determination of the water required. Two well-known American Engineering firms-- Michael Baker, Jr., Inc., and the Harza Engineering company-- were employed by the Jordanian Government to make a complete survey of the 943,000 dunums of land in Jordan within the valley. Standards somewhat lower than are customary in the United States and other countries were applied in this survey in consideration of Jordan's pressing need for agricultural land and the historic ability of the Arab farmer to grow crops under adverse conditions. While this resulted in the inclusion of land not normally considered to be adaptable for cultivation, it was found that about 520,000 dunums were arable.

From the gross figure of about 520,000 dunums, the Baker-Harza engineers deducted 3 percent, as an allowance for land that would be used for roads, ditches, drains, houses, farm buildings, etc. This provided a net irrigable area of 504,200 dunums. When water requirements were computed on the basis of standard formulas and a reasonable cropping pattern, the total water requirement for Jordan became 760 mcm. This figure, together with the allocations given above for Lebanon and Syria, were included in the memorandum prepared jointly by Ambassador Johnston and representatives of Lebanon, Syria and Jordan, at Beirut on February 19, 1995. The memorandum represented an understanding which, while not binding on any of the parties, provided a basis for Ambassador Johnston's report to his own Government and his further negotiations with the Israelis.

In a thorough examination by United States experts of the positions set forth in the Beirut Memorandum, it became clear that the allocation of 760 mcm for Jordan could not be effectively defended in one important respect. The amount of arable land in Jordan could be defended, since it was supported by a detailed physical survey. The requirement for water per dunum could also be defended because it was computed through the use of generally accepted scientific formulae. The allowance of only 3 percent for lands put to non-irrigation uses could not be effectively defended since it represented only a judgment that could not be supported by evidence. A review of a number of irrigation projects n various parts of the world showed that from 15 to 25 percent of the irrigable land area is always out of cultivation at any one time. In addition to land used for roads, farmsteads and other purposes which take it permanently out of cultivation, a significant portion is always unused because of social factors. The absence of the farmer, his death, the transfer of title and other similar problems often result in the temporary non-use of land. Under no circumstances was it possible to conceive that all of these factors would be accurately reflected by an allowance as small as 3 percent for non-crop use.

It was concluded that a more reasonable allowance, taking into consideration the special circumstances in the Valley was 8 percent -- only about half the reduction customarily made. This reduction decreased the computed water requirement of the Valley by 40 mcm per year, to a more realistic but entirely adequate total requirement of 720 mcm of this requirement, 243 mcm can be met from wadis and wells. The Yarmuk River can supply 377 mcm after making due allowance for the allocation to Syria of 90 mcm on the upper Yarmuk, an allocation of 25 mcm to Israel for uses in the Jordan-Yarmuk triangle, and allowing 14 mcm for evaporation and other losses form the Yarmuk reservoir. The remaining 100 mcm needed by Jordan must therefore be derived from the upper Jordan River. The present Plan would assure this amount.

A technical factor affecting the division of the water however is the need to assure that the quality of the water in the system as a whole is maintained at a usable standard of salinity and yet permit as much as possible of all the waters to be used. As Lebanon, Syria and Israel use more and more of the natural flow of the upper Jordan River the volume of water reaching Lake Tiberias will decrease. The lake will then tend to become more salty, since it is fed, in part, by a number of saline springs and since some 300 mcm is lost annually through evaporation form the Lake surface. Under the circumstances, it is prudent to plan ahead of the possibility that some of these springs around the edge of the Lake might be collected and prevented form flowing into the Lake as a means of keeping the salinity of the Lake within usable limits.

The best available information indicates that 1 or 2 mcm of the flow from these springs contains nearly one-third of the salt load and that this highly saline water can be diverted and disposed of. About 30 mcm of less saline spring water would be readily usable for crops when mixed with the fresh water of the Yarmuk. If conditions develop making it desirable to keep this water from reaching the Lake, the Plan proposes that up to 15 mcm be considered as part of the 100 mcm allocated to Jordan from the upper Jordan River. The increase in the salinity of Jordan's allocation would be negligible, while the loss of the 15 mcm to the system as a whole would be prevented.

4. Supervision

The Plan proposes the creation of an impartial engineering Board together with a Watermaster for the purpose of supervising operation of the water system and compliance of the parties. The Engineering Board would consist of 3 eminent engineers who would be selected from a list prepared by the Secretary General of the United nations. One of the engineers would be selected by the participating Arab States and one would be selected by Israel. Being selected from the proposed list, these two engineers would be expected to perform their functions as engineers and not as spokesmen or representatives of the sides that chose them. The two engineers so selected would choose a third, who would serve as chairman. None of the members of the Engineering Board nor the Watermaster could be a national of any Arab State or of Israel, nor be in their employ.

The Engineering Board would have certain functions which are stipulated in the Plan. It would review the engineering designs of the diversion structures and other project features comprising the total development program. This would be only for the purpose of ensuring that the structures and project features would not be inconsistent with the implementation of the agreement on the division and distribution of water and on the operation of the Jordan River system within the accepted framework of the plan. It would also establish patterns for withdrawals, releases and deliveries of the water of the river system. Flow patterns are suggested because of steady or scheduled delivery of water at any pint not only assures more effective operation of the dams and the irrigation system but also lessens the suspicions which might arise if the flows of water are irregular and not pre-determined. Flow patterns will also ensure the promptest detection of any possible violation of the agreement.

The Engineering Board world appoint a Watermaster whose stipulated duties would include supervising the deliveries and withdrawals of water and all regulatory and gauging facilities. He would make the calculations required in connection with the releases of Jordan's water form Lake Tiberias. He would also keep the necessary records and make the necessary reports concerning the operation of the system. Most important of all, in the event of a violation or threat of a violation of the Plan, he would take such steps as might be necessary to enforce compliance or remedy the violation.

As indicated above, through the means of established fixed flow patterns and through regulatory and gauging facilities, it would be, in fact, quite easy to anticipate or detect a violation of the Plan. Since the Watermaster is to have unimpeded access to all project features and all points on the watercourse it would be impossible for the any State to make an unauthorized or clandestine diversion of any substantial quantity of water. The proposed system for the delivery of water would also be to a large extent self- enforcing since any substantial or complete stoppage of water by Israel anywhere in the system could be countered by Jordanian stoppage of deliveries of water to the Yarmuk Triangle. Obviously the converse is true.

Give the erection of suitable diversion features and other necessary physical control mechanisms and given established flow patterns ensuring the prompt detection of violations, there would be no need for numerous and omnipresent supervisory personnel. With these enforcement procedures and with the Watermaster able to appeal to the United Nations as a last resort, the supervisory system as now proposed in the Plan would be efficient and economical, ensure the strictest impartiality, and impinge to the minimum possible extent upon the functions and authority of the governments concerned. The Plan proposes that the precise duties, rights and responsibilities of the Engineering Board ad the Watermaster will be defined in detail through a subsidiary agreement to the negotiated with the Arab States on the one hand and with Israel on the other either by U.S. or some other appropriate intermediary. The supervisory system proposed in the plan would obviate the need for direct contact or negotiation between the Arab States and Israel.

Appendix 4

Water Allocations and Requirements

Time Profile of Water Allocation to Jordan, MCM

Source	Johnston Oct. 53 (TVA)	Johnston, Feb. 55	Baker Harza July 55	Johnston Sept. 55	Minimum Arab Position Sept. 55
Average Contribution From Yarmouk Directly		288	305	297	312
Yarmouk Contribution via Lake Tiberias		92	79	80*	80*
Local Resources		240	227	243	243
Subtotal		620	611	620	635
Tiberias Water Required to Meet Total Irrigation Requirements		140	149	85+15(saline)	85
Total	774	760	760	720**	720**

* This amount could be stored in Lake Tiberias or any other economic location. Decision left to Neutral Board and deferred for five years.

** Jordan's allocation was progressively decreased with time.

The water requirements for the irrigation of the basin lands inside the Hashemite Kingdom was calculated by Baker-Harza Joint Venture. The total area was calculated to be 504,200 dunums requiring 744.9 mcm for a certain cropping pattern (Pattern A), and 759.3 mcm for another (Pattern B). To this total, and area of 9,500 dunums was added in Wadi Fara'a requiring 15.1 mcm of water from Wadi Fara'a to have it irrigated. Johnston's Mission adopted

Pattern A for 760 mcm per year, and had it reduced to 720 mcm per year by increasing the area of lands left fallow.

The table below shows the land areas and the diversion requirements as calculated by Baker-Harza, and as reduced by Johnston Mission.

Summary of Irrigable Area, in dunums, And Water Requirements (mcm/yr).

Area and Sub-Area	Net Irrigable Area, B&H	Net Irrigable Area, Johnston	Diversion Requirement, B&H	Diversion Requirement Johnston *
East Ghor (North)	169,500	160,762	213.5	202.5
East Ghor (South)	183,600	174,136	291.5	276.5
Total East Ghor	353,100	334,898	505.0	479.0 **
West Ghor	151,100	143,311	239.9	227.0
Total East and West Ghors	504,200	478,210	744.9	706.0
W. Fara'a	9,500	9,010	15.1	14.0
Grand Total	**513,700**	**487,220**	**760.0**	**720.0**

* As agreed by the Arab Technical Committee in late September 1955.

** Water requirements of the East Ghors, i.e, Jordan's share after the disengagement in 1988, and the share of the West Bank is the balance: (227+14)= 241 mcm

Sources of Water for the East and West Ghors (MCM/Yr).[358]

	East Bank	West Bank
Usable Flow from Side Wadis	175+ 8 (wells)	52 + 8 (wells)
Yarmouk Contribution	296	81 (*)
Jordan River	-	100
Total	**479**	**241**

* Johnston's calculations based on net depletion in Syria amounting to 59 mcm (31 mcm return flow). Total available from the Yarmouk to Jordan calculated at 377 mcm with the historic flow in the Yarmouk equaling 467 MCM. The Arab Technical Committee, in its minimum position estimated the Yarmouk's discharge at 475 mcm (the Main/ TVA figure), and a net depletion in Syria equaling 52 mcm. It estimated the flow available to Jordan at 392 mcm from the Yarmouk, and counted on 85 mcm from the Jordan River at Tiberias (no saline water of 15 mcm). In effect, the Arab Committee inflated the Yarmouk and would thus assign to the West Bank 96 MCM instead of the 81 under the final Johnston figure. The total for the West Bank does not change. In effect, the East Bank dependency on "international water" remained on the Yarmouk at 296 mcm per year, and the West Bank dependency stayed at 181 mcm, drawn from the remainder in the Yarmouk and from the Jordan River.

[358] Wells were estimated by the Main/TVA Plan to produce 10 mcm on the East Bank, and another 10 mcm on the West Bank (see Water and Power by Lowi, pp. 84-85). After debates, Johnston's Mission and the Arab Technical Committee agreed in February 1955 to consider 16 mcm for both banks.

Appendix 5

Cable of American Envoy to the President
Thursday, June 01, 1967

"Eyes only for President and Secstate From Robert Anderson

1. There follows a summary of my talk with President Nasser. Unless otherwise indicated, I will be trying to express his point of view to me.
2. After exchanging of pleasantries, Nasser said he became worried and afraid of Israeli attack because of speeches and his own intelligence of mobilization by Israel. [deleted sentence] As an example, he stated that 13 brigades were mobilized near Syria.
3. Nasser explained that he did not want repetition of 1956 affair when he was reluctant to believe that an attack had begun and was slow in moving troops to Sinai only to be caught between the Israelis in the north and the British at Port Said. He said he felt he had no choice but to mobilize and send troops to Sinai, which he did, and request the removal of UN forces. While he did not say so, I believe he was surprised at the rapidity of the removal of UN troops because he said they were only a token force and would have created no real obstacle.
4. He was asked specifically if he intended to begin any conflict and he said to please explain to my Government that he would not begin any fight but would wait until the Israelis had moved. This was qualified by saying that he did not know what the Syrians might do and had worried all day [Wednesday] for fear that the Syrians might start something out of anger because of the pact which he had made with Hussein. He also stated that, contrary to most public opinion, he did not have control over the radical elements of refugee organizations who were interested only in starting a conflict because they had no real responsibility for conduct of military affairs. He was asked if this conflict occurred, for example, if Syria should attack against his desires, whether he would respond, and he answered affirmatively by saying that it would bring response from him.
5. It was pointed out that if Israel felt she was virtually alone she might be motivated to strike first in order to secure a strategic advantage and that so long as she felt she had friends she might be restrained. Nasser replied that this was a risk which he would have to accept and that he thought the first Israeli target and main thrust of Israeli offensive would be against Egypt and Cairo. He said that elaborate plans had been made for instant retaliation, and that he was confident of the outcome of a conflict between Arabs and Israelis.

6. [This paragraph is deleted in its entirety, i.e., information is still classified.]

7. With reference to Gulf of Aqaba, Nasser stated that for eight years after 1948 the Straits had been closed to Israeli shipping and was open only by the illegal act of Israel, France and England, and he proposed merely to return to the status of 1956 which had been at least tolerated by all the nations for eight years. He explained that even we had deplored and opposed the act of the Israelis, British and French which changed the status quo in 1956. He stated that the Straits of Tiran were navigable only in a width of three miles which was clearly territorial waters and that he intended to maintain this position. He was asked specifically what commerce he would allow through the Straits under his concept and he replied by saying that the exclusions would be 1) Israeli ships, 2) Oil or any refined products, and 3) Arms for Israel. Here he stated that all countries claimed territorial waters to a greater distance offshore than he was asserting and further that he was at war with Israel and had been since 1948 with nothing existing between them except an armistice, and that under these circumstances he was entitled to assert jurisdiction.

8. He was asked if he would consider referring this matter of the Straits to either the United Nations or the World Court in view of the fact that four countries had borders on the gulf. He replied that he would not submit the question to the U.N. because the Israelis normally treated resolutions of the U.N. not favorable to them as "pieces of paper". He said that he did not have sufficient knowledge of the World Court to answer specifically about referring the matter to the World Court for decision but would consult his legal advisors. This was qualified by saying that he did not want to undertake any course of action that would take "years" to decide.

9. He also stated that even if he agreed on some other course of action, any other course of action would be strongly opposed by all Arab countries who were now his allies. On this point he seemed on the one hand adamant about the positions he had taken in the Straits and yet he did not rule out completely possibility of a World Court review if it could be done speedily. For the time being I think he will remain firm.

10. He was asked if he was not prepared to accept Israel as a matter of fact, even though he might have emotional and legal feelings concerning the establishment of the country in Palestine. Nasser replied by saying that he did not believe stable and lasting peace could be achieved without disposing of the refugee problem. He was asked if this could be done by compensation as well as some limited return of refugees. He replied that he thought practically all refugees

would return if permitted and that even if compensation were paid they would not be satisfied but would continue to agitate for return to Palestine. He went into long discourse on Arab mentality as it affects their feelings toward the place where they were born and reared.

11. Nasser stated that he had been prepared to sign an agreement with the Monetary Fund but had just received a letter saying that the Fund wished to review their relationships with Egypt further. He then stated he was glad he had not signed the agreement with the Fund because they were unreasonable and left him no flexibility. He emphasized that he did not want to be subject to economic pressure. It was explained to him that neither the Fund nor local American banks were in fact exerting pressure when they did not comply with national requests since they were all governed by strict rules that limited their own flexibility in making loans to countries that did not comply with all regulations.

12. Nasser expressed keen desire to have friendship of American people and American government explaining that under no circumstances was he a communist. On the other hand, he felt that U.S. policy was motivated largely by the large Jewish vote in U.S. and that American government would be reluctant to oppose this voting strength. He then called attention to the fact that Eisenhower had taken a strong position in 1956 against Israeli invasion and this had not hurt him politically.

13. He seemed anxious to have Zakaria Mohieddin explain his position directly to U.S. government and said he hoped we would take the long view because the Arab countries stretched from Morocco on the west to Pakistan on the east and that now he even had the support of Pakistan and India. He did not see how a minority in the U.S. could influence U.S. policy to oppose what such a vast region and such large numbers of people believed proper. It was explained to him that the U.S. government was not motivated by political considerations but was concerned essentially in maintaining peace and the integrity of countries.

14. At this time Nasser said that if the policy was for Arabs and Israelis to live together harmoniously then Israel should allow a million refugees to come back to Palestine, which would solve the refugee problem and still the Israelis would have two million of their own citizens in the same country, this, he said, would be true "living together."

15. He made it clear that he felt U.S. was taking the lead in peace efforts but that these efforts were oriented toward Israel, and not toward the Arab point of view. He kept reassuring me that he was not going to start a war but that he was not responsible for all groups and that he would intervene if any actual conflict began. He stated that under

present circumstances Jordanian troops insofar as the Israeli problem was concerned, were under UAR command. This of course is applicable to other troops such as Iraqis and Algerians who were reporting for duty.

16. This I think summarizes the basic points of our conversation on which I will elaborate further on my return.

17. For your general information I spent three days in Beirut before going to Cairo. During this visit I saw Saudi Arabs, Kuwaitis and Iraqis, as well as Lebanese. They are people who are generally moderate and have a tendency to oppose Nasser. At this time they were all applauding Nasser's action, insisting on the closing of the Gulf of Aqaba and taking a position that the U.S. was supporting a minority for political purposes. I am impressed more because of the quality of the people who made these assertions than the fact that they were made. Under the circumstances it would seem desirable that whatever international arrangements are thought proper it would be helpful if the initiative could be taken by some country other than U.S. and that U.S. be in a position of support of international efforts to secure peace rather than leadership which seems to be construed as favoring Israeli cause.

17. [Paragraph number duplicated.] During our conversation Nasser was relaxed, in sport clothes, and seemed confident both of his intelligence and of his military capability. We had no discussion Re Soviets except his assertion that he was not and would not be communist. I believe he would regard any effort to open the Straits of Tiran as hostile and any act of aggression, whether originating from Israel or resulting from actions in Syria by the terrorist groups, would bring response. He stated that his target system was prepared and that this time he would be ready.

18. I am proceeding to send message to Cairo through U.S. embassy to Nasser which will result in Zakaria Mohieddin arriving in New York presumably Sunday or early in week. I will return New York Saturday afternoon and will be available to come to Washington Sunday or thereafter. I can be reached through embassy here today and tomorrow morning, if desired.

19. Upon rereading this text I want to make clear as I understand it UAR has military command over its own troops, the Jordanian troops as related to any Israeli problem, the troops committed by Iraq, Algeria or any country sending troops, but does not include command over Syrian troops. It is because of this latter situation which I think bothers Nasser as to whether or not the Syrians might undertake unilateral action designed to force a confrontation. It was because of his concern on this subject that he was asked if he would intervene

even if the Syrians acted against UAR desires and the reply was affirmative."

Appendix 6

Letter of Invitation to the Bilateral Negotiations

[Sent to the parties by Secretary of State James Baker]

BEGIN MESSAGE:[359]

- AS YOU RECALL IN MADRID, I MADE IT CLEAR PUBLICLY AND PRIVATELY THAT I WAS PREPARED TO SEE THE PARTIES THEMSELVES TRY TO SORT OUT THE VENUE AND TIMING OF THE NEXT ROUND OF DIRECT BILATERAL NEGOTIATIONS.
- SINCE THEN, I CONVEYED TO YOU SEVERAL IDEAS FROM ISRAEL ON HOW TO RESOLVE THIS ISSUE. HOWEVER, THE ISRAELI SUGGESTIONS WERE NOT APPROVED BY ANY OTHER PARTY.
- I UNDERSTAND THE COMPLEXITIES INVOLVED IN SORTING OUT THE PROCEDURAL ASPECTS OF THIS ISSUE. I DO NOT MINIMIZE THEM, BUT NEITHER DO I BELIEVE THEY SHOULD BECOME. IN AND OF THEMSELVES, REASONS TO BLOCK FURTHER DISCUSSIONS BETWEEN THE PARTIES ON THE SUBSTANCE OF THE NEGOTIATIONS.
- IN THIS MESSAGE I WANT TO ADDRESS THREE ISSUES: FINDING AN ACCEPTABLE VENUE FOR THE NEXT ROUND OF NEGOTIATIONS; DETERMINING THE BEST DATE TO HOLD THE NEXT ROUND; AND BEGINNING TO FORESHADOW THE KINDS OF ISSUES THAT EACH OF YOU MIGHT DECIDE TO TAKE UP IN ORDER TO MAKE THE NEGOTIATIONS MOST PRODUCTIVE.

VENUE AND TIMING

- IT IS CLEAR THAT MADRID WILL NOT BE AN ACCEPTABLE VENUE FOR THE NEXT ROUND. IT IS EQUALLY CLEAR THAT THERE IS NO AGREEMENT YET TO HOLD TALKS IN THE REGION.
- I WANT TO MAKE CLEAR THE VIEW OF THE UNITED STATES THAT, OVER TIME, THERE IS NO REASON TO EXCLUDE HOLDING NEGOTIATIONS IN THE REGION.

[359] Letter copied by the author as is, in capital letters.

MANY SUCCESSFUL TALKS HAVE BEEN HELD IN THE REGION IN THE PAST AND A REGIONAL VENUE WOULD ALLOW CLOSE PROXIMITY FOR THE NEGOTIATORS TO CONSULT WITH THEIR POLITICAL LEADERSHIP. BUT I UNDERSTAND WE WILL NOT RESOLVE THIS ISSUE NOW.

- WE HAVE CONSULTED ON THIS MATTER WITH THE SOVIETS AND WE AGREE THAT THE MOST IMPORTANT ISSUE NOW IS TO RESUME THE DIRECT BILATERAL NEGOTIATIONS AS SOON AS POSSIBLE.
- ACCORDINGLY, I AM PLEASED TO INVITE EACH ONE OF YOU TO COME TO WASHINGTON TO START NEGOTIATIONS ON WEDNESDAY, DECEMBER 4.
- IN ORDER TO BEGIN PREPARATIONS FOR THESE NEGOTIATIONS, I WILL NEED YOUR AFFIRMATIVE RESPONSE NO LATER THAN MONDAY, NOVEMBER 25.
- THE UNITED STATES WILL ARRANGE THE SITES FOR THE TALKS AND OTHER ADMINISTRATIVE ISSUES; THE TALKS WILL TAKE PLACE HERE IN WASHINGTON.
- EACH DELEGATION WILL BE EXPECTED TO ASSUME ALL OTHER COSTS ASSOCIATED WITH THE NEGOTIATIONS SUCH AS TRANSPORTATION, LODGING, STAFF SUPPORT AND THE LIKE.
- WE WILL NOT BE IN A POSITION TO MEET DELEGATIONS ON ARRIVAL BUT, WITH ADVANCE NOTIFICATION ON ARRIVAL TIME, WE WILL ALERT AIRPORT AUTHORITIES TO YOUR PLANNED ARRIVAL.
- THE DELEGATIONS TO THESE DIRECT BILATERL NEGOTIATIONS NEED NOT BE THE SAME AS THOSE THAT ATTENDED THE MADRID CONFERENCE. IN ORDER TO ENSURE HOWEVER, THAT THE TERMS AND CONDITIONS AGREED UPON FOR THIS PROCESS REMAIN, WE NEED TO KNOW IN ADVANCE THE COMPOSITION OF EACH DELEGATION. OUR INTENTION WOULD BE TO NOTIFY THE COMPOSITION OF EACH DELEGATION TO ALL OTHER DELEGATIONS IN ADVANCE OF NEGOTIATIONS.
- IN ADDITION, WE ENVISAGE NO PARTICULAR CREDENTIALING FOR THE DELEGATIONS. HOWEVER, SINCE THE NEGOTIATIONS ARE LIKELY TO BE HELD IN U.S. GOVERNMENT BUILDINGS, ACCESS TO THOSE BUILDINGS WILL BE AVAILABLE ONLY TO THOSE DECLARED AS DELEGATES. OTHER ADVISERS AND STAFF WILL NOT BE PERMITTED AT THE SITES OF THE NEGOTIATIONS, BUT SHOULD PLAN TO REMAIN IN

WHATEVER OFFICES ARE SET UP BY THE DELEGATIONS THEMSELVES.

- SIMILARLY, WE PLAN TO PROVIDE NO PRESS FACILITIES, AND ANY PRESS ACTIVITY BY DELEGATES WILL HAVE TO BE ARRANGED BY THE DELEGATIONS THEMSELVES.

SUBSTANTIVE NEGOTIATIONS

- AS ALL OF US EXPECTED, THE FIRST ROUND IN MADRID COVERED LITTLE SUBSTANTIVE GROUND, AND THUS PROVIDED NO REAL DIRECTION AS TO WHERE THE NEGOTIATIONS MIGHT NOW HEAD.
- IT IS NOT THE INTENTION OF THE U.S OR THE SOVIET UNION, AS CO-SPONSORS TO SUGGEST TO YOU WHAT TO DO IN THE NEGOTIATIONS. HOWEVER, WE THOUGHT YOU MIGHT FIND IT USEFUL FOR US TO SHARE SOME PRELIMINARY THINKING ON WHICH APPROACHES TO NEGOTIATIONS MIGHT HELP GET THEM LAUNCHED SUCCESSFULLY.

ISRAEL-JORDANIAN/PALESTINIAN TALKS

- IN THE CASE OF THE NEGOTIATIONS BETWEEN ISRAEL AND THE JORDANIAN/PALESTINIAN DELEGATION, THERE ARE TWO DISTICNT SETS OF ISSUES THAT NEED TO BE DISCUSSED.
- JORDAN HAS NOTIFIED US AND PRESUMABLY ISRAEL, THAT THERE ARE A NUMBER OF BORDER ISSUES UNRELATED TO U.N. SECURITY COUNCIL RESOLUTION 242 WHICH NEED TO BE DISCUSSED. WE SEE NO REASON WHY THESE ISSUES SHOULD NOT BE RAISED EARLY IN THE NEGOTIATIONS SO THAT EXPERTS CAN GET TO WORK TO EXAMINE RESPECTIVE CLAIMS AND HISTORICAL BACKGROUND.
- AT THE SAME TIME, ISRAEL AND JORDAN COULD PROFITABLY DISCUSS A NUMBER OF OTHER BILATERAL ISSUES, SUCH AS THE NATURE OF PEACE, THE RESOLUTIONS OF MARITIME PROBLEMS IN THE GULF OF AQABA, JOINT MANAGEMENT OF WASTE WATER FACILITIES, JOINT PRODUCTION OF POTASH, TOURISM.

CIVIL AVIATION AND THE LIKE, WHICH COULD FORM THE ESSENTIAL BUILDING BLOCKS OF A FINAL SETTLEMENT, ONCE THE ELEMENTS OF A FINAL SETTLEMENT BEGIN TO FALL INTO PLACE, OR ALTERNATIVELY AS INTERIM ARRANGEMENTS OR UNILATERAL MEASURES TO SERVE MUTUAL INTERESTS AND IMPROVE THE ATMOSPHERE AND PROCESS OF NEGOTIATIONS.

- AS WE UNDERSTAND IT, THIS IS THE EXPECTATION OF BOTH SIDES THAT IN THESE NEGOTIATIONS BETWEEN ISRAEL AND JORDAN, PALESTINIANS FROM THE JOINT JORDANIAN/PALESTINIAN DELEGATION WOULD ALSO ATTEND IN ORDER TO MAINTAIN THE INTEGRITY OF THE JOINT DELEGATION IN NEGOTIATIONS WITH ISRAEL.
- WITH REGARD TO ISSUES RELATED TO THE WEST BANK AND GAZA IT IS SIMILARLY OUR UNDERSTANDING THAT PALESTINIANS WOULD TAKE THE LEAD BUT WOULD BE ACCOMPANIED BY JORDANIANS AS PART OF THE JOINT JORDANIAN/PALESTINIAN DELEGATION.
- IN THESE NEGOTIATIONS THERE IS ALREADY AGREEMENT THAT THE FIRST PHASE WILL FOCUS ON INTERIM SELF-GOVERNMENT ARRANGEMENTS.
- HAVING EXPERIENCED SEVERAL YEARS OF NEGOTIATIONS ON THESE ISSUES IN THE LATE 1980'S, IT IS OUR CONSIDERED VIEW THAT BOTH ISRAEL AND PALESTINIANS SHOULD AVOID AS MUCH AS POSSIBLE A PROTRACTED DEBATE ON SUCH PRINCIPLES AS THE "SOURCE OF AUTHORITY, NATURE OF THE INTERIM SELF GOVERNMENT AUTHORITY", AND THE LIKE.
- RATHER, ISRAEL AND THE PALESTINIANS MIGHT AGREE THAT EACH WOULD PRESENT IN THIS OR THE NEXT SESSION A PROPOSED MODEL OF INTERIM SELF-GOVERNMENT AUTHORITY.
- SUCH MODELS ARE LIKELY TO BE QUITE DIFFERENT; THEY WILL VARY WIDELY IN TERMS OF THE SCOPE OF AUTHORITY AND JURISDICTION THAT THEY ANTICIPATE EXTENDING TO PALESTINIANS. NONETHELESS, THEY WILL PROVIDE BOTH SIDES WITH SOME POTENTIALLY USEFUL STARTING POINTS TO BEGIN HAMMERING OUT THE POWERS AND RESPONSIBILITIES THAT WILL BE ASSUMED BY PALESTINIANS DURING THE TRANSITIONAL PERIOD, AS WELL AS THE ISSUES THAT NEED TO BE DEFINED AND NEGOTIATED DURING THE PERIOD AHEAD.

ISRAEL/ LEBANON TALKS

- WITH REGARD TO THE NEGOTIATIONS BETWEEN ISRAEL AND LEBANON THERE ARE CLEAR DIFFERENCES AS TO HOW TO PROCEED.
- IN OUR VIEW IT MIGHT BE MOST PRODUCTIVE FOR BOTH SIDES TO START ENGAGING ON THE MOST PRACTICAL ISSUE THAT THEY BOTH CONFRONT: NAMELY, HOW TO UNLOCK THE CURRENT STALEMATE IN JAZZINE.
- BY THIS, WE DO NOT MEAN IN ANY WAY A DEPARTURE FROM PRINCIPLES OF AN OVERALL SETTLEMENT TO SUBSTITUTE A JAZZINE SETTLEMENT FOR A RESOLUTION OF THE UNDERLYING ISSUES BETWEEN THE TWO COUNTRIES. HOWEVER, SINCE THE JAZZINE ISSUE INVOLVES THE COMPLEX OF POLITICAL AND SECURITY ISSUES THAT WILL BE REQUIRED IN ORDER TO DEAL EFFECTIVELY WITH THE PROBLEMS THAT EXTEND THROUGHOUT SOUTHERN LEBANON AND NORTHERN ISRAEL, IT CAN REPRESENT A USEFUL FIRST STEP AND IT CAN DEMONSTRATE TO BOTH PARTIES THAT NEGOTIATIONS CAN PRODUCE PRACTICAL SOLUTIONS TO PROBLEMS ON THE GROUND.

ISRAEL- SYRIA TALKS

- WITH REGARD TO THE NEGOTIATIONS BETWEEN ISRAEL AND SYRIA, THE FIVE HOUR TALKS HELD NOVEMBER 3 INDICATED THAT NEITHER SIDE WILL FIND IT EASY TO PROCEED UNTIL KEY ISSUES OF PRINCIPLE ARE RECOGNIZED BY THE OTHER SIDE. WHILE NOT DIMINISHING THE SIGNIFICANCE OF THESE PRINCIPLES TO EACH PARTY STRICT ADHERENCE TO THEM COULD LEAD TO EARLY STALEMATE IN THE NEGOTIATIONS.
- ONE WAY TO PROCEED, IN OUR VIEW, MIGHT BE TO PROBE THE OTHER SIDE'S POSITION IN CERTAIN HYPOTHETICAL CIRCUMSTANCES. FOR EXAMPLE, SYRIA MIGHT ASK, HYPOTHETICALLY, WHAT THE ISRAELI'S POSITION ON WITHDRAWAL WOULD BE IF SYRIA WERE PREPARED, AS PART OF A COMPREHENSIVE SETTLEMENT,

TO SIGN A PEACE TREATY WITH ISRAEL, EXCHANGE AMBASSADORS, AND WORK OUT MUTUALLY ACCEPTABLE SECURITY ARRANGEMENTS. ALTERNATIVELY, SYRIA MIGHT ASK WHETHER UNDER SUCH CIRCUMSTANCES ISRAEL WOULD EXCLUDE WITHDRAWAL AND RETURN TO SYRIAN SOVEREIGNTY OVER GOLAN.

- ISRAEL, FOR ITS PART, MIGHT ASK HYPOTHETICALLY WHAT SYRIA'S POSITION ON A PEACE TREATY, FULL NORMALIZATION AND DIPLOMATIC RELATIONS WOULD BE IF ISRAEL WERE PREPARED TO UNDERTAKE WITHDRAWAL. ALTERNATIVELY, ISRAEL MIGHT ASK WHETHER UNDER SUCH CIRCUMSTANCES SYRIA WOULD EXCLUDE A TREATY OF PEACE, FULL NORMALIZATION AND DIPLOMATIC RELATIONS.
- WHILE THIS KIND OF DIALOGUE DOES NOT ADEQUATELY OVERCOME DIFFERENCES OF PRINCIPLE, IT COULD PERMIT THE SIDES TO BEGIN EXPLORING SOME OF THESE ISSUES RAISED DURING THESE PRESENTATIONS. BAKER.

Appendix 7

Jordan's First Proposal for a Common Agenda

Informal Draft Agenda[360]
The Jordanian-Israeli Track

A. Goal
The achievement of a just, comprehensive and lasting peace.

B. Items
1) Implementation of Security Council Resolution 242 in all its aspects.
2) Israeli Settlements
3) Borders and Occupied Lands
4) Refugees and Other Displaced Persons
5) Water

In response, the Israeli delegation presented the Jordanian delegation with a proposal for a common agenda that read as follows:

Israel's Counterproposal

January 14, 1992

Informal Draft Agenda

Agenda

A. Goal of the negotiations

Israel-Jordan Treaty of Peace

B. Israel-Jordan peace negotiations

1. Peace Treaty [changed to read Vision of Peace]: components of peace
(1) Termination of the status of war and establishment of peace, including full diplomatic relations.
(2) Security.

---

360 Submitted to the Israeli Delegation in the afternoon of Tuesday, January 14, 1992.

(3) Borders.

(4) (a) Normalization, including various spheres (trade, civil aviation, culture etc).

(b) Areas of cooperation for special attention: water, energy, environment and economy.

2. Possible preliminaries for peace (during the course of negotiations):

(1) Liaison arrangements.

(2) Opening of borders and mutual visits.

(3) Liaison system between the two militaries.

(4) Non-political exchanges in spheres of mutual interest, including cooperation in the areas of water, energy and environment, and economy.

C, <u>Matters which may relate to both tracks and to the General Meeting</u>

1. Economic matters.
2. Legal matters.
3. Coordination Mechanism.
4. Rehabilitation of refugee camps in Jordan and in the territories.

C. <u>Possible Committees</u> (e.g.)

1. Israel-Jordan bilateral security arrangements.
2. Israel-Jordan civilian relationship-normalization.
3. Israel-Jordan cooperation in specific areas (water, energy, environment).
4. Legal matters.
5. Economic matters.

Appendix 8

Jordan-Israel Track
State Department
Monday, April 27, 1992

Jordan's Statement on Water

Presented by

Munther J. Haddadin

On

April 27, 1992

Thank you Dr. Majali,
Ambassador Rubinstein, Distinguished Audience,

We are at this time embarking on a crossroad at which we make a historic choice, the road to cooperation or that of continued conflict. This is why I saw it worth my while to compromise my religious duties[361] in favor of the obligations to set the grounds for peace in our region, obligations that carried me to Washington, D.C., before I could celebrate the resurrection of our Lord.

It is our view that water, basic to life and prosperity, should trigger cooperation. By its own nature, it is used to extinguish fires and not to ignite them. The lengthy disputes that are behind our meeting around this table did not provide the environment conducive to consultation or cooperation.

The good offices of the United States provided opportunities to avert hot conflicts over water. They helped diffuse tensions that otherwise could have gone the wrong way. This forum of bilateral negotiations provides the proper opportunity to compare notes, review the past developments on shared water resources, identify the heavy tolls that Jordan has been paying because of unilateral planning and use by Israel of the shared water resources; tolls

---

[361] Easter week started on 19 April and Easter Sunday fell on 26 April, according to the eastern calendar.

that have burdened Jordan with adverse economic, social and environmental impacts.

In our attempt to do that, we shall not claim that Israel is a water rich country, but hope to point out that Israel's efforts to alleviate its water shortage have been at the expense of water stressed Jordan. On the one hand, Jordan's reception of refugees and displaced persons from Palestine greatly distorted the balance in the water resources-population equation, and, on the other, Israel's unilateral use of Jordan's rightful share of water in the Jordan River basin further aggravated the situation.

Irrigation water in Jordan has been diverted to urban use, and has been replaced with treated wastewater with immense adverse environmental impacts and, at times, public health hazards. Water stress has cornered Jordan into over-pumping from its ground water aquifers to dangerous levels, and has never been able to supply more than 52 CM per capita for municipal use, at very high cost in capital and energy costs. Amman is being supplied with municipal water from the Jordan Valley.

The erosion of Jordan's rightful share in the waters of the Jordan basin left 35% of the Jordan Valley unproductive. Water shortages in the dry months have had their economic, social and environmental tolls.

The unilateral use of the Jordan waters by Israel and the transmission thereof to remote regions outside the basin contributed to the recession of the Dead Sea level with potential hazards posed on the environment.

It is our hope, on this second day of Easter, that we will succeed in:

a. Restoring Jordan's water shares in the Jordan River basin.
b. Mitigating the adverse impacts emanating from the use of Jordan's rightful water share, including the huge imbalance in the population-water resources equation.
c. Setting the stage for cooperation to control the level of the Dead Sea and reap the environmental benefits and benefits of hydro-power generation.

We stand prepared to proceed to achieve the above goals, and provide a boost for peace and cooperation.

Appendix 9

The Common Agenda[362]

"THE MIDDLE EAST PEACE PROCESS
The Bilateral Peace Negotiations
Jordan-Israel Track
Common Agenda

A. Goal:

The achievement of just, lasting and comprehensive peace between the Arab States, the Palestinians and Israel as per the Madrid invitation.

B. Components of Jordan-Israel Peace Negotiations:

1. Searching for steps to arrive at a state of peace based on Security Council Resolutions (242) and (338) in all their aspects.
2. Security:
   a. Refraining from actions or activities by either side that may adversely affect the security of the other or may prejudice the final outcome of negotiations. Threats to security resulting from all kinds of terrorism.
   b. i. Mutual commitment not to threaten each other by any use of force and not to use weapons by one side against the other including conventional and non-conventional mass destruction weapons.
   ii. Mutual commitment as a matter of priority and as soon as possible, to work towards a Middle East free from weapons of mass destruction, conventional and non-conventional weapons; this goal is to be achieved in the context of a comprehensive, lasting and stable peace characterized by the renunciation of the use of force, reconciliation and openness.

---

[362] As negotiated and agreed on Tuesday, 29 October, 1992. The Jordan side was represented by Majali, Haddadin and al Khasawneh; and the Israeli side by Rubinstein, Sabel and Ben Ari.

Note: The above (item b-ii) may be revised in accordance with relevant agreements to be reached in the multilateral Working Group on Arms Control and Regional Security.

c. Mutually agreed upon security arrangements and security confidence building measures.

3. Water:
   a. Securing the rightful water shares of the two sides.
   b. Searching for ways to alleviate water shortage.

4. Refugees and Displaced Persons:
   Achieving an agreed just solution to the bilateral aspects of the problem of refugees and displaced persons in accordance with international law.

5. Borders and Territorial Matters:
   Settlement of territorial matters and agreed definitive delimitation and demarcation of the international boundary between Jordan and Israel with reference to the definition under the Mandate, without prejudice to the status of any territories that came under Israeli Military Government control in 1967[363]. Both parties will respect and comply with the above international boundary.

6. Exploring the potentials of future bilateral cooperation, within a regional context where appropriate, in the following
   a. Natural Resources:
      - Water, energy and environment.
      - Rift Valley development.
   b. Human Resources:
      - Demography
      - Labor
      - Health
      - Education

---

[363] As amended in the ninth round of April-May, 1993. The original draft, agreed on 29 October 1992, read:
"...without prejudice to the territories presently under Israeli military administration."

- Drug control

c. Infrastructure:
   - Transportation: land and air.
   - Communication.

d. Economic areas including tourism.

7. Phasing the discussion, agreement and implementation of the items above including appropriate mechanisms for negotiations in specific fields.
8. Discussion on matters related to both tracks to be decided upon in common by the two tracks.

C. It is anticipated that the above endeavor will ultimately, following the attainment of mutually satisfactory solutions to the elements of this agenda, culminate in a peace treaty.

Appendix 10

Jordan's Position on Water[364]

Position Paper
On Water
Read by Dr. Majali

Bilateral Negotiations
Jordan - Israel Track
Monday, November 16, 1992

Gentlemen,

Let me at the outset reaffirm our intentions to engage with you in substantive and serious negotiations, in an attempt to arrive at successful resolutions of the issues of conflict outlined in our negotiations agenda as soon as it is concluded.

An exchange of technical views at this time over crucial issues is warranted in preparation for the upcoming negotiations.

The issue of water occupies a high priority for both parties. Not surprisingly, water is essential to life, development and prosperity. Its importance intensifies in the arid and fragile environment of our region, and the adverse impact of water shortages is exacerbated by demographic pressures to which our country has been vulnerable because of both natural and man made factors.

It is not a secret that Jordan has been exposed to a combination of factors that produced and intensified its water stress. Not only has Jordan been a primary host to the waves of Palestinian refugees and displaced persons, but also has been a target for denial of its rightful shares of the Jordan river waters. The adverse impact has thus been compounded, and water management in Jordan transformed to a continuous crisis management that has been striving to accommodate competition among the various uses domestically, and competition among the coriparians on the Jordan river system. The per capita share of water resources in Jordan is no where close to be envied by any other country, and amounts in fact to only 20% of the share recognized internationally as the share below which water poverty starts. It is

[364] Paper prepared by Dr. Munther J Haddadin, cleared by the officials including the Minister of Water and Irrigation, and some members of the High Committee for the Negotiations.

therefore not surprising to reveal that the domestic water consumption in Jordan has been constrained to modest levels, that the trade deficits in agricultural commodities is high, and that the abstraction of water from all of our groundwater aquifers have been exceeding their sustainable yields with very serious economic, social and environmental consequences.

Allow me, now, to point out that Israel has been behind Jordan's water agony. In 1946, the per capita share of renewable water resources was comfortable indicating a surplus in the population-water resources equation. That surplus abruptly diminished shortly afterwards with the arrival of the first wave of Palestinian refugees in 1948, and was eroded shortly there after.

We all are aware of our competing plans over the past decades to develop and use the waters of the Jordan river system. We all are further aware of the need to satisfy the needs of the people and the lands of the river basin before any notion of water transfer outside the basin can be entertained. And, finally, I think we are all aware of the efforts of the United States to reconcile the competing plans floated at the time by both parties to develop and use the waters of the Jordan river system. It is noteworthy that those efforts of reconciliation conducted in the mid 1950's never defied the notion that the basin requirements for water have to be satisfied first.

What do we have at hand today?

- We see Israel controlling and monopolizing the use of the upper Jordan, enhancing its water supplies to the detriment of water stressed Jordan.
- We see Israel seriously degrading the quality of water in the lower Jordan river.
- We see the waters of the Jordan transferred outside its basin that is left half parched.
- We have been witnessing Israel's desire to expand on the use of the Yarmouk river waters to Jordan's detriment.
- And we have at hand in Jordan more water shortages and deficits in the water budget than the country can afford.
- And, finally, a per capita water share in Jordan today reduced to less than 10% of what it was when the Kingdom was established in 1946, thus setting serious constraints on its economic growth, social well being and compromising its environmental integrity.

The above clarifies in general terms that Jordan can in no way afford the losses of its rightful water shares in the Jordan river, and justifies the efforts we are about to embark on to have Jordan's shares restored, a restoration that will undoubtedly help pave the way to cooperation in the field of water and other related fields.

The serious engagement we are about to embark on is the first such direct bilateral engagement in the decades of the conflict between us, our conduct of it will undoubtedly reinforce either of the two wider spread theories: is water in the Middle East an avenue for cooperation or is it one for continued conflict?

Let me assure you that we subscribe to the first. As evidenced by our conduct in the field throughout the decades of continued conflict, we have refrained from escalation of hostilities among the adversaries over water, and have substantially suffered in the process but never compromised our belief that water, by virtue of its nature, should be used to extinguish fires and not to ignite them.

Our hopes are high that, despite the difficulties, a resolution of the conflict over water will be within the reach of our negotiators, and that the avenue for cooperation to meet the challenges of the future will be opened. There are more attractions for success than there are deterrents. Meeting the challenges of water shortage on both sides awaits our collective and cooperative efforts. I am confident that positive attitudes on the part of all of us will enhance the chances for successful conflict resolution and the subsequent emergence of cooperative efforts.

Thank you.

Appendix 11

Israeli Proposal for Water Bilateral Sub-Agenda

Forwarded by Dr. Rosenthal
On
16 November 1992

During the bilateral talks held in Washington, experts from Jordan and Israel discussed the possibility of comprehensive approach to the water issues relevant to the two states based on a constructive spirit of cooperation rather than on mutually confrontational demands. This positive spirit was reflected in the relevant points referring to water issues included in the agreed Common Agenda of October 28, 1992:

"3. (a). Securing the rightful water shares of the two sides.
(b). Searching for ways to alleviate water shortages.

6. Exploring the potential of future bilateral cooperation within a regional context where appropriate, in the following:
A. Natural Resources:
- Water, energy and environment.
- Rift Valley development."

The following elements of settlement are proposed:

A) Cooperation on the exploitation of all water resources along the borders between the two countries for both agricultural and municipal uses.

B) Such cooperation will include transboundary transfer of water between the two countries.

C) During the negotiations process the two parties concerned shall refrain from erecting and putting in operation new water projects (such as the Qarame Reservoir), so as not to change the existing water arrangements between the two countries.

Once agreement will be reached, its elements shall be included in the Peace Treaty and thus become legal rights for all parties concerned.

<u>**Appendix 12**</u>

Water, Energy and Environment
Draft Final Sub Agenda
June 5, 1994

In their pursuit of achieving the goal of the Jordan-Israel track of bilateral negotiations, Jordan and Israel have identified the components of their negotiations and have listed them under Section B of the Common Agenda. Item B.7 addresses the phasing of discussion, agreement and implementation of the agenda items including appropriate mechanisms for negotiations in specific fields. It is understood that the two parties shall cooperate to achieve the goal.

Items of this sub-agenda listed hereunder respond to the intent of detailed discussions over the various issues in the subjects of water, energy, the environment, and the Jordan Rift Valley and provide practical steps to fulfill the objectives of items B.3, B.6-A and B.7 of the Common Agenda.

I. Surface Water Basins

A. Negotiation of mutual recognition of the rights of the two parties in waters from the Jordan and the Yarmouk rivers with mutually acceptable quality. Securing the unimpeded flow of the mutually recognized rightful shares of the Hashemite Kingdom of Jordan and Israel respectively in the above waters with agreed quality, through agreed conveyance systems.

B. Restoration of water quality in the Jordan River below Lake Tiberias to reasonably usable standards.

C. Protection of water quality.

II. Shared Groundwater Basins

A. Renewable freshwater aquifers- southern area between the Dead Sea and the Red Sea.

B. Fossil aquifers- between the Dead Sea and the Red Sea.

C. Protection of water quality of both.

III. Alleviation of Water Shortage Under Efficient Use

A. Development of water resources.

B. Municipal water shortages.

C. Irrigation water shortages.

IV. Potentials for Future Cooperation Within a Regional Context

A. General

1. Nature protection, natural reserves and bio-diversity.
2. Air quality control, including general standards, criteria and all types of manmade hazardous radiation, fumes and gases.
3. Marine environment and coastal resources management.
4. Waste management including hazardous wastes.
5. Pest control.
6. Abatement and control of hazards to the environment.
7. Desertification.
8. Public awareness and environmental education.
9. Environmental management tools.
10. Natural disasters.
11. Indigenous energy resources development.
12. Energy generation, conventional and non conventional, water needs and environmental considerations.
13. Electricity grid connection and oil pipelines.
14. Regional charters.

B. Jordan Rift Valley

1. Red Sea-Dead Sea Canal.
2. Dead Sea chemical industries.
3. Management of water basins.
4. Development of geothermal resources.
5. Industrial export processing zones.
6. Transportation facilities and infrastructure.
7. Agricultural development and operations.
8. Archaeology and tourism.
9. Inter-disciplinary activities in water, environment and energy.

Appendix 13

June 07, 1994

THE MIDDEL EAST PEACE PROCESS
THE BILATERAL PEACE NEGOTIATIONS
THE JORDAN-ISRAEL TRACK

COMMON SUB-AGENDA

Group A- Water Energy and Environment
Scope of Work

In their pursuit of achieving the goal of the Jordan-Israel track of negotiations, Jordan and Israel have identified the components of their bilateral negotiations and have listed them under section B of the Common Agenda. Item B.7 addresses the phasing of the discussion, agreement and implementation of the agenda items including appropriate mechanisms for negotiations in specific fields. It is understood that the two parties shall cooperate to achieve this goal.

Resolution of the issues of the conflict is to result in mitigation of the adverse impacts and damages resulting from the conflict.

Items of this sub-agenda listed hereunder respond to the intent of the detailed discussions over the various issues in the subjects of Water, Energy, the Environment and the Jordan Rift Valley, and provide practical steps to fulfill the objectives of items B.3, B.6-A and B.7 of the Common Agenda.

I. Surface Water Basins

A. Negotiation of mutual recognition of the rightful water allocations of the two sides in the Jordan River and Yarmouk River waters with mutually acceptable quality.

B. Securing the unimpeded flow of the mutually recognized rightful allocations of the Hashemite Kingdom of Jordan and Israel respectively in Jordan and Yarmouk River waters with the above agreed quality through agreed conveyance systems.

C. Restoration of water quality in the Jordan River below Lake Tiberias to reasonably usable standards.

D. Protection of water quality.

II. Shared Groundwater Aquifers

A. Renewable freshwater aquifers- southern area between the Dead Sea and the Red Sea.

B. Fossil aquifers- area between the Dead Sea and the Red Sea.

C. Protection of water quality of both.

III. Alleviation of Water Shortage

A. Development of water resources
B. Municipal water shortages
C. Irrigation water shortages

IV. Potentials for Future Cooperation Within a Regional Context Where Appropriate

A. General

1. Nature Protection, Natural Resources and Biodiversity.
2. Air quality control, including general standards, criteria and all types of man-made hazardous radiation, fumes and gases.
3. Marine environment and coastal resources management.
4. Waste management including hazardous wastes.
5. Pest control.
6. Abatement and control of hazards to the environment.
7. Desertification.
8. Public awareness and environmental education.
9. Environmental management tools.
10. Natural disasters.
11. Indigenous energy resources development.
12. Energy generation, conventional and non-conventional, water needs and environmental considerations.
13. Electricity grid connection and oil pipelines.
14. Regional charters.

B. Jordan Rift Valley Development

1. Red Sea-Dead Sea Canal.
2. Dead Sea chemical industries.
3. Management of water basins.
4. Development of geothermal resources.
5. Mining and mineral resources exploitation.
6. Industrial export processing zones.
7. Transportation facilities and infrastructure.
8. Agricultural development and operations.
9. Archaeology and tourism.
10. Inter-disciplinary activities in water, environment and energy.

For Israel	For Jordan
Initialed N.K (Noah Kinarti)	*Initialed MJH (MuntherJ Haddadin)*
Signed Elyakim Rubinstein	Signed Fayez Tarawneh

Done in Washington D.C at the Department of State on June 07, 1994.

Appendix 14

Wadi Araba Meetings of July 18-19, 1994

The Middle East Peace Process
The Bilateral Peace Negotiations
The Jordan Israel Track

WATER

The sub-commission for water, energy and the environment, having discussed the issue of water:

1. Agreed to mutually recognize, as a result of their negotiations, the rightful allocations of the two sides in Jordan River and Yarmouk River waters, and to fully respect and comply with the negotiated rightful allocations, in accordance with agreed acceptable principles, with mutually acceptable quality.
   In this framework, they will take into consideration their respective water shortages.

2. Stressed the importance and the need to minimize wastage of water resources throughout the chain of their uses.

3. Stressed the need to make more water available for use as soon as possible.

4. Agreed to share, where appropriate, the available data on shared water resources, and to review the potentials for enhancement of water resources development and use.

5. Accepted the principle that the water issues along their entire border must be dealt with in their totality, including the possibility of trans-boundary water transfers.

6. Accepted the principle of mutual prevention of harm to the water resources of the other party.

Initialed (Noah Kinarti) *Initialed* (Munther J Haddadin)

19.07.1994

ENVIRONMENT

Jordan and Israel acknowledge the importance of the ecology of the region, its high environmental sensitivity and the need to protect the environment and prevent danger and risks for the health and well-being of the region's population. They both recognize the need for conservation of natural resources, protection of biodiversity and the imperative of attaining economic growth based on sustainable development principles.

In light of the above, both parties agree, pursuant to Article B.6.a of the Common Agenda, to cooperate in matters relating to environmental protection in general and to those that may mutually affect them. Areas of such cooperation which deemed to reflect common understanding are detailed as follows, ad referendum:

1st Taking the necessary steps both jointly and/or severally to prevent damage and risks to the environment in general, and in particular those that may affect people, natural resources and environmental assets in the two countries respectively.

2nd Taking the necessary steps by both countries to cooperate in the following areas:

- Environmental planning and management including conducting Environmental Impact Assessment (EIA) and exchanging of data on projects possessing potential impact on their respective environments.
- Environmental legislation, regulations, standards and enforcement thereof.
- Research and applied technology.
- Emergency response, monitoring, related notification procedures and control of damages.
- Code of conduct through regional charters.

This may be achieved through the establishment of joint modalities and mechanisms of cooperation to ensure the exchange of information, communication and coordination regarding matters and activities of mutual environmental concern between their environmental administrations and experts.

3rd. Environmental subjects to be addressed, pursuant to the Common Subagenda signed on June 7, 1994:

1. Protection of nature, natural resources and biodiversity, including cooperation in planning and management of adjacent protected areas

along the common border, and protection of endangered species and migratory birds.

2. Air quality control, including general standards, criteria and all types of man made hazardous radiation's, fumes and gases.
3. Marine environment and coastal resources management.
4. Waste management including hazardous wastes.
5. Pest control including house flies and mosquitoes, and prevention of diseases transferred by pests, such as malaria and leishmaniosis.
6. Abatement and control of pollution, contamination and other man-made hazards to the environment.
7. Desertification. Combating desertification, exchange of information and research knowledge, and the implementation of suitable technologies.
8. Public awareness and environmental education, encouraging the exchange of knowledge, information, study materials, education programs and training through public actions and awareness campaigns.
9. Noise. Reducing noise pollution through regulation, licensing and enforcement based on agreed standards.
10. Potential cooperation in case of natural disasters.

4th. In accordance with the above, the two parties agree to cooperate in activities and projects in the following geographical areas:

I-The Gulf of Aqaba

I.1 The marine environment:

- Natural resources.
- Coastal reef protection.
- Marine pollution.
- Marine sources: e.g. Oil spills, Littering and waste disposal & other.
- Land-based sources: e.g. liquid waste, Solid waste and littering.
- Abatement including monitoring and emergency response actions.

I.2 Coastal Zone Management – The Littoral:

- Nature reserves and protected areas.
- Environmental protection of water resources.
- Liquid waste.
- Solid waste.
- Tourism and recreational activities.
- Ports.
- Transport.
- Industry and power generation.

- Air quality.
- Hazardous materials.
- Environmental assessments.

II-The Rift Valley

II.1 The Jordan River

Jordan and Israel agree to cooperate along their common boundaries in the following aspects:

- Ecological rehabilitation of the Jordan River.
- Environmental protection of water resources to ensure optimal water quality, at reasonably usable standards.
- Agricultural pollution control.
- Liquid waste.
- Pest control.
- Nature reserves and protected areas.
- Tourism and historical heritage.

II.2 The Dead Sea

- Nature reserves and protected areas.
- Pest control.
- Environmental protection of water resources, industrial pollution control.
- Tourism and historical heritage.

II.3 Area from the Red Sea to the Dead Sea

- Environmental protection of water resources.
- Nature reserves and protected areas.
- Pest control.
- Tourism and historical heritage.
- Agricultural pollution control.

Appendix 15

JOINT COMMUNIQUE BY U.S. SECRETARY OF STATE WARREN CHRISTOPHER, JORDANIAN PRIME MINISTER ABD AL-SALAM MAJALI AND ISRAELI FOREIGN MINISTER SHIMON PERES

DEAD SEA HOTEL JORDAN
JULY 20, 1994

The U.S. – Israel – Jordan Trilateral Economic Committee held its fifth meeting on July 20, 1994 at the Dead Sea Spa Hotel in Jordan. At this meeting, the American Delegation was headed by Secretary of State Warren Christopher, the Jordanian delegation by Prime Minister Abd al-Salam Majali, and the Isreali delegation by Foreign Minister Shimon Peres. The three parties expressed their sincere thanks and appreciation to the Government of Jordan for hosting this historic meeting.

The Trilateral Committee – established in October 1993 under the auspices of President Clinton by His Royyal Highness Crown Prince Hassan and Foreign Minister Peres – reviewed recent progress in the peace process. The Committee noted favorably the Israeli – Jordanian bilateral negotiations held in the region on July 18-19, 1994, and the intention to continue these discussions next month. The delegations reiterated their intention to energize efforts to promote further progress on the Israel – Jordan track, looking froward to the meeting between His Majesty King Hussein and Prime Minister Rabin in Washington on July 25, 1994.

The Trilateral Committee also reviewed recent progress on its work and reached agreement on a number of future activities. The delegation agreed to work in preparation for the meeting in the White House next week, on a Draft Master Plan for the Development of the Jordan Rift Valley. Meetings of experts earlier today indicated a substantial convergence of views on such a scheme and a common approach on proceeding. The experts will continue their meetings later today and tomorrow, and an intersectional meeting will be organized to unify the planning criteria and develop detailed terms of reference. The United States agreed to facilitate further the continued work on the Master Plan.

The Trilateral Committee also agreed to continue work on trade/ finance/ banking, civil aviation, tourism, and establishing a road link between the two countries. On trade, the parties agreed on the establishment of a set of principles concerning trade and commercial relationships between the two parties in the context of a peace treaty. On civil aviation, the parties agreed to establish a joint team to explore aviation routes serving the interests of both countries and flight safety. On tourism, the parties agreed to a travel and tourism arrangement and to establish a trilateral commission to facilitate cooperation in this area, specifically the opening of a crossing point in the Eilat-Aqaba area for tourists who are third country nationals. Finally, the parties agreed to conduct a preliminary site survey of a road linking Jordan, Israel and Egypt in the vicinity of Eilat and Aqaba.

The three ministers agreed to meet again periodically in the region, starting in the near future. Intersectional meeting of experts will also be organized to continue work on specific projects.

Appendix 16

Summary of Positions on Water

Summary Opinions of Kinarti and Haddadin on Tuesday 09 and Wednesday 10 August, 1994

KINARTI	HADDADIN
1. Jordan's allocation from Israel should based on municipal needs in Amman.	* (1) Allocation should be based on municipal needs and irrigation needs of perennial crops.
2. Jordan's allocation ranges between zero and 200 mcm/yr. The zero figure is not reasonable and the 200 mcm figure is equally unreasonable.	(2) I will be comfortable with 200 mcm/yr from the Jordan River.
3. Prefer not to mention the source of water at this time but to bring water to borders. Sources may be mentioned later.	(3) Sub agenda mentions sources explicitly (Jordan and Yarmouk).
4. Releasing water to Jordan will be to drinkable standards and to be compatible with the capability of transmission to Amman. Flexibility is there to adjust rates of release over the seasons (winter and summer).	(4) Account should also be made of the irrigation needs of perennial trees to save capital. Kinarti said Israel is concerned only with municipal needs in Amman.
5. All the above is contingent upon maintaining existing uses of Israel on the Yarmouk (65 mcm/yr.), and in Wadi Araba (50 mcm/yr.). Settlements in Wadi Araba should continue to cultivate the land and use the water resources regardless of the border line as finally demarcated.	(5) Israeli allocation from the Yarmouk should not exceed 25 mcm/yr. We have to know more about the water use in Wadi Araba. Nothing in any future arrangement shall prejudice the border's demarcating nor Jordan's sovereignty over its land and water resources.

6. We should cooperate in the North to make more water available for use by both parties. Such projects are envisaged on the Yarmouk and the Jordan. Most of the additional water would go to Jordan as part of the quantity requested under 2 above. Palestinians asked me not to forget them when I talk with you. If we don't come to agreement soon, maybe our above offer will change.	(6) NO DEAL on the objective of projects on the Yarmouk. I am not authorized to talk on behalf of Palestinians. We are here to resolve bilateral water issues. I repeat: Israel's allocation from the Yarmouk shall not exceed 25 mcm/yr.
7. Desalinating of brackish water and of sea water is the long term solution for both sides. I do not subscribe to water import or to the grand schemes designed for that. Assured that the activities of building a port in Turkey are not connected to Israel's planning.	(7) Agree desalination is one option. Another is water imports from the Euphrates after its waters are augmented by diversions from the north rivers of Turkey.

* With parenthesis means response to a point
Without parenthesis means initiation of the point

Haddadin & Kinarty
Lobby Discussion 15/8/1994

Kinarty :
Surface water in the North and groundwater in the South shall be dealt with as a package, The package will not be announced. He can not speak of figures before it is agreed to deal with it as a package.

Haddadin :
Let me assure you of one thing: There shall be no compromise on our sovereignty over our resources, land and water in the south or elsewhere. However, and this is purely personal, the important thing is the end result and not how you get there.

Kinarty :
I understand your position that you agree to a package deal along the borders but without losing sovereignty and without declaration.

Haddadin :
You are half correct! The other half is that it is not up to me to agree or refuse, it is our government's prerogative. I, personally will urge to look for the grapes and not to fight with the guard.

Kinarty :
Kinarty asked why Haddadin concludes that the Yarmouk belongs to Jordan. Kinarty also stated that the biggest change since the times of Johnston is that municipal water needs have become the most dominant. Haddadin agreed.

Haddadin :
Israel's share from the Yarmouk is to irrigate a part of the Yarmouk Triangle. Yarmouk Triangle share is to irrigate 16,000 dunums. The rest of the area should be irrigated from the Lake and the Jordan River.

Kinarty :
Is this based on your "Basin Principle"?

Haddadin :
Yes.

Appendix 17

ANNEX II

WATER RELATED MATTERS

Initialed N.K. (Noah Kinarti) *MJH (MuntherJ. Haddadin)*

ANNEX II

WATER RELATED MATTERS

Pursuant to Article 6 of the Treaty, Jordan and Israel agreed on following Articles on water related matters:

Article I: Allocation

1. *Water from the Yarmouk River*

a. Summer period – 15th May to 15th October of each year. Israel pumps (12) MCM and Jordan gets the rest of the flow.

b. Winter period – 16th October to 14th May of each year. Israel pumps (13) MCM and Jordan is entitled to the rest of the flow subject to provisions outlined herein below: Jordan concedes to Israel pumping an additional (20) MCM from the Yarmouk in winter in return for Israel conceding to transferring to Jordan during the summer period the quantity specified in paragraph (2.a) below from the Jordan River.

c. In order that waste of water will be minimized, Jordan and Israel may use, downstream of Adassiya Diversion/ point 121, excess flood water that is not usable and will evidently go to waste unused.

2. *Water from the Jordan River*

a. Summer Period – 15th May to 15th October of each year.
In return for the additional water that Jordan concedes to Israel in winter in accordance with paragraph (1.b) above, Israel concedes to transfer to Jordan in the summer period (20) MCM from the Jordan River directly upstream from Deganya gates on the river. Jordan shall pay the operation and maintenance cost of such transfer through existing systems (not including capital cost) and shall bear the total cost of any new transmission system. A separate protocol shall regulate this transfer.

b. Winter period – 16th October to 14th May of each year.
Jordan is entitled to store for is use a minimum average of (20) MCM of the floods in the Jordan River south of its confluence with the Yarmouk

(as outlined in Article II). Excess floods that are not usable and that will otherwise be wasted can be utilized for the benefit of the two parties including pumped storage off the course of the river.

c. In addition to the above, Israel is entitled to maintain its current uses of the Jordan River waters between its confluence with the Yarmouk and its confluence with Wadi Yabis/Tirat Zvi. Jordan is entitled to an annual quantity equivalent to that of Israel, provided however, that Jordan's use will not harm the quantity or quality of the above Israeli uses. The Joint Water Committee (outlined in Article VII) will survey existing uses for documentation and prevention of appreciable harm.

d. Jordan is entitled to an annual quantity of (10) MCM of desalinated water from the desalination of about (20) MCM of saline springs now diverted to the Jordan River. Israel will explore the possibility of financing the operation and maintenance cost of the supply to Jordan of this desalinated water (not including capital cost). Until the desalination facilities are operational, and upon the entry into force of the Treaty, Israel will supply Jordan (10) MCM of Jordan River water from the same location as in (2.a) above, outside the summer period and during dates Jordan selects, subject to the maximum capacity of transmission.

3. Additional Water

Jordan and Israel shall cooperate in finding sources for the supply to Jordan of an additional quantity of 50 MCM/yr of water drinkable standards. To this end, the Joint Water Committee will develop, within one year from the entry into force of the Treaty, a plan for the supply to Jordan of the above mentioned additional water. This plan will be forwarded to the respective governments for discussion and decision.

4. Operation and Maintenance

a. Operation and maintenance of the systems on Israeli territory that supply Jordan with water, and their electricity supply, shall be Israel's responsibility. The operation and maintenance of the new systems that serve only Jordan will be contracted at Jordan's expense to authorities or companies selected by Jordan.

b. Israel will guarantee easy unhindered access of personnel and equipment to such new systems for operation and maintenance. This subject will be further detailed in the agreements to be signed between Israel and the authorities or companies selected by Jordan.

Article II: Storage

1. Jordan and Israel shall cooperate to build a diversion/storage dam on the Yarmouk River directly downstream of the Adasiya Diversion/point 121. The purpose is to improve the diversion efficiency into the King Abdallah Canal of the water allocation of the Hashemite Kingdom of Jordan and, possibly, for the diversion of Israel's allocation of the river water. Other purposes can be mutually agreed.

2. Jordan and Israel shall cooperate to build a system of water storage on the Jordan River, along their common boundary, between it confluence with the Yarmouk River and its confluence with Wadi Yabis/ Tirat Zvi, in order to implement the provision of paragraph (2.b) of Article I above. The storage system can also be made to accommodate more floods; Israel may use up to 3 MCM/year of added storage capacity.

3. Other storage reservoirs can be discussed and agreed upon mutually.

Article III: Water Quality and Protection

1. Jordan and Israel each undertake to protect, within their own jurisdiction, the shared waters of the Jordan and Yarmouk rivers, and Araba/Arava groundwater, against any pollution, contamination, harm or unauthorized withdrawals of each others allocations.

2. For this purpose, Jordan and Israel will jointly monitor the quality of water along their boundary, by use of jointly established monitoring stations to be operated under the guidance of the Joint Water Committee.

3. Jordan and Israel will each prohibit the disposal of municipal and industrial wastewater into the courses of the Yarmouk or the Jordan rivers before they are treated to standards allowing their unrestricted agricultural use. Implementation of this prohibition shall be completed within three years from the entry into force of the Treaty.

4. The quality of water supplied from one country to the other at any given location shall be equivalent to the quality of the water used from the same location by the supplying country.

5. Saline springs currently diverted to the Jordan River are earmarked for desalination within four years. Both countries shall cooperate to ensure

that the resulting brine will not be disposed of in the Jordan River or in any of its tributaries.

6. Jordan and Israel will each protect water systems in its own territory supplying water to the other, against any pollution, contamination, harm or unauthorized withdrawal of each others' allocations.

Article IV: Groundwater in Wadi Araba/Emek Ha'arava

1. In accordance with the provisions or this Treaty, some wells drilled and used by Israel along with their associated systems fall on the Jordanian side of the borders. These wells and systems are under Jordan's sovereignty. Israel shall retain the use of these wells and systems in the quantity and quality detailed in Appendix 1, that shall be jointly prepared by December 31, 1994. Neither country shall take, nor cause to be taken any measure that may appreciably reduce the yields or quality of these wells and systems.

2. Throughout the period of Israel's use of these wells and systems, replacement of any well that may fail among them shall be licensed by Jordan in accordance with the laws and regulations then in effect. For this purpose, the failed well shall be treated as though it was drilled under license from the competent Jordanian authority at the time of its drilling. Israel shall supply Jordan with the log of each of the wells and the technical information about it to be kept on record. The replacement well shall be connected to the Israeli electricity and water systems.

3. Israel may increase the abstraction rate from wells and systems in Jordan by up to 10 MCM/yr. Above the yields referred to in paragraph 1 above, subject to a determination by the Joint Water Committee that this undertaking is hydro-geologically feasible and does not harm existing Jordanian uses. Such increase is to be carried out within five years from the entry into force of the Treaty.

4. Operation and Maintenance

    a. Operation and maintenance of the wells and systems on Jordanian territory that supply Israel with water, and their electricity supply shall be Jordan's responsibility. The operation and maintenance of these wells and systems will be contracted at Israel's expense to authorities or companies selected by Israel.

b. Jordan will guarantee easy unhindered access of personnel and equipment to such wells and systems for operation and maintenance. This subject will be further detailed in the agreements to be signed between Jordan and the authorities or companies selected by Israel.

Article V: Notification and Agreement

1. Artificial changes in or of the course of the Jordan and Yarmouk rivers can only be made by mutual agreement.

2. Each country undertakes to notify the other, six months ahead of time, of any intended projects which are likely to change the flow of either of the above rivers along their common boundary, or the quality of such flow. The subject will be discussed in the Joint Water Committee with the aim of preventing harm and mitigating adverse impacts such projects may cause.

Article VI: Co-operation

1. Jordan and Israel undertake to exchange relevant data on water resources through the Joint Water Committee.

2. Jordan and Israel shall cooperate in developing plans for purposes of increasing water supplies and improving water use efficiency, within the context of bilateral, regional or international cooperation.

Article VII: Joint Water Committee

1. For the purpose of the implementation of this annex, the parties will establish a Joint Water committee comprised of three members from each country.

2. The Joint Water Committee will, with the approval of the respective governments, specify its work procedures, the frequency of its meetings, and the details of its scope of work. The Committee may invite experts and/or advisors as may be required.

3. The Committee may form, as it deems necessary, a number of specialized subcommittees and assign them technical tasks. In this context, it is agreed that these subcommittees will include a northern subcommittee and a southern subcommittee, for the management on the ground of the mutual water resources in these sectors.

INDEX

GPSR Compliance

The European Union's (EU) General Product Safety Regulation (GPSR) is a set of rules that requires consumer products to be safe and our obligations to ensure this.

If you have any concerns about our products, you can contact us on ProductSafety@springernature.com

In case Publisher is established outside the EU, the EU authorized representative is:

Springer Nature Customer Service Center GmbH
Europaplatz 3
69115 Heidelberg, Germany

Batch number: 10755248

Printed by Printforce, the Netherlands